High-Quality Visual Experience

Marta Mrak, Mislav Grgic,
and Murat Kunt (Eds.)

High-Quality Visual Experience

Creation, Processing and Interactivity of
High-Resolution and High-Dimensional
Video Signals

 Springer

Dr. Marta Mrak
University of Surrey
Centre for Vision, Speech and
Signal Processing
GU2 7XH Guildford, Surrey
United Kingdom
E-mail: m.mrak@surrey.ac.uk

Prof. Mislav Grgic
University of Zagreb
Faculty of Electrical Engineering
and Computing
Department of Wireless
Communications
Unska 3/XII HR-10000 Zagreb
Croatia
E-mail: mislav.grgic@fer.hr

Prof. Murat Kunt
Signal Processing Laboratories
School of Engineering - STI
Swiss Federal Institute of Technology
EPFL STI IEL, Station 11
CH-1015 Lausanne
Switzerland
E-mail: murat.kunt@epfl.ch

ISBN 978-3-662-51927-1   ISBN 978-3-642-12802-8 (eBook)

DOI 10.1007/978-3-642-12802-8

Typesetting: Data supplied by the authors

Production & Cover Design: Scientific Publishing Services Pvt. Ltd., Chennai, India

Printed on acid-free paper

9 8 7 6 5 4 3 2 1

springer.com

Preface

Last few years have seen rapid acceptance of high-definition television (HDTV) technology around the world. This technology has been hugely successful in delivering more realistic television experience at home and accurate imaging for professional applications. Adoption of high definition continues to grow as consumers demand enhanced features and greater quality of content.

Following this trend, natural evolution of visualisation technologies will be in the direction of fully realistic visual experience and highly precise imaging. However, using the content of even higher resolution and quality is not straightforward as such videos require significantly higher access bandwidth and more processing power. Therefore, methods for radical reduction of video bandwidth are crucial for realisation of high visual quality. Moreover, it is desirable to look into other ways of accessing visual content, solution to which lies in innovative schemes for content delivery and consumption.

This book presents selected chapters covering technologies that will enable greater flexibility in video content representation and allow users to access content from any device and to interact with it. This book is divided into five parts:

(i)   Quality of visual information,
(ii) Video coding for high resolutions,
(iii) Visual content upscaling,
(iv) 3D visual content processing and displaying,
(v)   Accessing technologies for visual content.

Part I on quality of visual information introduces metrics and examples that are basics for evaluation of quality of high-resolution and high-dimensional visual data. This part of the book is addressed with three chapters. Chapter 1 introduces objective video quality assessment methods. In Chapter 2 the subjective video quality is addressed in terms of the quality of experience for specific case - digital cinema. Quality of 3D visual data is discussed in Chapter 3, where different quality metrics are evaluated for application on stereoscopic images.

The following part addresses the necessary technology that enables wide access to high resolution visual data - video compression methods. Chapter 4 brings an overview of state-of-the-art in video coding and introduces recent developments in ultra-high definition compression. Further details of current video coding standards and evaluation of their performance for high definition videos are given in Chapter 5. Two systems that support ultra-high definition are presented in

Chapters 6 and 7. In addition to the codecs for high resolutions, high frame rate videos are addressed in Chapter 8 where the related mathematical modelling method is given.

Methods for creation of content of higher frame rates and higher resolution are presented in Part III. Temporal content upscaling is in the focus of Chapters 9 to 11. Chapter 12 addresses the problem of natural image synthesis from low resolution images.

Technologies needed for 3D content creation, processing and displaying are presented in Part IV. Chapter 13 investigates the role of colour information in solving stereo correspondence problem. 3D object classification and segmentation methods are presented in Chapter 14. The techniques for generation and handling of videos combined from captured 3D content and computer graphics is proposed in Chapter 15. Discussion on this topic continues with Chapter 16 where a new approach for generation of 3D content is proposed. Chapter 17 brings detailed overview of 3D displaying technology, while Chapter 18 focuses on integral imaging.

Accessing technologies for visual content of high resolution and dimensions are presented in Part V. The techniques enabling video streaming with spatial random access are presented in Chapter 19. Chapter 20 addresses management of heterogeneous environments for enabling quality of experience. Transmission of 3D video is in the focus of Chapter 21 which presents a solution designed for wireless networks. Methods for retrieval of high-resolution videos are addressed in Chapter 22. Moreover, in Chapter 23 stereo correspondence methods are addressed in the context of video retrieval.

We believe that this collection of chapters provides balanced set of critical technologies that will facilitate development of future multimedia systems supporting high quality of experience. The refreshing perspectives of looking into visual data handling presented in this book complement current commercial visual technologies. Therefore, this book is essential to those whose interest is in futuristic high-quality visualisation systems.

January 2010   Marta Mrak
                                                                                    Mislav Grgic
                                                                                    Murat Kunt

Contents

Part I
Quality of Visual Information

Chapter 1
Automatic Prediction of Perceptual Video Quality: Recent Trends and Research Directions

Anush K. Moorthy and Alan C. Bovik

Abstract. Objective video quality assessment (VQA) refers to evaluation of the quality of a video by an algorithm. The performance of any such VQA algorithm is gaged by how well the algorithmic scores correlate with human perception of quality. Research in the area of VQA has produced a host of full-reference (FR) VQA algorithms. FR VQA algorithms are those in which the algorithm has access to both the original reference video and the distorted video whose quality is being assessed. However, in many cases, the presence of the original reference video is not guaranteed. Hence, even though many FR VQA algorithms have been shown to correlate well with human perception of quality, their utility remains constrained. In this chapter, we analyze recently proposed reduced/no-reference (RR/NR) VQA algorithms. RR VQA algorithms are those in which some information about the reference video and/or the distorting medium is embedded in the video under test. NR VQA algorithms are expected to assess the quality of videos without any knowledge of the reference video or the distorting medium. The utility of RR/NR algorithms has prompted the Video Quality Experts Group (VQEG) to devote resources towards forming a RR/NR test group. In this chapter, we begin by discussing how performance of any VQA algorithm is evaluated. We introduce the popular VQEG Phase-I VQA dataset and comment on its drawbacks. New datasets which allow for objective evaluation of algorithms are then introduced. We then summarize some properties of the human visual system (HVS) that are frequently utilized in developing VQA algorithms. Further, we enumerate the paths that current RR/NR VQA algorithms take in order to evaluate visual quality. We enlist some considerations that VQA algorithms need to consider for HD videos. We then describe exemplar algorithms and elaborate on possible shortcomings of these algorithms. Finally, we suggest possible future research directions in the field of VQA and conclude this chapter.

Anush K. Moorthy · Alan C. Bovik
Dept. of Electrical and Computer Engineering, The University of Texas at Austin,
Austin, Texas 78712, USA
e-mail: anushmoorthy@mail.utexas.edu, bovik@ece.utexas.edu

1 Introduction

Imagine this situation - you are given two videos, both having the same content but one of the videos is a 'low quality' (distorted) version of the other and you are asked to rate the low quality version vis-a-vis the original (reference) video on a scale of (say) 1-5 (where 1 is bad and 5 is excellent). Let us further assume that we collect a representative subset of the human populace and ask them the same question, and instead of just asking them to rate one pair of videos, we ask them to rate a whole set of such pairs. At the end of the day we now have a set of ratings for each of the distorted videos, which when averaged across users gives us a number between 1-5. This number represents the mean opinion score (MOS) of that video and is a measure of the perceptual quality of the video. The setting just described is called subjective evaluation of video quality and the case in which the subject is shown both the reference and the distorted video is referred to as a double stimulus study. One could imagine many possible variations to this technique. For example, instead of showing each video once, let us show each video twice so that in the first pass the human 'decides' and in the second pass the human 'rates'. This is a perfectly valid method of collecting subjective scores and along with a plethora of other techniques forms one of the possible methods for subjective evaluation of video quality. Each of these methods is described in a document from the International Telecommunications Union (ITU) [1] . If only we always had the time to collect a subset of the human populace and rate each video that we wish to evaluate quality of, there would have been no necessity for this chapter or the decades of research that has gone into creating algorithms for this very purpose.

Algorithmic prediction of video quality is referred to as objective quality assessment, and as one can imagine it is far more practical than a subjective study. Algorithmic video quality assessment (VQA) is the focus of this chapter. Before we delve directly into the subject matter, let us explore objective assessment just as we did with the subjective case. Imagine you have an algorithm to predict quality of a video. At this point it is simply a 'black-box' that outputs a number between (say) 1-5 - which in a majority of cases correlates with what a human would say. What would you imagine the inputs to this system are? Analogous to the double stimulus setup we described before, one could say that both the reference and distorted videos are fed as inputs to the system - this is full reference (FR) quality assessment . If one were to imagine practical applications of FR VQA, one would soon realize that having a reference video is infeasible in many situations. The next logical step is then truncating the number of inputs to our algorithm and feeding in only the distorted video - this is no reference (NR) VQA . Does this mean that FR VQA is not an interesting area for research? Surprisingly enough, the answer to this question is NO! There are many reasons for this, and one of the primary ones is that FR VQA is an extremely difficult problem to solve. This is majorly because our understanding of perceptual mechanisms that form an integral part of the human visual system (HVS) is still at a nascent stage [2, 3]. FR VQA is also interesting for another reason - it gives us techniques and tools that may be extended to NR VQA. Since FR VQA has matured over the years, we shall cease talking about it here. The interested reader is referred to [4, 5], for tutorial chapters on FR VQA .

Thinking solely from an engineering perspective one would realize that there exists another modality for VQA. Instead of feeding the algorithm with the reference and distorted videos, what if we fed it the distorted video and *some features* from the reference video? Can we extract features from the reference video and embed them into the video that we are (say) transmitting? If so, at the receiver end we can extract these reference features and use them for VQA. Such assessment of quality is referred to as reduced-reference (RR) VQA . RR and NR techniques for VQA form the core of this chapter.

In describing the RR technique, we have inadvertently stumbled upon the general system description for which most algorithms described in this chapter are designed. There exists a pristine reference video which is transmitted through a system from the source. At the receiver, a distorted version of this video is received whose quality is to be assessed. Now, the system through which the video passes could be a compression algorithm. In this case, as we shall see, measures of blockiness and bluriness are used for NR VQA. In case the system is a channel that drops packets, the effect of packet loss on quality may be evaluated. These concepts and many others are discussed in this chapter. Before we describe recent algorithms, let us briefly digress into how the performance of an algorithm is evaluated.

2 Performance Evaluation of Algorithms and Databases

At this stage we have some understanding of what a VQA algorithm does. We know that the aim of VQA is to create algorithms that predict the quality of a video such that the algorithmic prediction matches that of a human observer. For this section let us assume that we have an algorithm which takes as input a distorted video (and some reference features) and gives us as output a number. The range of the output could be anything, but for this discussion, let us assume that this range is 0-1, where a value of 0 indicates that the video is extremely bad and a value of 1 indicates that the video is extremely good. We also assume that the scale is continuous, i.e., all possible real-numbers between 0 and 1 are valid algorithmic scores. With this setup, the next question one should ask is, 'How do we know if these numbers generated are any good?'. Essentially, what is the guarantee that the algorithm is not spewing out random numbers between 0 and 1 with no regard to the intended viewer?

The ultimate observer of a video is a human and hence his perception of quality is of utmost importance. Hence, a set of videos are utilized for a subjective study and the perceptual quality of the video is captured in the MOS . However, picking (say) 10 videos and demonstrating that the algorithmic scores correlate with human subjective perception is no good. We require that the algorithm perform well over a wide variety of cases, and hence the database on which the algorithm is tested must contain a broad range of distortions and a variety of content, so that the stability of its performance may be assessed. In order to allow for a fair comparison of algorithms that are developed by different people, it is imperative that the VQA database, along with the subjective MOS be made publicly available. One such publicly available dataset for VQA is the popular Video Quality Experts Group (VQEG) FRTV Phase-I dataset [6] . The VQEG dataset consists of 20 reference videos, each subjected to 16 different distortions to form a total of 320 distorted videos. In [6], a study of various algorithms was conducted on this dataset and it was shown that none of the assessed algorithms were statistically better than peak signal-to-noise ratio (PSNR)[1] ! Over the years, many new FR VQA algorithms which perform well on this dataset have been proposed [9, 10] . However, the VQEG dataset is not without its drawbacks.

The dataset is dated, since the report of the study was released in the year 2000. Previous generation compression techniques such as MPEG [11] were used to produce distortions. Current generations compressions standards such as H.264/AVC [12] exhibit different perceptual distortions and hence a database that covers the H.264/AVC compression standard is relevant for modern systems. Further, the perceptual separation of videos in the VQEG dataset is poor, leading to inconsistent judgments for humans and algorithms. In order to alleviate many such problems associated with the VQEG dataset, researchers from the Laboratory for Image and Video Engineering (LIVE) have created two new VQA datasets . The LIVE databases are now available for non-commercial research purposes; information may be found online [13, 14]. The LIVE VQA datasets include modern day compression techniques such as the H.264/AVC and different channel induced distortions. Descriptions of the datasets and the evaluated algorithms may be found in [15] and [16].

Now that we have a dataset with subjective MOS and scores from an algorithm, our goal is to study the correlation between them. In order to do so, Spearman's Rank Ordered Correlation Coefficient (SROCC) [17] is generally used [6]. SROCC of 1 indicates that the two sets of data under study are perfectly correlated. Other measures of correlation include the Linear (Pearson's) correlation coefficient (LCC) and the root-mean-square error (RMSE) between the objective and subjective scores. LCC and RMSE are generally evaluated after subjecting the algorithms to a logistic function . This is to allow for the objective and subjective scores to be non-linearly related. For eg., figure 1 shows a scatter plot between MOS scores from the VQEG dataset and an FR VQA algorithm [18]. As one can see, the two are definitely correlated, only that the correlation is non-linear. Transformation of the scores using the logistic accounts for this non-linearity and hence application of LCC and RMSE make sense. It is essential to point out that application of the logistic in no way constitutes 'training' an algorithm on the dataset (as some authors claim). It is simply a technique that allows for application of the LCC and RMSE as statistical measures of performance. A high value (close to 1) for LCC and a low value (close to 0) for RMSE indicate that the algorithm performs well.

Having summarized how one would analyze a VQA algorithm, let us move on to the human visual system whose properties are of tremendous importance for developing VQA algorithms.

---

[1] Why PSNR is a poor measure of visual quality is described in [7] and [8].

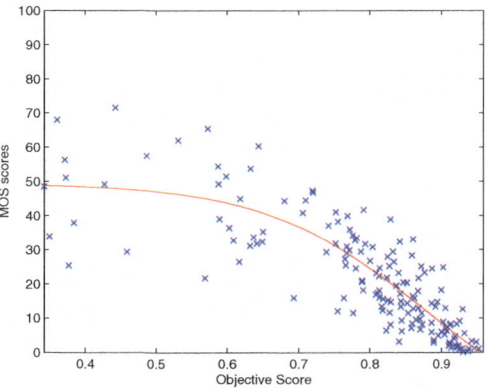

Fig. 1 Figure showing a scatter plot between MOS from the VQEG dataset and an FR VQA algorithm's scores. A non-linear correlation is evident. Figure also shows a best-fit-line through the scatter obtained using the logistic function proposed in [6].

3 A Brief Foray into the Human Visual System

You are currently staring at these words on a sheet of paper. Due to acquired fluency in English, it takes you a fraction of a second to view, process, understand and proceed along this page. But it is not language alone that guides you along. The human visual system (HVS) which processes all of the information incident upon the eye and renders it into a form recognizable by higher areas of the human brain for cognitive processes to occur has been one of the most actively researched areas of neuroscience.

The first stage of visual processing in the human are the eyes. This spherical mass is home to different kinds of photoreceptors - receptors that produce a response when incident with photons. The response of these receptors is fed through the retinal ganglion cells and then to the Lateral Geniculate Neucleus (LGN) which resides in the thalamus. The LGN is analogous to an 'active' switch - receiving and processing both feed-forward and feedback information. LGN responses are passed on to area V1 of the primary visual cortex (situated at the back of your head) which then connects to area V2, V4 as well as area V5/Middle-temporal (MT) and other higher areas in the brain. This kind of hierarchical structure is common in neural processing.

Each of the above described units is an interesting area of study, however we shall not pursue them in detail here. The interested reader is referred to [2] for overviews and descriptions. Here we shall look at these regions of processing using a system-design perspective. The first stage of processing is the human eye. The eye behaves akin to a low-pass filter since light at frequencies above 60 cycles per degree (cpd) are not passed on to the receptors at the back of the eye. Current research indicates that there are two kinds of photoreceptors - rods and cones, based on their response characteristics [3]. Rods are generally in use in low-light conditions while cones are used for vision under well-lit conditions and for color vision. There exist 3 types of cones and depending upon their response characteristics are classified as Long (L), Medium (M) and Short (S) wavelength cones. Another very important characteristic of the eye is the fact that not every region in the visual field is perceived with the same amount of acuity. For example, stare at any one word in this sentence and then try (without moving your eye) to read the beginning of this paragraph. You will notice that even though the word that you are staring at is extremely clear, as you move away from the word under focus, you start loosing resolution. This is referred to as foveation . If you haven't thought about this before, it may come as a surprise, since the world seems sharp in daily life. This is because the eye performs an efficient engineering solution (given the contraints). The HVS is designed such that the when viewing at a scene, the eye makes rapid movements called saccades interleaved with fixations. Fixations, as the name suggests, refers to the process of looking at a particular location for an extended period of time. Little to no information is gathered during a saccade and most information is gathered during a fixation. Using this strategy of eye movements where the region of maximum visual acuity (fovea) is placed at one location for a short period of time, and then moved to another, the HVS constructs a 'high resolution' map of the scene.

The reason why we described the process of foveation in some detail is because for HD videos, foveated video coding can help reduce bandwidth while maintaining picture quality. Indeed, foveation driven video coding is an active area of research [19].

VQA systems which seek to emulate the HVS generally model the first stage of processing using a point-spread-function (PSF) to mimic the low-pass response of the human eye. The responses from the receptors in the eye are fed to the retinal ganglion cells. These are generally modeled using center-surround filters, since ganglion cells have been shown to possess on-center off-surround structure [2]. Similar models are used for the LGN. The next stage of the HVS is area V1. The neurons in V1 have been shown to be sensitive to direction, orientation, scale and so on. A multi-scale, multi-orientation decomposition is generally used to mimic this. Better models for V1 involve using multi-scale Gabor filterbanks [9]. Since we are concerned with video, we skip areas V2 and V4 and move on to area V5/MT. This area is responsible for processing motion information. Motion estimates are of great importance for the human since they are used for depth perception, judging velocities of oncoming objects and so on. The engineering equivalent of this region is estimating optical flow [20] from frames in a video. A coarser approximation is block-based motion estimation [21]. As we shall see most NR VQA algorithms do not use this information and currently perform only a frame-based spatial computation.

In the HVS, the responses from MT/V5 are further sent to higher levels of the brain for processing. We do not discuss them here. The interested reader is referred to [2] for details.

We have now seen how the human visual system works and how algorithms that seek to evaluate quality akin to a human observer are evaluated. We also listed some considerations for HD video. Having said that, one should note that any NR/RR VQA technique that is proposed can be used for quality assessment of HD video. Additional considerations for HD may improve performance of these algorithms. In the rest of this chapter we shall discuss RR and NR algorithms for VQA. All discussed algorithms unless otherwise stated do not utilize color information - i.e., all processing is undertaken using luminance information alone. Color-distortions and their assessment are interesting areas of research [22].

4 Reduced-Reference Algorithms

Reduced reference (RR) VQA algorithms generally follow a model in which some amount of information from the source video is embedded in the video being transmitted. This embedded information is recovered at the receiver and compared with the same information extracted from the received video to assess quality. We classify these algorithms into two categories - one that relies on inserting a watermark in each frame and all other techniques including natural scene statistics, multivariate data analysis and so on. Even though some authors claim that the techniques based on watermarking are NR, based on our definition, embedding information at the sender in the video stream constitutes a RR scheme.

4.1 Techniques Based on Watermarking

Hidden markers for MPEG quality. Sugimoto *et. al.* proposed a technique for measuring MPEG quality based on embedding binary data in videos [23]. A particular previously agreed upon binary sequence (for example, a series of 1's or 0's or alternating 1's and 0's) is first generated. The AC components of each macroblock[2] in each frame are computed and its spectrum is spread using a pseudo-random sequence (PN) [24]. Embedding is then performed in the frequency domain using a set of rules based on the quantization parameter. At the receiver a reverse process is performed and the false detection rate is computed. The authors claim that this rate is linearly correlated with PSNR and hence with quality.

Tracing watermark and blind quality. A different PN matrix for each frame is multiplied with the same watermark, and this product is embedded in the mid-band frequencies from the discrete cosine transform (DCT) of each frame [25]. At the receiver a reverse procedure is carried out and the average mean squared error between the transmitted and the original watermark across the video is computed. Even though correlation with perceived quality is not evaluated, the authors demonstrate that mean squared error (MSE) correlates with the bit-error rate.

Fu-Zheng *et. al.* propose a similar method for VQA, however instead of using the whole frame, they use only certain sub-blocks is alternate frames [26]. Further, the watermark is weighted based on the quantization parameter and instead of using MSE, the number of correctly recovered watermark pixels are used as a measure of degradation. Correlation with PSNR is demonstrated.

---

[2] A macroblock is the default unit of coding for video compression algorithms. Refer to [21] for details.

Video Quality Assessment based on Data Hiding. Farias *et. al.* proposed a VQA algorithm based on data hiding in [27]. Even though the authors claim that the method is NR, a method that incorporates any information at the source end which is utilized at the receiver end for QA, as per our definition, is RR VQA. The proposed algorithm is based on the technique of water-marking. At the source end, a 8×8 block Discrete Cosine Transform (DCT) of the frame is performed. A binary mask is multiplied by an uncorrelated pseudo-random noise matrix, which is rescaled and added to the medium frequency coefficients from the DCT. A block is selected for embedding only if the amount of motion (estimated using a block motion estimation algorithm) exceeds a certain threshold (T_{mov}). At the receiver end an inverse process is performed and the mark is extracted. The measure of degradation of the video is then the total squared error between the original mask and the retrieved mask. The authors do not report statistical measures of performance as discussed before.

Data hiding in perceptually important areas. Carli *et. al.* proposed a block-based spread-spectrum method for RR VQA in [28]. The proposed method is similar to that in [27], and only differs in selecting where to place this watermark. Regions of perceptual importance are computed using motion information, contrast and color. The watermark is embedded only in those areas that are perceptually important, since degradation in these areas are far more significant. A single video at different bit-rates is used to demonstrate performance.

Limitations to the use of watermarking include the fact that the use of squared error (which is generally computed at the receiver as a measure of quality) does not relate to human perception, and that the degradation of a watermark may not be proportional to (perceptual) video degradation.

4.2 Other Techniques

Low Bandwidth RR VQA. Using features proposed by the authors in [29], Wolf and Pinson developed a RR VQA model in [30]. A spatio-temporal (ST) region consisting of 32 pixels $\times$ 32 pixels $\times$ 1 second is used to extract three features. Further a temporal RR feature which is essentially a difference between time-staggered ST regions is also computed. At the receiver the same set of parameters are extracted and then a logarithmic ratio or an error ratio between the (thresholded) features is computed. Finally a Minkowski pooling is undertaken to form a quality score for the video. The authors claim that the added RR information contributes only about 10 kbits/s of information.

Multivariate Data Analysis based VQA. Oelbaum and Diepold utilized a multivariate data analysis approach, where the HVS is modeled as a black-box, with some input features [31, 32]. The output of this box is the visual quality of the video. The authors utilize previously proposed features for NR IQA including blur , blocking and video 'detail'. They also extract noise and predictability based on simple techniques. Edge, motion and color continuity form the rest of the features. Features are extracted on a frame-by-frame basis and the mean is utilized for processing. A set of test videos and multivariate data analysis [33] are used to compute a feature matrix **F**. A multiplicative signal correction (MSC) is performed using linear regression to account for correlations between features to obtain **F'**. Further, partial least squares regression (PLSR) is used to map the feature vectors onto subjective ratings. Up to this point the method described is NR. However, the authors use a quality estimate from the original video in order to improve NR VQA performance thus creating a RR VQA algorithm. The authors demonstrate high correlation with human perception. The use of the original video for the NR to RR transition is non-standard. Further, the innovativeness of the algorithm hinges on the use of multivariate data analysis, since the features used have been previously proposed in literature.

Neural Network based RR VQA. Le Callet *et. al.* proposed a time-delay neural network (TDNN) [34] based RR VQA index in [35]. The algorithms follows the general description of an RR algorithm, with extracted features borrowed from previous works - including power of frame differences, blocking and frequency content measures. Their main contribution is to utilize a TDNN to perform a temporal integration of these indicators without specifying a particular form for temporal pooling. A small test to evaluate performance is undertaken and decent performance is demonstrated.

Foveation-based RR VQA. Meng *et. al.* proposed an algorithm based on features extracted from spatio-temporal (ST) regions from a video for HD RR VQA [36]. The features extracted and the ST regions are based on the ideas proposed by Wolf and Pinson [29]. Extracted features from the original video are sent to the receiver over an ancillary channel carrying RR information. Based on the fact that the human perceives regions within the fovea with higher visual acuity (a fact that is very pertinent for HD video), the authors divide the video into foveal, parafoveal and peripheral regions, where the ST regions are computed with increasing coarseness. The authors claim that the use of these different regions increases performance, however, analysis of performance is lacking.

Quality Aware Video. Extending the work in [37] for images, Hiremath *et. al.* proposed an algorithm based on natural video statistics for RR VQA in [38]. A video is divided into a group of pictures (GOP) and each frame in the GOP is decomposed using a steerable pyramid [39] (an overcomplete wavelet transform). Subbands at same orientation and scale but from different frames are then aligned to obtain $H(s, p, t)$, where s is the scale, p is the orientation (translation factor for the wavelet) and t represents the frame. The authors then compute $L_2(s, p) = \sum_{n=0}^{2} (-1)^n \binom{N}{n} \log H(s, p, t + n\Delta t)$. The histogram of L_2 appears to be peaked at zeros with heavy tails and this is fitted with a four parameter logistic function. The four parameters of the fit and the KL divergence [40] between the fit and the actual distribution for each subband in each GOP form the RR features. Further, marginal distributions in each subband is fitted using a generalized Gaussian model , which are additional RR features. The RR features are embedded in the

DCT domain before transmission. At the receiver side, a similar process is undertaken to estimate the same parameters on the degraded video. The KL divergence between the original parameters and the distorted parameters is computed and averaged across subbands to form a distortion measure for the video. The proposed algorithm is not tested on a public dataset, but instead a small set of videos are used for evaluation.

Representative-Luminance based RR VQA. The essence of the idea proposed by Yamada *et. al.* in [41] is to estimate PSNR at the receiver using luminance information embedded in the transmitted video stream. Block variance of each 16×16 block is evaluated and the representative luminance of a frame is chosen from a subset of the blocks which have variance equal to the median variance of the frame. The authors claim that this captures the luminance of pixels in the medium frequency range. PSNR is computed at the receiver using this additional information and is used as the quality metric.

RR VQA based on Local Harmonic Strength. Gunawan and Ghanbari proposed a RR VQA algorithm in [42] based on local harmonic strength. First a Sobel filter is used to produce a gradient image [43]. This image is then segmented into blocks and a harmonic analysis is applied on each of these blocks. Harmonic analysis consists of applying the 2-D fast Fourier transform (FFT) [44] on a block-by-block basis and computing the magnitude of the transform at each pixel location. The local harmonic strength is the sum of the magnitudes of the transform at particular locations within a block. The local harmonic feature is used as the RR feature. A similar analysis is performed at the receiver on the distorted video and harmonic gains and losses are computed as differences between the harmonic features of the reference and distorted videos. A motion correction factor obtained from the mean of motion vectors (computed using a block-based motion estimation algorithm) is then applied to obtain the corrected harmonic gain/loss. The quality measure of the sequence is a linear combination of these corrected harmonic features. The parameters of the combination are obtained using a small training set.

Distributed Source Coding based estimation of channel-induced distortion. In [45], each macroblock in a frame is rasterized (i.e., converted into a vector) $\mathbf{x}^{(\mathbf{k})}$ and then a RR feature vector $\mathbf{y}$ is computed, where each entry of the feature vector is

$$y_i = \mathbf{a}^T \mathbf{x}^{(\mathbf{k})}$$

where $\mathbf{a}$ is s pseudo random vector with $||\mathbf{a}|| = 1$. This vector $\mathbf{y}$ is then subjected to a Wyner-Ziv encoding [40], in order to reduce the bit-rate. At the receiver, a similar process is carried out using the transmitted pseudo random vector and an estimate of the mean square error (MSE) is obtained between the transmitted RR feature vector and the one computed at the receiver. The authors claim that the method estimates MSE well with a small increase in transmission bit-rate (for the RR features).

RR Video Structural Similarity. The Structural SIMilarity index (SSIM) was proposed as a FR image quality assessment technique in [46] and extended to VQA in [47]. Albonico *et. al.* proposed a RR VQA in [48] where the extracted features were similar to those proposed for video SSIM. Using 16×16 blocks the mean $\mu_x(i,n)$ and standard deviation $\sigma_x(i,n)$ for the ith macroblock in the nth frame is computed at the source and are transmitted as the RR feature vector. At the receiver end, a distortion estimation is undertaken using the technique described in [49] and reviewed in the next section to produce $\widehat{D}(i,n)$. Further, the mean $\mu_{\widehat{x}}(i,n)$ and standard deviation $\sigma_{\widehat{x}}(i,n)$ are estimated at the receiver end from the received frame. A covariance estimate is then formed as:

$$\sigma_{x\widehat{x}}(i,n) = 0.5 \times [\sigma_{\widehat{x}}(i,n)^2 + \sigma_x(i,n)^2 + (\mu_{\widehat{x}}(i,n) - \mu_x(i,n))^2 - \widehat{D}(i,n)]$$

This allows for computation of the SSIM index for each frame. The authors state that the index estimated in such a fashion fails to match-up to the original SSIM value and hence a lookup-table based approach is used to eliminate a *bias* in the estimated values. The authors claim that this allows for an estimate of the SSIM index and hence of video quality.

5   No-Reference Algorithms

No-reference algorithms are those that seek to assess quality of a received video without any knowledge of the original source video. In a general setting, these algorithms assume that the distorting medium is known - for example, compression, loss induced due to noisy channel etc. Based on this assumption, distortions specific to the medium are modeled and quality is assessed. By far the most popular distorting medium is compression and blockiness and bluriness are generally evaluated for this purpose. We classify NR algorithms as those based on measuring blockiness, those that seek to model the effect of multiple artifacts (for example, blocking and blurring) and other techniques based on modeling the channel or the HVS .

5.1   Blockiness-Based Techniques

Blockiness measure for MPEG-2 Video. Tan and Ghanbari [50] used a harmonic analysis technique for NR VQA which was also used in [42] (for RR VQA). A Sobel operator is used to produce a gradient image, which is then subjected to a block FFT . The ratio of sum of harmonics to sum of all AC components within a block is computed in both horizontal and vertical directions. The phase of harmonics across the frame are then histogrammed. The authors suggest that a smaller standard deviation in the (empirical) probability density function (PDF) indicates greater blockiness. Based on certain pre-set thresholds on the harmonic ratio of the magnitudes and the phase of the harmonics a block is considered to be 'blocky'. The authors however test their technique on I and P-frames and state that the method does not function well for B-frames.

Detecting blocking artifacts in compressed video. Vlachos proposed an algorithm based on phase correlation to detect blockiness in MPEG coded SD videos [51]. A set of sub-sampled images from each frame is cross-correlated to produce a block-iness measure. The sampling structure chosen is such that each sub-image consists of a particular pixel from one of the 8×8 block used for MPEG compression. Even though the authors do not test this measure on a database and use an informal subjective test for one video, Winkler *et. al.* compare this measure with two other measures in [52]. However, Winkler *et. al.* also choose to use a part of the VQEG dataset to examine performance. Vlachos' measure does not seem to perform too well.

Perceptually significant block-edge impairment metric. Suthaharan proposed the perceptually significant block-edge impairment metric (PS-BIM) in [53] that uses luminance masking effects to improve performance. The measure is a ratio of two terms, each of which is expressed as a linear combination of horizontal and vertical blocking measures. The horizontal and vertical blocking measures are weighted sums of simple luminance differences where the weights are based on luminance masking effects. The authors used I-frames from coded sequences to demonstrate performance. No subjective evaluation was undertaken to evaluate performance.

NR blocking measure for adaptive video processing. Muijs and Kirenko compute a normalized horizontal gradient $D_{H,norm}$ at a pixel as the ratio of the absolute gradient and the average gradient over a neighboring region in a frame of a video [54]. They then sum $D_{H,norm}$ over the rows to produce a measure S_h as a function of the column. Blocking strength is then a ratio of the mean value of S_h at block boundaries to the mean value of S_h at intermediate positions. A small study was used to evaluate subjective quality and the algorithm was shown to perform well.

5.2 Multiple Artifact Measurement Based Techniques

No-reference objective quality metric (NROQM). Caviedes and Oberti computed a set of features including blocking, blurring , and sharpness from the degraded video in order to assess its quality [55]. Blocking is computed as weighted pixel differences between neighboring blocks. Ringing [3] is computed using a combination of edge detection and 'low-activity' area detection in regions around edges. Clipping - saturation at low/high pixel values due to numerical precision - is evaluated based on grey level values of pixels. Noise is measured using a block-based approach. A histogram-based approach is utilized to compute contrast and a kurtosis based approach is used for sharpness measurement. A training set of videos (with subjective MOS) is used to set parameters in order to combine these measures into a single score. A separate set of videos was used for testing and high correlation with human perception was demonstrated.

---

[3] Ringing artifacts are spurious signals that appear around regions of sharp-transitions. In images, these are seen as 'shimmering' rings around edges.

Quantifying blockiness and packet-loss. Babu et al. proposed two NR VQA metrics - one for blocking and another for packet-loss - in [56]. In essence, a block is said to be blocky if the edge strength does not have 'enough' variance. Overall blockiness is then the ratio of blocky blocks to total blocks in a frame. To quantify packet loss binary edge images are formed using row differences and for each macroblock, the measure of packet loss is the sum of the absolute differences between the edge images. For each frame, a squared sum is then computed as the final measure. A comparison of the proposed metrics is undertaken with others proposed in literature, however this comparison does not involve any subjective correlations.

NR VQA based on artifact measurement. Farias and Mitra proposed a metric that is based on measurement of blockiness , bluriness and noisiness [57]. Blockiness is measured using a modification of Vlachos' algorithm [51]. Width of edges (computed using a Canny edge detector) in a frame is used as a measure of blur. For each of the 8×8 blocks, variances of each of the 9 overlapping 3×3 regions within a block is computed, and the average of the lowest 4 variances is the block noise variance. A histogram of these averages is then used to compute a measure of the frame noisiness. A weighted Minkowski sum of these artifacts is a measure of quality. Parameters for this sum are estimated using subjective data. Using subjective data from a small study, the algorithm was shown to perform well.

NR VQA based on HVS. Massidda *et. al.* proposed an NR metric for blur detection, specifically for 2.5G/3G systems [58]. They computed blockiness, bluriness and moving artifacts to evaluate quality. Blockiness is evaluated using 8×8 non-overlapping blocks. Within each block they define 4 regions, and sums of horizontal and vertical edges (obtained using a Sobel filter) are computed over each of these regions. These values are then collapsed using Minkowski summation to form a single value (B_{Sob}). Blur is evaluated using the approach proposed in [59]. Mean and variance of gradients from two consecutive frames are computed and pooled to obtain a measure of moving artifacts-based distortion. The final quality index is is then a weighted combination of these three artifacts. The authors evaluate the quality index as a function of the quantization parameter, instead of using subjective scores.

Prototype NR VQA system. Dosselmann and Yang propose an algorithm that estimates quality by measuring three types of impairments that affect television and video signals - noise, blocking and bit-error based color impairments [60]. To measure noise, each frame is partitioned into blocks and inter-frame correlations between (a subset of) blocks are computed and averaged to form an indicator for that frame. Using a spatio-temporal region spanning 8 frames, the variance σ_η^2 is computed and the noise measure is $\eta = 1 - (1 - \sigma_\eta^2)^{p_1}$, where p_1 is experimentally set to 2048. The final noise metric is an average of 256 such values. An alignment based procedure is used to quantify blocking. Measurement of channel induced color error is performed by inspecting the R, G and B values from the RGB color space of a frame and thresholding these values. Even though these measures are not pooled the authors demonstrate how these measures can be incorporated into a system. Correlation with subjective assessment is not studied.

NR VQA based on frame quality measure. Kawayoke and Horita proposed a model for NR VQA consisting of frame quality measure and correction, asymmetric tracking and mean value filtering [61]. The frame quality measure is simply neighboring pixel differences with and without edge preservation filtering. This is histogrammed, pooled and corrected to obtain the final measure. Asymmetric tracking accounts for the fact that humans tend to perceive poorer regions with greater severity than good ones [62]. Finally mean value filtering removes high frequency ingredients from the measure to produce the quality index. Evaluation on a small dataset shows good correlation with perception.

5.3 Other Techniques

NR VQA incorporating motion information. Yang *et. al.* proposed a NR VQA which incorporates motion information in [63]. Block-based motion estimation is applied on a low-pass version of the distorted video. Translation regions of high spatial complexity are identified using thresholds on variance of motion vectors and luminance values. Using the computed motion vectors for these regions, sum of squared error is computed between the block under consideration and its motion compensated block in the previous frame, which is then low-pass filtered to give a spatial distortion measure. A temporal measure is computed using a function of the mean of the motion vectors. A part of the VQEG dataset is used to train the algorithm in order to set the thresholds and parameters in the functions. Testing on the rest of the dataset, the algorithm is shown to correlate well with human perception of quality.

NR Blur measurement for VQA. Lu proposed a method for blur evaluation to measure blur caused by video compression and imaging processes [64] . First a low pass filter is applied to each frame in order to eliminate blocking artifacts. Only a subset of pixels in a frame are selected on the basis of edge intensity and connectivity for blur measurement, in order to process only that 'type' of blur we are interested in (for example, blur due to compression as against blur due to a low depth of field). Blur is then estimated using a combination of an edge image and gradients at the sample points. The authors demonstrated that their algorithm correlates well with PSNR and the standard deviation of the blurring kernel for three videos.

NR Fluidity Measure. In this chapter, we have so far discussed NR measures which generally do not model frame-drops. The measure proposed by Pastrana-Vidal and Gicquel in [65] covers this important aspect of NR VQA. Other works along the same lines that we haven't discussed here include [66], [67] and [68]. In [65] the discontinuity along the temporal axis, which the authors label fluidity break is first isolated and its duration is computed. Abrupt temporal variation is estimated using normalized MSE between luminance components in neighboring frames. Based on experimental results, a series of transformations and thresholds are applied to produce a quality index for the video. A subjective study was carried out and good correlation with results were demonstrated.

The authors extend the framework proposed in [65] for fluidity measurement in [69] where a clearness-sharpness metric is introduced to quantify perceptually significant blurring in video. The two measures are pooled using a multiplicative term (in order to account for approximate spatio-temporal separability of the HVS) and a final quality metric is produced. A set of videos were used to test the proposed index.

Perceptual Temporal Quality Metric. Yang *et. al.* compute the dropping severity as a function of the number of frames dropped using timestamp information from the video stream [70]. They then segment the video temporally into cuts/segments [71], and determine the motion activity of each such segment using the average of (thresholded) motion vectors from that segment. The dropping severity is then mapped onto a perceptually significant dropping factor based on this motion information. A temporal pooling using a temporal window based approach is then undertaken. This value is then subjected to a non-linearity and the parameters are estimated using a fit to subjective data, with hardcoded thresholds. Further weighting and pooling of this temporal indicator leads to the final quality score. Using a mix of expert and non-expert viewers, a subjective study was undertaken and good performance on a small set of videos was demonstrated.

NR VQA based on error-concealment effectiveness. Yamada *et. al.* defined an error concealment process to be ineffective for a block if the absolute sum of motion vectors for that block (obtained from motion vector information in the encoded video stream) is greater than some threshold [72]. Luminance discontinuity is then computed at the erroneous regions as the mean of absolute differences between correctly decoded regions and regions where error concealment has been applied. Another threshold indicates if this region has been concealed effectively. No evaluation with respect to subjective perception is carried out, but effectiveness of the measures are evaluated on a test-set based on packet-loss ratio and number of impairment blocks.

NR modeling of channel distortion. Naccari *et. al.* proposed a model for channel induced distortion at the receiver for H.264/AVC [21] coded videos in [49]. The model seeks to estimate the mean-squared error between the received and transmitted videos - which can also be expressed as the mean distortion induced by all the macroblocks in a frame. Hence the quantity they wish to estimate is the distortion induced by the ith macroblock in frame n - $\widehat{D}_n^i$. In order to do this, they consider two cases depending upon whether the macroblock under consideration was lost or correctly received. In the former case, the predicted distortion is modeled as a sum of distortions arising from motion vectors, prediction residuals and distortion propagation. For the latter case, the distortion is simply due to error propagation from the previous frame. The de-blocking filter in H.264/AVC [21] is further modeled using attenuation coefficients. Correlation between estimated channel distortion and measured channel distortion is estimated for performance evaluation.

NR VQA for HD video. Keimel *et. al.* proposed an NR VQA algorithm specifically for HD video [73]. Features extracted for this purpose include blur, blockiness, activity and predictability using previously proposed methods. Different pooling strategies are used for each of these features, followed by principal component analysis (PCA) for dimensionality reduction and partial least squares regression (PLSR) to map the features to visual quality. A correction factor is then incorporated based on a low-quality version of the received video to find the final quality score. The astute reader would have observed that the proposed method is pretty general and does not specifically address HD video. The proposed algorithm is tested on HD video however and decent performance is demonstrated.

Video quality monitoring of streamed videos. Ong *et. al.* model jerkiness between frames using absolute difference between adjacent frames and a threshold [74]. Picture loss is similarly detected with another threshold. Blockiness is detected using using a technique similar to those proposed previously. The test methodology is non-standard and requires users to identify number of picture freezes, blocks and picture losses in the videos. Perceptual quality is not evaluated however.

Other techniques for NR VQA include the one proposed in [75] which we do not explore since a patent on the idea was filed by the authors. Further, many of the metrics discussed here were evaluated for their performance using a variety of criteria in [76, 77]. Hands *et. al.* provide an overview of NR VQA techniques and their application in Quality of Service (QoS) [78]. Kanumri *et. al.* model packet loss visibility in MPEG-2 video in [79] to assess quality of video.

6   Conclusion

In this chapter we began with a discussion of video quality assessment and introduced datasets to evaluate performance of algorithms. We then went on to describe the human visual system briefly. A summary of recent reduced and no reference algorithms for quality assessment then followed.

We hope that by now the reader would have inferred that NR VQA is a difficult problem to solve. It should also be amply clear that even though a host of methods have been proposed (most of which are listed here) there does not seem to emerge an obvious winner. Our arguments on the use of a common publicly available dataset for performance evaluation are hence of importance. The reader would have observed that most authors tend to select a particular kind of distortion that affects videos and evaluate quality. Any naive viewer of videos will testify to the fact that distortions in videos are not singular. In fact, compression - which is generally assumed to have a blocking distortion, also introduces blurring and motion-compensation mismatches, mosquito noise, ringing and so on [80]. Given that there exist a host of distortions that may affect a video, one should question the virtue of trying to model each individual distortion. Further, if one does choose to model each distortion individually, a method to study the effect of multiple distortions must be undertaken. Again, this is a combinatorially challenging problem.

A majority of algorithms seek to model spatial-distortions alone and even though some methods include elementary temporal features, a wholesome approach to NR VQA should involve a spatio-temporal distortion model. Further, in most cases a majority of the design decisions are far removed from human vision processing. It is imperative as researchers that we keep in mind that the ultimate receiver is the human and hence understanding and incorporating HVS properties in an algorithm is of essence. Finally, even though we listed statistical measures to evaluate performance, researchers are working on alternative methods to quantify performance [81, 82, 83].

References

1. Bt-500-11: Methodology for the subjective assessment of the quality of television pictures. International telecommuncation union
2. Sekuler, R., Blake, R.: Perception. McGraw Hill, New York (2002)
3. Wandell, B.: Foundations of vision. Sinauer Associates (1995)
4. Seshadrinathan, K., Bovik, A.C.: Video quality assessment. In: Bovik, A.C. (ed.) The Essential Guide to Video Processing. Academic Press, London (2009)
5. Moorthy, A.K., Seshadrinathan, K., Bovik, A.C.: Digital Video Quality Assessment Algorithms. Springer, Heidelberg (2009)
6. Final report from the video quality experts group on the validation of objective quality metrics for video quality assessment,
   http://www.its.bldrdoc.gov/vqeg/projects/frtv_phaseI
7. Girod, B.: What's wrong with mean-squared error? In: Watson, A.B. (ed.) Digital images and human vision, pp. 207–220 (1993)
8. Wang, Z., Bovik, A.C.: Mean squared error: Love it or leave it? - a new look at fidelity measures. IEEE Signal Processing Magazine (2009)
9. Seshadrinathan, K.: Video quality assessment based on motion models. Ph.D. thesis, The University of Texas at Austin (2008)
10. Wang, Z., Li, Q.: Video quality assessment using a statistical model of human visual speed perception. Journal of the Optical Society of America 24(12), B61–B69 (2007)
11. Generic coding of moving pictures and associated audio information - part 2: Video, ITU-T and ISO/IEC JTC 1. ITU-T Recommendation H.262 and ISO/IEC 13 818-2 (MPEG-2) (1994)
12. Advanced video coding, ISO/IEC 14496-10 and ITU-T Rec. H.264 (2003)
13. Live wireless video database (2009),
    http://live.ece.utexas.edu/research/quality/live_wireless
    live_wireless_video.html
14. Live video database (2009),
    http://live.ece.utexas.edu/research/quality/
    live_video.html
15. Seshadrinathan, K., Soundararajan, R., Bovik, A.C., Cormack, L.K.: Study of subjective and objective quality assessment of video. IEEE Transactions on Image Processing (to appear)

16. Moorthy, A.K., Bovik, A.C.: Wireless video quality assessment: A study of subjective scores and objective algorithms. IEEE Transactions on Circuits and Systems for Video Technology (to appear)
17. Sheskin, D.: Handbook of parametric and nonparametric statistical procedures. CRC Pr. I Llc., Boca Raton (2004)
18. Moorthy, A.K., Bovik, A.C.: A motion-compensated approach to video quality assessment. In: Proc. IEEE Asilomar Conference on Signals, Systems and Computers (2009)
19. Wang, Z., Lu, L., Bovik, A.: Foveation scalable video coding with automatic fixation selection. IEEE Transactions on Image Processing 12(2), 243 (2003)
20. Beauchemin, S., Barron, J.: The computation of optical flow. ACM Computing Surveys (CSUR) 27(3), 433–466 (1995)
21. Richardson, I.: H. 264 and MPEG-4 video compression: video coding for next-generation multimedia. John Wiley & Sons Inc., Chichester (2003)
22. Winkler, S.: A perceptual distortion metric for digital color video. In: Proc. SPIE, vol. 3644(1), pp. 175–184 (1999)
23. Sugimoto, O., Kawada, R., Wada, M., Matsumoto, S.: Objective measurement scheme for perceived picture quality degradation caused by MPEG encoding without any reference pictures. In: Proceedings of SPIE, vol. 4310, p. 932 (2000)
24. MacWilliams, F., Sloane, N.: Pseudo-random sequences and arrays. Proceedings of the IEEE 64(12), 1715–1729 (1976)
25. Campisi, P., Carli, M., Giunta, G., Neri, A.: Blind quality assessment system for multimedia communications using tracing watermarking. IEEE Transactions on Signal Processing 51(4), 996–1002 (2003)
26. Fu-zheng, Y., Xin-dai, W., Yi-lin, C., Shuai, W.: A no-reference video quality assessment method based on digital watermark. In: 14th IEEE Proceedings on Personal, Indoor and Mobile Radio Communications, PIMRC 2003, vol. 3 (2003)
27. Farias, M., Carli, M., Neri, A., Mitra, S.: Video quality assessment based on data hiding driven by optical flow information. In: Proceedings of the SPIE Human Vision and Electronic Imaging IX, San Jose, CA, USA, pp. 190–200 (2004)
28. Carli, M., Farias, M., Gelasca, E., Tedesco, R., Neri, A.: Quality assessment using data hiding on perceptually important areas. In: IEEE International Conference on Image Processing (2005)
29. Wolf, S., Pinson, M.: Video quality measurement techniques. National Telecommunications and Information Administration (NTIA) Report 02-392 (2002)
30. Wolf, S., Pinson, M.: Low bandwidth reduced reference video quality monitoring system. In: First International Workshop on Video Processing and Quality Metrics for Consumer Electronics, Scottsdale, AZ, USA (2005)
31. Oelbaum, T., Diepold, K.: Building a reduced reference video quality metric with very low overhead using multivariate data analysis. In: International Conference on Cybernetics and Information Technologies, Systems and Applications, CITSA 2007 (2007)
32. Oelbaum, T., Diepold, K.: A reduced reference video quality metric for avc/h.264. In: Proc. European Signal Processing Conference, pp. 1265–1269 (2007)
33. Hair, J.: Multivariate data analysis. Prentice Hall, Englewood Cliffs (2006)
34. Haykin, S.: Neural networks: a comprehensive foundation. Prentice Hall, Englewood Cliffs (2008)
35. Le Callet, P., Viard-Gaudin, C., Barba, D.: A convolutional neural network approach for objective video quality assessment. IEEE Transactions on Neural Networks 17(5), 1316 (2006)
36. Meng, F., Jiang, X., Sun, H., Yang, S.: Objective Perceptual Video Quality Measurement using a Foveation-Based Reduced Reference Algorithm. In: IEEE International Conference on Multimedia and Expo., pp. 308–311 (2007)

37. Wang, Z., Wu, G., Sheikh, H., Simoncelli, E., Yang, E., Bovik, A.: Quality-aware images. IEEE Transactions on Image Processing 15(6), 1680–1689 (2006)
38. Hiremath, B., Li, Q., Wang, Z.: Quality-aware video. In: IEEE International Conference on Image Processing, ICIP 2007, vol. 3 (2007)
39. Simoncelli, E., Freeman, W., Adelson, E., Heeger, D.: Shiftable multiscale transforms. IEEE Transactions on Information Theory 38(2), 587–607 (1992)
40. Cover, T., Thomas, J.: Elements of information theory. Wiley-Interscience, Hoboken (2006)
41. Yamada, T., Miyamoto, Y., Serizawa, M., Harasaki, H.: Reduced-reference based video quality-metrics using representative-luminance values. In: Third International Workshop on Video Processing and Quality Metrics for Consumer Electronics, Scottsdale, AZ, USA (2007)
42. Gunawan, I., Ghanbari, M.: Reduced-reference video quality assessment using discriminative local harmonic strength with motion consideration. IEEE Transactions on Circuits and Systems for Video Technology 18(1), 71–83 (2008)
43. Gonzalez, R., Woods, R.: Digital image processing. Prentice-Hall, Inc., Upper Saddle River (2002)
44. Oppenheim, A., Schafer, R.: Discrete-time signal processing. Prentice-Hall, Inc., Upper Saddle River (1989)
45. Valenzise, G., Naccari, M., Tagliasacchi, M., Tubaro, S.: Reduced-reference estimation of channel-induced video distortion using distributed source coding. In: Proceeding of the 16th ACM international conference on Multimedia (2008)
46. Wang, Z., Bovik, A.C., Sheikh, H.R., Simoncelli, E.P.: Image quality assessment: From error measurement to structural similarity. IEEE Signal Processing Letters 13(4), 600–612 (2004)
47. Wang, Z., Lu, L., Bovik, A.C.: Video quality assesssment based on structural distortion measurement. Signal Processing: Image communication (2), 121–132 (2004)
48. Albonico, A., Valenzise, G., Naccari, M., Tagliasacchi, M., Tubaro, S.: A Reduced-Reference Video Structural Similarity metric based on no-reference estimation of channel-induced distortion. In: IEEE International Conference on Acoustics, Speech, and Signal Processing, ICASSP (2009)
49. Naccari, M., Tagliasacchi, M., Pereira, F., Tubaro, S.: No-reference modeling of the channel induced distortion at the decoder for H. 264/AVC video coding. In: Proceedings of the International Conference on Image Processing, San Diego, CA, USA (2008)
50. Tan, K., Ghanbari, M.: Blockiness detection for MPEG2-coded video. IEEE Signal Processing Letters 7(8), 213–215 (2000)
51. Vlachos, T.: Detection of blocking artifacts in compressed video. Electronics Letters 36(13), 1106–1108 (2000)
52. Winkler, S., Sharma, A., McNally, D.: Perceptual video quality and blockiness metrics for multimedia streaming applications. In: Proceedings of the International Symposium on Wireless Personal Multimedia Communications, pp. 547–552 (2001)
53. Suthaharan, S.: Perceptual quality metric for digital video coding. Electronics Letters 39(5), 431–433 (2003)
54. Muijs, R., Kirenko, I.: A no-reference blocking artifact measure for adaptive video processing. In: Proceedings of the 13th European Signal Processing Conference, EUSIPCO 2005 (2005)
55. Caviedes, J., Oberti, F.: No-reference quality metric for degraded and enhanced video. In: Proceedings of SPIE, vol. 5150, p. 621 (2003)
56. Babu, R., Bopardikar, A., Perkis, A., Hillestad, O.: No-reference metrics for video streaming applications. In: International Packet Video Workshop (2004)

57. Farias, M., Mitra, S.: No-reference video quality metric based on artifact measurements. In: IEEE International Conference on Image Processing, vol. 3, pp. 141–144 (2005)
58. Massidda, F., Giusto, D., Perra, C.: No reference video quality estimation based on human visual system for 2.5/3G devices. In: Proceedings of SPIE, vol. 5666, p. 168 (2005)
59. Marziliano, P., Dufaux, F., Winkler, S., Ebrahimi, T.: Perceptual blur and ringing metrics: Application to JPEG 2000. Signal Processing: Image Communication 19(2), 163–172 (2004)
60. Dosselmann, R., Yang, X.: A Prototype No-Reference Video Quality System. In: Fourth Canadian Conference on Computer and Robot Vision, CRV 2007, pp. 411–417 (2007)
61. Kawayoke, Y., Horita, Y.: NR objective continuous video quality assessment model based on frame quality measure. In: 15th IEEE International Conference on Image Processing, ICIP 2008, pp. 385–388 (2008)
62. Moorthy, A.K., Bovik, A.C.: Visual importance pooling for image quality assessment. IEEE Journal of Selected Topics in Signal Processing, Issue on Visual Media Quality Assessment 3(2), 193–201 (2009)
63. Yang, F., Wan, S., Chang, Y., Wu, H.: A novel objective no-reference metric for digital video quality assessment. IEEE Signal processing letters 12(10), 685–688 (2005)
64. Lu, J.: Image analysis for video artifact estimation and measurement. In: Proceedings of SPIE, vol. 4301, p. 166 (2001)
65. Pastrana-Vidal, R., Gicquel, J.: Automatic quality assessment of video fluidity impairments using a no-reference metric. In: Proc. of Int. Workshop on Video Processing and Quality Metrics for Consumer Electronics (2006)
66. Pastrana-Vidal, R., Gicquel, J., Colomes, C., Cherifi, H.: Sporadic frame dropping impact on quality perception. In: Proceedings of SPIE, vol. 5292, p. 182 (2004)
67. Pastrana-Vidal, R., Gicquel, J., Colomes, C., Cherifi, H.: Frame dropping effects on user quality perception. In: 5th International Workshop on Image Analysis for Multimedia Interactive Services (2004)
68. Lu, Z., Lin, W., Seng, B., Kato, S., Ong, E., Yao, S.: Perceptual Quality Evaluation on Periodic Frame-Dropping Video. In: Proc. of IEEE Conference on Image Processing, pp. 433–436 (2007)
69. Pastrana-Vidal, R., Gicquel, J.: A no-reference video quality metric based on a human assessment model. In: Third International Workshop on Video Processing and Quality Metrics for Consumer Electronics, VPQM, vol. 7, pp. 25–26 (2007)
70. Yang, K., Guest, C., El-Maleh, K., Das, P.: Perceptual temporal quality metric for compressed video. IEEE Transactions on Multimedia 9(7), 1528–1535 (2007)
71. Cotsaces, C., Nikolaidis, N., Pitas, I.: Video shot detection and condensed representation. a review. IEEE signal processing magazine 23(2), 28–37 (2006)
72. Yamada, T., Miyamoto, Y., Serizawa, M.: No-reference video quality estimation based on error-concealment effectiveness. In: Packet Video 2007, pp. 288–293 (2007)
73. Keimel, C., Oelbaum, T., Diepold, K.: No-Reference Video Quality Evaluation for High-Definition Video. In: Proceedings of the International Conference on Image Processing, San Diego, CA, USA (2009)
74. Ong, E., Wu, S., Loke, M., Rahardja, S., Tay, J., Tan, C., Huang, L.: Video quality monitoring of streamed videos. In: IEEE International Conference on Acoustics, Speech and Signal Processing, pp. 1153–1156 (2009)
75. Knee, M.: A single-ended picture quality measure for MPEG-2. In: Proc. International Broadcasting Convention, pp. 7–12 (2000)
76. Leontaris, A., Reibman, A.: Comparison of blocking and blurring metrics for video compression. In: Proceedings of the IEEE International Conference on Acoustics, Speech, and Signal Processing, ICASSP 2005, vol. 2 (2005)

77. Leontaris, A., Cosman, P.C., Reibman, A.: Quality evaluation of motion-compensated edge artifacts in compressed video. IEEE Transactions on Image Processing 16(4), 943–956 (2007)
78. Hands, D., Bourret, A., Bayart, D.: Video QoS enhancement using perceptual quality metrics. BT Technology Journal 23(2), 208–216 (2005)
79. Kanumuri, S., Cosman, P., Reibman, A., Vaishampayan, V.: Modeling packet-loss visibility in MPEG-2 video. IEEE transactions on Multimedia 8(2), 341–355 (2006)
80. Yuen, M., Wu, H.: A survey of hybrid MC/DPCM/DCT video coding distortions. Signal Processing 70(3), 247–278 (1998)
81. Wang, Z., Simoncelli, E.P.: Maximum differentiation (MAD) competition: A methodology for comparing computational models of perceptual quantities. Journal of Vision 8(12), 1–13 (2008)
82. Charrier, C., Knoblauch, K., Moorthy, A.K., Bovik, A.C., Maloney, L.T.: Comparison of image quality assessment algorithms on compressed images. In: SPIE conference on Image quality and System Performance (to appear, 2010)
83. Charrier, C., Maloney, L.T., Cheri, H., Knoblauch, K.: Maximum likelihood difference scaling of image quality in compression-degraded images. Journal of the Optical Society of America 24(11), 3418–3426 (2007)

Chapter 2
Quality of Experience for High Definition Presentations – Case: Digital Cinema

Andrew Perkis, Fitri N. Rahayu, Ulrich Reiter, Junyong You, and Touradj Ebrahimi

Abstract. World-wide roll out of Digital Cinema (D-Cinema) is pushing the boundaries for Ultra High Definition content to all of us – the users. This poses new challenges on assessing our perception of high quality media and also understanding how quality can impact the business model. We use Quality of Experience (QoE) as the term used to describe the perception of how usable or good the users think the media or services are. In order to understand QoE we explore the Human Visual System (HVS) and discuss the impact of HVS on designing a methodology for measuring subjective quality in D-Cinema as a use case. Following our methodology, we describe our laboratory set up at the NOVA kino – a 440 seat fully digitized screen in full cinema production 24/7. This setup is used for subjective assessment of D-Cinema content, applying a test methodology adopted from existing recommendations. The subjective quality results are then used to design and to validate an objective metric. Our experiments demonstrate that traditional quality metrics cannot be adopted in D-Cinema presentations directly. Instead we propose a HVS-based approach to improve the performance of objective quality metrics, such as PSNR, SSIM, and PHVS, in quality assessment of D-Cinema setups. Finally we conclude by discussing how quality impacts the business model, using empirical data from a pilot D-Cinema roll out in Norway – the NORDIC project. This pilot showed that quality can influence users and create more loyalty and willingness to pay a premium if the high quality is used to enhance their experience. Examples are the widespread reintroduction of 3D as well as alternative content such as music, sports, and business related presentations.

1 Introduction

The future world of ubiquitous, interconnected media services and devices can be analyzed from two perspectives: the perspective of the media production/delivery

Andrew Perkis · Fitri N. Rahayu · Ulrich Reiter · Junyong You · Touradj Ebrahimi
Centre for Quantifiable Quality of Service in Communication Systems (Q2S),
Norwegian University of Science and Technology, Trondheim, Norway
e-mail: andrew@iet.ntnu.no, fitri@q2s.ntnu.no,
reiter@q2s.ntnu.no, junyong.you@q2s.ntnu.no,
touradj.ebrahimi@epfl.ch

chain – the INDUSTRY, and the perspective of the media consumption chain – the USER. Behind these two perspectives, there is a technology drive and a business drive. The technology drive is to a large extent summarized through efforts such as the European framework initiative Networked Media and is thus well covered through research and development. However, Networked Media is not complete without analysis and understanding of business perspectives [1]. In this context:

- The media production/delivery chain represents the business perspectives and opportunities as well as the needs for investments.
- The media consumption chain represents the buyers of the media and thus represents the income potential.

Together, these two perspectives create the media value chain. The relative importance of each perspective in the media value chain is a chicken and egg problem. In our analysis, we choose to put the user at the centre of attention considering that the industry is there to serve the user and must understand their needs and perception in offered products and services.

The expectations and technical comfort levels of the users have evolved in terms of complexity, as users are increasingly embracing advanced technologies which fit their lifestyle (leisure, work and education). In this direction, a typical user today utilizes an array of digital media providers, services and devices.

The framework to assess the user's behavior and the necessary technology management is based on assessing the user experience in a consistent way, and rewarding the user's loyalty through innovative packages and new engaging services, and content delivered through their device of choice whenever and wherever they want it. These assessments are crucial for the industry and drive their innovations and investments in future new digital media and services.

Under these new conditions, the ultimate measure of media communications and the services they offer is how users perceive the performance and especially the quality of the media, in technical terms denoted by Quality of Experience (QoE). QoE is typically the term used to describe the perception of a specific consumer of a given product or service, in terms of added value, usability, and satisfaction.

In this light, QoE can be defined as the characteristics of the sensations, perceptions and expectations of people as they interact with multimedia applications through their different perceptual sensors (mostly restricted to vision and hearing in today's audiovisual context). QoE is a combined/aggregated metric that tells if Quality is good, acceptable or bad. The impact on the business model is obvious although not always taken into consideration.

A poor QoE will result in unsatisfied users, leading to a poor market perception and ultimately, brand dilution. Being able to quantify the QoE as perceived by end-users can play a major role in the success of future media services, both for the companies deploying them and with respect to the satisfaction of end-users that use and pay for the services [2].

In our definition, we will view the media sector as consisting of four major industries:

- The printed media industry (newspapers and magazines)
- The broadcast industry (radio and TV)
- The electronic media industry (web, mobile and media on the move)
- The motion picture industry (cinema)

Both the electronic media industry and especially the motion picture industry strive for High Definition. In this chapter we will focus on the motion picture industry, as this at the moment is perceived as producing the highest quality media to users.

Going to the movies is the end product of a long process involving a complex value chain. This value chain has developed and operated in the same manner for over 100 years. Innovations have evolved and refined the process. This process includes a few major revolutions such as going from silent movies to sound and more recently being the last of the entertainment industries to go digital [3]. This innovation bears in it a revolution and a dramatic change for some of the players in the value chain. The innovation itself is technical, which also results in organizational and managerial changes.

The open question still remains and is whether quality plays a role in the development, is an integral part of the business model, and that the focus on the innovation is on the user.

Digital Cinema (D-Cinema) requires a complete change of infrastructure in all screens worldwide. The traditional 35mm film projector needs to be replaced with a D-Cinema server and a digital projector. The process of the change is referred to as the D-Cinema roll out which results in the exhibitors adopting and starting to use the new technology. The way the roll out is done is governed by a business model where the actors in the value chain have to agree on a financial model to cover investments, and novel organization and business models to deliver the film. The predominant financial model used so far is a so called Virtual Print Fee (VPF), where studios pay the exhibitors a contribution towards the investment based on long term agreements to screen their digital films. The associated business model has been through setting up businesses to purchase D-Cinema equipments and to lease them to exhibitors. These businesses are referred to as 3rd party integrators.

So far quality has not been used explicitly to drive the roll out, although it is an important factor. The motivations for the change are complex and not solely based on quality, and not all benefits are seen by the user.

One of the fundamental problems of D-Cinema is that the exhibitors themselves, the owners of the screens, are those in the value chain with the least benefit of the digitization. The highest benefits are for the content owners, the studios producing the film, and possibly on the end-user, the cinema goers. This means that the roll out has to be communicated and motivated between all players in the value chain in order for the adoption to take place. It is crucial that the financial and business model considers this for D-Cinema to succeed and grasp all players. D-Cinema roll out opens for the use of change agents, arenas, people and technologies trying to convince and to persuade the exhibitors to change despite the fact that the pain is greater than the gain.

The rest of this chapter is organized as follows. Section 2 provides an introduction to the Human Visual System and how this relates to quality metrics and evaluations. Section 3 outlines our evaluation set up and methodology for doing subjective quality assessments in a real cinema on D-Cinema quality content. Section 4 provides our work on developing perceptual image metrics for D-Cinema content and correlates the results to the subjective assessment. Section 5 brings us back to the business perspective and provides results from the NORDIC D-Cinema trial which was run in Norway from 2005 to 2008.

2 Role of Human Visual System in the Perception of Visual Quality

The Human Visual System (HVS) seldom responds to direct stimulation from a light source. Rather, light is reflected by objects and thus transmits information about certain characteristics of the object. The reflected ray of light enters the eyeball through the cornea. The cornea represents the strongest part of the refracting power of the eye, providing about 80% of the total eye's refracting capacity. After passing through the cornea and the watery aqueous humor, the photon beam enters the inner eye through the pupil, which regulates the amount of light allowed to enter. The lens focuses the light on the sensory cells of the retina [4].

The internal layer of the eyeball is made up of a nervous coat called retina. The retina covers the inner back part of the eyeball and is where the optical image is formed by the eye's optical system. Here, a photochemical transduction occurs: nerve impulses are created and transmitted along the optic nerve to the brain for higher cortical processing. The point of departure of that optic nerve through the retina does not have any receptors, and thus produces a "blind spot". The retina consists of two different types of light-sensitive cells, rods and cones. There are about 6.5 million cones in each eyeball, most of them located in the middle of the retina, in a small area about 1.mm in diameter called the fovea or fovea centralis. Fovea is the center of the eye's sharpest vision and the location of most color perception, performing in bright light, but being fairly insensitive at lower light levels. Located around the fovea centralis are about 120 million rods. They are mainly responsible for vision in dim light and produce images consisting of varying shades of black and white. The acuity over most of that range is poor, and the rods are multiply connected to nerve fibers, so that a single nerve fiber can be activated by any one of about a hundred rods. In contrast, cones in the fovea centralis are more individually connected to nerve fibers [5].

The eyeball is situated in the orbital cavity, a location that protects it and provides a rigid bony origin for the six extrinsic muscles that produce ocular movement. When the visual system focuses on a certain object, then the optical axes of both eyes are adjusted toward it. According to Snell and Lemp, the sensation of tension in the muscles serves as an indicator of the distance the object is away. The direction of a visually perceived object corresponds directly to the position of its image on the retina. In the determination of an object's distance to the eye, there are a number of potential sources or cues of depth.

2.1   Cortical Processing of Stimuli

The optic nerves, transmitting sensory information from each eye, proceed posterior and medially to unite at the optic chiasm. There, fibers from the nasal halves of each retina cross over to the opposite hemisphere. Fibers from the temporal halves project to the hemisphere on the same side. The result is that signals from the same regions of the visual field are projecting to the same hemisphere; thus, the left half of the visual field projects to the right half of retina, which in turn sends neural signals to the right hemisphere.

There is a point-to-point relation between retina, lateral geniculate nucleus, and the primary visual cortex. Impulses from the upper and lower halves of the visual field are located in different parts of the optic radiation, and consequently also project into different areas of the primary visual cortex V1. This is called retinotopic projection, as the distribution of stimulation on the retina is preserved in the cortex.

The primary visual cortex V1 seems to separate the pattern of light falling on the retina into discrete features. Apparently, these are retinal location, orientation, movement, wavelength, and the difference between the two eyes. In subsequent cortical processing these features are further differentiated. Therefore, the primary visual cortex has the task of sorting visual information and distributing it to other, more specialized cortical areas.

Two visual streams have been identified by Ungerleider and Mishkin in 1982 that originate from the primary visual cortex: the dorsal or parietal stream, and the ventral or temporal stream. Apparently, the first correlates more to location, depth and movement, whereas the latter is more connected to color, spatial detail and form [6, 7]. Goldstein, on the basis of the experiments performed by Ungerleider and Mishkin, suggests that perceiving the location of an object is attributed to the dorsal stream, whereas the ventral stream determines the object's identity – the *where* and *what* dimensions of vision [8].

Although the basic mechanisms of sensory information transmission are well understood, a detailed understanding of how visual input is processed and interpreted is still missing. Especially the transition from neuronal reaction to perception, i.e. the process of attaching a meaning to the stimulus, remains unexplained. Based on the findings introduced above, it seems that for the perception of visual quality, the ventral stream is of higher importance than the dorsal stream. How this affects visual perception remains unclear.

2.2   Perception and Attention

From experience we know that perceived visual quality is highly context- and task-dependent. This is related to the way we generally perceive stimuli: Neisser's model of the Perceptual Cycle describes perception as a setup of schemata, perceptual exploration and stimulus environment [9]. These elements influence each other in a continuously updated circular process (see Fig. 1). Thus, Neisser's model describes how the perception of the environment is influenced by background knowledge which in turn is updated by the perceived stimuli.

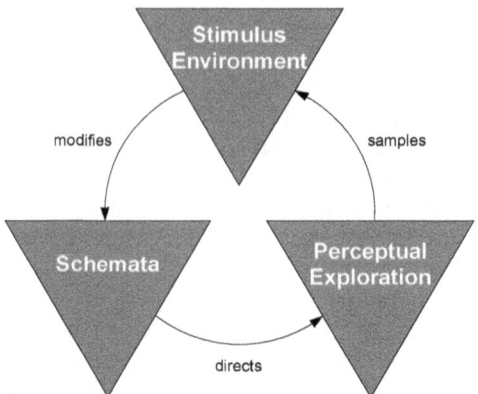

Fig. 1 Neisser's perceptual cycle

In Neisser's model, schemata represent the knowledge about our environment. They are based on previous experiences and are located in the long term memory. Neisser attributes them to generate certain expectations and emotions that steer our attention in the further exploration of our environment. The exploratory process consists, according to Neisser, in the transfer of sensory information (the stimulus) into the short-term memory. In the exploratory process, the entirety of stimuli (the stimulus environment) is compared to the schemata already known. Recognized stimuli are given a meaning, whereas unrecognized stimuli will modify the schemata, which will then in turn direct the exploratory process The differences in schemata present in the human individual cause the same stimulus to provoke different reactions between subjects. Following Neisser's model, especially new experiences (those that cause a modification of existing schemata) are likely to generate a higher load in terms of processing requirements for the percepts.

The schemata therefore also control the attention that we pay toward stimuli. A large number of studies have tried to identify and describe the strategy that is actually used in the human perceptual process. Pashler gives an overview and identifies two main concepts of attention [10]: attention as based on exclusion (gating) or based on capacity (resource) allocation. The first concept defines the mechanism that reduces processing of irrelevant stimuli to be attended. It can be regarded as a filtering device that keeps out stimuli from the perceptual machinery that performs the recognition. Attention is therefore identified with a purely exclusionary mechanism. The second concept construes the limited processing resource (rather than the filtering device) as attention. It suggests that when attention is given to an item, it is perceptually analyzed. When attention is allocated to several items, they are processed in parallel until the capacity limits are exceeded. In that case, processing becomes less efficient.

Neither of the two concepts can be ruled out by the many investigations performed up to now. Instead, assuming either, the gating or the resource interpretation, all empirical results can be accounted for in some way or other. As a result, it must be concluded that both capacity limits and perceptual gating characterize the human perceptual processing. This combined concept is termed Controlled Parallel Processing (CPP). CPP claims that parallel processing of different objects is achievable, but optional. At the same time, also selective processing of a single object is possible, largely preventing other stimuli from undergoing full perceptual analysis.

For the evaluation of perceived visual quality in the D-Cinema context this means that subjective assessment methodologies and objective metrics found to be applicable for other types of visual presentations may not be valid here. In fact, screen size and resolution in the D-Cinema context differ greatly from those set-ups for which recommendations exist. At the same time, it can be assumed that cinema goers will expect the highest quality possible (or perceivable), given that the industry has claimed to provide exactly this.

3 Subjective Image Quality Assessment in D-Cinema

In the field of subjective evaluation, there are many different methodologies and rules to design a test. The test recommendations described by the ITU have been internationally accepted as guidelines for conducting subjective assessments. Recommendation ITU-R BT500-11 [11] provides a thorough guideline for the testing methods and the test conditions of subjective visual quality assessments. Important issues in the guidelines include characteristics of the laboratory set up, stimulus viewing sequence, and rating scale. Although recommendation ITU-R BT500-11, is the guidelines intended for assessing picture quality of traditional television broadcast, it still provides relevant guidelines for conducting subjective visual quality assessment of recent and enhanced services such as internet based multimedia applications, HDTV broadcasting, etc. Several issues described in the ITU-R BT500-11 are relevant to subjective quality assessment in digital cinema environment. Another important guideline relevant to this work is recommendation ITU-R BT.1686 [12]; it provides recommendations on how to perform on-screen measurements of the main projection parameters of large screen digital imagery applications, based on presentation of programs in a theatrical environment.

3.1 Laboratory Set Up

The evaluation described here has been conducted at a commercial digital cinema in Trondheim, Norway. The DCI-specified cinema set up is considered to provide ideal viewing conditions. Figure 2 shows a view of the auditorium. Table 1 gives the specifications of the movie theatre.

The digital cinema projector used is a Sony CineAlta SRX-R220 4K projector, one of the most advanced projectors in digital cinema installations around the world (for more details on this projector see [13, 14]). Projector installation, calibration, and maintenance have been performed by Nova Kinosenter, Trondheim Kino AS. Therefore, it did not seem necessary to perform any additional measurement of contrast, screen illumination intensity and uniformity, or any other measurements recommended in [12].

Table 1 Specification of the movie theater

DISPLAY			HALL		PROJECTOR	
Screen (H x W)	5 x 12 m		Number of Seats	440	Type	Sony SRX-R220 4K
Projection Distance	19 m		Number of Wheelchair Seats	3		
Image Format	WS 1:1.66		Width	18.3 m		
	WS 1:1.85		Floor area	348 m^2		
	CS 1:2.35		Built Year	1994		

Fig. 2 Liv Ullman Auditorium of Nova Kinosenter (NOVA 1)

In order to reproduce a movie theatre experience, the assessment was conducted in the same conditions as when watching a feature film, i.e. in complete darkness. To illuminate the subject's scoring sheets during the subjective assessment without affecting the projected images perception, small low-intensity lights were attached to the clipboard used for voting by each subject.

The physical dimensions of the screen are 5 meters by 12 meters (H x W); as a result the observation at 1H is equal to observation at 5 meters from the screen. To get a viewing distance of 1H, subjects must be seated in the front rows of the theatre. However, this location is not optimal because the point of observation is too close to the lower border of the screen, and is uncomfortable for the subjects. For this reason, a viewing distance of 2H was selected. Consequently, the test subjects' seats were located in the 6th row from the screen as illustrated by the cross mark in Fig. 3. In order to ensure a centralized viewing position, only five seats located in the 6th row from the screen were used by subjects during the evaluation. The location of these seats is illustrated by the cross marks in Fig. 4.

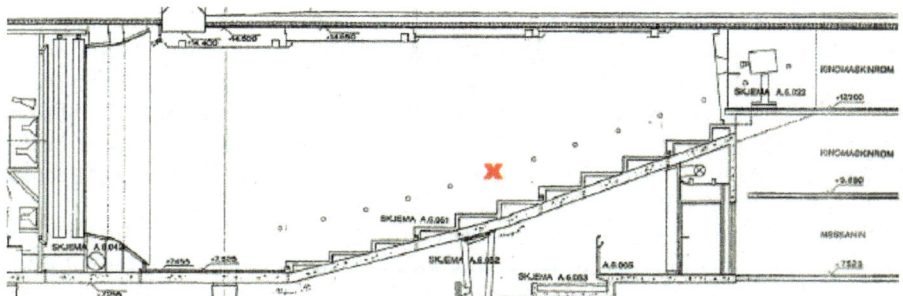

Fig. 3 Liv Ullman auditorium of Nova Kinosenter (NOVA 1) (side view)

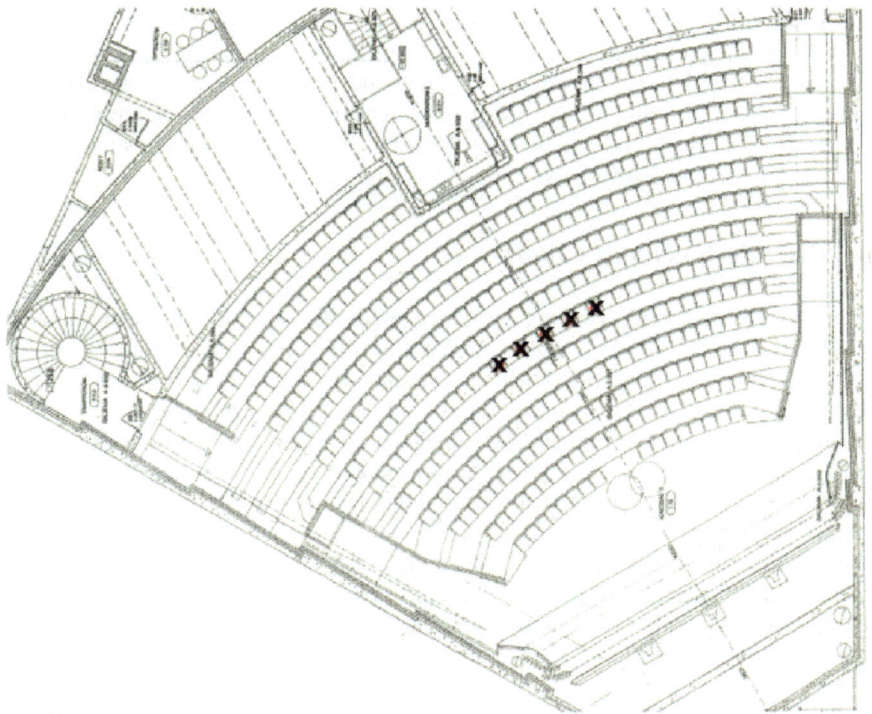

Fig. 4 Liv Ullman auditorium of Nova Kinosenter (NOVA 1) (top view)

3.2   Test Materials

The digital cinema specification [15] provides guidance for selection of test materials for the subjective assessments' stimuli. Digital cinema is based on 2K or 4K imagery, which is a significantly higher quality in terms of larger pixel counts per image when compared to standard and high definition content, respectively. In order to comply with the DCI specifications, the stimuli used in the assessment were images taken from the DCI Standard Evaluation Material (StEM) [16]. From these, six 2K images were selected. Because we only take into account the luminance component of images in this study, the luminance component was extracted from each image resulting in six gray scale 2K images.

The subjective assessment was performed by examining a range of JPEG 2000 compression errors introduced by varying bit rates. In the design of a formal subjective test, it is recommended to maintain a low number of compression conditions in order to allow human subjects an easier completion of their evaluation task. Accordingly, 8 different conditions were applied to create 8 processed images from each source image. The selected conditions covered the whole range of quality levels, and the subjects were able to note the variation in quality from each quality level to the next. This was verified prior to the subjective quality assessment with a pilot test that involved expert viewers in order to conclude the selection of the final 8 bit rates. As a result of the pilot test, the selected bit rates were in the range of 0.01 to 0.6 bits/pixel. To create 48 processed gray scale images, 6 source images were compressed using the KAKADU software version 6.0, with the following settings: codeblock size of 64x64 (default), 5 decomposition levels (default), and switched-off visual frequency weighting.

3.3   Test Methods and Conditions

There are several stimuli viewing sequence methods described in Recommendation ITU-R BT.500-11 [11]. They can be classified into two categories: single stimulus (subjects are presented with a sequence of test images and are asked to judge the quality of each test image) and double stimulus (subjects are presented with the reference image and the test image before they are asked to judge the quality of the test image). The presentation method of single stimulus is sequential, whereas the presentation method of double stimulus can be sequential and simultaneous (side by side). The decision on which test method to use in a subjective assessment is crucial, because it has a high impact on the difficulty of the test subjects' task. The pilot test prior to the main subjective assessment was also conducted to compare sequential presentation and simultaneous presentation. Differentiating between levels of high quality images requires a test method that possesses a higher discriminative characteristic. Our pilot test indicated that the simultaneous (side by side) presentation had a higher discriminative characteristic than the sequential presentation order. Therefore, the subjective quality assessment used the Simultaneous Double Stimulus test method, in which the subjects are presented with the reference image and the distorted test image displayed side by side on the screen. Figure 5 illustrates the display format in this method.

Fig. 5 Display format of Simultaneous Double Stimulus

The reference image is always shown on the left side of the image and the distorted image is shown on the right side. Test subjects grade the quality of the distorted image on the right hand side by comparing it to the reference image on the left.

The quality scale is the tool that the human subjects utilize to judge and to report on the quality of the tested images. One of the most popular quality scales in the subjective quality assessment research field is the 5 point quality level. Here, a 10 point quality scale was chosen, because the pilot test had shown that eight different quality levels could be clearly differentiated. Also, selecting a finer scale seemed to be advantageous due to the higher quality of test images used, in which a finer differentiating quality is suitable. The test used a discrete quality grading scale, which implies that the subjects are forced to choose one of the ten values and nothing in between. The quality grading scale, which is illustrated in Fig. 6, refers to "how good the picture is".

The test was conducted as a single session. Each of the 48 processed images and the 6 references were presented for a period of 10 seconds; subjects evaluate each presented image once. Subjects then needed to vote on their questionnaire sheet before the next image was presented, and they were given 5 seconds to cast their vote. The presentation structure of the test is illustrated in Fig. 6. The total session length was 15 minutes. Prior to the main session, a training session was conducted. Subjects were informed about the procedure of the test, how to use the quality grading scale, and the meaning of the designated English term related to the distortion scale of the image. During the training session, a short pre-session was run in which 19 images were shown to illustrate the range of distortions to be expected. The order of the main session was randomized, meaning that the six images and eight processing levels were randomized completely. Four to five subjects participated at the same time, and six such rounds were needed to include all subjects (see next section). The images presentation orders for each six rounds were different.

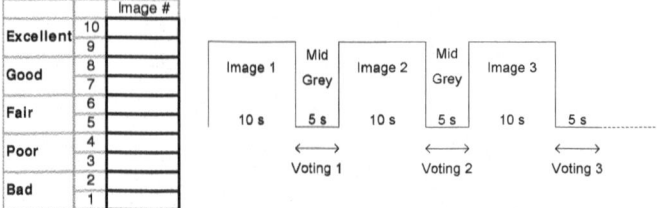

Fig. 6 Ten point quality scale and presentation structure of the test

3.4 Subjects

A proper evaluation of visual quality requires human subjects with good visual acuity and high concentration, e.g. young persons such as university students. 29 subjects (10 female, 19 male) participated in the evaluation tests performed in this work. 27 of them were university students. Some of the subjects were familiar with image processing. Their age ranged from 21 to 32 years old. All subjects re-ported that they had normal or corrected to normal vision.

3.5 Subjective Data Analysis

Before processing the resulting data, post-experiment subject screening was conducted to exclude outliers using a method described by VQEG [17]. In addition to using this method, the scores of each subject on reference images were also examined. As a result, one subject was excluded because he/she showed randomness due to scoring low for the quality of reference images. Then the consistency level for each of the remaining 28 subjects was verified by comparing his/her scores for each of the 48 processed images to the corresponding mean scores of those images over all subjects. The consistency level was quantified using Pearson's correlation coefficient r, and if the r value for one subject was below 0.75, this subject was excluded [17]. Here, the value of r for each subject was ≥ 0.9. Hence, data from all remaining 28 subjects was considered.

All data was then processed to obtain the Mean Opinion Score (MOS) by averaging the votes for all subjects. Figure 7 illustrates the MOS results. In addition, the Standard Deviation and the 95% Confidence Intervals (CI) were computed (based on a normal distribution assumption).

The behavior of a codec is generally content dependent, and this can be observed in Fig. 7. As an example, for the lowest bit rate subjects scored higher for Images 1 and 5 when compared to other images; these two images show a close up face, which typically has low spatial complexity characteristics. Furthermore, Image 2, which depicts a crowd and has high spatial complexity, tends to have the lowest score of all the images except for the highest bit rate.

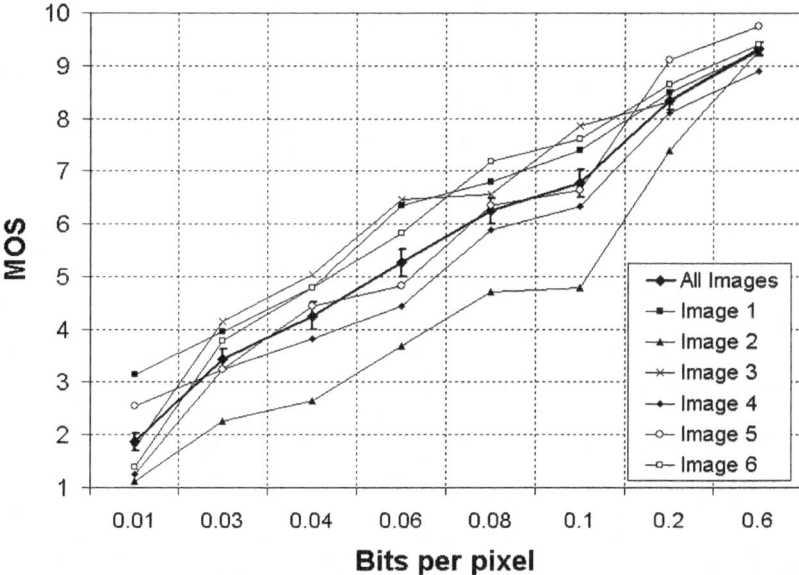

Fig. 7 MOS score vs. bit rate

4 Objective Image Metrics in D-Cinema

Although the subjective quality assessment is considered to be the most accurate method to evaluate image quality, it is time-consuming. Many objective image quality metrics have been proposed. Peak signal-to-noise ratio (PSNR) is a traditional metric usually used in evaluating the quality of a visual presentation suffering from compression and transmission errors. However, it was found that its performance is not credible for measuring the perceived quality because it does not take the characteristics of the human visual system (HVS) into account [18]. Subsequently, a number of researchers have contributed significant research in the design of image quality assessment algorithms, claiming to have made headway in their respective domains. For example, under an assumption that the HVS is highly adapted for extraction of structural information from a scene, Wang *et al.* have proposed a structural similarity (SSIM) measure to predict the image quality based on luminance, contrast, and structure comparisons between reference and distorted images [19]. Additionally, a multi-scale SSIM (MSSIM) measure has been proposed based on the SSIM algorithm [20]. MSSIM iteratively applies a low-pass filter to reference and distorted images and down-samples the filtered images by a factor of 2. At each image scale, the contrast comparison and the structure comparison are calculated, respectively. The luminance comparison is computed only at the highest scale. An overall MSSIM measure is obtained by combining the measures at different scales.

Although these objective quality metrics have been validated to be accurate in evaluation of quality degradations for normal size images [21], there is no guarantee that they will perform as well for a digital cinema system because of the difference between conventional visual presentations and visual presentation in a digital cinema setup. To our best knowledge, no rigorous analysis of the performance of objective image quality assessment specific to digital cinema has been studied so far. Here, we will study the difference of image quality assessment between a normal size image and a high-resolution image displayed on a large screen, from the viewpoint of the HVS. We will adopt related characteristics of the HVS to analyze this difference and propose an HVS-based approach to improve the performance of PSNR and SSIM in image quality assessment for digital cinema. In addition, the human vision cannot perceive all distortions in an image, especially in those regions with high-frequency, because of the contrast sensitivity of the HVS that varies with different frequencies, and the existence of masking effects [22]. Therefore, a modified PSNR by taking into account the contrast masking effect and removing the imperceptible distortion from the quality computation has been proposed in [23] and will be employed in this study. The experimental results with respect to the subjective quality evaluation demonstrate that the proposed approach can evidently improve the performance of existing metrics in image quality assessment for digital cinema systems.

4.1 Visual Characteristics in D-Cinema and Its Application to Image Quality Metrics

The most significant difference between digital cinema images and traditional images is that the former have a much higher resolution and are shown on much larger screens. Therefore, the quality assessment of the image presentation in a digital cinema setup is accordingly different from that in other controlled laboratories on a normal size display, such as a TV or computer monitor.

In the human visual system, eye movement is typically divided into fixation and saccades. Fixation is the maintaining of the visual gaze on a single location. Saccade refers to a rapid eye movement. Humans do not look at a scene in fixed steadiness, instead, the fovea sees only the central 2° of visual angle in the visual field and fixed on this target, then moves to another target by saccadic eye movement [24]. Saccades to an unexpected stimulus normally take about 200 milliseconds to initiate, and then last about 20-200 milliseconds, depending on their amplitude (20-30 milliseconds is typical in reading). In image quality assessment, quality evaluation takes place during eye fixation when the fovea can perceive the visual stimulus with maximum acuity [25]. Thus, when viewing an image on a large screen in the digital cinema, subjects cannot see the entire image at once and evaluate distortions in all regions. Even though the fovea might not be able to perceive an entire image simultaneously on a normal size display, we believe the situation in this case is significantly different from that in a digital cinema setup.

Many physiological and psychological experiments have demonstrated that human attention is not allocated equally to all regions in the visual field, but focused on certain regions known as salient regions or attention regions [26]. We have proposed a visual attention based perceptual quality metric by modeling the visual attention based on a Saliency model and visual content analysis [27]. However, it was found that the visual attention does not have an evident influence on the image quality assessment. The NTIA video quality model (VQM) assumes that the HVS is more sensitive to those regions with severe distortions [28We firstly tested the performance of two image quality metrics: PSNR and SSIM, on image quality assessments for normal size images and high-resolution images in a digital cinema setup. PSNR is a traditionally used metric based on Mean Square Error (MSE), computed by averaging the squared intensity differences between the distorted and reference image pixels, and defined as follows:

$$PSNR = 10 \cdot \log_{10}(\frac{P^2}{MSE})$$
(1)

where P denotes the peak value of the image. Although PSNR does not always correlate well with subjective image quality assessment, it is still widely used for evaluation of the performance of compression and transmission systems. PSNR and MSE are appealing because they are simple to compute, have clear physical meanings, and are mathematically convenient in the context of optimization.

Based on the hypothesis that the HVS is highly adapted for extraction of structural information, Wang *et al.* have developed a measure of structural similarity to estimate image quality by comparing local patterns of pixel intensities that have been normalized for luminance and contrast. The SSIM measure is calculated based on three components: luminance, contrast, and structure comparison, defined as follows:

$$SSIM(x, y) = \frac{(2\mu_x\mu_y + C_1)(2\sigma_{xy} + C_2)}{(\mu_x^2 + \mu_y^2 + C_1)(\sigma_x^2 + \sigma_y^2 + C_2)}$$
(2)

where μ and σ denote mean and standard deviation on the luminance component, and C_1 and C_2 are two small constants to avoid instability when $(\mu_x^2 + \mu_y^2)$ or $(\sigma_x^2 + \sigma_y^2)$ are very close to zero. In addition, the authors of SSIM calculated the SSIM measure within a local square window, moving pixel-by-pixel over the entire image. A mean of SSIM indices over all windows is computed after applying a circular-symmetric Gaussian weighting function to the reference and the distorted images to eliminate blocking artifacts.

In this study, we used our subjective quality assessment in Section 4 to evaluate the performance of PSNR and SSIM for digital cinema applications. In addition, the subjective quality results on JPEG 2000 compressed images with normal sizes were extracted from the LIVE image quality dataset [29], where 29 reference images have been compressed using JPEG 2000 at different bit rates ranging from 0.028 bits per pixel (bpp) to 3.15 bpp. After calculating the quality values using PSNR and SSIM on these distorted images, a nonlinear regression operation between the metric results (*IQ*) and the subjective scores (DMOS), as suggested in [30], was performed using the following logistic function:

$$DMOS_P = \frac{a_1}{1+\exp(-a_2 \cdot (IQ-a_3))} \tag{3}$$

The nonlinear regression function is used to transform the set of metrics results to a set of predicted DMOS values, $DMOS_P$, which are then compared against the actual subjective scores (DMOS) and resulted in three evaluation criteria: root mean square error (RMSE), Pearson correlation coefficients, and Spearman rank order correlation coefficient. RMSE and Pearson correlation express the prediction accuracy of a quality metric, and Spearman rank order correlation provides information about the prediction monotonicity of the metric [30]. In addition, the reference [30] also suggests another criterion, the outlier ratio, that relates to prediction consistency based on standard errors of the subjective quality values. However, the LIVE dataset does not provide such standard errors for computing the outlier ratio.

Table 2 Performance evaluation of PSNR and SSIM in digital cinema setup and LIVE dataset

Criteria	Digital Cinema		LIVE Dataset	
	PSNR	SSIM	PSNR	SSIM
RMSE	1.00	1.13	7.45	5.71
Pearson	0.914	0.888	0.888	0.936
Spearman	0.913	0.875	0.890	0.931

Table 2 gives the performance evaluation of PSNR and SSIM on image quality assessment in our digital cinema setup and LIVE image dataset, respectively. It is noticed that the RMSE values are strongly related to score ranges in a subjective quality assessment, which is why the RMSE values between the digital cinema scenario and LIVE dataset are quite different. However, according to the comparison on the correlation coefficients, we can find that the performance of SSIM is worse than PSNR. In our opinion, the reason is that subjects do not compare the entire structural information between the distorted image and the reference image in a digital cinema setup, because the image size is too large.

According to the above performance comparison of PSNR and SSIM between two different scenarios, as well as the analysis of the characteristics of the HVS in a digital cinema setup, we think that an image quality metric developed for normal size images cannot be adopted as is to predict the image quality in digital cinema applications. The main reason is that a subject cannot see an entire image at once when he/she evaluates the quality of this image in digital cinema. Thus, we propose to first divide the image into different blocks, and then perform image quality metrics on each block. The overall quality of this image can be derived from the metric results in all blocks or parts of the image blocks. In this study, square blocks are employed. The block size (S) is determined based on several factors, including the human fovea acuity angle (α), image size (S_1), screen size (S_2), and viewing distance (D), as follows:

$$S = \frac{S_1}{S_2} \cdot D \cdot \text{atan}(\frac{\alpha}{180}) \qquad (4)$$

In the subjective quality assessment introduced in Section 4, the reference and distorted images were shown on the screen simultaneously (Fig. 3), therefore, we used the height values of the image and the screen as their sizes S_1 and S_2.

The reference and the distorted images have been divided into different blocks whose size are $S{\times}S$, and then PSNR and SSIM values were calculated in each block between the reference and the distorted images. Actually, it is unnecessary to perform the computations for all blocks. One reason is that some blocks with severe distortions may dominate the overall quality of an image [28]. Another reason is that subjects usually estimate their judgment of the quality of an image based on evaluation of a subset of regions or blocks in that image, and they may not have enough time to examine all blocks or regions in a short viewing duration, such as 10 seconds in our subjective experiments. Furthermore, we found that subjects paid more attention to those image blocks with higher contrast when assessing the image quality. Hence, the standard deviation in each divided block was computed to express the contrast information, and all the blocks were sorted in a descending order according to their standard deviation values. Blocks with lower contrast levels were excluded from the quality calculation, where a threshold T for distinguishing the blocks was set. This threshold (T) was estimated, as in equation (5), according to the saccadic and fixation time of eye movement and the viewing duration in the subjective quality assessment:

$$T = \frac{10(\text{s})/30(\text{ms})}{M} \qquad (5)$$

where M denotes the number of all divided blocks in an image, 10(seconds) is the viewing duration, and 30(milliseconds) is the saccadic and fixation time.

PSNR and SSIM measures are computed in the candidate blocks between the reference and distorted images whose contrast levels exceed the threshold, and then mean values of PSNR and SSIM over these blocks were used as the quality of that image measured by PSNR and SSIM, respectively. Our experimental results demonstrate that the performance of this approach is better than the original PSNR and SSIM methods.

As aforementioned, we computed the quality values in those blocks with high contrast levels. However, the distortions introduced to image blocks, especially the blocks with high contrast levels, are not perceived by the HVS totally, because of the contrast sensitivity of the HVS that varies with different frequencies and the existence of masking effects. Contrast sensitivity is a measure of the ability to discern between luminance of different levels in an image. In addition, when an image has high activity, there is a loss of sensitivity to errors in those regions. This is the masking effect of the image activity. Many approaches have been proposed to model the contrast sensitivity and masking effects in order to compute a visually optimal quantization matrix for a given image in compression algorithms. The Discrete Cosine Transform (DCT) has usually been used in contrast making due to its suitability for certain applications and accuracy in modeling the cortical neurons [31]. In this study, we employ a DCT based approach to model the contrast sensitivity function (CSF) and masking effect as in [23]. The quality value is computed according to a modified PSNR by excluding the imperceptible distortion because of the contrast sensitivity and masking effect from the computation of PSNR. This method is called PHVS, and readers can refer to [23] for details.

4.2 Performance Evaluation of Image Quality Metrics in D-Cinema

According to the sizes of tested images and the screen in our digital cinema setup, as well as the human fovea acuity angle (2°), the block size in this study was set to 16 according to equation (4), and the threshold T was 10% calculated by equation (5). A reference image and its distorted image (JPEG 2000 compressed) were divided into different blocks with sizes 16×16. Subsequently, all blocks in the reference image were sorted in a descending order according to their standard deviations, and 10% of all blocks with highest standard deviations were selected to compute the quality of distorted image. A quality value in each candidate block was calculated by PSNR, SSIM, or PHVS. The mean of the quality values over all the candidate blocks was taken as an overall quality of the distorted image.

   To evaluate the performance of the proposed approach, four evaluation criteria were used. As aforementioned, a nonlinear mapping operation in equation (3) was performed between the metric results and the subjective DMOS values. RMSE, Pearson correlation coefficient, and Spearman rank order correlation coefficient can be computed between the mapped metric results and the DMOS values. In addition to these three criteria, another criterion, outlier ratio relating to the prediction consistency of a metric, can be obtained, because our subjective quality experiment provided standard errors of the subjective quality results. The outlier ratio is defined as the ratio of the number of outlier point images compared to the total number of the tested images, in which an outlier point image is detected if it satisfies the following condition:

$$|DMOS_P - \text{DMOS}| > 2 \cdot SE \tag{6}$$

where SE denotes the standard error value.

   In our experiments, the original methods of PSNR, SSIM, and PHVS were performed with respect to the subjective image quality assessment in the digital cinema setup, and four evaluation criteria were computed. To validate the proposed approach, we used two methods as follows:

1) The first method was to divide the image into different 16×16 blocks, and these three metrics were computed in all blocks. The mean over all blocks in the image was taken as the quality metric for this image.

2) The image was still divided into different 16×16 blocks in the second method, however the metrics were only performed for those blocks with high contrast levels, as described above. The image quality was computed by the mean over these

blocks, while the remaining blocks with lower contrast levels were excluded from the quality computation.

Table 3 gives the evaluation results of the original metrics and the above two methods. According to the evaluation results, the proposed approach can evidently improve the performance of the original metrics in image quality assessment of digital cinema, especially for SSIM. The first method can improve the performance of these metrics, especially for SSIM and PHVS, which indicates subjects formulate their judgment on the image quality in a digital setup based on a all blocks, rather than the global image. Further, the second method has better performance when compared to the first method. It indicates that subjects pay more attention to those image regions with higher contrast levels when assessing the image quality. This observation will be useful to develop a quality metric for digital cinema applications.

Table 3 Evaluation results of different methods on image quality assessment in digital cinema setup

Criteria	Original metrics			First method			Second method		
	PSNR	SSIM	PHVS	PSNR	SSIM	PHVS	PSNR	SSIM	PHVS
RMSE	1.00	1.13	0.85	1.00	1.00	0.77	0.73	0.53	0.56
Pearson	0.914	0.888	0.938	0.914	0.913	0.949	0.954	0.976	0.974
Spearman	0.913	0.875	0.941	0.913	0.904	0.952	0.956	0.974	0.976
Outlier ratio	0.063	0.063	0.021	0.042	0.063	0.021	0.021	0.021	0

Table 4 Evaluation results of Minkowski summation for spatial pooling

Criteria	PSNR	SSIM	PHVS
RMSE	0.71	0.54	0.57
Pearson	0.958	0.976	0.973
Spearman	0.961	0.974	0.976
Outlier ratio	0.021	0.021	0.021

In the above experiments, we used the direct mean of quality values over the blocks. In addition, another pooling scheme, Minkowski summation, is widely used in some quality metrics to pool the quality values over different spatial regions, such as the perceptual distortion metric (PDM) proposed by Winkler *et al.* [32]. Therefore, we also tested the Minkowski summation with different exponents on those blocks with higher contrast levels. The experimental results demonstrated that the best performance was achieved when the exponent in the Minkowski summation was set to 2, and the results indicated that the Minkowski summation can also be used in pooling the quality values over different blocks in the proposed approach, as shown in Table 4.

Finally, we also employed the same approach in this chapter to evaluate these three metrics for normal size images using the LIVE image quality database.. We tested the JPEG 2000 compression distortion type, as well as three other distortion types: Gaussian blur, JPEG compression and white noise. The performance was, however, worse than the original methods either for JPEG 2000 compression or other distortion types. We believe that the cause is due to the fact that the mechanisms are different when subjects assess the image quality between a normal size monitor and a large screen in a digital cinema setup. The future work will be focused on an in depth analysis of the differences of the visual quality assessment between a normal scenario with normal size displays and a digital cinema system, and development of more suitable metrics for digital cinema.

5 Quality as Part of the Business Plan

For a media presentation to be possible, the most important process is of course the creative one, created by the artist and represented by the content provider or content aggregator. One of the predominant slogans recently in the media industry has been "Content is king", which, certainly, has been used as one of the driving forces for digitization and development of new services and applications.

Even more recently, following the rapid deployment of wireless services, more focus has been put on the delivery. As slogan goes, the novelty is represented by "Connectivity is Queen".

Our claim, however, is contrary to this, and assumes all providers have one goal in common, the satisfied and loyal customer, buying and consuming their services and applications regardless of technology, simply meaning that the "User is president". The hypothesis is that a satisfied user gets his/her expectations fulfilled and that expectations are connected to the experience of the media as represented by the QoE. Although QoE currently is not an objectively measurable parameter, we will assume that this is represented by our broader understanding of the term *Quality*. High quality gives more satisfied customers and a better experience than poor quality.

5.1 Case: D-Cinema

The motivations for D-Cinema were at least three fold:

- To reduce distribution costs (benefits for studios)
- To reduce piracy (benefits for studios)
- To enhance Quality of Experience (benefits for cinema goers – the users)

Several other motivating factors have become more and more important as the innovators and early adopters are implementing D-Cinema. Some of these were planned; some again are spin-off innovations. Examples of these are 3D and Other Digital Stuff – ODS (Alternative Content).

D-Cinema roll out has hit a crucial moment in time since reaching the famous chasm of diffusion of innovation simultaneously with the global financial crisis.

The uncertainty and ambiguity of the Virtual Print Fee as sole financial model for overcoming the investments are also somewhat preventing the early majority to follow. These are indeed exciting times, and finding the ways of crossing the chasm is important as ever. Two technical reinventions are acting as major change agents, the rebirth of 3D and the possibility for exhibitors to screen alternative content – the so called Other Digital Stuff – ODS. Together they share an optimistic prospect. They embrace the possible realization of a complete transition from 35mm film to D-Cinema, although slower than formerly anticipated by the founders of the Digital Cinema Initiative. There are no easy and obvious solutions, however, the D-Cinema arenas as well as respected opinion leaders are working side by side, utilizing the change agents, to define and communicate the future direction, potentially crossing the chasm.

Digital Cinema Initiatives, LLC [15] was created in March, 2002, and is a joint venture of Disney, Fox, Paramount, Sony Pictures Entertainment, Universal and Warner Bros. Studios. DCI's primary purpose is to establish and document voluntary specifications for an open architecture for digital cinema that ensures a uniform and high level of technical performance, reliability and quality control.

On July 20, 2005, DCI released its first version of the final overall system requirements and specifications for D-Cinema. Based on many existing SMPTE and ISO standards, it explains the route to create an entire Digital Cinema Package (DCP) from a raw collection of files known as the Digital Cinema Distribution Master (DCDM), as well as the specifics of its content protection, encryption, and forensic marking. The specification also establishes standards for the decoder requirements and the presentation environment itself.

The case study uses results obtained through the NORDIC projects (NORDIC and NORDIC 2.0) – Norway's Digital Interoperability in Cinemas [3]. The NORDIC projects brings together Norway's leading experts in the D-Cinema field, including Midgard Media Lab at NTNU, telecom and pay-television operator Telenor, installation and service company Hjalmar Wilhelmsen, and digital cinema advertising pioneers Unique Promotions/Unique Digital/Unique Cinema Systems, as well as major exhibitors and cinemas across Norway. The main achievements of the NORDIC projects are the lessons learned in order to provide advice to the Norwegian cinema industry on when and how to shift the whole of Norway to Digital Cinema. A complete overview of the NORDIC models, results and discussions has recently been published [3].

Michael Karagosian recognized the chasm for D-Cinema and published it in September 2007 [33]. The chasm is well documented by Rogers and Moore and follows a fairly classic path, very much comparable to many of the cases reviewed by the two authors. D-Cinema is currently facing a situation where we see:

- A market where there are few or no new customers. This is identified by reports from all vendors that sales are down and reports from the D-Cinema arenas that the D-Cinema roll out seems to have stalled. This means that not everyone is automatically following.
- The innovators and early adopters still have their D-Cinema equipment and films are being screened, but not in the volume foreseen, and 35mm is still dominating at most exhibitors. The disappointment of current customers is

coming to the surface. This can be identified by discussions on the D-Cinema arenas. This shows that not all players have the same understanding of D-Cinema and its process. There especially seems to be a gap between the studios, distributors and exhibitors, representing a role each in the value chain.

- Competition is tightening between the vendors as alliances are building. Vendors from Asia offering more price competitive products are entering the market. Classically we would have seen aggressive competition from already established products in the market. However, since D-Cinema is new, no existing products were in the market.
- The investors are getting worried and some are losing interest. The global financial crisis is severely increasing this effect, putting the existing business models to their hardest test ever.

None of these reasons are due to the users' reaction to D-Cinema. As a matter of fact, the NORDIC project has shown that the users make little or no distinction whether it's a digital screening or not. This shows that the perceived quality is stable and withheld during the digitization. The real advantages are in the added functionalities offered by D-Cinema such as 3D and alternative content.

5.2 Does Quality Matter?

The ability to move from 2D to 3D screening was once viewed as important as the transition from silent to sound movies by the motion picture industry. Although its popularity has come and gone, it is definitely back again and seen as one of the most important change agents for D-Cinema.

The revenue upside for 3D is clear. Estimates suggest that exhibitors can charge a 20% premium on a ticket for a digital 3D (non-Imax) movie. Like with "Journey to the center of the earth" 3D versions also tend to outperform 2D on a per-screen basis. And 3D films are viewed as less susceptible to piracy. There is very little material and numbers currently available to be able to substantiate such a claim. However, Hollywood sees a future, documented by the more than 30 3D productions in the pipeline and the enormous interest gathering around 3D.

Other Digital Stuff – the reference term for Alternative Content – covers all content screened in a cinema that is not a feature film. This includes advertising, live events (sports, music), educational and gaming. ODS has the potential of diversifying exhibitor offering and provide new revenue streams.

As technology, ODS came as the unexpected side effect of reinventions. By digitizing screens one realized the huge potential the equipment (server and projector) pose for other users requiring ultra high definition. The exploitations have ranged from Opera to Laparoscopic surgery.

The D-Cinema business is about creating a digital media asset (the feature or motion picture) and screening this on as many screens as possible in order to optimize the profit of the studios owning the asset. For more than 100 years a value chain consisting of the studios, the distributors and the exhibitors has managed this. Digitization has the potential of dramatically changing the value chain, the roles of the players in the value chain, and opens for new business models and possibilities.

D-Cinema began as a purely technological change, but is continuing by involving all the processes in the motion picture industry affecting the complete value chain. The change is destined to transform the whole business of cinema exhibition into something different from what we know today. Change agents operating on the D-Cinema arena, respected opinion leaders, and the industry itself influence this.

35mm film is the longest lasting and most universally adopted and used audio-visual format of all time. Even audio technologies that have found global acceptance (78rpms, 45rpm, 33rpm, tapes, CD, DAT, SACD/DAD-A) have never lasted as long as the century old cinema standard. While the road to D-Cinema has been a long process, it is only correct that the replacement for such a universal medium should take great care in crafting something that both surpasses it and finds as much universal acceptance.

D-Cinema completely changes the roles of the players and stakeholders in the media value chain. The NORDIC projects have shown that the changes in technology are manageable and largely solved by achieving interoperability. However, the change on the business models is severe and not solved.

6 Discussions and Conclusions

Given the potential impact on quality introduced by D-Cinema and the quest for quality in the media consumption by most users, the transition from 35 mm to D-Cinema should happen as soon as possible. The higher quality of D-Cinema at the relatively low cost can potentially increase the ticket sales and be the winner in the battle for the eyeball – proving quality impacts the business model.

To assess the quality and utilize the result in the business models we need to be able to measure and to model Quality of Experience (QoE). QoE always puts the end-user at the centre of attention. This makes QoE research in the D-Cinema environment a strong change agent for enforcing and necessitating the D-Cinema roll out – enhancing the fact that the motion picture industry exists to serve the user, the moviegoers. Our quality assessment investigates how humans consume and perceive digital media presentation in the cinema environment. The purpose is to define objective parameters and models that can lead to a definite answer on whether customers are indeed satisfied.

QoE is affected by several factors including perception, sensations, and expectations of users as they consume digital content presented to them using their perceptual sensors. In D-Cinema the main perceptual sensors are sight and hearing. Hence perceived visual quality and perceived audio quality in D-Cinema are integral in QoE research. Nevertheless, it can be hypothesized that the most interesting and prospective aspect of understanding the perceived visual and audio quality in D-Cinema is not by viewing and treating these two factors separately but together to investigate how perceived quality is influenced by multimodal factors. The most striking difference by using a cinema for subjective assessments than just regular screens is the impressive sensation caused by the D-Cinema environment. Even though some may be content with watching movies on their home digital cinema equipment, there are still many moviegoers who buy tickets and watch movies loyally in the theatre. For these, watching a movie in the theatre provides them with a totally different experience than viewing in their own home. The sensation while consuming digital presentation in a darkened large auditorium with the large screen contributes greatly to the user experience. Investigating how these factors influence the QoE is important. Subjective experiments using human subjects are one of the key stages in the QoE research for D-Cinema. In order to give realistic results, the subjective experiment must be conducted in a realistic environment such as the DCI specified cinema in Trondheim—Nova Kinosenter—to provide the same viewing condition as regular moviegoers have. However, it is important to note that subjective experiments using human subjects are complex and must be in controlled environments using proper methodologies. The data from the subjective experiments is then used as a foundation to develop objective models to determine quality in the context of QoE.

We have presented the subjective test methodology and setup in NOVA 1 using characteristics of the HVS that are related to image quality assessment in a D-Cinema setup. Based on an intensive analysis on the mechanism of image quality assessment in a digital cinema setup, we proposed an approach for improving the performance of three image quality metrics. The images were divided into different blocks with a given size, and metrics were performed in certain blocks with high contrast levels. The mean of quality values over these blocks was taken as the image quality. The experimental results with respect to the subjective quality results in the D-Cinema setup and LIVE dataset demonstrated the promising performance of the proposed approach in improving the image quality metrics for digital cinema applications.

Our work in perceived visual quality in D-Cinema is a starting point in QoE research. Our study showed that due to different and unique digital image content and viewing conditions of D-Cinema, quality research of D-Cinema especially in the context of QoE is not really in the same category as any other application. Our future work will study the mechanisms of QoE in D-Cinema in depth.

The current problem still remains, how do we motivate for a total roll out of D-Cinema and how can quality contribute in the process. Many stakeholders are trying to influence, ranging from individual to organizations and a few novel reinventions of the technology that were not initially spotted. The most important of these are the ability to screen 3D content using the existing D-Cinema equipment. In addition, the industry is exploiting how to benefit from the possibility of screening other digital stuff in times where the screen is not used for traditional film screenings. Opinion leaders, experts and the industry itself are constantly communicating these benefits and using their powers as change agents to finally be able to convince and reach the early majority creating a mass market for D-Cinema with real competition and inertia. Our hope is that improved QoE will prevail and become the most important change agent of them all.

References

1. Perkis, A.: Innovation and change. Case: Digital Cinema. Master thesis: Master of Technology Management (MTM), NTNU (2008)
2. Perkis, A.: Does quality impact the business model? Case: Digital Cinema. In: Proc. IEEE QoMEX, San Diego, USA (2009)
3. Perkis, A., Sychowski, P.V., et al.: NORDIC – Norway's digital interoperability in cinemas. Journal St. Malo: NEM - Networked and Electronic Media (2008), ISBN 978-3-00-025978-4
4. Murch, G.M.: Visual and Auditory Perception, 4th printing. Bobbs-Merrill Co., Indianapolis (1973)
5. Snell, R.S., Lemp, M.A.: Clinical Anatomy of the Eye, 2nd edn. Blackwell Science Inc., Malden (1998)
6. Wade, N.J., Swanston, M.T.: Visual Perception: An Introduction, 2nd edn. Psychology Press, Hove (2001)
7. Nolte, J.: The Human Brain - An Introduction to Its Functional Anatomy, 5th edn. Mosby Inc., St. Louis (2002)
8. Goldstein, E.B.: Wahrnehmungspsychologie (German language). In: Ritter, M., Akadem, S. (eds.), 2nd edn. Verlag, Berlin (2002), ISBN 3-8274-1083-5
9. Neisser, U.: Cognition and Reality. Freeman, San Francisco (1976)
10. Pashler, H.E.: The Psychology of Attention, 1st paperback edn. The MIT Press, Cambridge (1999)
11. Recommendation ITU-R BT.500-11: Methodology for the subjective assessment of the quality of television pictures. ITU-R, Geneva, Switzerland (1974-2002)
12. Recommendation ITU-R BT.1686: Methods of measurement of image presentation parameter for LSDI programme presentation in a theatrical environment. ITU-R, Geneva, Switzerland (2004)
13. Sony: 4K Digital Cinema Projectors SRX-R220/SRX-R210 Media Blok LMT-100 Screen Management System LSM-100. Technical Specification (2007)
14. Sony: SRX-R220 (2009),
    `http://www.sony.co.uk/biz/view/ShowProduct.action?product=SRX-R220&site=biz_en_GB&pageType=Overview&imageType=Main&category=D-Cinema`
15. DCI: Digital Cinema System Spesification version 1.2. Digital Cinema Initatives (2008)
16. DCI: StEM Access Procedures (2009),
    `http://www.dcimovies.com/DCI_StEM.pdf`
17. VQEG: VQEG Multimedia Group Test Plan Version 1.21 (2008)
18. Girod, B.: What's wrong with mean-square error. In: Watson, A.B. (ed.) Digital Images and Human Vision, pp. 207–220. MIT Press, Cambridge (1993)
19. Wang, Z., Bovik, A.C., et al.: Image quality assessment: from error visibility to structural similarity. IEEE Trans. Image Processing 13(4), 600–612 (2004)
20. Wang, Z., Simoncelli, E.P., et al.: Multi-scale structural similarity for image quality assessment. In: Proc. IEEE Asilomar Conference Signals, Systems and Computers (2003)
21. Sheikh, H.R., Sabir, M.F., et al.: A Statistical Evaluation of Recent Full Reference Image Quality Assessment Algorithms. IEEE Trans. Image Processing 15(11), 3440–3451 (2006)

22. Wu, H.R., Zhao, K.R.: Digital video image quality and perceptual coding. CRC / Taylor & Francis Press (2006)
23. Ponomarenko, N., Silvestri, F., et al.: On between-coefficient contrast masking of DCT basis functions. In: CD-ROM Proc. of 3rd Int Workshop VPQM (2007)
24. Carpenter, G.A.: Movements of the eyes. Pion Press, London (1988)
25. Burr, D., Morrone, M.C., et al.: Selective suppression of the magnocellular visual pathway during saccadic eye movements. Nature 371(6497), 511–513 (1994)
26. Itti, L., Koch, C.: Computational modeling of visual attention. Nature Review: Neuroscience 2(3), 194–203 (2001)
27. You, J., Perkis, A., et al.: Perceptual quality assessment based on visual attention analysis. In: Proc. of ACM Int. Conference on Multimedia (2009)
28. Pinson, M., Wolf, S.: A new standardized method for objectively measuring video quality. IEEE Trans. Broadcasting 50(3), 312–322 (2004)
29. LIVE database, http://live.ece.utexas.edu/research/quality
30. VQEG: Final report from the Video Quality Experts Group on the validation of objective models of video quality assessment, Phase II, FR-TV 2 (2003), http://www.vqeg.org
31. Wang, Z., Sheikj, H.R., et al.: The handbook of video database: Design and applications. CRC Press, Boca Raton (2003)
32. Winkler, S.: Digital video quality: vision models and metrics. John Wiley & Sons Press, Chichester (2005)
33. Karagosian, M.: Report on the Progress of Digital Cinema in 2007: SMPTE Motion Imaging Journal 2007 Progress Report (2007), http://mkpe.com/publications/D-Cinema/reports/May2007_report.php

Chapter 3
Quality of Visual Experience for 3D Presentation – Stereoscopic Image

Junyong You, Gangyi Jiang, Liyuan Xing, and Andrew Perkis

Abstract. Three-dimensional television (3DTV) technology is becoming increasingly popular, as it can provide high quality and immersive experience to end users. Stereoscopic imaging is a technique capable of recoding 3D visual information or creating the illusion of depth. Most 3D compression schemes are developed for stereoscopic images including applying traditional two-dimensional (2D) compression techniques, and considering theories of binocular suppression as well. The compressed stereoscopic content is delivered to customers through communication channels. However, both compression and transmission errors may degrade the quality of stereoscopic images. Subjective quality assessment is the most accurate way to evaluate the quality of visual presentations in either 2D or 3D modality, even though it is time-consuming. This chapter will offer an introduction to related issues in perceptual quality assessment for stereoscopic images. Our results are a subjective quality experiment on stereoscopic images and focusing on four typical distortion types including Gaussian blurring, JPEG compression, JPEG2000 compression, and white noise. Furthermore, although many 2D image quality metrics have been proposed that work well on 2D images, developing quality metrics for 3D visual content is almost an unexplored issue. Therefore, this chapter will further introduce some well-known 2D image quality metrics and investigate their capabilities in stereoscopic image quality assessments. As an important attribute of stereoscopic images, disparity refers to the difference in image location of an object seen by the left and right eyes, which has a significant impact on the stereoscopic image quality assessment. Thus, a study on an integration of the disparity information in quality assessment is presented. The experimental results demonstrated that better performance can be achieved if the disparity information and original images are combined appropriately in the stereoscopic image quality assessment.

Junyong You · Liyuan Xing · Andrew Perkis
Centre for Quantifiable Quality of Service in Communication Systems (Q2S), Norwegian University of Science and Technology, Trondheim, Norway
e-mail: junyong.you@ieee.org, liyuan.xing@q2s.ntnu.no, andrew@iet.ntnu.no

Gangyi Jiang
Ningbo University, Ningbo, China
e-mail: jianggangyi@nbu.edu.cn

1 Introduction

Networked three-dimensional (3D) media services are becoming increasingly feasible through the evolution of digital media, entertainment, and visual communication. Three-dimensional television (3DTV), one of the popular media services, can provide a dramatic enhancement in user experience, compared with the traditional black-and-white and color television. Although David Brewster introduced the stereoscope, a device that could take photographic pictures in 3D, in 1844, it was not until 1980s that experimental 3DTV was presented to a large audience in Europe. However, although various recent technological developments combined with an enhanced understanding of 3D perception have been achieved, many important topics related to 3D technology are almost unexplored [1]. A networked 3DTV service consists of an entire chain from content production and coding schemes for transmitting through communication channels to adequate displays presenting high quality 3D pictures. During this chain, the quality of a 3D presentation may be degraded at each stage. In this chapter, we will focus on perceptual quality assessment of visual experience for stereoscopic images. Human factors as well as typical artefacts in 3D presentations that may affect the quality of visual experience will be reviewed, and we will mainly focus on coding and transmission artefacts. To study the relationship between perceived quality and distortion parameters for stereoscopic images, a subjective quality assessment has been conducted. Furthermore, accurate metrics which can predict stereoscopic image quality will be proposed based on two-dimensional (2D) image quality metrics and disparity information.

Before going to the detailed discussions on quality assessment at each stage in the chain, understanding the human factors that can affect the quality of visual experience is necessary. The relationship among some psycho-visual factors, such as sensation of depth, perceived sharpness, subjective image quality, and relative preference for stereoscopic over non-stereoscopic images, was investigated in [2]. The main finding is that viewers usually prefer a stereoscopic version rather than a non-stereoscopic version of image sequences, given that the image sequences do not contain noticeable stereo distortions, such as exaggerated disparity. Perceived depth is rated greater for stereoscopic sequences than that for non-stereoscopic ones, whereas perceived sharpness of stereoscopic sequences is rated same or lower compared to non-stereoscopic sequences. Subjective rating on stereoscopic image quality is influenced primarily by apparent sharpness of image sequences, whereas the influence of perceived depth is not evident. As early as in 1993, technological requirements for comfortable viewing in 3D display were studied [3]. To reduce visible image distortion and visual strain, a basic requirement (image size), visual noise requirements (disparity range, disparity resolution), and motion parallax (viewpoint sampling, brightness constancy, registration tolerance, and perspective interpolation) are required. Stelmach et al. [4] found that the sensation of depth, image quality and sharpness are affected differently by different spatial resolutions and temporally filtering schemes. The overall sensation of depth is not affected by low-pass filtering, and larger spatial resolutions usually make more contribution to the rating of quality and sharpness. Field averaging and dropped-and-repeated frame conditions may result in images with poor quality and sharpness, even though the perceived depth is relatively unaffected. Ijsselsteijn et al. [5] investigated appreciation-oriented measures on perceived quality and naturalness with parameters of displaying duration. The experimental results on 5s versus 10s displaying durations demonstrated that there is no significant influence of displaying duration on the perceived quality. However, a small yet significant shift between the naturalness and quality was found for these two duration conditions. The experimental results with displaying durations ranging from 1s to 15s also showed a small yet significant effect of displaying duration on the perceived quality and naturalness. Besides, longer displaying durations do not have a negative impact on the appreciative scores of optimally reproduced stereoscopic images. However, observers usually give lower judgments to monoscopic images and stereoscopic images with unnatural disparity values as displaying duration increases. Meegan et al. [6] were opinion of that the binocular vision assigns a greater weight to an un-degraded image than a degraded one. In addition, compared to blockiness, blurring has greater influence on the perceived quality. Therefore, a quality metric for a stereo-pair images can be developed by assigning a greater weight to the un-degraded or less degraded image in a stereo-pair images while a smaller weight to the other one. The same rule can be applied to the situation between blockiness and blurring. These subjective experiments tried to investigate the physiological process in quality assessment for 3D presentations, and the corresponding observations have been used in practical systems already. For example, inspired by the work done by Meegan et al., asymmetric coding schemes for multi-view videos have been proposed. In addition, effects of camera-base distance and JPEG-coding (Symmetric versus Asymmetric) on an overall image quality, perceived depth, perceived sharpness and perceived eye-strain were investigated in [7]. Bounds of an asymmetric stereo view compression scheme by H.264/AVC and its relationship with eye-dominance were examined based on a user study [8].

Considering the capture and visualization distortions, we will focus on shooting, viewing condition and representation related degradation. The geometry relation-ship between shooting and viewing conditions was formulated. Woods et al. [9] analyzed the geometry of stereoscopic camera and display systems to show their effect on image distortions such as depth plane curvature, depth non-linearity, depth and size magnification, shearing distortion, and keystone distortion. In addition, Yamanoue et al. [10] clarified that the shooting and viewing conditions and conditions under which the puppet-theater effect and cardboard effect occur, in geometrical terms. Generally, crosstalk is a significant factor of the most annoying distortions in 3D display. Inter-channel crosstalk for the auto-stereoscopic displays utilizing slanted lenticular sheets was modeled in [11]. Although the cause of most introduced distortions has been studied, their influence on the perceived quality is still an unexplored issue. Seuntiens et al. [12] designed an experiment to investigate the relevant subjective attributes of crosstalk, such as perceived image distortion, perceived depth, and visual strain. The experimental results indicated that image distortion ratings show a clear increasing trend when increasing crosstalk level and camera base distance. Especially, a higher crosstalk level is visible more clearly in a longer camera base distance. Ratings of visual strain and perceived depth increase only when increasing the camera base distance. However, if the crosstalk level increases, visual strain and perceived depth might not change accordingly. Furthermore, Kim et al. [13] has proposed an objective metric by taking into account both acquisition and display issues. For instance, using a multiple cameras structure may cause impairment such as misalignment. The experimental results demonstrated that a depth map is a useful tool to find out implied impairments, in which the depth map was obtained by estimating disparity information from stereoscopic videos. By using the depth map, the depth range, vertical misalignment and temporal consistency can be modeled separately to exhibit different viewing aspects. They are then integrated into a metric by a linear regression, which can predict the levels of visual fatigue.

Existing work on perceptual quality evaluation for both video-plus-depth and multi-view video 3D presentation is mostly focused on assessing the quality degradation caused by compression errors. Currently, most 3D compression schemes are developed for stereoscopic images or videos that consist of two views taken from a lightly different perspective in a 3D scene. Since one image (target) in a stereo-pair images can be restored from the disparity information and the other one image (reference), the reference image is in general coded with a traditional 2D compression scheme whereas the target image can be represented by disparity vectors. Stereoscopic coding schemes using the disparity estimation can be classified into: 1) intensity-based methods and 2) feature-based methods [1]. Although many quality metrics for 2D image quality assessment have been proposed, the quality models on stereoscopic images have not been widely studied. Hewage et al. [14] tested the performance of three quality metrics, including peak signal-to-noise ratio (PSNR), video quality model (VQM) proposed by NTIA [15], and structural similarity model (SSIM) [16], with respect to a subjective quality experiment on a series of coded stereoscopic images. The experimental results demonstrated that VQM is better than other two metrics while its performance is still not promising. Similar work has been done in [17]. Four metrics, as well as three approaches, called average approach, main eye approach, and visual acuity approach, were tested for evaluating the perceptual quality of stereoscopic images. Further, disparity information was integrated into two metrics for the quality assessment [18]. It was found that the disparity information has a significant impact on stereoscopic quality assessment, while its capability has not been studied adequately. In [19], only absolute disparity was used. It was found that added noise on the relatively large absolute disparity has greater influence than on other disparity. Subsequently, a metric called stereo sense assessment (SSA) based on the disparity distribution was proposed.

In addition, some special metrics that take into account advantage of the characteristics of 3D images have been proposed. Boev et al. [20] combined two components: a monoscopic quality component and a stereoscopic quality component, for developing a stereo-video quality metric. A cyclopean image for monoscopic quality, a perceptual disparity map, and a stereo-similarity map for stereoscopic quality were defined. These maps were then measured using SSIM in different scales and combined into a monoscopic quality index and a stereoscopic quality index, respectively. The experimental result demonstrated that the proposed method is better than signal-to-noise ratio (SNR). Additionally, an artifact distribution of coding schemes at different depth layers within a 3D image was modeled in a single metric [21]. The metric included three steps. Firstly, a set of 2D image pairs were synthesized at different depth layers using an image based rendering (IBR) scheme. Secondly, pixels that can be discerned to belong to each depth layer were identified. Finally, the image pairs were masked and the coding artifact at each depth layer was evaluated using the multi-scale SSIM. Three coding schemes were studied including two H.264 based pseudo video coding schemes and JPEG 2000. The experimental results showed a high correlation between the coding artifacts and their distribution in different depth layers. Gorley et al. [22] used a new Stereo Band Limited Contrast (SBLC) algorithm to rank stereoscopic pairs in terms of image quality. SBLC took into account the sensitivity to contrast and luminance changes in image regions with high spatial frequency. A threshold for evaluating image quality produced by SBLC metric was found to be closely correlated to subjective measurements. Sazzad et al. [23] assumed that perceived distortion and depth of any stereoscopic images are strongly dependent on the spatial characteristics in certain image regions, such as edge regions, smooth and texture regions. Therefore, a blockiness index and zero crossing rates within these regions in an image were then evaluated. They were finally integrated into a single value using an optimization algorithm according to subjective quality evaluation results. With respect to a quality database on JPEG coded stereoscopic images, the model performed quite well over a wide rang of image contents and distortion levels.

Transport methods of 3DTV were surveyed from early analog systems to most recent digital technologies in [24]. Potential digital transport architectures for 3DTV include the digital video broadcast (DVB) architecture for broadcast, and the Internet Protocol (IP) architecture for wired or wireless streaming. Motivated by a growing impact of IP based media transport technologies, Akar et al. mainly focused on the ubiquitous Internet by using it as a choice of the network infrastructure for future 3DTV systems in [24]. To our best knowledge, the quality evaluation issues for 3D presentations with transmission errors have almost not addressed so far. However, transmission related factors in IP based architecture have a non-neglectable impact on the perceived quality of 3D presentations. For example, different network protocols, such as Datagram Congestion Control Protocol (DCCP) and Peer-to-Peer Protocols, different transmission control schemes, e.g. effective congestion control, packet loss protection and concealment, video rate adaptation, and network/service scalability will definitely affect the quality of visual experience. The artefacts introduced by the transmission errors (such as packet loss and jitter) are quite different from the coding artefacts (such as blurring). Therefore, an accurate quality metric should also take into account the characteristics of the transmission artefacts.

Finally, for the video-plus-depth technique, depth image based rendering (DIBR) is inevitable and the influence of DIBR on the perceptual quality should also be taken into consideration. Although depth perception can provide more comfortable experience for end-users, the rendering process of current display technology also introduces degradation on the image quality. Barkowsky et al. tried to achieve a trade-off between such increased comfort and the introduced distortion [25]. However, their experimental results indicated that the distortion caused by the depth rendering process is usually greater than the comfort provided by the depth perception, especially for certain types of visual contents. Another objective metric for DIBR was proposed lately in [26]. This metric consisted of Color and Sharpness of Edge Distortion (CSED) measures. The color measure evaluated luminance loss of the rendered image compared against the reference image. The sharpness of edge distortion measure calculated a proportion of the remained edge in the distorted image regarding the edge in the reference image, whilst taking into account the depth information. To validate the performance of the proposed metric, a subjective assessment of DIBR techniques with five different hole-filling methods (Constant color filling, Horizontal interpolation, Region of interest filling, Horizontal extrapolation and Horizontal and vertical interpolation) was performed. The experimental results indicated the promising performance of this metric.

The rest of this chapter is organized as follows. In Section 2, a subjective quality experiment on stereoscopic images is briefly summarized and an analysis on the relationship between the perceived quality and distortion parameters is performed. Section 3 introduces some well-known 2D image quality metrics and investigates their capabilities in evaluating the stereoscopic image quality; an integration of disparity information into objective quality metrics is proposed based on an intensive analysis of the disparity information on stereoscopic quality evaluation. Finally, conclusions are drawn in Section 4.

2 Subjective Stereoscopic Image Quality Assessment

The quality of a visual presentation that is meant for human consumption (the user) can be evaluated by showing it to a human observer and asking the subject to judge its quality on a predefined scale. This is known as subjective assessment and is currently the most common way to evaluate the quality of image, video, and audio presentations. Generally, the subjective assessment is also the most reliable method as we are interested in evaluating quality as seen by the human eye. In this section, we will present a subjective quality assessment on stereoscopic images, which can be exploited for understanding perception of stereoscopic images and providing data for designing objective quality metrics [27]. We mainly focus on distortion types introduced by image processing but ignore the influence of a display device. The relationship between distortion parameters and the perceived quality will be investigated, and we will also validate whether PSNR can be used in predicting the stereoscopic image quality.

2.1 Experimental Materials and Methodology

The source stereo-pair images were collected from the Internet [28]. Ten pairs of high resolution and high quality color images that reflect adequate diversity in image contents were chosen. Figure 1 shows parts of the right eye images in the data set. For each pair of the source images, the right eye image was distorted with four different distortion types that may occur in real-world applications while the left eye image was kept undistorted. The distortion types included:

- Gaussian blurring: The R, G, and B color components were filtered using a circular-symmetric 2-D Gaussian kernel of standard deviation σB pixels. Three color components of an image were blurred using the same kernel, and σB values ranged from 0.2 to 100 pixels.
- JPEG compression: The distorted images were generated by compressing the reference images (full color) with JPEG at different bit rates ranging from 0.15 bits per pixel (bpp) to 3.34 bpp. The compression was implemented by a MATLAB's toolbox function (imwrite.m).
- JPEG2000 compression: The distorted images were generated by compressing the reference images (full color) with JPEG2000 at different bit rates ranging from 0.003 bpp to 2 bpp. Kakadu version 2.2 [29] was used to generate the JPEG2000 compressed images.
- White noise: White Gaussian noise with standard deviation σN was added to RGB color components of the images after scaling these three color components between 0 and 1. The used values of σN were between 0.012 and 2.0. The distorted components were clipped between 0 and 1, and then re-scaled to a range of [0-255].

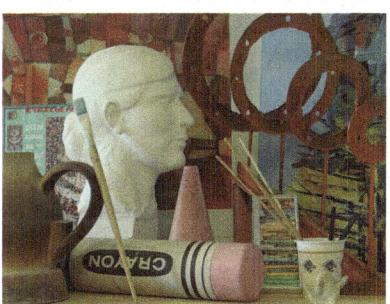

Fig. 1 Examples of stereoscopic images (top left: *Art*; top right: *Bowling*; bottom left: *Dwarves*; bottom right: *Moebius*)

These distortions reflected a broad range of types of image impairment, from smoothing to structured distortion, image-dependent distortions, and random noise. The distortion levels covered a wide range of quality degradation from imperceptible levels to high levels of impairment.

A subjective human trial, based on the ITU-R recommendation BT.500 [30], assessed the quality of the generated distorted stereoscopic images against the original images. A double-stimulus continuous quality-scale (DSCQS) method for stereoscopic image assessment was employed. The DSCQS method was cyclic, in which subjects viewed a pair of pictures with same content, i.e. a distorted image and an original image, and were asked to assess the qualities of both two images. The subjects were presented with a series of stereoscopic image pairs in a random order.

In the experiment, polarized glasses were worn in order to separate the left and right images on a single screen to the appropriate eyes. The experiment was conducted in a dark room, with constant minimal light levels. Twenty non-expert subjects participated in the quality evaluation, whose ages vary from 20 to 25 with a mean of 23 years. All the subjects participating in this experiment met the minimum criteria of acuity of 20:30 vision, stereo-acuity at 40 sec-arc, and passed a color vision test. The participants were not aware of the purpose of this experiment or that one of the stereo-pair images was undistorted. Before starting the experiment, all the subjects received instructions and completed a training trial of the stereo display. This training trial contained four sets of stereo-pair images viewed and rated in a same way to that in the actual test, but the results from the training trial were not included in the result analysis. The observers then completed the experimental trials for each distortion category. They were asked to be as accurate as possible to judge the image quality. The experiment lasted 50 minutes including short breaks after each distortion type.

In each trial the images were rated on a sliding scale of Excellent, Good, Fair, Poor, and Bad. The participants were asked to assess the overall quality of each stereo-pair images by filling in an answer sheet. After the quality evaluation for all the images was finished, difference mean opinion scores (DMOS) were calculated using ITU-R Recommendation BT.500 on a scale of 0-100 as follows. Raw scores were firstly converted to raw quality difference scores as following:

$$D_{ij} = R_{iref(j)} - R_{ij} \tag{1}$$

where R_{ij} denotes the raw score of the j-th stereoscopic image given by the i-th subject, and $R_{iref(j)}$ denotes the raw quality score assigned by the i-th subject to the reference image against the j-th distorted image. Then, the raw difference scores D_{ij} were converted into $DMOS_j$ by averaging the raw difference scores.

2.2 Experimental Results and Analysis

Two phenomena were observed in the experiment. Firstly, the luminance of the stereoscopic display device can affect the eye strain when subjects evaluated image quality. The dominant eye will feel uncomfortable when the luminance values of two projectors are same to each other. This uncomfortableness can be abated by turning down the luminance of the dominant projector. Secondly, the difference between the resolutions of the stereoscopic images and the stereoscopic display device also has influence on the stereoscopic quality assessment.

Subsequently, we will investigate the relationship between the perceived quality and the distortion types and parameters. Firstly, the DMOS values on Gaussian blurred images are shown in Fig. 2. The perceived quality values of stereoscopic images show a decreasing trend as the distortion levels on right view images increasing. But all the DMOS values are below 20 regarding the whole range of quality score being from 0 to 100. In our opinion, the stereoscopic image quality degradation caused by Gaussian blur is less affected by the poor quality presented to the right eye images because subjects pay more attention to the left eye images that have no distortion. In addition, no obvious difference between the DMOS values on different image contents was found for the blurring distortion.

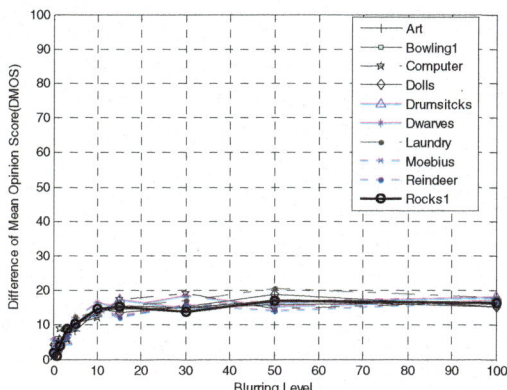

Fig. 2 DMOS values of Gaussian blurred stereoscopic images

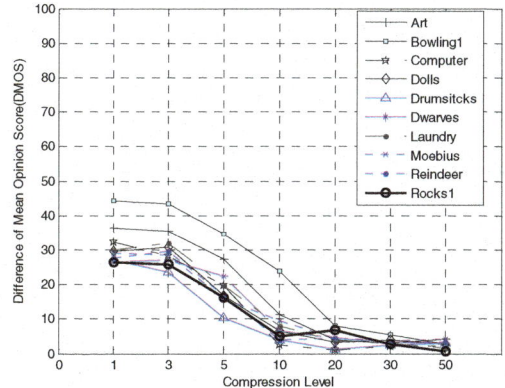

Fig. 3 DMOS values of JPEG compressed stereoscopic images

For the distortion caused by JPEG compression, the fluctuation of the perceived stereoscopic image qualities is more drastic than that of the blurring distortion, as shown in Fig. 3. The x-axis of Fig. 3 represents the JPEG Q-parameter which determines the quality of the compressed image. In the figure, the quality scores show an increasing trend as the bit-rates for right view images increasing, and the maximal DMOS value is below 50. According to the figure, we can find that the perceived quality of JPEG compressed image is content and texture dependent. The smoother areas the image contains, the more blocking artifacts are visible. For example, the *Bowling* image is relatively smoother than other images, as shown in Fig. 1, and the perceived distortion on this image introduced by blockiness artifact is, therefore, more visible than on other images, as shown in Fig. 3.

In reference [1], it is stated that for blockiness artifact, the quality of stereoscopic images is approximately a mean of qualities of the images presented to the left and right eyes; while for blurring artifacts, the image quality is less affected by the poor quality presented to one eye because more weight is given to the input that has the sharper image, therefore, low-pass filtering (blur) of the images for one eye is a more effective method for reducing bandwidth than quantization. However, our experiment indicated that the perceived quality of images with blockiness artifact is content and texture dependent, and the depth perception degrades when the blurring level is increased. Compared with watching the JPEG compressed images with a similar perceived quality, all participants felt more eye strain and uncomfortable when viewing the blurred images. Thus, the use of low-pass filtering instead of quantization for processing stereoscopic images is required to be explored further.

For white noise distortion, the quality scores show a linear increasing trend as the noise added to right eye images increasing and the maximal DMOS value is below 50. According to Fig. 4, we can find that the image content and texture information have no significant influence on the perceived quality.

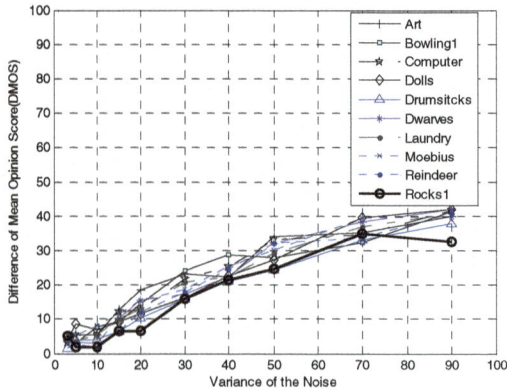

Fig. 4 DMOS values of stereoscopic images with white noise

As shown in Fig. 5, the bit rate in JPEG2000 compression schemes has less impact on the perceived quality of stereoscopic images when the bit rate is more than a threshold, such as 0.1bpp, and the DMOS values is usually in a range of [0-10]. In other words, non-experts usually feel less uncomfortable when watching these distorted images. But when the bit rate is less than 0.1bpp, the quality decreases dramatically as the bit rate decreasing. In addition, the perceived quality is also dependent on image contents. The DMOS value of the *Bowling* image that contains more smooth areas is smaller than that of other images when the bit-rate is fixed. The reason might be that JPEG2000 compression scheme discards higher frequency components when encoding images with plenty of texture, compared with the images with more smooth areas.

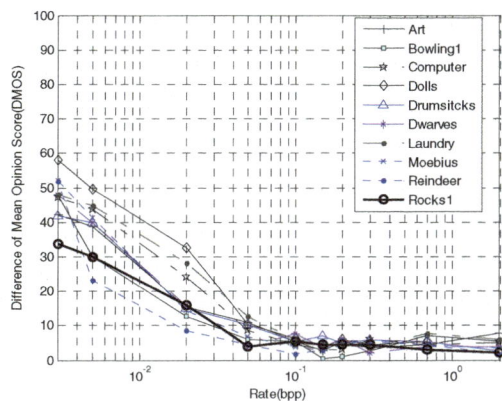

Fig. 5 DMOS values of JPEG2000 compressed stereoscopic images

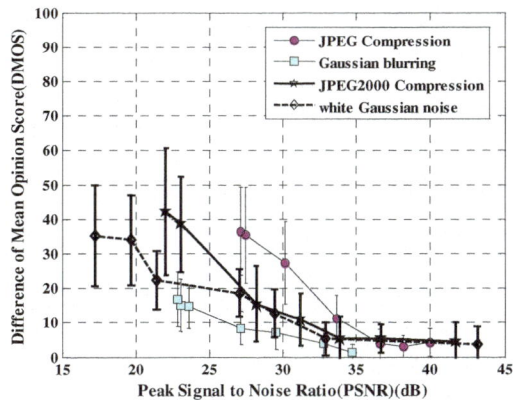

Fig. 6 Scatter plot of *Art* image: DMOS values versus PSNR

According to the analysis on the relationship between the perceived stereo-scopic quality and distortion conditions, the influence of different distortion types and image contents on the perceived quality is different. Subsequently, we will simply analyze the performance of PSNR on the quality prediction for stereo-scopic images. Figure 6 and Figure 7 show the mean of DMOS values and 95% confidence interval plotted against PSNR (dB) from each distortion type for the *Art* image and *Bowling* image, respectively. As expected, the PSNR values in-crease as DMOS values decreasing for both images. However, it is found that the difference of the DMOS values of the *Art* image between the Gaussian blurring and JPEG2000 compression is less than that of the *Bowling* image for the same PSNR value. This phenomenon indicates that the performance of PSNR is de-pendent on the distortion type and image content, and it might not be an appropri-ate metric for evaluating the quality of stereoscopic images. Therefore, we will propose more suitable objective quality metrics for the stereoscopic image based on some 2D image quality metrics and the disparity information. The extensive analysis on the experimental results above is a fundament work and provides a solid basis for designing objective quality metrics.

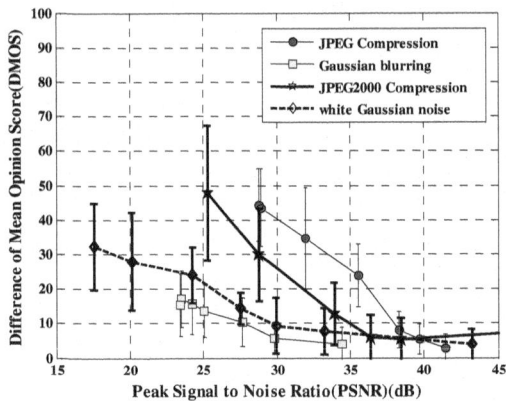

Fig. 7 Scatter plot of *Bowling* image: DMOS values versus PSNR

3   Perceptual Stereoscopic Image Quality Metric Based on 2D Image Quality Metrics and Disparity Analysis

Accurately predicting stereoscopic quality is an important issue for improving the ability and feasibility of compression and transmission schemes for stereoscopic images. Although many 2D image quality metrics (IQMs) have been proposed that work well on 2D images, developing quality metrics for 3D presentations is almost an unexplored issue. As indicated in Section 2, PSNR is not appropriate for evalu-ating the quality of stereoscopic images. Therefore, in this section, we will intro-duce some well-known 2D IQMs and investigate their capabilities in stereoscopic image quality assessment. Furthermore, as disparity is an important attribute of stereopsis, we will try to improve the performance of IQMs on stereoscopic image quality assessment by integrating disparity information into the IQMs. We will mainly focus on the full reference (FR) metrics in this study, which means that the undistorted images are required for evaluating the quality of the distorted images.

3.1 Introduction to 2D Image Quality Metrics

Over the years, a number of researchers have contributed significant research in the design of full reference image quality assessment algorithms, claiming to have made headway in their respective domains [31]. In this study, eleven IQMs that are summarized in Table 1 were employed and explained in detail as follows.

Table 1 Descriptions of image quality metrics

IQM	Descriptions
PSNR	Peak signal-to-noise ratio
SSIM	Single scale structural similarity
MSSIM	Multi-scale structural similarity
VSNR	Visual signal-to-noise ratio
VIF	Visual information fidelity
UQI	Universal quality index
IFC	Information fidelity criterion
NQM	Noise quality measure
WSNR	Weighted signal-to-noise ratio
PHVS	Modified PSNR based on HVS
JND	Just noticeable distortion model

PSNR is a traditionally used metric for visual quality assessment and still widely used in evaluating the performance of compression and transmission schemes. Although the performance of PSNR is worse than many other image quality metrics in certain distortion types and respective domains, it is still appealing because it is simple to compute, has clear physical meanings, and is mathematically convenient in the context of optimization.

SSIM (Structural SIMilarity) [15] is to compare structural information between the reference and distorted images. Under an assumption that the human visual system is highly adapted for extracting structural information from a scene, a similarity measure can be constructed based on luminance comparison, contrast comparison, and structure comparison between the reference and distorted images.

MSSIM (Multi-scale SSIM) [32] is an extension of SSIM. MSSIM iteratively applies a low-pass filter in the reference and distorted images and down-samples the filtered images by a factor of 2. At each image scale j after j-1 iterations, the contrast comparison and the structure comparison are calculated, respectively. The luminance comparison is computed only at the highest scale. The overall MSSIM measure is obtained by combining the measures at different scales.

VSNR (Visual Signal-to-Noise Ratio) [33] operates via a two-stage approach. In the first stage, contrast thresholds for detection of distortions in the presence of natural images are computed by wavelet-based models of visual masking and visual summation. The second stage is applied if the distortions are suprathreshold, which operates based on low-level visual property of perceived contrast and mid-level visual property of global precedence. These two properties are measured by the Euclidean distance in a distortion-contrast space of multi-scale wavelet decomposition. VSNR is computed based on a simple linear sum of these distances.

VIF (Visual Information Fidelity) [34] is to quantify loss of image information to the distortion process based on natural scene statistics, the human visual system, and an image distortion model in an information-theoretic framework.

UQI (Universal Quality Index) [35] is similar to SSIM, and it is to model image distortions as a combination of three factors: loss of correlation, luminance distortion, and contrast distortion.

IFC (Information Fidelity Criterion) [36] is a previous work of VIF. IFC is to model the natural scene statistics of the reference and distorted images in wavelet domain using steerable pyramid decomposition [37].

NQM (Noise Quality Measure) [38] is a measure aiming at the quality assessment of additive noise by taking into account variation in contrast sensitivity, variation in local luminance, contrast interaction between spatial frequencies, and contrast masking effects.

WSNR (Weighted Signal-to-Noise Ratio) [38] is to compute a weighted signal-to-noise ratio in frequency domain. The difference between the reference image and distorted image is transformed into the frequency domain using a 2D Fourier transform and then weighted by the contrast sensitivity function.

PHVS (PSNR based on the Human Visual System) [39] is a modification of PSNR based on a model of visual between-coefficient contrast masking of discrete cosine transform (DCT) basis functions. This model can calculate the maximal distortion that is not visible at each DCT coefficient due to the between-coefficient contrast masking.

JND (Just Noticeable Distortion) [40] model integrates spatial masking factors into a nonlinear additivity model for masking effects to estimate the just noticeable distortion. A JND estimator applies to all color components and accounts for a compound impact of luminance masking, texture masking and temporal masking. Finally, a modified PSNR is computed by excluding the imperceptible distortions from the computation of the traditional PSNR.

Because four typical distortion types were adopted in the subjective quality assessment on the stereoscopic images in Section 2, we will also investigate the performance of these IQMs on 2D images with the same distortion types. The source 2D images and corresponding subjective evaluation results were collected from the LIVE image quality database [15, 31, 41], and the distortions are as following:

- Gaussian blur: The R, G, and B color components were filtered using a circular-symmetric 2-D Gaussian kernel of standard deviation σB pixels. These three color components of the image were blurred using the same kernel. The values of σB ranged from 0.42 to 15 pixels.

- JPEG compression: The distorted images were generated by compressing the reference images (full color) using JPEG at different bit rates ranging from 0.15 bpp to 3.34 bpp. The implementation was performed by the imwrite.m function in MATLAB.
- JPEG2000 compression: The distorted images were generated by compressing the reference images (full color) using JPEG2000 at different bit rates ranging from 0.028 bits per pixel (bpp) to 3.15 bpp. Kakadu version 2.2 was used to generate the JPEG2000 compressed images.
- White noise: White Gaussian noise of standard deviation σN was added to the RGB components of the images after scaling the three components between 0 and 1. The same σN was used for the R, G, and B components. The values of σN used were between 0.012 and 2.0. The distorted components were clipped between 0 and 1, and then rescaled to the range of [0-255].

Basically, the distortion types and the generation in the stereoscopic image quality assessment are very similar to those in the 2D image quality assessment. Therefore, these two image data sets and the corresponding subjective assessment results can provide a fair comparison of the IQMs between the stereoscopic and 2D image quality assessments. The performance comparison and analysis will be performed with respect to the LIVE database and the subjective stereoscopic image quality experiment described in Section 2, respectively.

3.2 Performance Analysis of IQMs on 2D and Stereoscopic Image Quality Assessment

We performed the 11 IQMs on the 2D and the stereoscopic images, respectively. As some IQMs use the luminance component only, while others can employ the color components as well, we transformed all the color images into gray images firstly, and then computed the image quality using these IQMs. After obtaining the metric results, a nonlinear regression operation between the metric results (*IQ*) and the subjective scores (DMOS), as suggested in [42], was performed using the following logistic function:

$$DMOS_P = \frac{a_1}{1 + \exp(-a_2 \cdot (IQ - a_3))} \tag{2}$$

The nonlinear regression function was used to transform the set of metric results to a set of predicted DMOS values, $DMOS_P$, which were then compared against the actual subjective scores (DMOS) and result in two evaluation criteria: root mean square error (RMSE) and Pearson correlation coefficient. The evaluation results of these eleven IQMs on the quality assessment and the LIVE 2D image data set are given in Table 2 and 3. According to the evaluation results, some general conclusions can be drawn as follows.

Table 2 RMSE of IQMs on LIVE data set

IQM	Blurring	JPEG	JPEG2000	Noise	All
PSNR	9.78	8.43	7.45	2.71	9.60
SSIM	7.50	5.97	5.71	3.89	8.50
MSSIM	5.25	5.43	4.84	4.16	7.11
VSNR	5.94	5.78	5.52	3.35	7.47
VIF	4.39	6.49	5.13	2.97	6.53
UQI	5.09	8.46	8.59	5.53	8.77
IFC	4.99	7.51	7.55	5.50	7.37
NQM	7.55	6.31	6.00	2.79	7.45
WSNR	6.30	6.57	6.97	3.52	7.79
PHVS	6.41	5.81	5.52	2.56	7.71
JND	5.99	6.87	6.11	3.28	8.83
Average	*6.29*	*6.69*	*6.31*	*3.66*	*7.92*

Table 3 Pearson correlation coefficients of IQMs on LIVE data set

IQM	Blurring	JPEG	JPEG2000	Noise	All
PSNR	0.783	0.850	0.888	0.986	0.801
SSIM	0.879	0.928	0.936	0.970	0.848
MSSIM	0.943	0.941	0.954	0.943	0.896
VSNR	0.926	0.932	0.940	0.926	0.885
VIF	0.960	0.914	0.949	0.960	0.913
UQI	0.946	0.849	0.848	0.946	0.837
IFC	0.949	0.883	0.885	0.949	0.888
NQM	0.877	0.919	0.929	0.877	0.885
WSNR	0.916	0.912	0.903	0.916	0.874
PHVS	0.913	0.932	0.940	0.913	0.877
JND	0.925	0.901	0.927	0.930	0.835
Average	*0.911*	*0.906*	*0.918*	*0.938*	*0.867*

- The performance of IQMs has significant difference for different distortion types. For example, PSNR is suitable for evaluating the quality degradation caused by noise, while its performance on blurring distortion is not promising. UQI and IFC have excellent performance in predicting the compression degradation.
- Statistically speaking, these IQMs have promising performance on a single distortion type, while the robustness to the change of the distortion types is not very strong.

Table 4 RMSE of IQMs on stereoscopic images

IQM	Blurring	JPEG	JPEG2000	Noise	All
PSNR	1.97	5.97	5.09	2.74	7.64
SSIM	3.00	7.70	8.91	2.43	9.28
MSSIM	1.91	4.94	4.97	2.51	7.62
VSNR	2.58	5.71	5.42	3.54	8.63
VIF	1.84	4.39	6.45	3.41	7.78
UQI	1.89	3.96	6.44	4.06	7.13
IFC	1.78	3.55	6.37	3.98	8.61
NQM	2.17	3.53	4.23	4.36	8.70
WSNR	2.02	6.74	6.09	3.85	9.07
PHVS	2.10	5.62	5.27	2.65	8.06
JND	2.18	6.97	5.73	3.55	8.58
Average	*2.13*	*5.38*	*5.91*	*3.37*	*8.29*

Table 5 Pearson correlation coefficients of IQMs on stereoscopic images

IQM	Blurring	JPEG	JPEG2000	Noise	All
PSNR	0.939	0.882	0.950	0.978	0.795
SSIM	0.851	0.793	0.824	0.983	0.677
MSSIM	0.943	0.920	0.948	0.981	0.797
VSNR	0.893	0.893	0.948	0.962	0.731
VIF	0.947	0.938	0.916	0.966	0.788
UQI	0.943	0.950	0.913	0.950	0.825
IFC	0.951	0.962	0.929	0.954	0.734
NQM	0.925	0.961	0.963	0.942	0.726
WSNR	0.935	0.846	0.924	0.955	0.696
PHVS	0.930	0.896	0.949	0.979	0.769
JND	0.924	0.836	0.937	0.961	0.738
Average	*0.926*	*0.898*	*0.927*	*0.965*	*0.752*

Subsequently, we performed the IQMs on the right eye images in the stereo-scopic image quality assessment, as there were no distortions on the left eye images. Table 4 and Table 5 give the evaluation results. In addition, we can use a constant as the quality value on the left eye images, e.g. 1, for SSIM, MSSIM, UQI, etc. Then, the significance of the interaction effects between the quality values of the left eye images, the right eye images, and the overall qualities was tested by performing a two-way ANOVA (ANalysis Of Variance) on the results. The results of the ANOVA show that the quality of the right eye image dominates the overall quality and there is almost no influence of the left eye image on the overall quality. In addition, according to the evaluation results, the situation of the performance of these IQMs in evaluating the stereoscopic image quality is similar to that in predicting the 2D image quality. A better IQM on the 2D image quality assessment usually has better performance on the stereoscopic quality assessment. However, according to the averages of different IQMs, as shown in the last rows in the Tables, the robustness of these IQMs to the change of distortion types in stereoscopic image quality assessment is much worse than that in 2D image quality assessment. The performance of these IQMs on the entire distortion types in the 2D image quality assessment is much better than that in the stereoscopic image quality assessment. In our opinion, the reason is that the perceived quality is not only affected by image content, but other attributes of stereopsis, such as disparity, have significant influence on the quality evaluation of the stereoscopic images as well.

3.3 Perceptual Stereoscopic Quality Assessment Based on Disparity Information

Human eyes are horizontally separated by about 50-75 mm (interpupillary distance) depending on each individual. Thus, each eye has a slightly different view of the world. This can be easily seen when alternately closing one eye while looking at a vertical edge. At any given moment, the lines of sight of the two eyes meet at a point in space. This point in space projects to the same location (i.e. the center) on the retinae of the two eyes. Because of different viewpoints observed by the left and right eyes however, many other points in space do not fall on corresponding retinal locations. Visual binocular disparity is defined as the difference between the points of projection in the two eyes and is usually expressed in degrees as the visual angle. The brain uses binocular disparity to extract depth information from the two-dimensional retinal images in stereopsis. In computer stereo vision, binocular disparity refers to the same difference captured by two different cameras instead of eyes [43]. Generally, one image of stereo-pair images can be restored from the disparity and the other one image. Therefore, we believe that the disparity between the left and right eye images has an important impact on visual quality assessment. In this subsection, we apply the disparity information in the stereoscopic image quality assessment [44].

In this work, we do not intend to study the estimation methods of disparity map between a stereo-pair images and their impact on the quality assessment. We chose a belief propagation based method to estimate the disparity map [45]. Because of the distorted regions, the disparity of the original stereo-pair images is different from that of the distorted stereo-pair images, even though the relative positions of the objects in the image pair do not change at all. Figure 8 shows examples of an original *Art* image (right eye), distorted images, and the corresponding disparity maps. Because the real objects in the image do not change during the distortion process, changes between two disparity images (one is original disparity and another is the disparity between the left eye image and the distorted right eye image) are usually located at those positions where the distortions are clearly visible, such as noise added regions, regions with blockiness. Consequently, we can compare the disparity images to obtain a quality prediction for the distorted stereoscopic images.

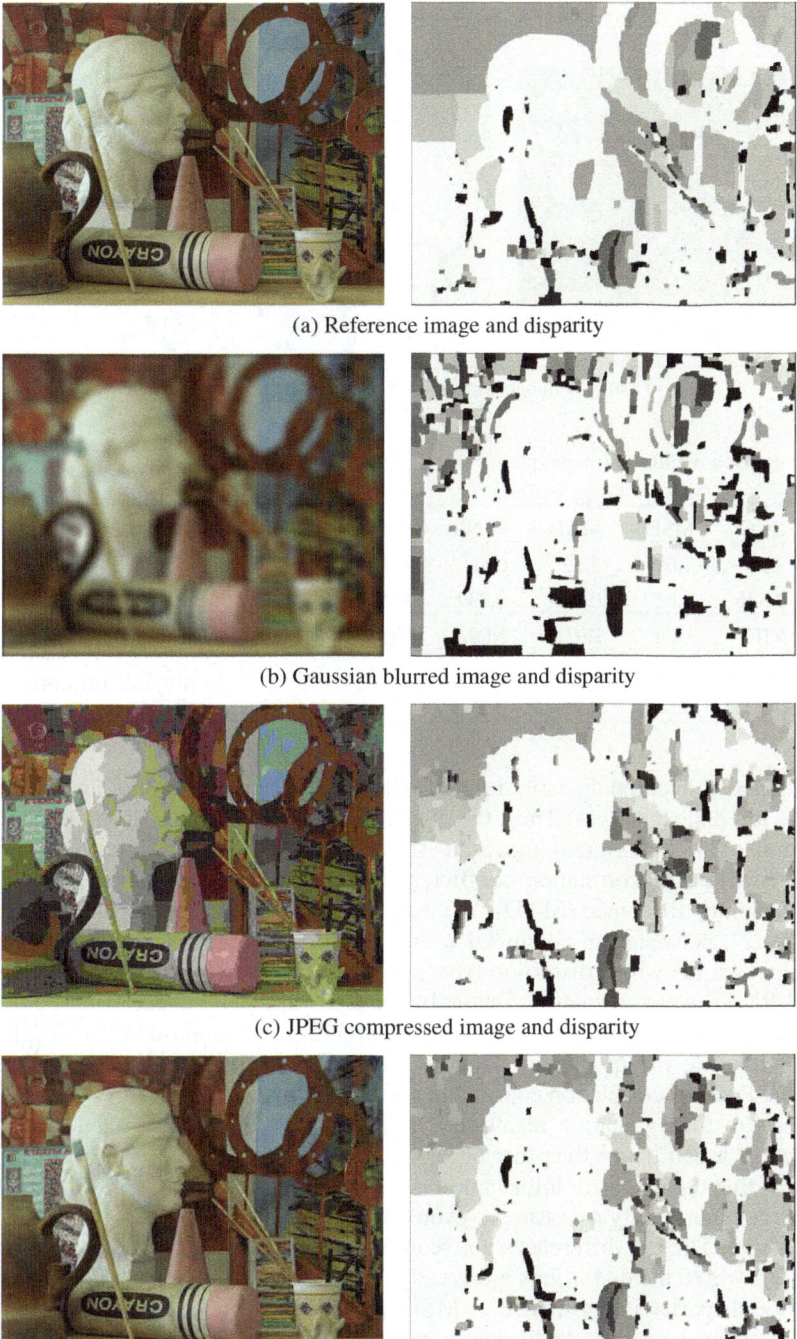

(a) Reference image and disparity

(b) Gaussian blurred image and disparity

(c) JPEG compressed image and disparity

(d) JPEG2000 compressed image and disparity

Fig. 8. *Art*: reference, distorted images, and corresponding disparity images

Fig. 8 *(Cont.)*

(e) Distorted image with white noise and disparity

Table 6 Evaluation results of image quality metrics on disparity images

Criteria	GCC	MSE	MAD	PSNR	SSIM	MSSIM	VSNR
RMSE	7.11	7.09	7.22	6.86	6.40	6.99	6.85
Pearson	0.826	0.827	0.820	0.839	0.862	0.832	0.840
Criteria	VIF	UQI	IFC	NQM	WSNR	PHVS	JND
RMSE	8.31	6.37	8.08	7.94	7.20	6.87	7.59
Pearson	0.755	0.863	0.769	0.776	0.821	0.839	0.816

As explained above, the disparity refers to the difference in location of an object seen by the left and right eyes. Thus, the disparity images have quite different modalities compared to the original images, as shown in Fig. 8. Firstly, we tested three simple metrics: global correlation coefficient (GCC), mean square error (MSE), and mean absolute difference (MAD). We performed the same fitting operation, as in Equation (2), between the computed results obtained by these metrics and the DMOS values on the whole distortion types, and then the Pearson correlation coefficient and RMSE were calculated. Secondly, we also validated the performance of IQMs on the disparity images, even though these IQMs were supposed to be developed for predicting the quality of natural images. Table 6 gives the evaluation results of the Pearson correlation coefficient and RMSE using these metrics.

According to the evaluation results of the IQMs on the disparity images, the performance is much better than that on original images. This observation probably indicates that the disparity information is more important than the original images for perceptual quality assessment, even though the disparity does not contain any real objects. The big differences between two disparity images usually appear in the regions where the distortions are greatly annoying. Thus, even a very simple metric on the disparity images, such as MSE, performs better than a complicated IQM on the original images. Additionally, we found that SSIM and UQI have the best performance within all the IQMs. We believe that this is because these two metrics are based on comparing the structural information, and the disparity can express such structural information of the original images.

Since the disparity images have significant influence on the stereoscopic image quality assessment, we naturally suppose that the combination of the disparity images and the original images can perform better than using either the disparity or the original images solely. Subsequently, we used three approaches to combine the disparity and original images to compute the stereoscopic image quality.

Table 7 Evaluation results of global combination between image quality and disparity quality on stereoscopic image quality assessment

DQ	IQ										
	PSNR	SSIM	MSSIM	VSNR	VIF	UQI	IFC	NQM	WSNR	PHVS	JND
GCC	0.869	0.867	0.840	0.830	0.831	0.835	0.836	0.837	0.828	0.833	0.839
MSE	0.887	0.878	0.838	0.830	0.828	0.844	0.843	0.829	0.847	0.828	0.846
MAD	0.888	**0.899**	0.853	0.828	0.825	0.841	0.833	0.829	0.838	0.830	0.851
PSNR	0.876	0.887	0.848	0.836	0.837	0.847	0.874	0.842	0.840	0.839	0.829
SSIM	0.858	0.859	0.870	0.862	0.858	0.870	0.861	0.866	0.856	0.859	0.866
MSSIM	0.857	0.865	0.837	0.832	0.836	0.846	0.853	0.833	0.840	0.834	0.815
VSNR	0.850	0.842	0.844	0.841	0.837	0.860	0.834	0.838	0.833	0.845	0.863
VIF	0.817	0.819	0.804	0.741	0.779	0.826	0.730	0.732	0.730	0.766	0.778
UQI	0.855	0.859	0.865	0.862	0.858	0.868	0.863	0.868	0.855	0.857	0.864
IFC	0.814	0.807	0.793	0.764	0.775	0.822	0.760	0.762	0.760	0.778	0.780
NQM	0.847	0.856	0.829	0.770	0.784	0.827	0.774	0.764	0.763	0.796	0.775
WSNR	0.865	0.878	0.852	0.817	0.831	0.838	0.840	0.821	0.818	0.823	0.818
PHVS	0.853	0.879	0.845	0.813	0.818	0.843	0.823	0.817	0.813	0.818	0.825
JND	0.839	0.876	0.827	0.833	0.806	0.852	0.795	0.836	0.827	0.815	0.839

The first approach, called global combination, was to compute two quality values of the distorted image and the distorted disparity firstly, denoted as IQ and DQ, respectively. IQ was computed by IQMs on the original images, and DQ by GCC, MSE, MAD, and the IQMs. Then, an overall quality which was taken as the quality of the stereoscopic image was calculated using the following function with different coefficients and exponents:

$$OQ = a \cdot IQ^d + b \cdot DQ^e + c \cdot IQ^d \cdot DQ^e \qquad (3)$$

In this study, we employed Levenberg-Marquardt algorithm to find the optimum parameters in Equation (3). Although the optimum parameters may change if different initial values were used, we found that the highest correlation coefficient between OQ and DMOS values is 0.899. For example, one set of the optimum parameters is $a=3.465$, $b=0.002$, $c=-0.0002$, $d=-1.083$, and $e=2.2$. In this experiment, we used the direct correlation between OQ and DMOS values while the fitting operation in Equation (2) was not performed because we have performed an optimization operation between OQ and DMOS values in Equation (3). We report the highest correlation for different combinations in Table 7 while the corresponding optimum parameters are omitted for the sake of clarity.

According to the experimental results, it was found that appropriate combinations of the image quality and the disparity quality perform better than using the quality of either the original images or the disparity images solely. In addition, we also found that the combination of SSIM and MAD, i.e. SSIM was used to compute IQ and MAD was used to compute DQ, always obtains the best performance within all the possible combinations. Furthermore, SSIM has a promising performance in the combinations either for measuring the original image quality or for computing the disparity image quality. This result indicates that a good metric for predicting the stereoscopic image quality can be developed if appropriate methods are found to combine the original image quality and the disparity image quality.

The second approach is called local combination. Some IQMs, e.g. PSNR (based on MSE), SSIM, MSSIM, UQI, PHVS, and JND, compute a quality map between the reference image and the distorted image to depict the distribution of quality degradation at image pixels directly or indirectly, and the overall quality of the distorted image is usually computed as a mean over all the pixels in the quality map. Furthermore, we can also compute a quality map of the disparity image which can reflect an approximate distribution of the degradation on the distorted disparity image. In this study, four methods were used to compute the quality map on the disparity image as following:

$$DDQ = \begin{cases} (D-\overline{D})^2 \\ |D-\overline{D}| \\ 1-\dfrac{\sqrt{D^2-\overline{D}^2}}{255} \\ IMQ(D,\overline{D}) \end{cases} \tag{4}$$

where D and $\overline{D}$ denote the original disparity image and the distorted disparity image, respectively, and $IMQ(D,\overline{D})$ denote the quality map using the corresponding IQMs (including PSNR, SSIM, MSSIM, UQI, PHVS, and JND) between the original disparity image and the distorted disparity image. After computing the quality maps of the original image and the disparity image, Equation (3) was used to pool each pixel pair on the quality maps, and then the mean over all pixels was taken as the overall quality of the stereoscopic image. Table 8 gives the Pearson correlation coefficients between the quality values and the DMOS values by using the local combination, where the highest correlation coefficients were reported.

According to the evaluation results, it was found that the performance improvement by using the local combination is not as significant as if the global combination was performed. Some combinations even reduced the correlation between the overall quality values and the subjective DMOS values. However, we found that SSIM and UQI algorithms on the original images and disparity images have the best performance for local combination, regardless of what kinds of combination are used. For example, let *IQM* and *DQM* be quality maps of the original image and disparity image computed by UQI and SSIM, respectively, and the combination at each pixel pair be $OQM = \sqrt{IQM} + \sqrt{DQM} + \sqrt{IQM \cdot DQM}$, the Pearson correlation coefficient between the predictive qualities and the subjective results is 0.899. Therefore, not all metrics are suitable for the local combination, and an appropriate method is needed to explore the relationship between the image quality map and the disparity quality map.

Table 8 Evaluation results of local combination between image quality map and disparity quality map on stereoscopic image quality assessment

DQ	IQ					
	PSNR	SSIM	MSSIM	UQI	PHVS	JND
$(D - \overline{D})^2$	0.866	0.832	0.776	0.833	0.837	0.770
$\left| D - \overline{D} \right|$	0.840	0.849	0.743	0.853	0.806	0.803
$1 - \dfrac{\sqrt{D^2 - \overline{D}^2}}{255}$	0.807	0.792	0.815	0.898	0.815	0.795
PSNR	0.799	0.805	0.786	0.842	0.776	0.782
SSIM	0.799	0.821	0.800	**0.899**	0.832	0.816
MSSIM	0.762	0.801	0.774	0.859	0.769	0.802
UQI	0.798	0.822	0.831	0.895	0.841	0.839
PHVS	0.795	0.826	0.804	0.846	0.765	0.728
JND	0.804	0.823	0.781	0.835	0.774	0.807

Finally, the third approach was to integrate the local combination into the global combination by the following three steps:

- Two quality maps were computed firstly using appropriate metrics on the original image and disparity image, respectively;
- These two maps were combined locally and the mean was taken as an intermediate quality of the distorted image;
- The final step was to combine the intermediate quality and the quality of the disparity image and then obtained the overall quality of the stereoscopic images.

In our experiment, the highest correlation coefficient (*0.91*) was achieved when UQI was used in computing the quality maps of the original image and disparity image, and the local combination on the quality maps was then combined with the MAD of the disparity image again. Figure 9 gives the scatter plot of the subjective DMOS values versus the optimum predictive quality values. According to the experiment results, the proposed model has better performance on predicting perceptual quality of the stereoscopic images with lower impairments than that on the images with higher impairments. Therefore, improving the robustness of the quality metric to different impairment levels is also an important task in the future work.

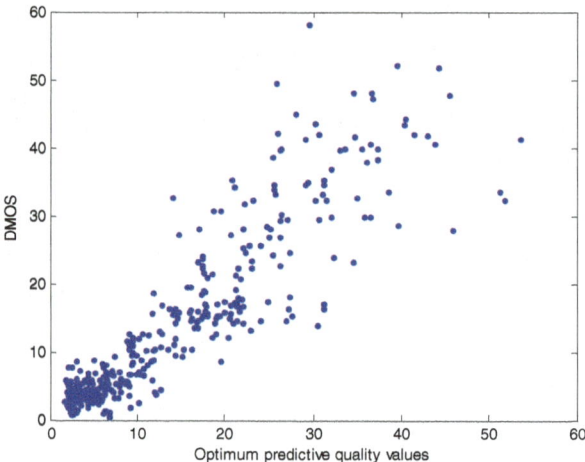

Fig. 9 Scatter plot of DMOS versus optimum predictive quality values

In summary, our experiments indicate that 2D IQMs can not be adopted in evaluating the stereoscopic image quality directly, and the disparity information has a significant impact on the perceived quality. The future work is needed to explore the relationship between the original image, the disparity information, and the quality assessment in depth.

4 Conclusions

In this chapter, we have investigated related issues in visual quality assessment for 3D presentations, especially the stereoscopic image. Typical distortion types on 3D presentations introduced from content capture, coding schemes, to transmission through communication channels, and in displaying the 3D presentation on an auto-stereoscopic display, were reviewed. We mainly focused on an analysis of the quality degradation caused by coding errors. To study the relationship between the perceived quality and distortion conditions for the stereoscopic images, a subjective quality assessment was conducted. Four typical distortion types: Gaussian blur, JPEG compression, JPEG2000 compression, and white noise, were introduced to some popular stereoscopic images, and the subjective quality evaluation was conducted in a controlled laboratory. We performed an intensive analysis on the relationship between the perceived quality and distortion conditions on the stereoscopic images. It was found that the perceived quality is dependent strongly on the distortion type and image content. The performance of PSNR in predicting the stereoscopic image quality was evaluated with respect to the subjective results. However, it was found that PSNR is not an appropriate metric for the stereoscopic image quality assessment. Therefore, we investigated the capabilities of some well-known 2D image quality metrics, including SSIM, MSSIM, VSNR, VIF, UQI, IFC, NQM, WSNR, PHVS, and JND model, in the stereoscopic image quality assessment. The experimental results indicated that 2D image quality metrics can not be adopted in evaluating the stereoscopic image quality directly. Furthermore, as an important factor in stereopsis, the disparity was taken into account in the stereoscopic image quality assessment. The experimental results demonstrated the promising performance by using the disparity information in evaluating the stereoscopic quality, and the best performance can be achieved when the disparity information and the original image are combined appropriately.

Although some tentative work on developing objective quality metrics for stereoscopic images has been done in the literature and this chapter, we are still a long way from 3D quality metrics that are widely applicable and universally recognized. The future work is to understand the fundamental of 3D presentation impact on the human visual and perceptual system. Determining how to model such impact and the relationship between the characteristics of 3D presentations and quality assessment is another critical issue for evaluating the quality of visual experience in 3D visual presentations.

References

1. Meesters, L.M.J., Ijsselsteijn, W.A., Seuntiëns, P.J.H.: A survey of perceptual evaluation and requirements of three-dimensional TV. IEEE Trans. Circuits and Systems for Video Technology 14(3), 381–391 (2004)
2. Tam, W.J., Stelmach, L.B., Corriveau, P.J.: Psychovisual aspects of viewing stereoscopic video sequences. In: Proc. SPIE Stereoscopic Displays and Virtual Reality System V, San Jose, CA, USA (1998)
3. Pastoor, S.: Human factors of 3D displays in advanced image communications. Displays 14(3), 150–157 (1993)
4. Stelmach, L., Tam, W.J., Meegan, D., Vincent, A.: Stereo image quality: effects of spatio-temporal resolution. IEEE Trans. Circuits and Systems for Video Technology 10(2), 188–193 (2000)
5. Ijsselsteijn, W.A., de, R.H., Vliegen, J.: Subjective evaluation of stereoscopic images: effects of camera parameters and display duration. IEEE Trans. Circuits and Systems for Video Technology 10(2), 225–233 (2000)
6. Meegan, D.V.: Unequal weighting of monocular inputs in binocular combination: implications for the compression of stereoscopic imagery. Journal Experimental Psychology: Applied 7(2), 143–153 (2001)
7. Seuntiens, P., Meesters, L., Ljsselsteijn, A.: Perceived quality of compressed stereoscopic images: Effects of symmetric and asymmetric JPEG coding and camera separation. ACM Trans. Applied Perception 3(2), 95–109 (2006)
8. Kalva, H., Christodoulou, L., Mayron, L.M., Marques, O., Furht, B.: Design and evaluation of a 3D video system based on H.264 view coding. In: Proc. ACM Network and operating systems support for digital audio and video, Newport, Rhode Island, USA (2006)
9. Woods, A., Docherty, T., Koch, R.: Image distortions in stereoscopic video systems. In: Proc. SPIE Stereoscopic Displays and Applications IV, San Jose, CA, USA (1993)

10. Yamanoue, H., Okui, M., Okano, F.: Geometrical analysis of puppet-theater and card-board effects in stereoscopic HDTV images. IEEE Trans. Circuits and Systems for Video Technology 16(6), 744–752 (2006)
11. Boev, A., Gotchev, A., Eqiazarian, K.: Crosstalk Measurement Methodology for Auto-Stereoscopic Screens. In: Proc. IEEE 3DTV, Kos, Greece (2007)
12. Seuntiens, P.J.H., Meesters, L.M.J., Ijsselsteijn, W.A.: Perceptual attributes of crosstalk in 3D images. Displays 26(4-5), 177–183 (2005)
13. Kim, D., Min, D., Oh, J., Jeon, S., Sohn, K.: Depth map quality metric for three-dimensional video. In: Proc. SPIE Stereoscopic Displays and Applications XX, San Jose, CA, USA (2009)
14. Hewage, C.T.E.R., Worrall, S.T., Dogan, S., Kondoz, A.M.: Prediction of stereoscopic video quality using objective quality models of 2-D video. Electronics Letters 44(16), 963–965 (2008)
15. Pinson, M., Wolf, S.: A new standardized method for objectively measuring video quality. IEEE Trans. Broadcasting 50(33), 312–322 (2004)
16. Wang, Z., Bovik, A.C., Sheikh, H.R., Simonselli, E.P.: Image quality assessment: from error visibility to structural similarity. IEEE Trans. Image Processing 13(4), 600–612 (2004)
17. Campisi, P., Callet, P.L., Marini, E.: Stereoscopic images quality assessment. In: Proc. 15th European Signal Processing Conference (EUSIPCO), Poznan, Poland (2007)
18. Benoit, A., Callet, P.L., Campisi, P., Cousseau, R.: Quality Assessment of Stereo-scopic Images. EURASIP Journal Image and Video Processing (Article ID 659024) (2008), doi:10.1155/2008/659024
19. Jiachen, Y., Chunping, H., Zhou, Y., Zhang, Z., Guo, J.: Objective quality assessment method of stereo images. In: Proc. IEEE 3DTV, Potsdam, Germany (2009)
20. Boev, A., Gotchev, A., Eqiazarian, K., Aksay, A., Akar, G.B.: Towards compound ste-reo-video quality metric: a specific encoder-based framework. In: Proc. IEEE South-west Symposium Image Analysis and Interpretation, Denver, CO, USA (2006)
21. Olsson, R., Sjostrom, M.A.: Depth dependent quality metric for evaluation of coded integral imaging based 3D-images. In: Proc. IEEE 3DTV, Kos, Greece (2007)
22. Gorley, P., Holliman, N.: Stereoscopic image quality metrics and compression. In: Proc. SPIE Stereoscopic Displays and Applications XIX, San Jose, CA, USA (2008)
23. Sazzad, Z.M.P., Yamanaka, S., Kawayoke, Y., Horita, Y.: Stereoscopic image quality prediction. In: Proc. IEEE QoMEX, San Diego, CA, USA (2009)
24. Akar, G.B., Tekalp, A.M., Fehn, C., Civanlar, M.R.: Transport Methods in 3DTV: A Survey. IEEE Trans. Circuits and Systems for Video Technology 17(11), 1622–1630 (2007)
25. Barkowsky, M., Cousseau, R., Callet, P.L.: Influence of depth rendering on the quality of experience for an autostereoscopic display. In: Proc. IEEE QoMEX, San Diego, CA, USA (2009)
26. Hang, S., Xun, C., Er, G.: Objective quality assessment of depth image based render-ing in 3DTV system. In: Proc. IEEE 3DTV, Potsdam, Germany (2009)
27. Wang, X., Yu, M., Yang, Y., Jiang, G.: Research on subjective stereoscopic image quality assessment. In: Proc. SPIE Multimedia Content Access: Algorithm and Sys-tems III, San Jose, CA, USA (2009)
28. Middlebury Stereo Vision Page, http://vision.middlebury.edu/stereo
29. Taubman, D.S., Marcellin, M.W.: JPEG 2000: Image Compression Fundamentals, Standards and Practice. Kluwer, Norwell (2001)

30. ITU-R Recommendation. BT.500-10 Methodology for the subjective assessment of the quality of television. ITU-R, Geneva, Switzerland (2002)
31. Sheikh, H.R., Sabir, M.F., Bovik, A.C.: A Statistical Evaluation of Recent Full Reference Image Quality Assessment Algorithms. IEEE Trans. Image Processing 15(11), 3440–3451 (2006)
32. Wang, Z., Simoncelli, E.P., Bovik, A.C.: Multi-scale structural similarity for image quality assessment. In: Proc. IEEE Asilomar Conference Signals, Systems and Computers, Pacific Grove, CA, USA (2003)
33. Chandler, D.M., Hemami, S.S.: VSNR: A wavelet-based visual signal-to-noise ratio for natural images. IEEE Trans. Image Processing 16(9), 2284–2298 (2007)
34. Sheikh, H.R., Bovik, A.C.: Image information and visual quality. IEEE Trans. Image Processing 15(2), 430–444 (2006)
35. Wang, Z., Bovik, A.C.: A universal image quality index. IEEE Signal Processing Letters 9(3), 81–84 (2002)
36. Sheikh, H.R., Bovik, A.C.: An information fidelity criterion for image quality assessment using natural scene statistics. IEEE Trans. Image Processing 14(12), 2117–2128 (2005)
37. Simoncelli, E.P., Freeman, W.T.: The steerable pyramid: A flexible architecture for multi-scale derivative computation. In: Proc. IEEE ICIP, Washington, DC, USA (1995)
38. Damera-Venkata, N., Kite, T.D., Geisler, W.S., Evans, B.L., Bovik, A.C.: Image quality assessment based on a degradation model. IEEE Trans. Image Processing 9(4), 636–650 (2000)
39. Ponomarenko, N., Battisti, F., Egiazarian, K., Carli, M., Astola, J., Lukin, V.: On between-coefficient contrast masking of DCT basis functions. In: CD-ROM Proc. VPQM, Scottsdale, Arizona, USA (2007)
40. Yang, X.K., Ling, W.S., Lu, Z.K., Ong, E.P., Yao, S.S.: Just noticeable distortion model and its applications in video coding. Signal Processing: Image Communication 20(7), 662–680 (2005)
41. Sheikh, H.R., Wang, Z., Cormack, L., Bovik, A.C.: LIVE Image Quality Assessment Database Release 2, http://live.ece.utexas.edu/research/quality
42. VQEG: Final report from the Video Quality Experts Group on the validation of objective models of video quality assessment (2000), http://www.vqeg.org
43. Qian, N.: Binocular Disparity and the Perception of Depth. Neuron 18, 359–368 (1997)
44. You, J., Xing, L., Perkis, A., Wang, X.: Perceptual quality assessment for stereoscopic images based on 2D image quality metrics and disparity analysis. In: Proc. VPQM, Scottsdale, Arizona, USA (2010)
45. Felzenszwalb, P.F., Huttenlocher, D.P.: Efficient belief propagation for early vision. Int. Journal Computer Vision 70(1), 41–54 (2006)

Part II
Video Coding for High Resolutions

Chapter 4
The Development and Standardization of Ultra High Definition Video Technology

Tokumichi Murakami

1 Introduction

Video technology has evolved from analog to digital and SD (Standard Definition) to HD (High Definition). However, to provide a visual representation with high quality that satisfies the full range of human visual capabilities it requires further advances in video technology. One important direction is ultra high resolution video. Although UHD (Ultra High Definition) has already been standardized as a video format with spatial resolution 3840x2160 and 7680x4320 in an ITU recommendation (ITU-R BT.1769), actual deployment of UHD services have not yet been realized.

In order to realize UHD video services, the basic technologies that support UHD video, such as high quality camera, display, storage and transmission infrastructure, are indispensable. Presently, these technologies have accomplished remarkable progress, and the video and the visual equipments with 4Kx2K (4K) or 8Kx4K (8K) resolutions exceeding HD are shown at many trade shows and exhibitions. Also, several cameras corresponding to 4K have already been announced, and there are a variety of displays, such as liquid crystal displays (LCD), plasma display panels (PDP) and projectors, which can render 4K video. Moreover, organic electroluminescence (organic EL) equipped with thinness, power saving and high resolution is also promising as a UHD display. Furthermore, the Japan Broadcasting Corporation (NHK) has developed a 33 million pixel camera for a Super Hi-Vision system with 7680x4320 resolution, and is demonstrating a projector and a liquid crystal panel with 8K resolution. Thus, the realization of UHD video service is within reach.

Tokumichi Murakami
Mitsubishi Electric Corporation
Research and Development Center
5-1-1 Ofuna, Kamakura City, Kanagawa, 247-8501 Japan
Tel.: +81-467-41-2801
e-mail: Murakami.Tokumichi@eb.MitsubishiElectric.co.jp

High performance video coding technology is another indispensable element to realize UHD video. At present, the main standards for video coding are MPEG-2 and MPEG-4 AVC (Advanced Video Coding)/H.264 (AVC/H.264). However, the development of a new video coding technology is necessary for UHD video applications since the video must be compressed further to be transmitted within current systems, while still keeping the high quality of the original source as much as possible. In response to such environmental conditions and demands, the standardization activity of a next generation video coding for UHD video is getting underway.

In this chapter, the history and international standardization of video coding technology are described. Then, the fundamental constituent factors of video coding are introduced. Next, the requirements for the video coding technology towards the realization of UHD video are described, and the progress of the supporting UHD video technologies is surveyed. Finally, the challenges toward the technical development of a next generation video coding and the view of future video coding technology are discussed.

2 Progress of Digital Video Technique

Looking back upon the history of video technology, it is evident that video coding is one of the most important elements when considering the progress of digital video technology. During these two decades, MPEG (Moving Picture Experts Group) has occupied a central position in the international standardization of video coding technology. In this section, the results of MPEG standardization and current activities are surveyed.

2.1 History of Video Coding

2.1.1 Before AVC/H.264

Television broadcasting first started as analog in the 1940s and spread generally and widely. In the telecommunications area, video transmission was realized as TV phone service at the beginning, which was sending semi-video in addition to voice through the telephone line. However, it was not practical at that time to allocate wide range of the bandwidth of the communication line for video transmission. Although research and development for efficient video transmission were conducted during the 1950s and 1960s, most of them were video transmission systems made use of analog technology.

In the 1970s, since the digital signal processing began to evolve into practical and more matured technology, the hierarchy of the digital telecommunications network was specified in the communications field. In the beginning of 1980s, the practical development on high efficiency digital video compression had come accelerated. As a result, innovative video coding technology was introduced into the video conference systems for business use, by KDD, NTT and Mitsubishi etc., in Japan. In the middle of 1980s, it became possible to simulate the video coding algorithms more easily with improved workstation capability, and practical research was greatly advanced. As it was at the dawn of a new age of digital communication line based on Integrated Services Digital Network (ISDN), the development of the products of pioneering video conference system was carried out. The CCITT (Consultative Committee for International Telephone and Telegraph), now known as ITU-T (International Telecommunication Union – Telecommunication sector), began to consider and discuss the needs of interconnectivity and interoperability for video transmission assuming TV phone, video conference, remote surveillance, etc. In 1990, CCITT has recommended H.261 for video coding scheme at the transmission rate of px64 Kbit/s (p=1, 2, ...) for the communication of video and audio on ISDN [1]. In H.261, a hybrid coding system using the combination of motion compensated prediction coding and transform coding was adopted, and many of the current video coding systems are derived from the hybrid coding system from H.261. In this period, VTR became more widespread in the home because of its tendency of lower-pricing. Under such a situation, MPEG, which has been a working group under ISO/IEC, started the development of an international standard for video coding that aimed at consumer appliances in 1988.

Since the 1980s, the history of video coding technology has been deeply related to the international standardization. MPEG specifically aimed at the development of audio and video coding methods for CD (Compact Disc) which began to spread rapidly with the advent of digital music. MPEG-1 specified the coding of video with up to about 1.5 Mbit/s; this standardization was completed in 1993 and was adopted for video CD and CD karaoke [2].

Subsequently, MPEG-2 was standardized as MPEG-2/H.262 [3, 4] and aimed at the coding of SDTV and HDTV; this standardization was completed through cooperation between ISO/IEC and ITU-T in 1994. The standardization of MPEG-2 triggered the roll-out of digital broadcasting. Video coding technology provided a means to satisfy constraints on the communication bandwidth and storage capacity for transmitting and storing video, respectively. Until MPEG-2 was standardized, the video was always treated with lower resolution than standard television broadcasting. However, with the advent of MPEG-2, video coding technology was able to realize high quality video services. In 1995, NHK and Mitsubishi had jointly developed an HDTV codec conforming to MPEG-2 specification, and conducted a verification experiment on digital HDTV broadcasting, This became a turning point to accelerate digital TV broadcasting. After then, MPEG-2 began to be adopted as a video coding scheme for digital broadcasting in Japan, Europe and the United States. HDTV digital broadcasting began to be a full-fledged service world-widely in early 2000s, and the spread of LCD displays brought the realistic video experience of 1920 scanning lines to home.

On the other hand, the combination of the Internet and PC had grown greatly as a platform for multimedia services since Mosaic, which was an Internet browser, first released in 1993. In that period, many proprietary coding methods were developed outside the conventional standardization organizations. In spite of this situation, H.263 [5] and MPEG-4 [6] were still used in many applications. H.263 was recommended for the transmission of VGA (Video Graphic Array) video from tens of Kbit/s to several Mbit/s in 1995. MPEG-4 was a successor of H.263 and was completed its standardization in 1999. MPEG-4 was utilized for 3G mobile video phones with 64 kbit/s, portable video players with up to 2-3 Mbit/s, as well as the animation function of digital still camera, etc.

In the 2000s, the development of video coding technology progressed rapidly due to an increase in processing speed of devices, user demand for higher quality and an abundance of video services. AVC/H.264 [7, 8] is the standard which was developed based on the coding techniques examined under the H.26L project in ITU-T/SG16/Q.6 known as VCEG (Video Coding Experts Group). A collaborative team known as the JVT (Joint Video Team) was formed between MPEG and VCEG in 2001. AVC/H.264 which achieved twice as much compression ratio of MPEG-2 was standardized in 2003. The AVC/H.264 standard was then adopted as the coding method for mobile TV broadcasting called One-seg in Japan and for Blu-ray with HD resolution, and continues to extend the scope of its applications.

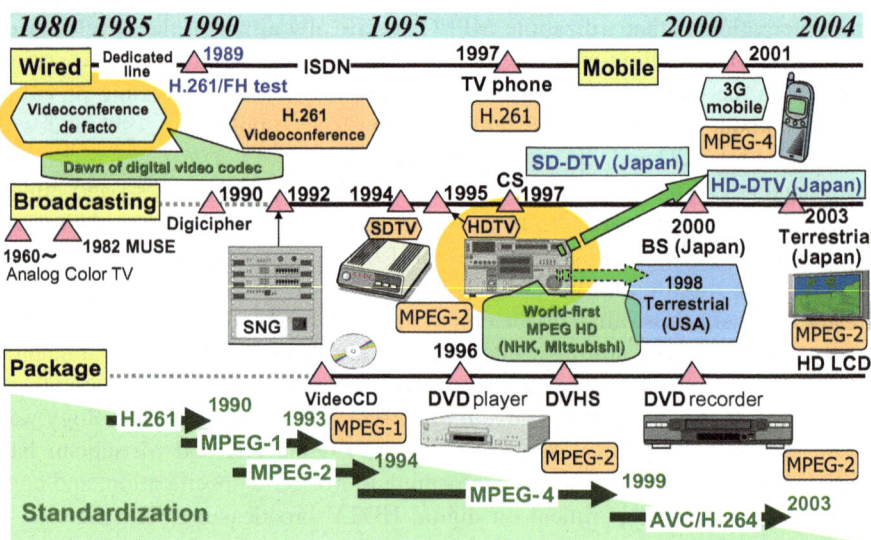

Fig. 1 History of Video Technology (From Visual Communication to Digital Broadcasting)

2.1.2 After AVC/H.264

AVC/H.264 continued to be improved for high quality video after the recommendation in 2003. Baseline Profile including 4x4 transform was standardized in 2003 and High Profile which employs 8x8 transform and individual quantization matrix for HDTV was recommended in 2005. High Profile has been adopted and deployed in home AV equipments including Blu-ray disc players and recorders. In 2007, some additional coding tools to support high quality video coding were added, including the support for coding of video in 4:4:4 format and high definition levels, according to the proposals from Mitsubishi, etc. [9].

In response to this progress in video coding technology, television broadcasting, which is the most familiar video media to the public, has started shifting from SDTV to HDTV in a digital form. Video recorders for home use and small handheld camcorders are also operated with HDTV quality; such devices realize not only a small size but a low price as well.

Thus, while HD video is becoming the norm, the development of UHD video technology such as 4K with 4 times the resolution of HD is progressing steadily. Visual equipments with 4K resolution are now being exhibited at shows. Several cameras corresponding to 4K resolution have already been announced, and LCD, PDP and projectors which can display 4K image can be seen. Moreover, the organic EL equipped with thinness, power saving and high resolution is also promising as a display of UHD video. With respect to practical use, digital cinemas with 4K resolution have been specified and their use for digital cinema including distribution to theaters has already started [10]. Furthermore, NHK is planning for advanced television broadcasting with 8K resolution from the year of 2025, and already has developed a 33 million pixel camera with 8K resolution and a projector with 8192x4320 resolution. Video standardization of 4K and 8K resolutions is being progressed by ITU-R (Radiocommunications sector) and SMPTE (Society of Motion Picture and Television Engineers) which is responsible for production standards used by the cinema and television industries. Next-generation video coding standards, including UHD video as a target, are also in the process of starting in response to these environmental conditions and expectations.

In 2009, MPEG invited the public to submit evidence of new video coding technologies that fulfill the conditions for UHD video, and evaluated the technologies considering the emerging application requirements [11]. As a result of this study, sufficient evidence was obtained and MPEG is now planning a new standardization initiative to meet these goals. The current schedule is to collect proposals in 2010, and to recommend an international standard in 2012-2013. MPEG and VCEG are likely to cooperate on this activity.

Thus, information projected on a screen will be diversified in the future when UHD video technology for 4K and 8K resolutions is realized. For example, we will be able to enjoy realistic and immersive video on a large screen TV that is over 100 inches diagonal, and display web browsers simultaneously with UHD video contents on the screen. We may also use a photogravure TV of A3 size like an album of photographs.

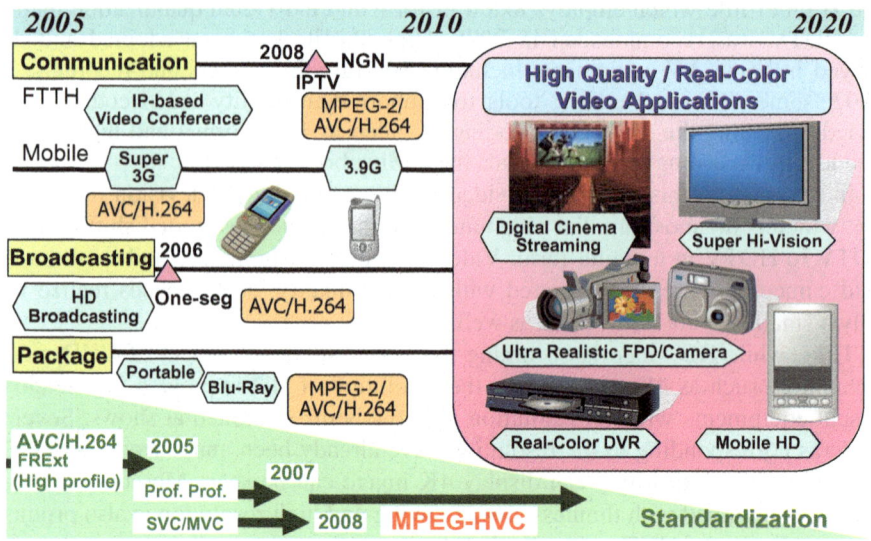

Fig. 2 History of Video technology (Nowadays and Future)

2.2 Technical Standardization for Video Coding

The history of video coding technology has been deeply related to international standardization since the 1980s. Because interoperability is very important for the widespread utilization of video contents, it is important that coded video contents should conform to an international standard. In the following, international standards for video coding focusing on MPEG are surveyed.

2.2.1 International Organizations of Video Coding Standards
Video coding technology has been playing an important role in the progress of video technology and the spread of video contents and applications. Standardization organizations responsible for digital video related technologies and their mutual relationships are shown in Fig. 3.

ISO/IEC and ITU-T are the primary organizations in the world that engage in the international standardization of video coding. MPEG is one of the working groups, formally known as ISO/IEC JTC1/SC29/WG11, which belongs to a Joint committee of ISO (International Standardization Organization), which promulgates industrial and commercial standards, and IEC (International Electro-technical Commission) which treats international standards for all electrical, electronic and related technologies. On the other hand, ITU-T is an international standardization organization for telecommunications which was formerly called CCITT, and it has been coordinating many standards in the field of telecommunications.

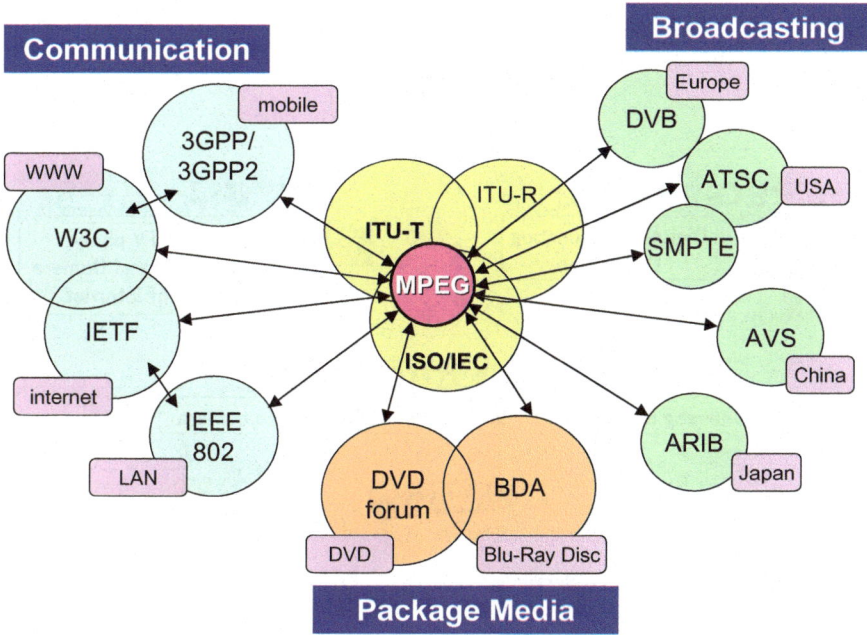

Fig. 3 Standardization Organizations and their Relationships

Among the digital video related standardization organizations, MPEG has been and continues to occupy the central position of video coding technology, and MPEG standards are considered as one of the essential technologies for digital video services. Since its inauguration in 1988, MPEG has standardized MPEG-1, MPEG-2/H.262, MPEG-4 and AVC/H.264, and has been promoting the development and standardization of multimedia technologies including video coding. MPEG standards specifies the technologies to compress video data with compression ratios in the range from 30:1 to 100:1 as well as the technologies for transmission and storage of video and audio contents, and offers open specifications and compatibilities.

Furthermore, MPEG has cooperated on international standardization of video coding with VCEG (Video Coding Experts Group) which is affiliated with ITU-T/SG16. Experts from both MPEG and VCEG committees formed the JVT (Joint Video Team) for the development of AVC/H.264.

2.2.2 Improvement in Compression Ratio by MPEG

The compression ratio of video data has improved through MPEG standardization. For example, AVC/H.264 can perform twice as much compression ratio as MPEG-2. High resolution and multi-channel were attained by the improvement of the compression ratio of video coding by MPEG.

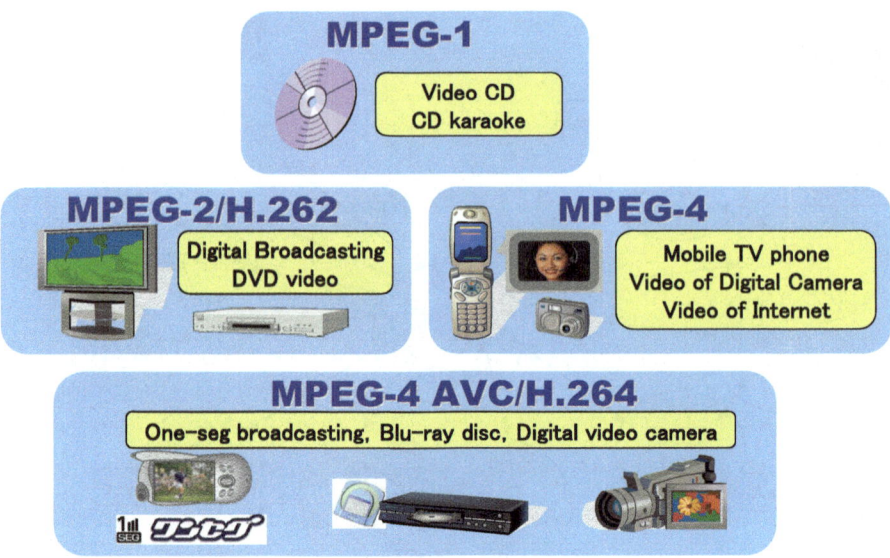

Fig. 4 MPEG and Digital Video Services

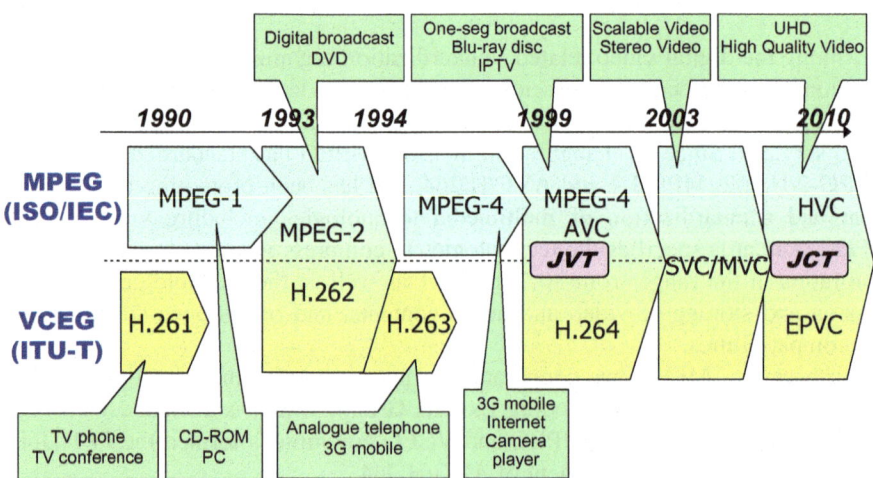

Fig. 5 Cooperative Relationship between MPEG and VCEG

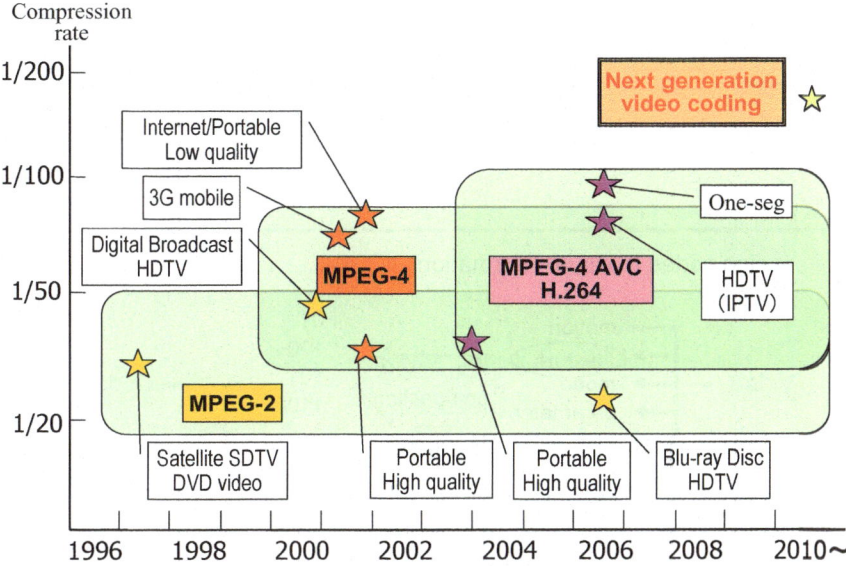

Fig. 6 Improvement in Compression Ratio by MPEG

3   Video Coding Technology

3.1   The Shannon Theory and Video Coding Technology

Video coding aiming at digital information compression began its development from the necessity of transmitting a vast quantity of digital video data through communication line with narrow band. Therefore, it is possible to draw an analogy between the information and telecommunication model of Shannon [12, 13] and the composition of digital video coding transmission system (refer to Fig. 7). In the sending side of a digital video transmission system, the analog video signal acquired from a camera is digitally sampled and quantized. Format conversion is performed using various filters and a sequence of digital images is generated [14]. Then, prediction, transform, quantization and entropy coding are applied to the image sequence to produce a compressed bitstream (Source Coding) [15]. The compressed bitstream then undergoes encryption, multiplexing, error detection/correction, modulation, etc., and is transmitted or recorded according to the characteristic of a transmission line or a recording medium (Channel Coding). On the other hand, in the receiving side, the video signal is reproduced by the inverse operations performed in the sending side and the video is displayed on a screen.

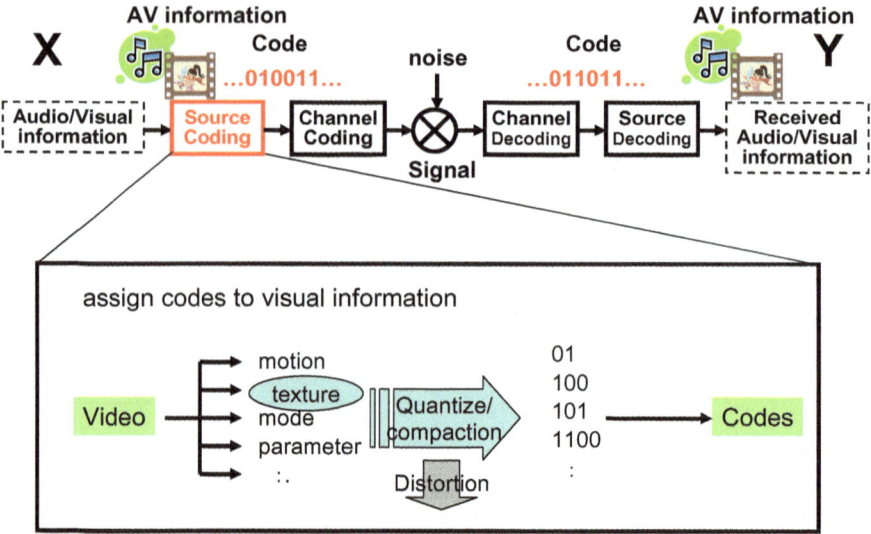

Visual information is distorted by both source coding and channel error.
→ Quality is measured by PSNR versus Bitrate.

Fig. 7 Video Coding in Shannon's model

3.2 Main Components of Video Coding

Video coding usually consists of four main components including prediction, transform, quantization and entropy coding. Prediction reduces relative redundancy exploiting correlation within a picture or across several pictures. The residual signal that represents the difference between the original and the predicted signal is encoded. Transform is a process for energy compaction of the signal to reduce the correlation of the symbols. In practice, the signal is transformed from a spatial domain to a frequency domain. There are several transforms that have been used in typical image and video coding standards including Discrete Cosine Transform (DCT) and Discrete Wavelet Transform (DWT). Quantization is a technique that reduces the amount of information directly. There are two main methods of quantization including Scalar Quantization and Vector Quantization. Entropy Coding is a reversible coding method based on statistical characterization of the symbols to be encoded. Huffman coding and arithmetic coding are typical examples of entropy coding schemes.

3.2.1 Prediction

A picture has high correlation between neighboring pixels in both spatial and temporal directions. Consequently, the amount of information can be reduced by the combination of the prediction between pixels and the coding of the prediction error (residual signal). The prediction exploiting spatial correlation within a picture is known as Intra prediction, while the prediction exploiting temporal correlation across two or more pictures is known as Inter prediction. A method of further exploiting correlation between frames is to utilize motion prediction, which is referred to as Motion Compensated Prediction. Fig. 8 shows the difference of power between several signals in a typical video coding system.

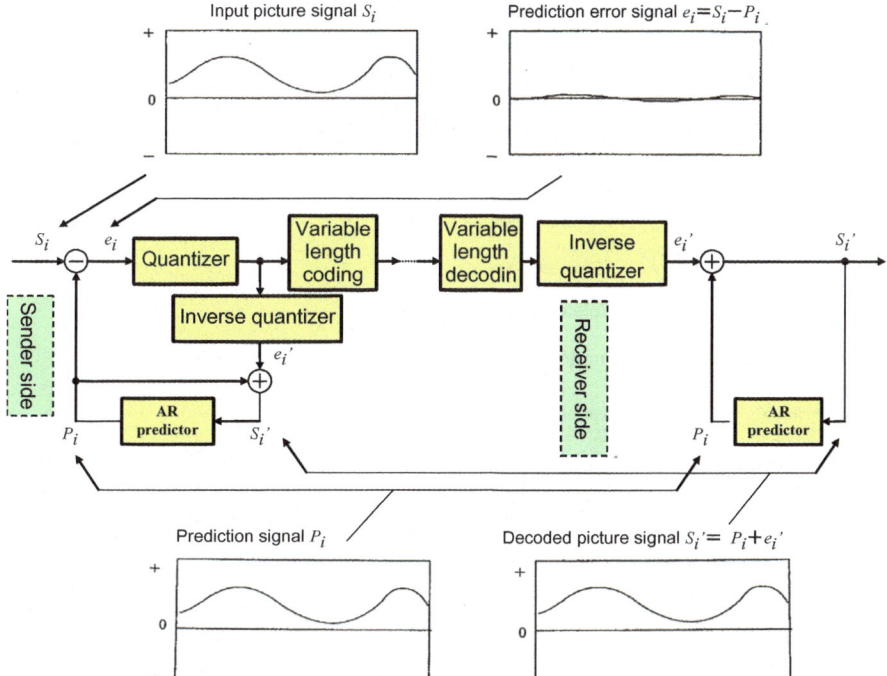

Fig. 8 Predictive Coding Scheme

3.2.1.1 Intra Frame Prediction

Intra Frame Prediction is a prediction technique that uses the neighboring pixels within a frame. Three prediction methods including Previous-sample Prediction, Matrix Prediction and Plane Prediction are shown as examples of Intra Frame Prediction in Fig. 9. Previous-sample Prediction uses neighboring pixels in the horizontal direction as a prediction pixel, Matrix Prediction uses neighboring pixels in both horizontal and vertical directions, and Plane Prediction uses neighboring pixels in horizontal direction and subtracts the pixel values at the same positions on the former line.

3.2.1.2 Motion Compensated Prediction

Motion Compensated Prediction is a technique which creates a prediction image that resembles the current image by linear translation of a block within a reference picture which is already transmitted and decoded. Compression is achieved by

coding the difference between the predicted and original pictures. The principle of
Motion Compensated Prediction is shown in Fig. 10.

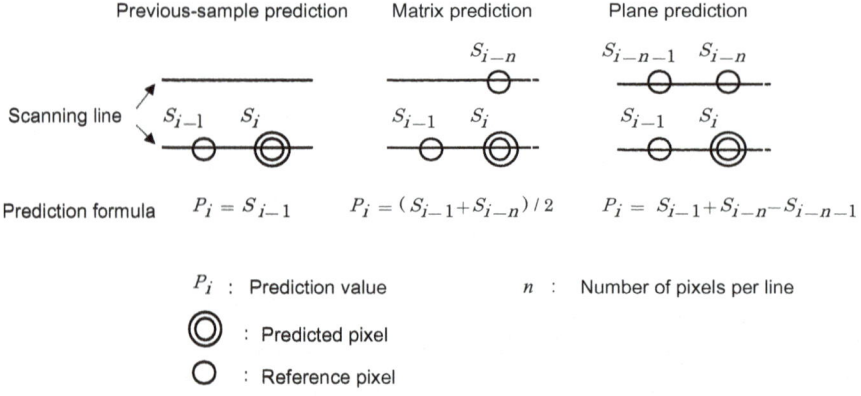

Fig. 9 Examples of Intra Frame Prediction

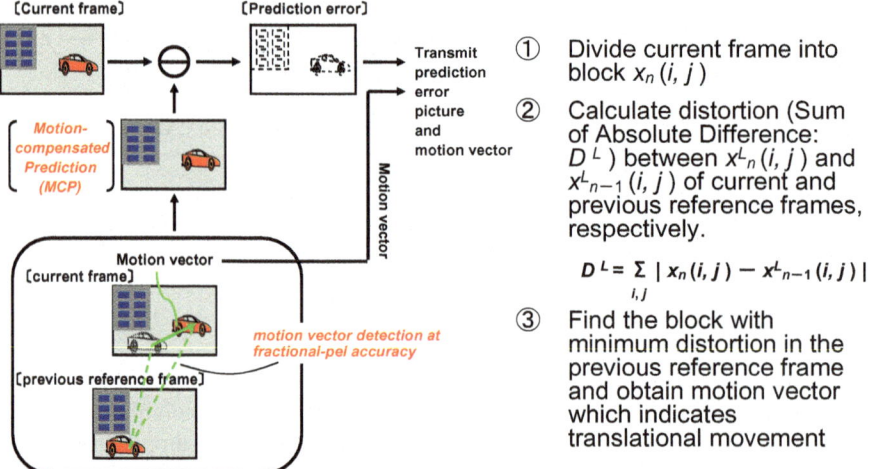

① Divide current frame into block $x_n(i, j)$

② Calculate distortion (Sum of Absolute Difference: D^L) between $x^L_n(i, j)$ and $x^L_{n-1}(i, j)$ of current and previous reference frames, respectively.

$$D^L = \sum_{i,j} | x_n(i, j) - x^L_{n-1}(i, j) |$$

③ Find the block with minimum distortion in the previous reference frame and obtain motion vector which indicates translational movement

Fig. 10 Principle of Motion Compensated Prediction

Motion Compensated Prediction can reduce the energy of the residual signal compared with the simple difference between frames. Fig. 11 shows an example that compares the simple difference signal between frames and the Motion Compensated Prediction difference signal. It is clear that the difference signal decreases dramatically when Motion Compensated Prediction is utilized.

Fig. 12 shows an example of the signal characteristics of the original, Intra Frame Prediction, Inter Frame Prediction and Motion Compensated Prediction pictures of a HDTV picture with their entropy and signal power. It is shown that signal power decreases sharply when Motion Compensated Prediction is utilized.

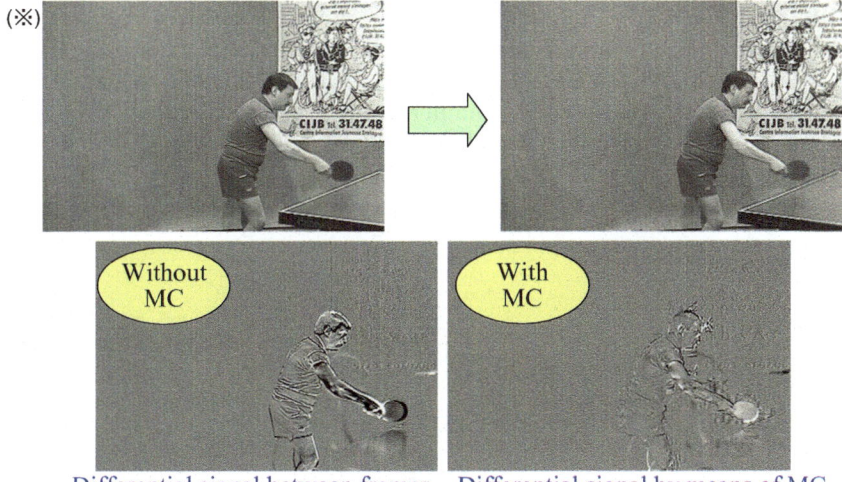

Differential signal between frames   Differential signal by means of MC

(※) "Table Tennis" (MPEG standard image)

Fig. 11 Effect of Motion Compensated Prediction

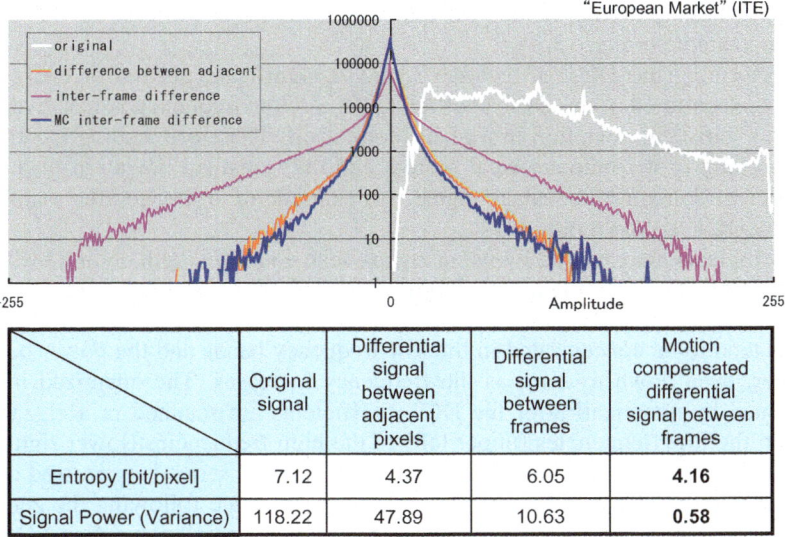

	Original signal	Differential signal between adjacent pixels	Differential signal between frames	Motion compensated differential signal between frames
Entropy [bit/pixel]	7.12	4.37	6.05	**4.16**
Signal Power (Variance)	118.22	47.89	10.63	**0.58**

Fig. 12 Characteristics of Picture Signal

3.2.2 Transform

Transform is the method of converting an image signal into another signal domain, and centralizing signal power to specific frequency bands. There exist DCT and DWT for this purpose, which are used in the current picture coding standards.

3.2.2.1 DCT

DCT converts the spatial domain signal into the frequency domain using a window with fixed width for the transformation. Usually, a picture is divided into NxN pixel blocks (N pixels width both horizontal and vertical directions) and the transform is performed for each pixel block. The DCT is expressed as follows,

$$F(u, v) = \frac{2}{N} C(u) C(v) \sum_{x=0}^{N-1}\sum_{y=0}^{N-1} f(x, y) \cos\left[\frac{(2x+1)u\pi}{2N}\right] \cos\left[\frac{(2y+1)v\pi}{2N}\right]$$

where

$$C(u), C(v) = \begin{cases} \dfrac{1}{\sqrt{2}} & (u, v = 0) \\ \\ 1 & (u, v \neq 0) \end{cases}.$$

On the other hand, the inverse transform (IDCT) reconverts a transformed signal to the spatial domain and is expressed as follows,

$$f(x, y) = \frac{2}{N} \sum_{u=0}^{N-1}\sum_{v=0}^{N-1} C(u) C(v) F(u, v) \cos\left[\frac{(2x+1)u\pi}{2N}\right] \cos\left[\frac{(2y+1)v\pi}{2N}\right].$$

The transform basis patterns of the two dimensional DCT in the case of 8x8 is shown as an example in Fig. 13.

After performing the DCT of a video signal, a significant portion of energy tends to be concentrated in the DCT coefficients in the low frequency bands, even if there is no statistical deviation in a pixel block itself. Therefore, coding is performed according to the human visual system and the statistical deviation in the DCT coefficient domain of an image signal. An example of an image after transformation by DCT is shown in Fig. 14.

DCT coefficients are encoded by using zigzag scan and run length coding technique after quantization. Run length coding is a method of coding the combination of (number, length) of the same kinds of continuous symbols. Higher power DCT coefficients tend to be concentrated in the low frequency bands and the power becomes lower, even down to zero, as the frequency increases. The quantized indexes obtained by quantization of the DCT coefficients are scanned in a zigzag pattern from the low frequencies (upper left) to the high frequencies (lower right) and are rearranged into a one dimensional series. The signal series is expressed as a pair of the number of zeros (zero run) and a non-zero value following the zero series (level). When the last non-zero value is reached, a special sign called EOB (End of block) is assigned to reduce coding signals. By following this process, the statistical nature of the signal series can be exploited. Namely, symbols that have a large level will typically have a short zero run and symbols that have a long zero run are typically associated with a small level. In this way, a variable length code can be assigned to the combination of (zero run, level) to be compressed with shorter codes assigned to more probable symbols and longer codes assigned to less probable ones. The example of a zigzag scan and run length coding adopted in MPEG-2 are shown in Fig. 15.

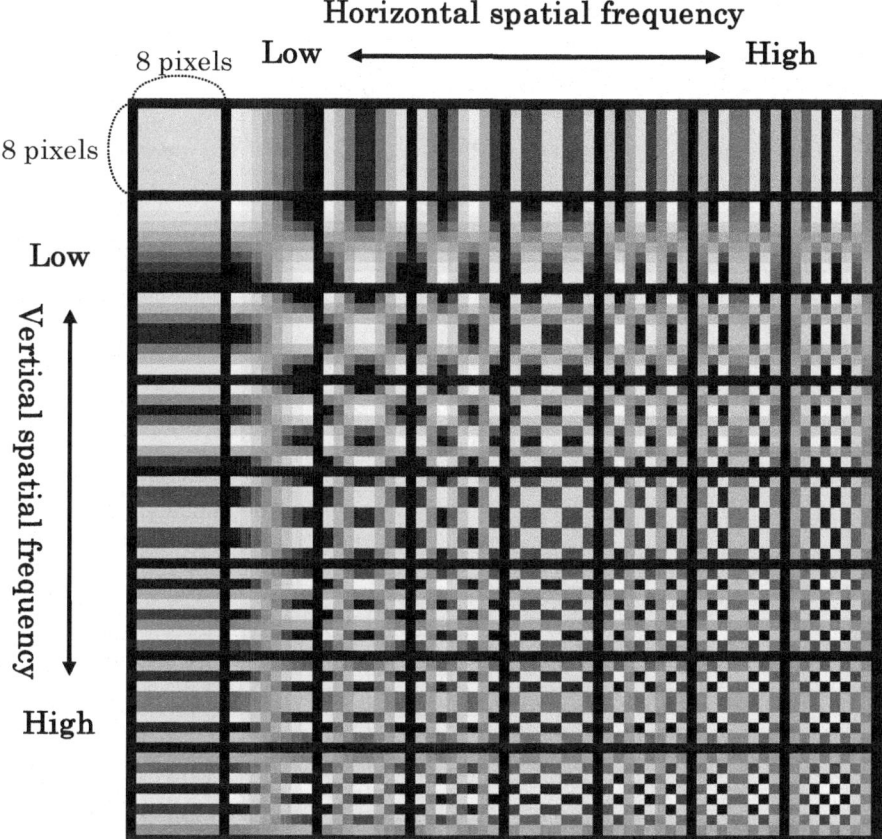

Fig. 13 Transform Basis Patterns of Two Dimensional 8x8 DCT

3.2.2.2 DWT

DWT is one of the transform methods using the transform basis made by the operation of expanding and moving a function localized in frequency domain. DWT allows using windows whose sizes are different according to frequencies, and has the feature of high response for both low-frequency and high-frequency portions of signals. DWT has also the following features,

(1) The correspondence for local waveform change is high by using flexible transform windows for unsteady signals

(2) Block noise which is often present in DCT transform does not occur inside the window width for the conversion of lowest frequency

(3) Hierarchical coding can be realized easily.

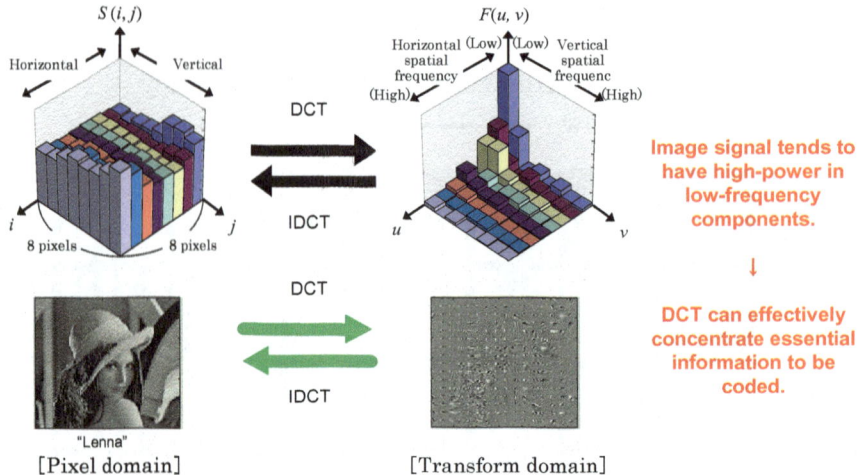

Fig. 14 Example of transformation by DCT

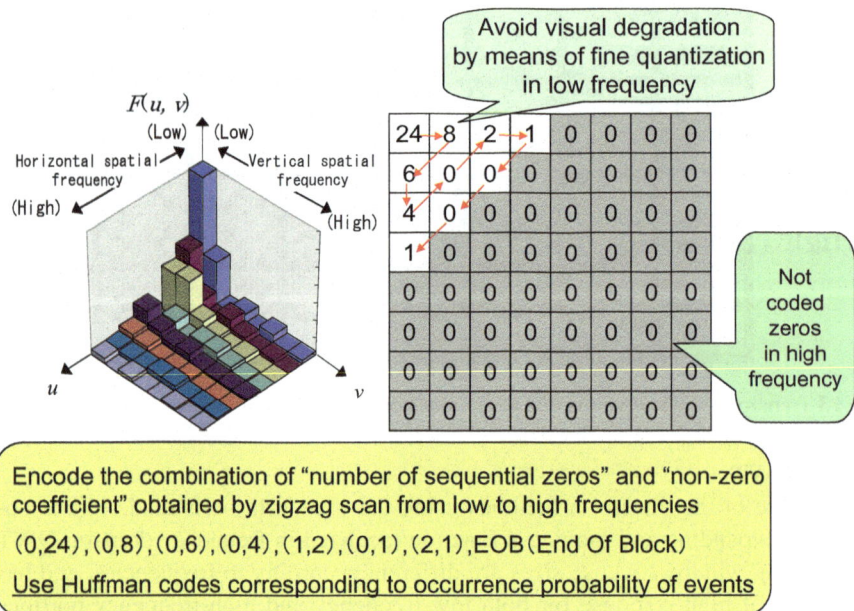

Fig. 15 Example of Zigzag Scan and Run Length Coding

3.2.3 Quantization

Quantization is a technique of reducing the amount of information directly, and there are mainly two methods well-known for video compression, which are Scalar Quantization and Vector Quantization. Scalar Quantization is an operation of making an input signal correspond to one of k kinds of values which are represented as $q_1, ..., q_k$ as shown in Fig. 17.

Sub-band decomposition in horizontal and vertical directions

Recursive divisions of only LL elements

Fig. 16 Discrete Wavelet Transform

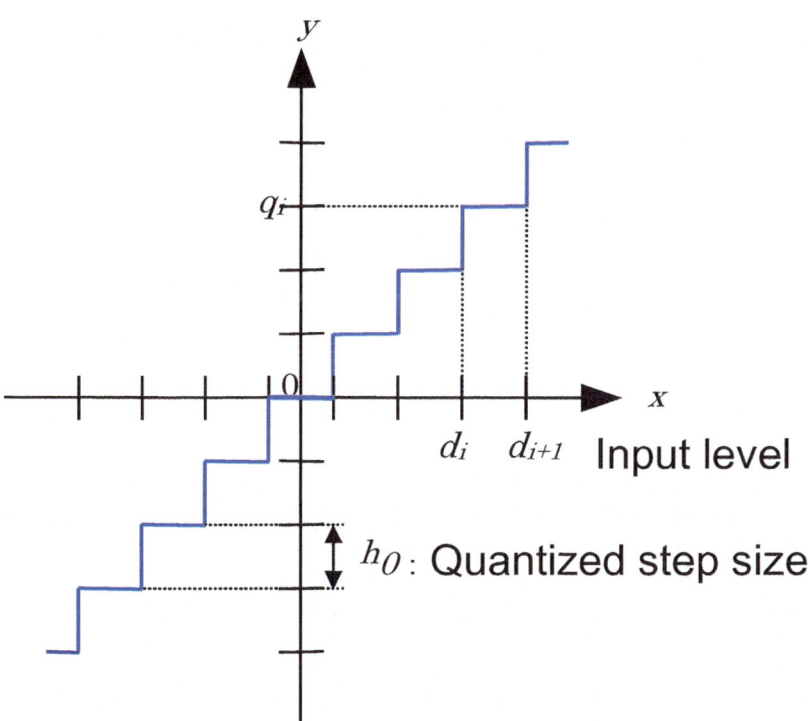

Fig. 17 Example of Scalar Quantization

Vector Quantization is an operation which quantizes several samples at the same time and expresses them with a representative vector which gives the best approximation of the samples [16, 17]. The sources of information which consist of many dimensions are quantized by one of the representative points of a multi-dimension space by Vector Quantization. Therefore, Vector Quantization has the following advantages,

(1) Coding efficiency can be raised by adopting the correlation and the dependency between the vectorized samples in the quantization mechanism.
(2) Even if the vectorized samples are completely independent, the multi-dimensional signal space can be divided into its quantized sections.
(3) Samples can be coded with non-integer word size by assigning a quantized represented vector or its codeword.

The key map and the principle of Vector Quantization are shown in Fig. 18 and Fig. 19, respectively.

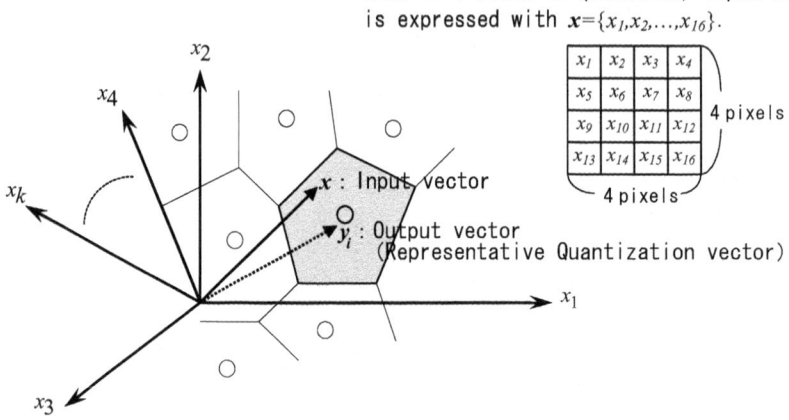

Fig. 18 Key map of Vector Quantization

3.2.4 Entropy Coding

Entropy Coding is a method of describing the mode information, motion vector information, quantized values, etc. as a series of binary signals which consists of only 0 and 1 (binarization). The total amount of codes is reducible by assigning coded words according to the occurrence probability of symbols. Huffman coding and arithmetic coding are typical entropy coding methods used in video coding. Huffman coding is a method of designing and using a variable length code table which associates symbols and code-words. This method can shorten the average code length by assigning short codes to symbols with high occurrence probability and long codes to symbols with low occurrence probability. An example of Huffman coding is shown in Fig. 20.

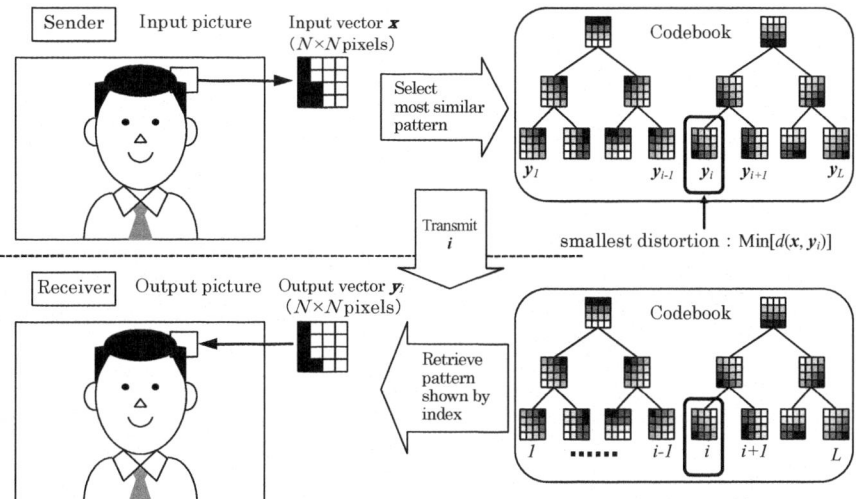

Fig. 19 Principle of Vector Quantization

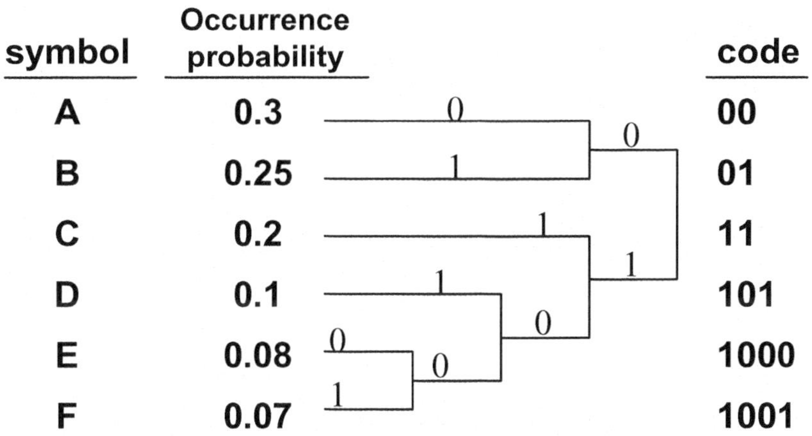

symbol	Occurrence probability		code
A	0.3		00
B	0.25		01
C	0.2		11
D	0.1		101
E	0.08		1000
F	0.07		1001

Fig. 20 Example of generation of Huffman Coding

Arithmetic coding is a method of coding the divided section of an interval of number line and generates a codeword one-by-one according to the occurrence probability of symbols. Moreover, the code length of a non-integer bit can be assigned to a symbol. The concept of Arithmetic coding is shown in Fig. 21.

3.2.5 Hybrid Coding Architecture
The Hybrid Coding Architecture which combines Prediction and Transform techniques is adopted in video coding standards such as H.26x and MPEG. The brock diagram of the typical Hybrid Coding Architecture is shown in Fig. 22.

Probability

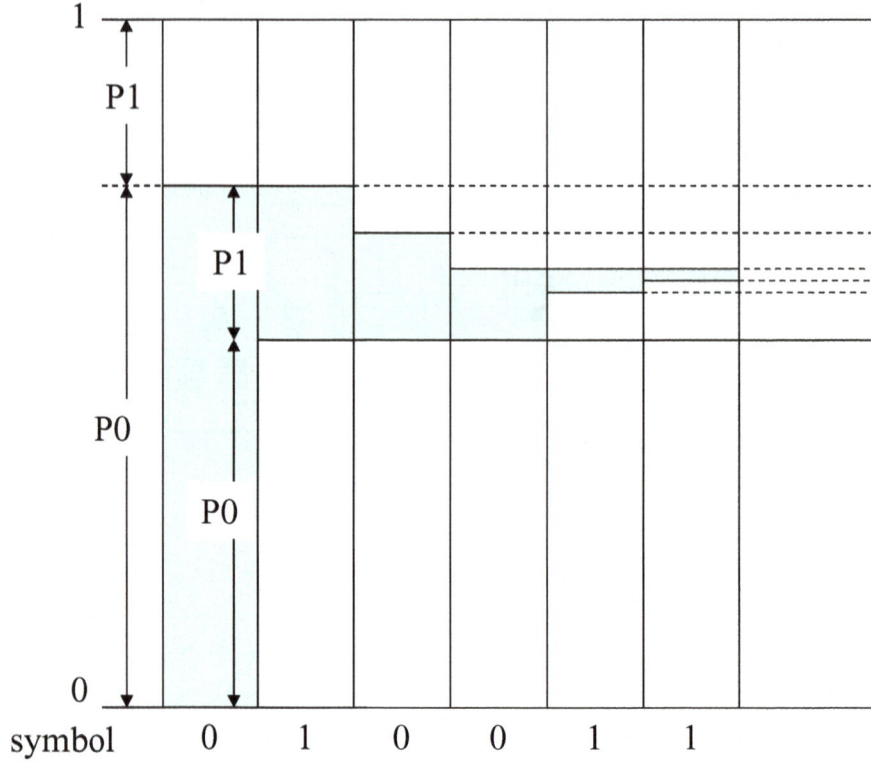

Fig. 21 Process of Arithmetic Coding

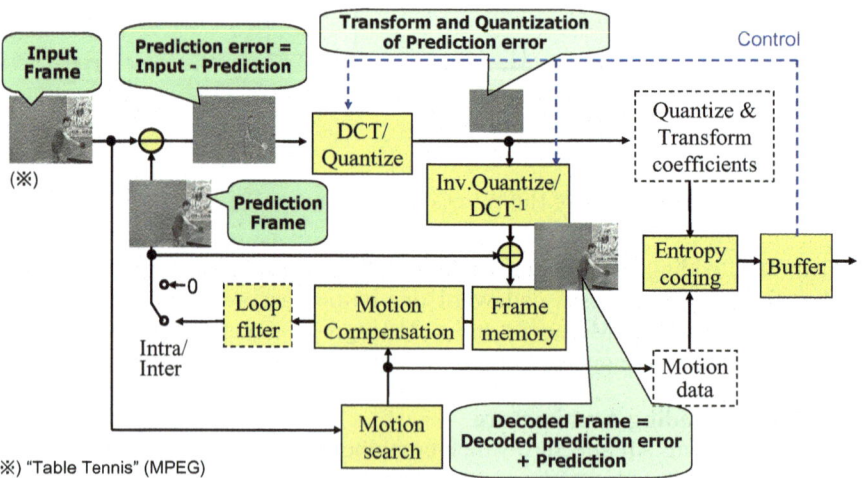

Fig. 22 Brock Diagram of Typical Hybrid Coding Architecture

3.3 MPEG Coding Methods

The features of the major MPEG video coding standards such as MPEG-2, MPEG-4 visual and AVC/H.264 are introduced [18, 19].

3.3.1 MPEG-2

MPEG-2 is an international standard that specifies video coding for the purpose of digital television broadcasting with high quality. Various coding tools are included to support the coding of interlaced video signals. MPEG-2 is used in satellite and terrestrial digital broadcastings and recording medias such as DVD, and is the mainstream coding method for video at present. The main features of MPEG-2 video coding are described below.

First, it adopts a hierarchical structure of video format. That is, MPEG-2 video format has a layered structure composed of Sequence, Group of Picture (GOP), Picture, Slice and Macroblock (MB) as shown in Fig. 23.

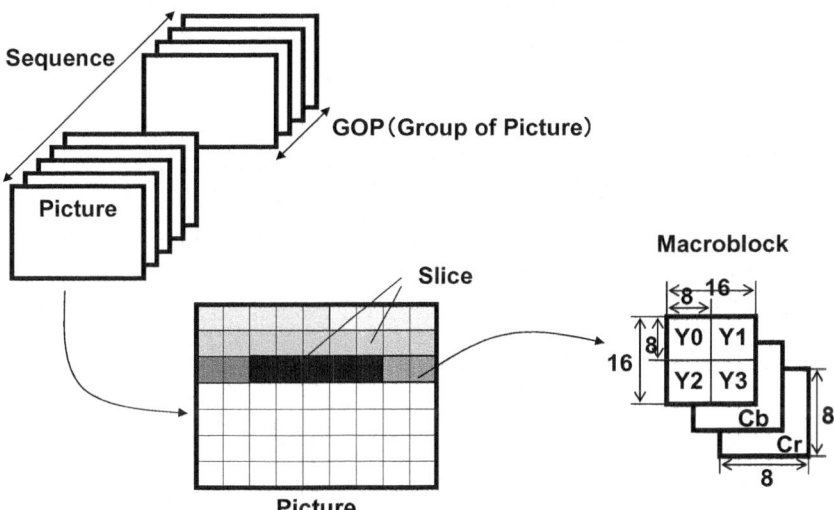

Fig. 23 Hierarchical structure of MPEG-2 video format

Secondly, three kinds of fundamental picture types and various prediction methods are adopted for Motion Compensated Prediction. Prediction efficiency is increased by adopting forward, backward, and bi-directional predictions. Forward prediction is a method of predicting the present frame from the past frame in time. Backward prediction is a method of coding the future frame and predicting a past frame from it. Bi-directional prediction is a method of using both the past and the future frames for prediction. Then, three kinds of pictures such as I, P, and B pictures are defined for using these prediction methods. I pictures are predicted within itself and do not refer to any other pictures. P pictures are predicted with only forward prediction. B pictures are predicted by choosing the most effective prediction among forward, backward and bi-directional predictions. A classification of picture types and the prediction methods of MPEG-2 are shown in Fig. 24. This structure makes it possible to realize random access of picture and also improves the coding performance for package media.

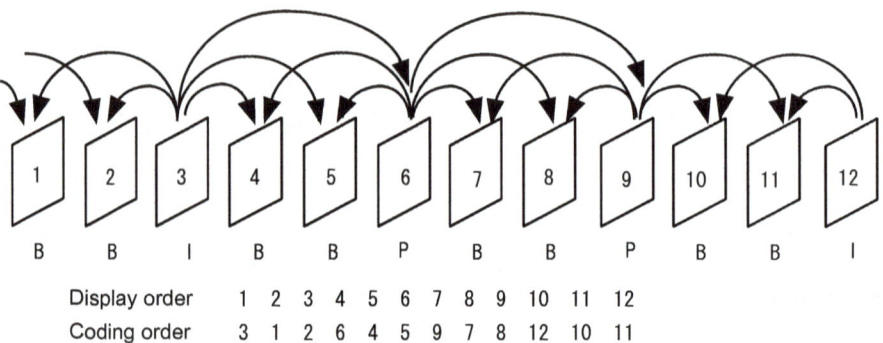

* Each arrow is pointed from reference picture to predicted picture.

Fig. 24 Picture Classification and Prediction Methods

In addition, prediction with half-pel accuracy is defined in MPEG-2, whereby the unit of displacement is expressed with a motion vector pointing to half a pixel position (middle position of adjacent pixels). Since the value of half a pixel position does not actually exist, it is virtually generated by interpolation from neighboring pixels. The concept of half-pel accuracy prediction and the half-pel calculation method are shown in Fig. 25 and Fig. 26, respectively.

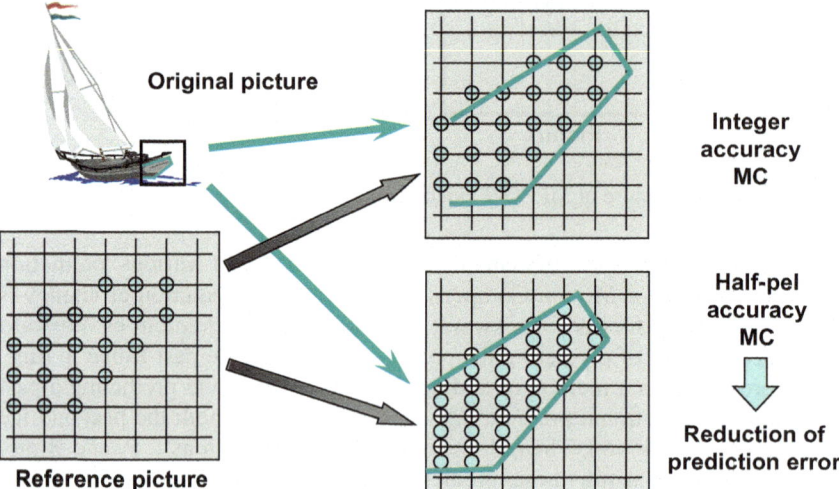

Fig. 25 Half-pel Accuracy Prediction

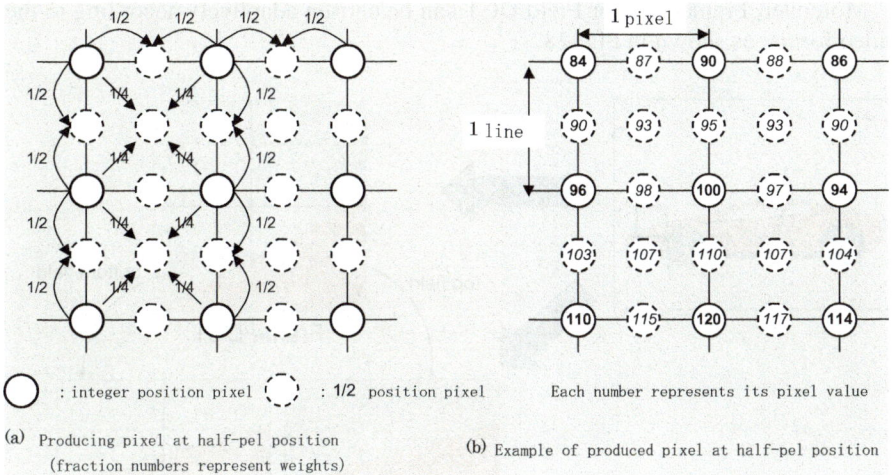

(a) Producing pixel at half-pel position
(fraction numbers represent weights)

(b) Example of produced pixel at half-pel position

Fig. 26 Half a pixel Calculation Method

MPEG-2 has adopted various coding tools to support the efficient coding of Interlace video signals. First, Frame prediction, Field prediction or Dual prime prediction can be selected adaptively in order to perform optimal prediction according to the movement of the objects in video as shown in Fig. 27.

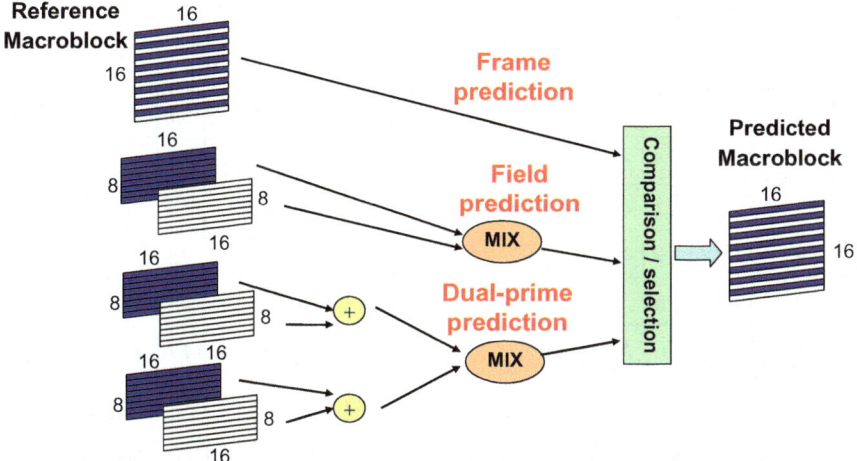

Frame picture MC is performed by comparison and selection of best matching macroblock among frame, field and dual prime predictions.

Fig. 27 Frame/field Adaptive Prediction

Moreover, Frame DCT or Field DCT can be chosen adaptively according to the video format as shown in Fig. 28.

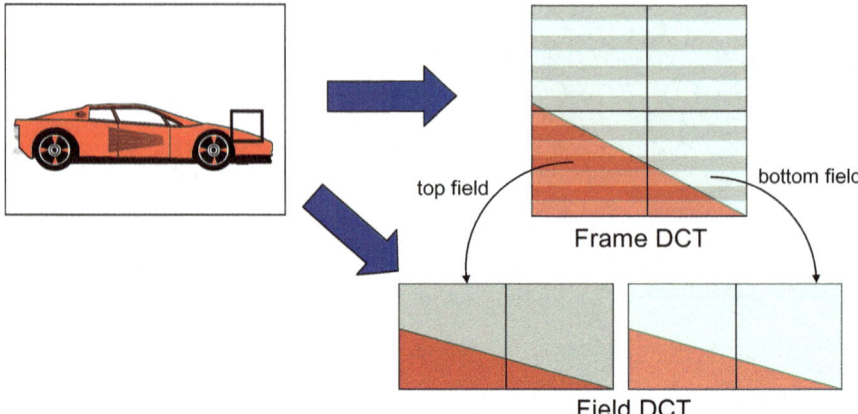

top field   bottom field

Frame DCT

Field DCT

MPEG-2 DCT can adaptively performed with frame or field structure block.

Fig. 28 Frame/field Adaptive DCT

Furthermore, scanning order can be switched adaptively according to frame/field DCT transform as shown in Fig. 29.

0	1	5	6	14	15	27	28
2	4	7	13	16	26	29	42
3	8	12	17	25	30	41	43
9	11	18	24	31	40	44	53
10	19	23	32	39	45	52	54
20	22	33	38	46	51	55	60
21	34	37	47	50	56	59	61
35	36	48	49	57	58	62	63

(a) zigzag scan

0	4	6	20	22	36	38	52
1	5	7	21	23	37	39	53
2	8	19	24	34	40	50	54
3	9	18	25	35	41	51	55
10	17	26	30	42	46	56	60
11	16	27	31	43	47	57	61
12	15	28	32	44	48	58	62
13	14	29	33	45	49	59	63

(b) alternate scan

Fig. 29 Adaptive scan change

The composition of MPEG-2 video coding is shown in Fig. 30.

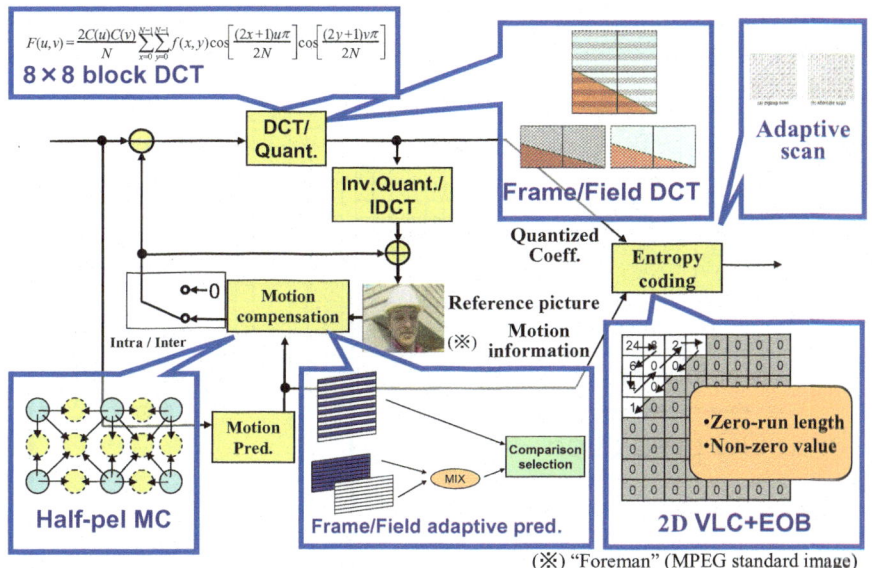

(※) "Foreman" (MPEG standard image)

Fig. 30 MPEG-2 Video Coding

3.3.2 MPEG-4 visual

MPEG-4 visual is an international standard for the purpose of the coding at low bit rate for mobile equipments. It is used in mobile devices such as cellular phones and portable video players. MPEG-4 mainly performs coding and transmission for progressive video signal from QCIF (176 pixels x 144 lines) to VGA (640 pixels x 480 lines) at lower bit rates of about 1 - 3 Mbit/s. The main features of MPEG-4 visual are shown below.

First, Intra Frame Prediction is performed for DC and AC data of transformed and quantized coefficients. The entropy of a symbol which should be coded as DCT coefficients can be reduced by prediction since the DC coefficient is equivalent to the average value in a block and AC coefficients including low frequency harmonics have high spatial correlations. The outline of Intra Frame Prediction in MPEG-4 is shown in Fig. 31.

Next, in addition to half-pel accuracy of Motion Compensated Prediction, MPEG-4 also supports quarter-pel accuracy prediction which uses the virtual samples between half-pel pixels as a candidate of the prediction. Additionally, a 16x16 pixel macroblock domain can be equally divided into four 8x8 sub-blocks and Motion Compensated Prediction can be adaptively performed in a 16x16 macroblock unit or the unit of an 8x8 sub-block. With this technique, the performance of prediction for complicated motions within a macroblock can be improved. Furthermore, three dimensional VLC is performed on the quantization indexes after transform and quantization. This method includes the information (LAST), which indicates that the coefficient to be coded is the last non-zero coefficient in a block, into the set of (zero run, non-zero value). Then, the set of (LAST, zero run, non-zero value) is coded.

B
C
D
A
X
Y
Macroblock
Intra DC prediction
B
C
D
A
X
Y
Macroblock
Intra AC prediction

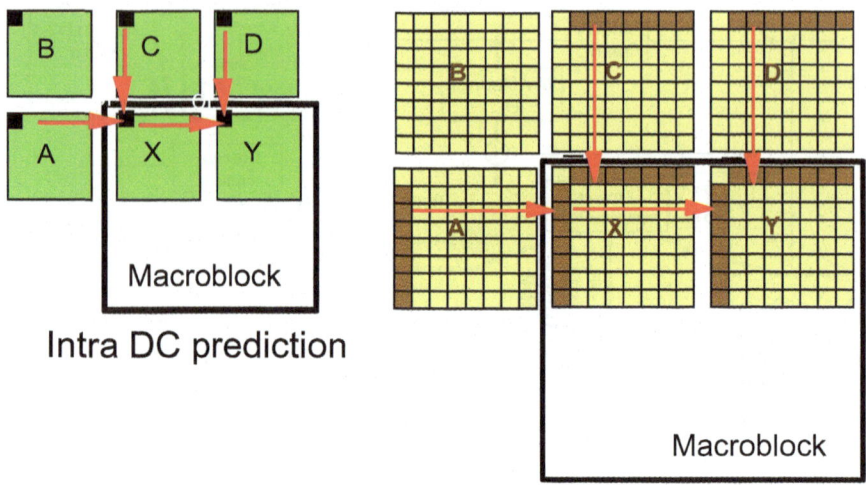

Fig. 31 Intra Frame DC/AC Prediction

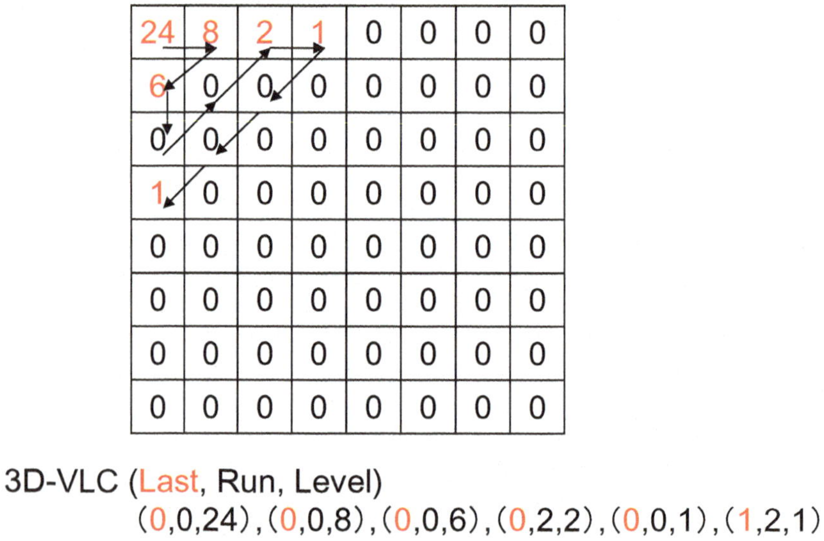

24 8 2 1 0 0 0 0
6 0 0 0 0 0 0 0
0 0 0 0 0 0 0 0
1 0 0 0 0 0 0 0
0 0 0 0 0 0 0 0
0 0 0 0 0 0 0 0
0 0 0 0 0 0 0 0
0 0 0 0 0 0 0 0
3D-VLC (Last, Run, Level)
(0,0,24),(0,0,8),(0,0,6),(0,2,2),(0,0,1),(1,2,1)

Fig. 32 Three dimensional VLC

The composition of MPEG-4 coding is shown in Fig. 33.

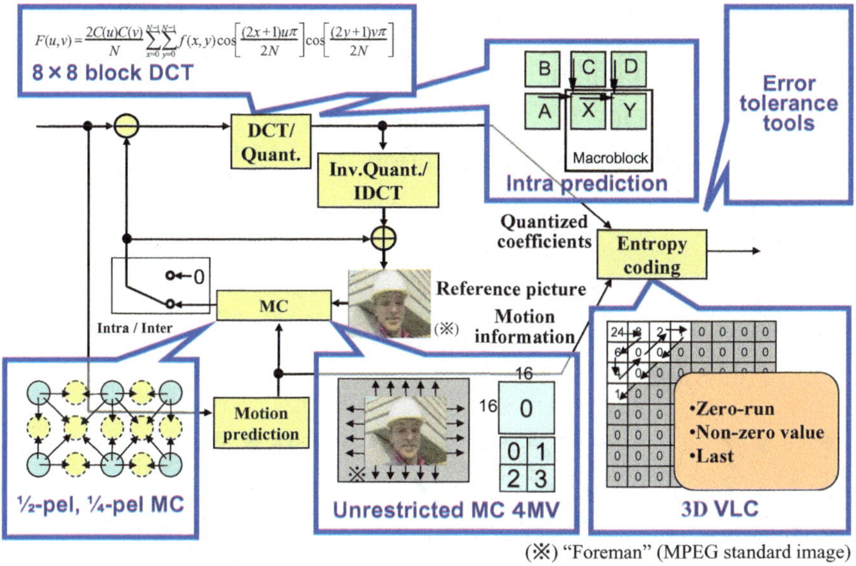

(※) "Foreman" (MPEG standard image)

Fig. 33 MPEG-4 Visual coding

3.3.3 AVC/H.264

The AVC/H.264 standard is specified as MPEG-4 Part 10 by ISO/IEC as well as Recommendation H.264 by ITU-T. The improvement in coding efficiency is taken into consideration as the top priority when AVC/H.264 standardization was performed. It has been reported that AVC/H.264 has the twice as much compression efficiency of MPEG-2. In AVC/H.264, a multi-directional prediction in the spatial domain (pixel domain) is adopted as Intra Frame Prediction in order to reduce the amount of video information. Several prediction methods are defined for luminance and chrominance signals; 16x16 and 4x4 intra predictions for luminance are introduced below. Intra 16x16 prediction for luminance is a method which chooses either of four prediction modes shown in Fig. 34 per macroblock to predict a 16x16 pixel macroblock.

On the other hand, Intra 4x4 prediction for luminance divides a 16x16 pixel macroblock into 16 blocks which consist of 4x4-pixel blocks and chooses one of nine prediction modes as shown in Fig. 35 per block.

Moreover, an adaptive block size partition is adopted for Motion Compensated Prediction of AVC/H.264. Since seven block size partitions including 16x16, 16x8, 8x16, 8x8, 8x4, 4x8, and 4x4 are defined for Motion Compensation "Prediction, the size of prediction can be chosen per macroblock (16x16) or subblock (8x8). In addition, subblock partitioning can be used for four 8x8 blocks independently.

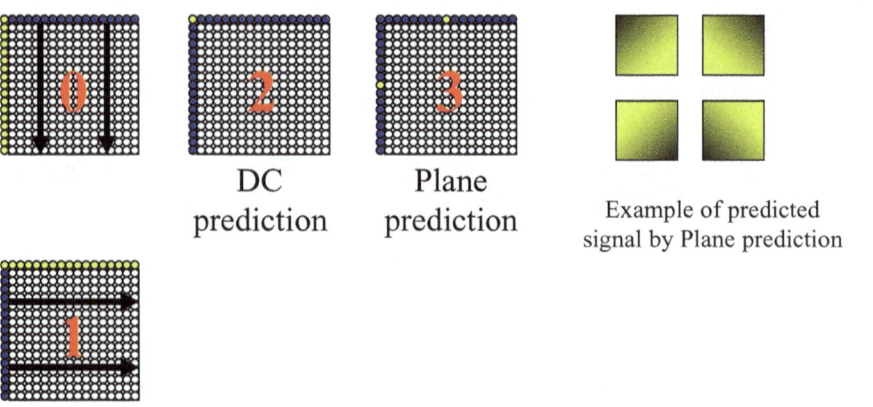

DC
prediction

Plane
prediction

Example of predicted
signal by Plane prediction

Fig. 34 Intra 16x16 prediction

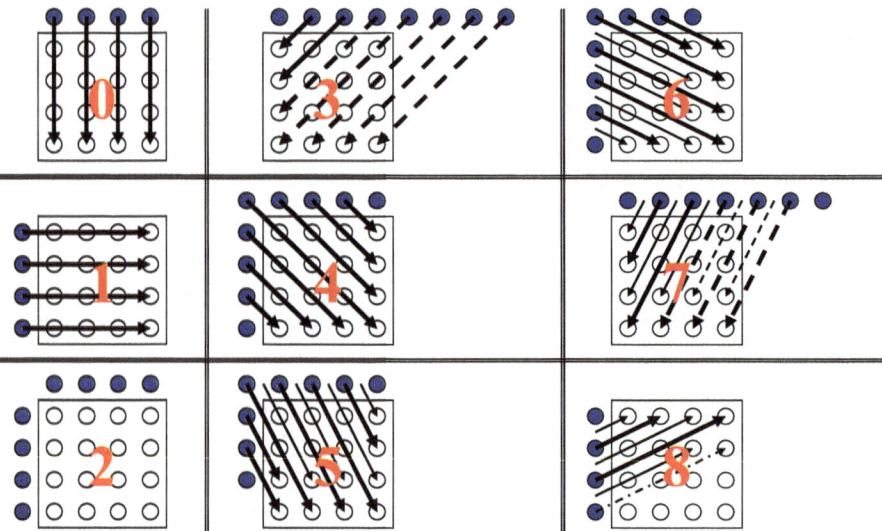

Fig. 35 Intra 4x4 prediction

In AVC/H.264, Motion Compensated Prediction can be performed by referring two or more reference frames. That is, the frames of the past and the future are stored in the frame memory, and can be chosen as reference frames for each block partition greater than 8x8 sub-blocks. B slices in AVC/H.264 support prediction from two reference pictures and the combination of the two pictures can be freely chosen. In contrast to MPEG-2, it is possible to perform bi-prediction even from two past pictures or two future pictures.

In AVC/H.264, 4x4 and 8x8 Integer Transform has been adopted for the conversion from spatial domain to frequency domain; the integer transform ensures that there is no mismatch between encoder and decoder. However, some parts of the transform process are included in quantization and de-quantization processing.

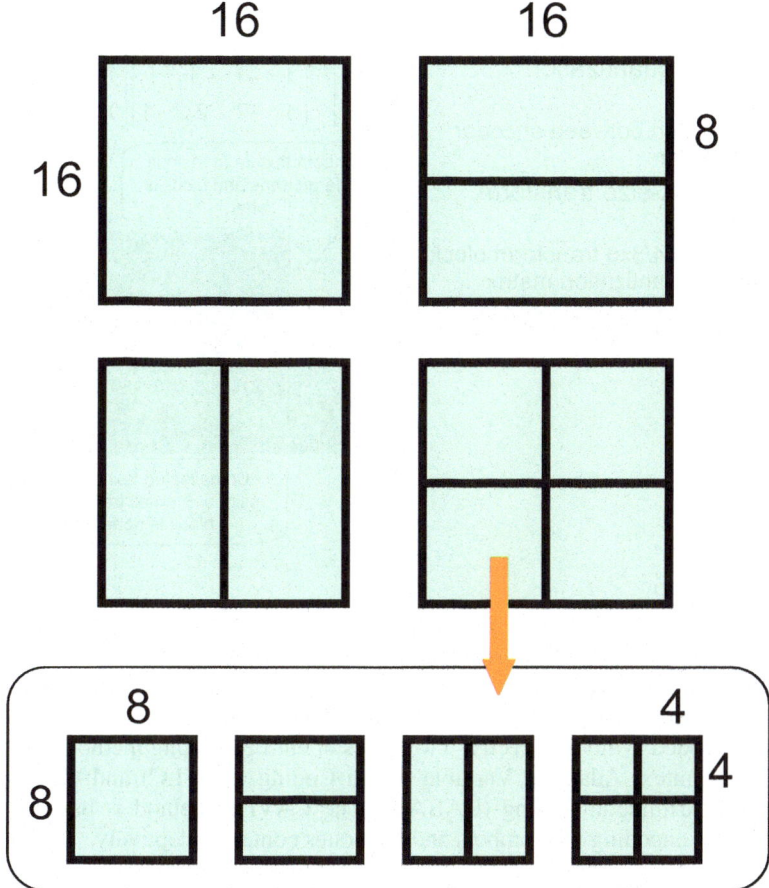

Fig. 36 Block size partitions of Motion Compensated Prediction

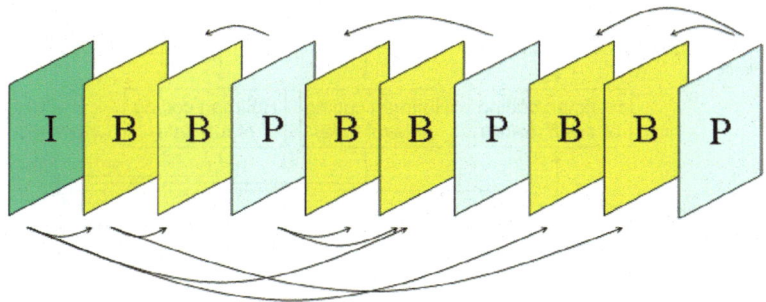

Fig. 37 Multi reference frame prediction

- Integer approximation of DCT together with quantization process
  - No mismatch between encoder and decoder
- Adaptive block-size transform (High Profile)
  - Adaptive 4x4/8x8 transform block size with quantization matrix
    - Allow better adaptation to local signal statistics of HDTV signal

$$\begin{bmatrix} Y0 \\ Y1 \\ Y2 \\ Y3 \end{bmatrix} = \begin{bmatrix} 1 & 1 & 1 & 1 \\ 2 & 1 & -1 & -2 \\ 1 & -1 & -1 & 1 \\ 1 & -2 & 2 & -1 \end{bmatrix} \begin{bmatrix} X0 \\ X1 \\ X2 \\ X3 \end{bmatrix}$$

Smooth texture in residue → large transform block is better

Complicated texture in residue → small transform block is better

Fig. 38 Integer Transform

In entropy coding of AVC/H.264, high compression is achieved by encoding the symbols adaptively and using the knowledge (Context) in connection with coding states such as the information on the surrounding block data as well as variable length coding of coded symbols directly. Two types of entropy coding methods exist in AVC/H.264: Context Adaptive Variable length Coding (CAVLC) and Context Adaptive Binary Arithmetic Coding (CABAC). The CAVLC method is based on Huffman tables for encoding the symbols and generates contexts adaptively.

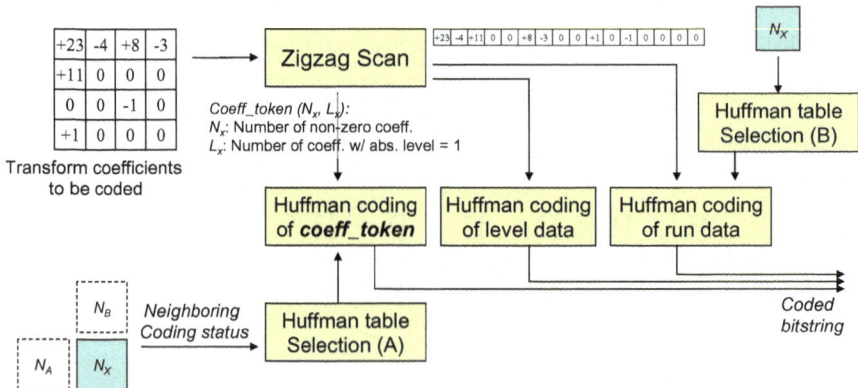

CAVLC encodes zigzag scanned transform coefficients with adaptive VLC table selections.

Fig. 39 CAVLC

On the other hand, CABAC converts symbols (coding mode, motion vector, transform coefficients, etc.) into a binary code series based on the rule defined by the standard (binarization), then chooses an occurrence stochastic model based on a context model (context modeling) and finally performs a binary arithmetic coding based on the selected occurrence stochastic model (binary arithmetic coding). In addition, an occurrence stochastic model is updated based on the result of coding (probability estimation).

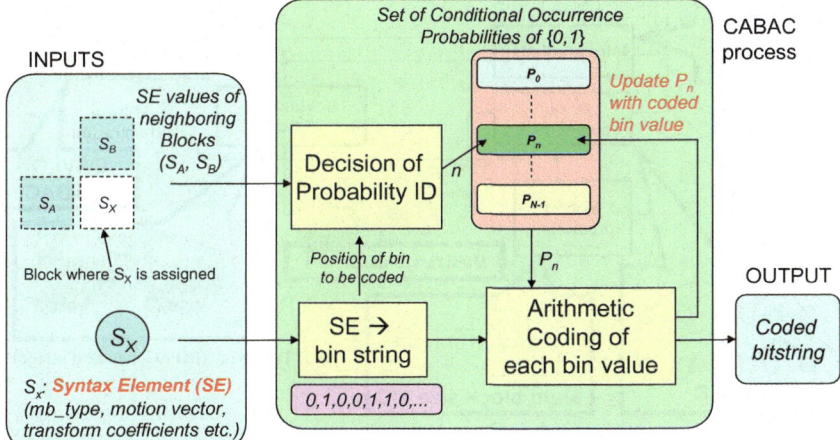

CABAC encodes binarized syntax elements through selecting probability models for each syntax element according to element's context, adapting probability estimates based on local statistics and using arithmetic coding.

Fig. 40 CABAC

The composition of AVC/H.264 video coding is shown in Fig. 41.

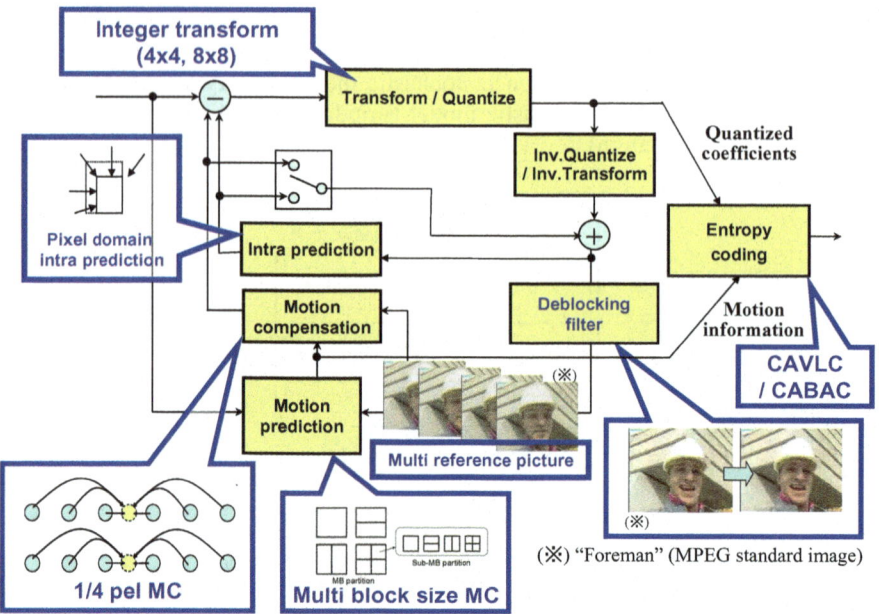

Fig. 41 AVC/H.264 video coding

3.3.4 Technical Achievements of MPEG Video Standards

The technical achievments of MPEG-2, MPEG-4 visual and AVC/H.264 are summarized in Table 1.

Table 1 Technical achievements of MPEG Video Standards

	Data Structure	Motion Compensation	Texture Coding (Intra and Inter)	Entropy Coding
MPEG-2	GOP including Interlace	I,P,B prediction / Half-pel MC / Interlace adaptation	Alternate Scanning / DCT mismatch control	2D Huffman
MPEG-4 visual	Video Object/ Sprite	16x16/8x8 MC / Quarter-pel MC / Unrestricted MC	Intra DC/AC Prediction Adaptive Scanning	3D Huffman
AVC/H.2 64	NAL Unit	16x16~4x4 Multi-blocksize MC / Multiple Reference Pictures	Pixel-domain Intra Prediction / Integer Transform / Adaptive Transform size	CAVLC / CABAC

Coding performance has been improved by the standard evolution of MPEG. The improvement of coding performance according to the progress of coding methods is shown in Fig. 42.

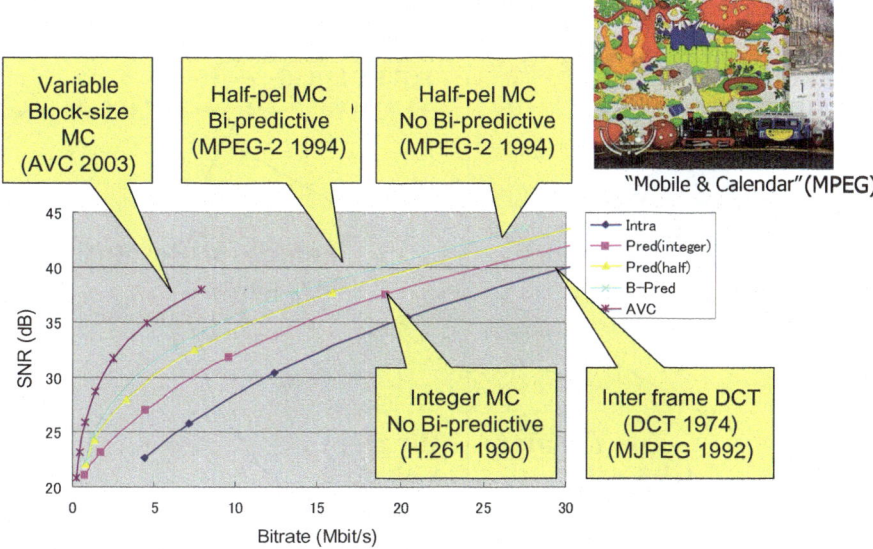

Fig. 42 Improvement of coding performance

On the other hand, the operation load in terms of complexity of video coding continues to increase with the standard evolution of MPEG. By improving Intra and Inter prediction accuracy, the complexity of AVC/H.264 encoding process has increased by 5 to 10 times compared with that of MPEG-2 or MPEG-4 visual. The complexity of the decoding process of AVC/H.264 is double for that of MPEG-4 visual by adopting context adaptive entropy coding and in-loop deblocking filter. However, the progress of semiconductor technology including LSI, processor and large scale storage has supported the realization of evolution of video compression technology.

4 Requirement for Quality of UHD Video System

4.1 Required Specifications for UHD Video Service

HDTV was realized by MPEG-2 as a digital broadcasting service for the home, and HDTV broadcasting has promoted both the thinness and enlargement of television displays. Camcorders and video recorders also support HDTV. Since the visible difference of video quality between SDTV and HDTV on a large screen became clear for users, the merits of HDTV have been validated. When UHD video, which is expected as the next generation video service, will be realized, it is necessary to improve each parameter, which influences video quality including spatial and temporal resolutions, gradation, and color space of HDTV. An illustration that re-examines the various factors to be qualified into UHD Video is shown in Fig. 43.

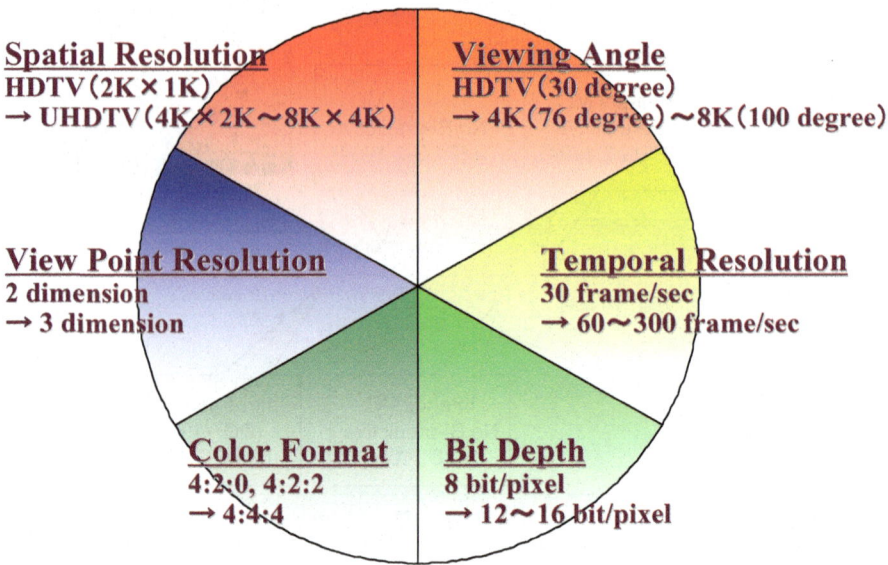

Fig. 43 Factors to be qualified into UHD Video

The specifications for UHD formats are planned to be standardized in ITU-R by 2012.

In the next section, the requirements about these resolutions and modes of the expression are considered, and the increase in the amount of information by fulfilling the requirements is measured.

4.1.1 Requirement for UHD Video

4.1.1.1 Spatial Resolution

Based on the fact that the angle of resolution of the human visual system is one minute degree, an HD image covers about 30 degrees of useful visual field, which can be recognized only with eyeball motion and without moving ones head. When the viewing angle is extended to 60 degrees of gazing viewing angle in which objects can be recognized only with little movement of ones head, 4K image can cover the angle. Furthermore, when a viewing angle is extended to 100 degrees of guidance viewing angle in which the existence of objects within the angle can be felt, 8K image can cover the angle. Although the video with 4K and 8K of UHD exceeding HD are already standardized as a video format of 3840x2160 and 7680x4320 in ITU-R Recommendation [20], video services with such resolution have not yet been deployed.

4.1.1.2 Temporal Resolution

30 fps (frame per second) used by the present television broadcasting was selected since it is the limit of the human visual system for flicker detection. However, it is

expected that at least 60 fps and up to 120 fps of temporal resolution is probably needed for UHD video since the present television broadcasting adopts an interlace signal and 60 fields per second is used in practice. If the temporal interval between frames is shortened, Motion Prediction of video coding will become more effective, and it should result in increased compression efficiency

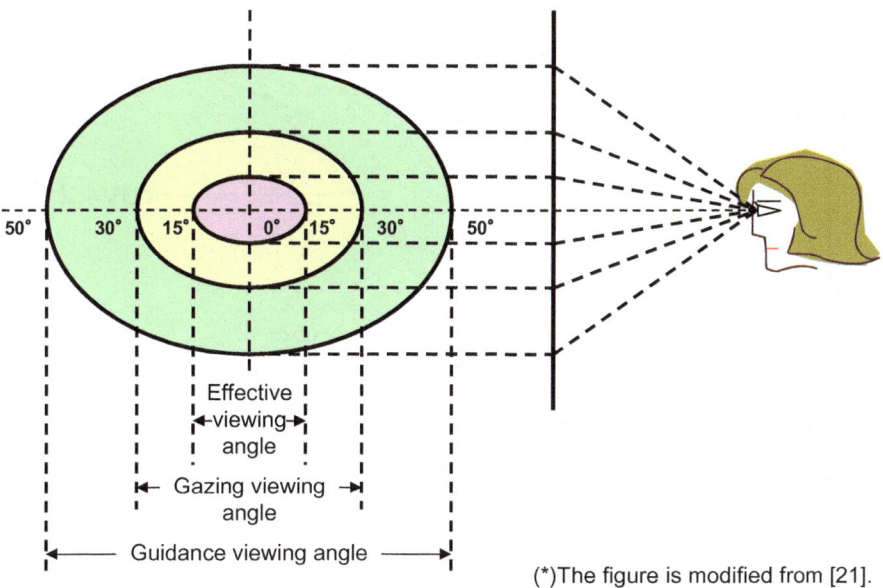

(*)The figure is modified from [21].

Fig. 44 The information acceptance characteristic within view [21]

4.1.1.3 Gradation per pixel
If the gradation per pixel increases, the contrast in the dark portions of a screen can be more visible. Moreover, there is also an advantage which can perform high precision calculations in the filter processing and the sub pixel processing for Motion Prediction in video coding. Therefore, the quality improvement of video is expected to be improved by increasing bit depth from the present 8 bpp to 10-12 bpp.

4.1.1.4 Color space
Although the amount of information becomes 1.5-2 times by increasing the color sampling format from 4:2:0/4:2:2 to 4:4:4, the possibility to use 4:4:4 format will increase if the compression efficiency of video coding is improved. Since the number of the signal elements of 4:2:0 format is one fourth of that of 4:4:4 format for chrominance signal, there is a trade-off in quality for video coding. At low bit rates, 4:2:0 video is preferred over 4:4:4 video since it contains fewer pixels. On the other hand, at higher bit rates, 4:4:4 video is preferred over 4:2:0 video. As a result, there is a cross-over point on performance between 4:2:0 and 4:4:4 video coding. For example, even if the cross-over point between 4:2:0 and 4:4:4 formats occurs at a higher bit rate than that of the practical use with existing video coding technology,

the realization of high quality video applications with 4:4:4 format will be attained since the cross-over point is achieved to shift to a lower rate by the development of a new video coding technology with higher coding performance[22].

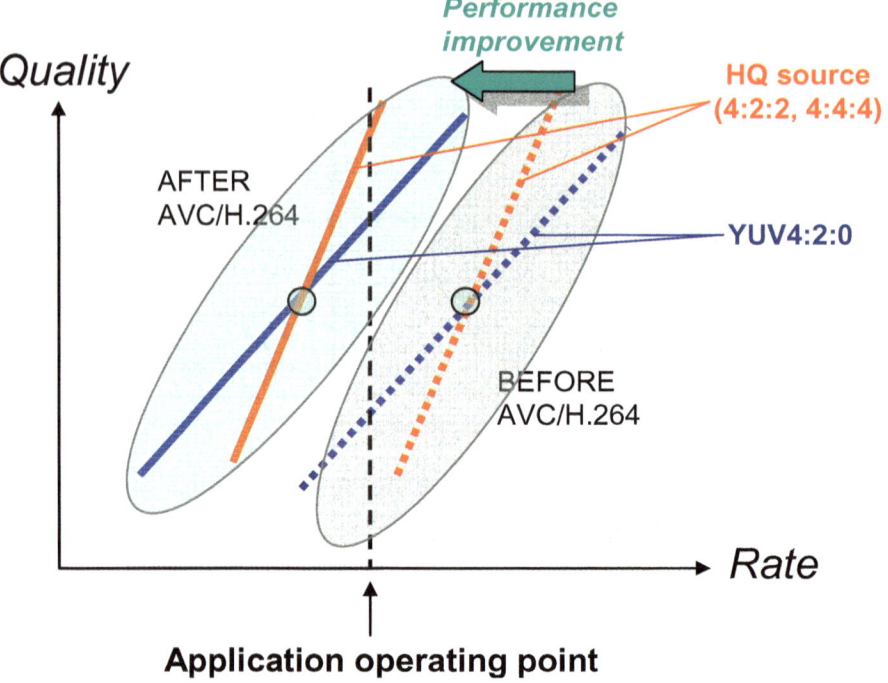

Fig. 45 Recent Progress of Video Coding Technology on Rate-Distortion Characteristic [13]

Moreover, most of video cameras and displays can treat RGB signal directly now. Then, if RGB and 4:4:4 formats are treated directly also in video coding, the degradation of the quality by means of color conversion does not occur and high quality video can be consistently provided.

4.1.2 The Amount of Information of UHD Video

Realizing new video expression which fulfills the requirements of UHD video is simultaneously accompanied with the steep increase of the amount of information. For example, the uncompressed rate of HD video (1920x1080/8bpp/4:2:0/30fps) is around 1 Gbit/s. This uncompressed rate increases to 3 to 18 Gbit/s for 4K video (3840x2160/8-12bpp/YUV4:2:0-4:4:4/30-60fps) and 12 to 72 Gbit/s for 8K video (7680x4320/8-12bpp/YUV4:2:0-4:4:4/30 - 60fps). Even if compared by the same bit length (8bpp), video format (4:2:2) and frame rate (30fps), 4K video has around four times more information and 8K has about 16 times more information than HD. With exponential increase of video data band width, the transmission rate for the interface between display/camera and storage/codec should also be exponentially increased as shown in Fig. 47. To satisfy the needs of I/O interface for 4K UHD Video, the standardization of 25Gbps serial optical interface for studio use is currently under consideration in SMPTE 32-NF-30.

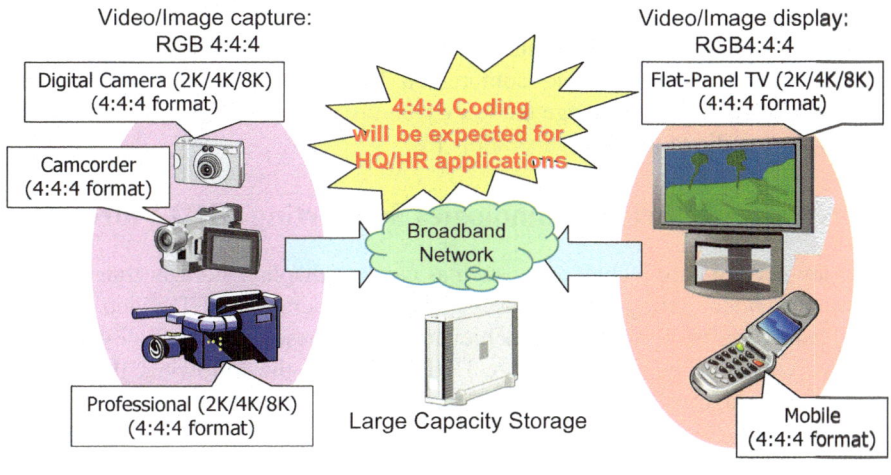

Fig. 46 High Quality and High Resolution/real-Color Video Applications [13]

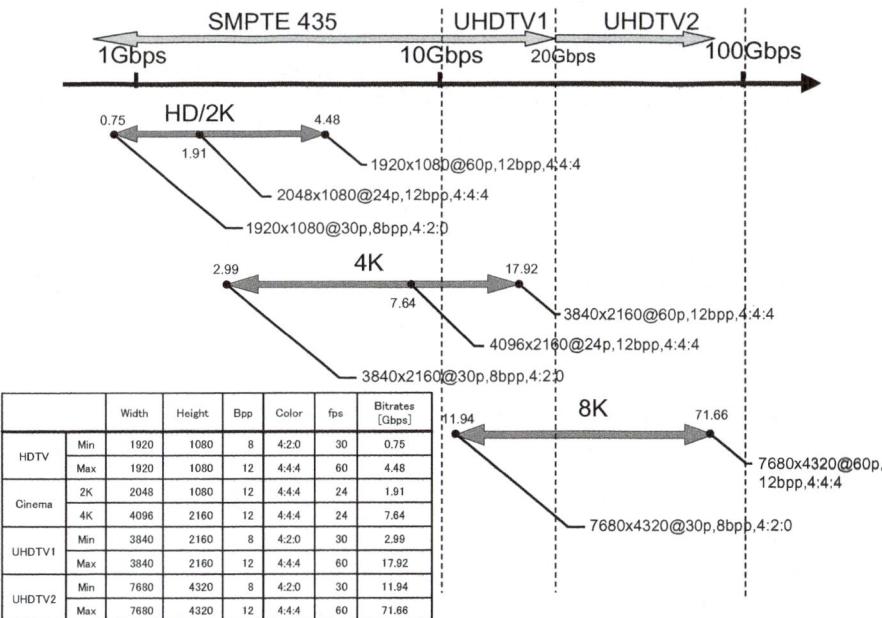

		Width	Height	Bpp	Color	fps	Bitrates [Gbps]
HDTV	Min	1920	1080	8	4:2:0	30	0.75
	Max	1920	1080	12	4:4:4	60	4.48
Cinema	2K	2048	1080	12	4:4:4	24	1.91
	4K	4096	2160	12	4:4:4	24	7.64
UHDTV1	Min	3840	2160	8	4:2:0	30	2.99
	Max	3840	2160	12	4:4:4	60	17.92
UHDTV2	Min	7680	4320	8	4:2:0	30	11.94
	Max	7680	4320	12	4:4:4	60	71.66

Fig. 47 Bitrates for UHD Video Source [23]

4.2 Expectation for New Video Coding Technology

In order to utilize the UHD video with 4K and 8K resolutions, it is insufficient to use the present standard (AVC/H.264) which was standardized for the purpose of the coding video with the resolution up to HD. New video coding technology will be required, which compresses video greater while maintaining the high quality of an original video as much as possible. New video coding technology has a possibility to change SD/HD video compressed by MPEG-2 or AVC/H.264 into HD/UHD video and to exchange the television broadcasting which is the most familiar video media to the next generation TV.

5 Progress of Device Technologies Supporting UHD Video

There are several base technologies such as camera, display, storage, transmission system and video coding, which can realize next generation UHD video technology. There have been remarkable achievements in these areas until now. Therefore, we will soon be able to realize UHD video. In the following, the present status of camera, display, storage and digital network, is surveyed.

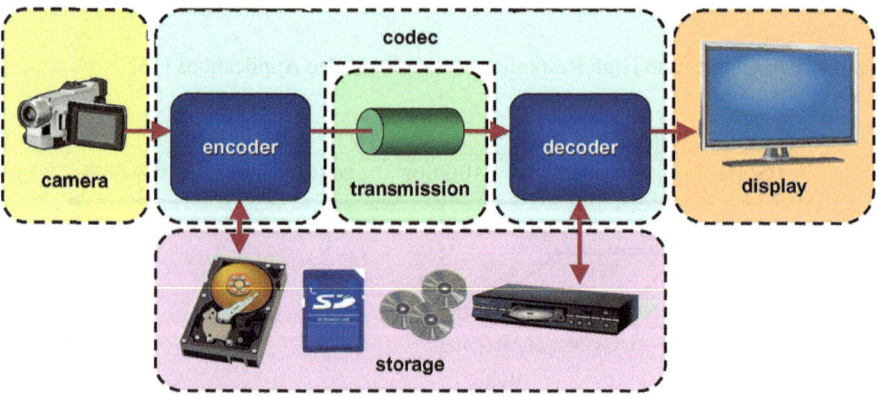

Fig. 48 Device technologies for realizing UHD video

5.1 Video Capture Device

Pixel size reduction of the image sensors for digital video cameras, in which high speed pick-up is possible, has progressed and the realization of high resolution cameras is ready.

With regards to consumer cameras, small camcorders that fit in the hand also support HDTV resolution with remarkably low price. Moreover, in professional use, several cameras with support of 4K resolution have already been announced and cameras with 8K resolution have also been developed. Since digital cinema services have started, the development of 4K cameras for digital cinema is progressing. For example, RED (RED Digital Cinema Camera Company) is manufacturing a 4K camera called RED ONE which is equipped with a CMOS sensor of 12 Mpixel and has the resolution of 4520x2540. DALSA Origin with 4096x2048 resolution (8 million pixels) has been developed by DALSA and the camera is characterized with 16 bit/pixel of high gradation. On the other hand, Octavision with 3840x2160 resolution (8 million pixels) has been developed by Olympus. A CMOS 4K camera which has a frame rate of 60fps with the resolution of 3840x2160 has been announced by Victor. Furthermore, an 8K video camera with 7680x4320 resolution (33 million pixels) captured by a high speed CMOS sensor, 60fps and 12 bit/pixel has been also announced by the Japan Broadcasting Corporation (NHK).

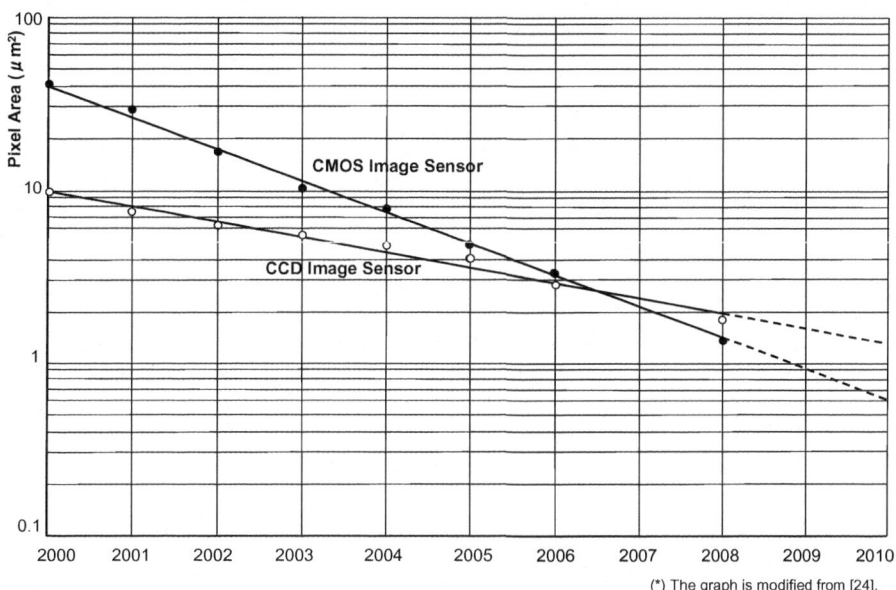

(*) The graph is modified from [24].

Fig. 49 Transition of image sensor pixel improvement [24]

5.2 Display

Flat panel TVs such as LCD and PDP have been progressing with larger screen and higher resolution. These TVs corresponded to the spread of digital contents including digital broadcasting. A screen size of the 40 inches has become popular and support for full HD resolution which can display images from television broadcasting is becoming typical.

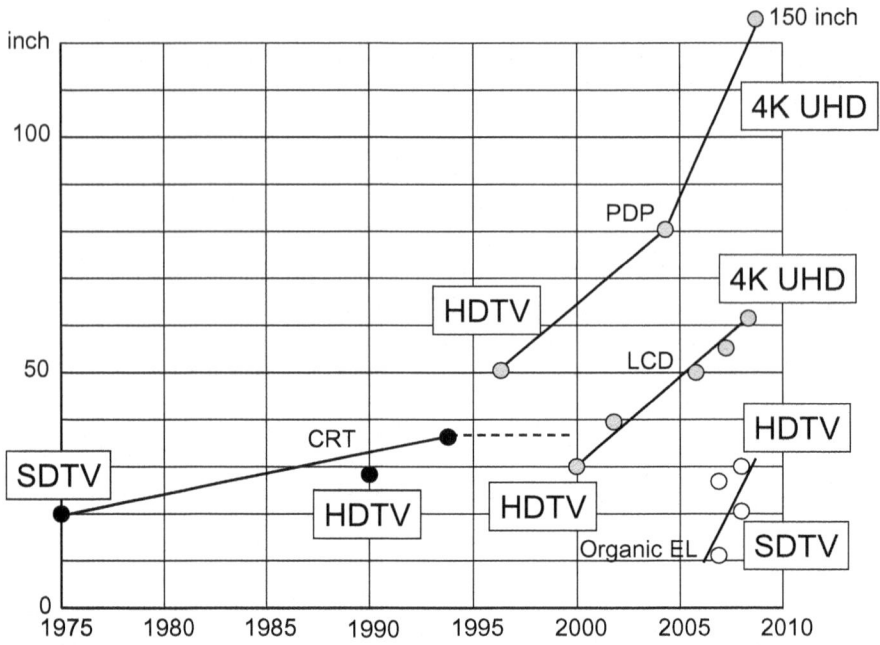

Fig. 50 Progress of TV screen size in Japan [25]

Moreover, the development of 4K television with 4 times as many pixels compared to full HD is progressing steadily and the possibility of its appearance in the home from 2010 to 2012 has also increased. There are trial products of LCD and PDP with 4K resolution including 3840x2160 and 4096x2160. The former expands Full HD twice horizontally and vertically, while the latter corresponds to digital cinema resolution. The LCD, PDP and projector which can display 4K video have been exhibited at several shows, and some of them are produced commercially. Several examples of 4K displays and 4K/8K projectors are shown in Table 2 and Table 3.

Table 2 Example of 4K and 8K digital cameras

Product, Manufacturer	Resolution	Frame Rate (fps)	notes
RED One	4520 × 2540	~30(4K) ~60(3K) ~120(2K)	2007~ CMOS, RAW data 12bit/pixel
DALSA Origin II	4096 × 2048	~30	2003~(Origin), 2007~(Origin II) CCD, RAW data, 16bit/pixel
Olympus Octavision	3840 × 2160	24, 30	2005~ HDTV CCD, 4:2:2 format
JVC-Victor	3840 × 2160	60	2009 CMOS, RAW data 12bit/pixel
NHK	7680 x 4320	60	2009~ CMOS, RAW data 12bit/pixel
Vision Research Phantom 65	4096 x 2440	~125	2006~ CMOS

Table 3 Examples of 4K liquid crystal and plasma display

Manufacturer	Resolution	Thickness (inch)	System
SAMSUNG	3840 x 2160	82	Liquid Crystal (2008)
SAMSUNG SDI	4096 x 2160	63	Plasma (2008)
ASTRO design	3840 x 2160	56	Liquid Crystal(2007)
SHARP	4096 x 2160	64	Liquid Crystal (2008)
Panasonic	4096 x 2160	150	Plasma (2008)
NHK+Panasonic	3840 x 2160	103	Plasma (2009)
MITSUBISHI	3840 x 2160	56	Liquid Crystal(2007)

On the other hand, organic EL equipped with thinness, power saving and high resolution is very promising as a UHD video display. Organic EL displays have also been introduced at the various trade shows and some of them are produced commercially. Several organic EL displays are shown in Table 5.

Table 4 Examples of 4K and 8K Projectors

Product, Manufacturer	Resolution	notes
Victor DLA-SH4K	4096 x 2400	D-ILA(2007)
MERIDIAN	4096 x 2400	D-ILA(2009) House Use
SONY SRX	4096 x 2160	SXRD(2005)
NHK+Victor	7680 x 4320	D-ILA(2004)
Victor	8192 x 4320	D-ILA(2009) RGB 12bit/pixel 60fps

Table 5 Examples of organic EL displays

Product, Manufacturer	Resolution	Thickness (inch)	notes
SAMSUNG	1920 x 1080	31	(2008)
SAMSUNG SDI	1920 x 1080	40	(2005/2008)
LG	1280 x 720	15	0.85mm (2009)
SONY	1920 x 1080	27	10mm(2007)
SONY XEL-1	960 x 540	11	3mm, (2007) ¥200,000

5.3 Storage

The speed at which storage capacity increases for HDD, SSD and other card type storage is remarkable. Storage capacity of a Terabyte has already arrived. Fig. 51 and Fig. 52 shows the improvement of recoding bit rate and strange density of typical storage media.

In the space of SD memory cards, SDHC with a capacity of 32 GB and a speed of 25MB/s has been realized, and SDXC with a capacity of 2 TB and a speed of 50-300MB/s has also been realized.

5.4 Digital Network for Communication and Broadcasting

The performance in the speed of optical line has progressed and some transmission experiments of the 4K resolution uncompressed video have been also

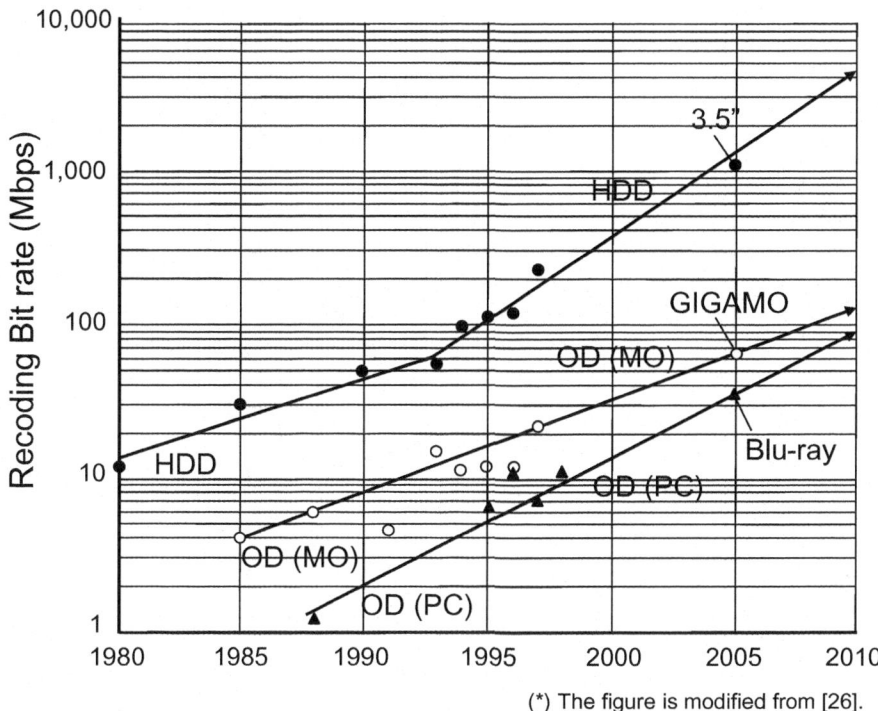

Fig. 51 Trend of Recording bit rate of hard disk and optical disk [26]

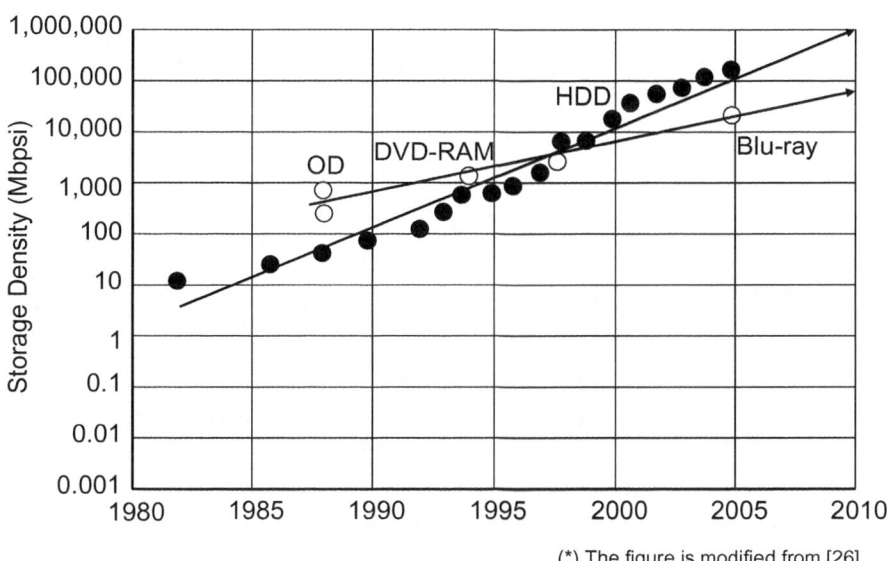

Fig. 52 Trend of Storage Density of Hard Disk and Optical Disk [26]

conducted. In Japan, the commercial service of broadband Ethernet with 1 Gbit/s has already been carried out, and a future service of 100 Gbit/s is also planned. Furthermore, UHD video will be able to be sent via broadband by NGN.

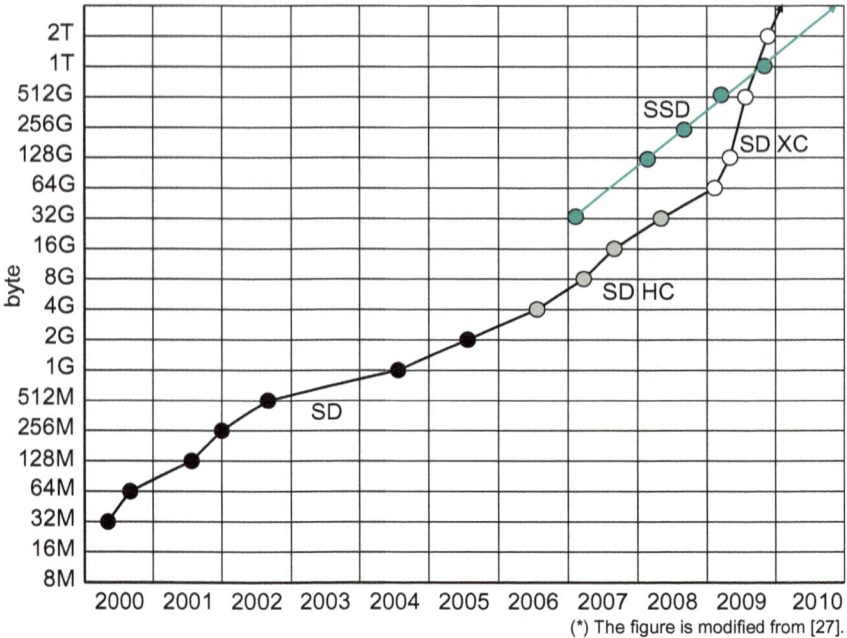

Fig. 53 Memory size of SD and SSD [27]

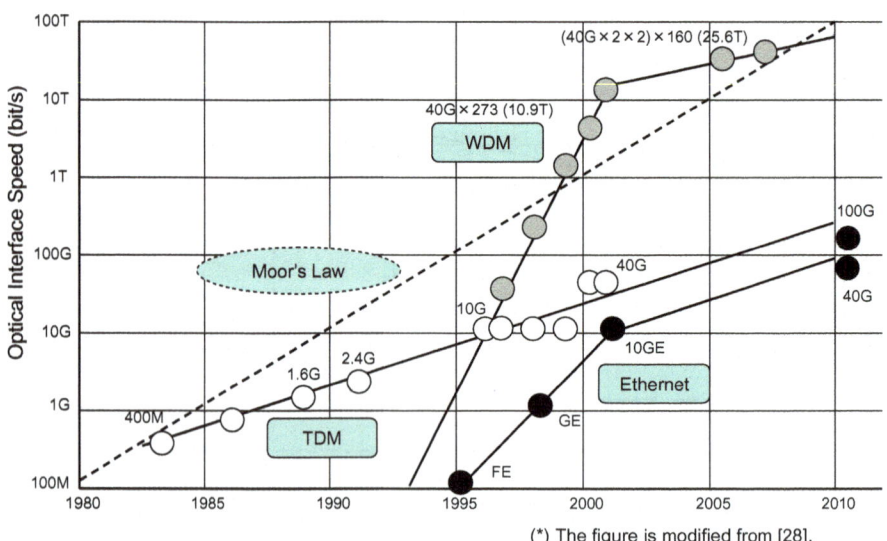

Fig. 54 Advance of optical transmission technology [28]

6   Standardization of UHD Video Coding Technology

6.1   Realization of UHD Video Application

When UHD video is considered, video coding technology is required to reduce the huge amount of information. An improvement in a compression ratio is needed while maintaining a high quality and ultra high resolution. Furthermore, from the scale and the cost of the video coding technology realized, UHD video technology is required to be a system which is more conscious of implementation. UHD video requires different inputs and outputs of a sensor and a device compared to present systems. Therefore, it is necessary to take the characteristics of those devices into consideration. Even if UHD video technology follows the framework of existing video encoding methods, its high resolution may change the optimal coding parameters. This change may result in the replacement of some coding tools in the current system. There is also the possibility that techniques which could be realized only in simulation are not able to be utilized in practical codecs because of memory integration, calculation complexity and device cost.

6.2   Next Generation Video Coding Standard

JVT have moved the principal axis of their activities toward the addition of other functionalities such as SVC (Scalable Video Coding) and MVC (Multi-view Video Coding) after the standardization of AVC/H.264 High Profile. However, since the development of UHD video devices has become remarkable from 2008 to 2009 and the necessity for UHD video coding was appealed from Japan, the argument of the standardization of a next generation video coding has become active toward the development of HVC (High performance Video Coding) in MPEG since April 2009. Then, MPEG invited the public to submit evidence of new video coding technologies that fulfill the conditions for UHD video, and evaluated the proposed technologies at the meeting in June-July 2009 [11]. As a result of this study, MPEG decided to work towards issuing a formal call for proposals and initiating the development of a new video coding standard. The current plan is to collect proposals in 2009, and to issue an international standard in 2012 or 2013. At the same time, VCEG is considering the improvement of video coding efficiency and the reduction of the complexity based on AVC/H.264 as KTA (Key Technical Areas). KTA has started to be discussed since 2005 and has continued to be added new coding tools. Then, the examination of NGVC (Next Generation Video Coding) was started in 2009, which is assumed to be used for UHD video, HD broadcasting, HD video conference, mobile entertainment, etc. A draft document describing the collection of test sequences and the requirement of NGVC was created during the Yokohama meeting in April 2009. The requirement was updated and the working title was changed from NGVC to EPVC (Enhanced Performance Video Coding) during the Geneva meeting in July 2009. In addition, the cooperation between MPEG and VCEG is likely to be established. The requirement conditions of MPEG HVC [29] and VCEG EPVC [30] are shown below.

(1) Video coding performance

To realize 50% reduction of coding bits with subjective quality equivalent to AVC/H.264 High Profile.

(2) Correspondence to a high quality video

(2.1) Resolution

From VGA to 4K (also 8K)

(2.2) Chroma format

From 4:2:0/4:2:2 to 4:4:4

(2.3) Bit length

8 - 14 bit per pixel

(2.4) Frame rate

24 - 60 fps or more

(2.5) Complexity

(i) To be possible of implementation at its standardization period.

(ii) To be possible to control the trade off between complexity and coding performance.

(iii) To be possible of parallel processing.

MPEG HVC and VCEG EPVC are going to publish a joint Call for Proposals on the next video coding standard in January 2010, and the target bitrates for UHD and HDTV are currently defined as shown in Table 6.

Table 6 Target bitrates for Call for Proposals on HVC/EPVC

Class	Resolution	Frame rate	Target bitrate for evaluation
A	2560 x 1600p cropped from 4K	30 [fps]	2.5[Mbps], 3.5[Mbps], 5.0[Mbps], 8.0[Mbps], 14.0[Mbps]
B	1920 x 1080p	24 [fps]	1.0[Mbps], 1.6[Mbps], 2.5[Mbps], 4.0[Mbps], 6.0[Mbps]
		50/60 [fps]	2.0[Mbps], 3.0[Mbps], 4.5[Mbps], 7.0[Mbps], 10.0[Mbps]

6.3 Challenge toward UHD Video Coding

6.3.1 Trial to improve AVC/H.264

Within VCEG, experiments have shown the improvement in video coding performance of AVC/H.264 using KTA and continual improvements of the reference software. Many proposals for Evidence collection of MPEG used the tools extracted from KTA. Therefore, it can be said that KTA at present is a benchmark for the improvement of video coding efficiency. The following tools have been incorporated into the KTA software and have been found to yield gains in coding efficiency.

(1) Extension of bit depth
(2) 1/8-pel accuracy Motion Compensated Prediction
(3) Prediction direction adaptive transform
(4) Adaptive prediction error coding
(5) Adaptive quantization matrix
(6) Adaptive quantization rounding
(7) RD optimal quantization
(8) Adaptive interpolation filter
(9) High precision interpolation filter
(10) Adaptive loop filter

The block diagram of encoder with these tools is shown in Fig. 55.

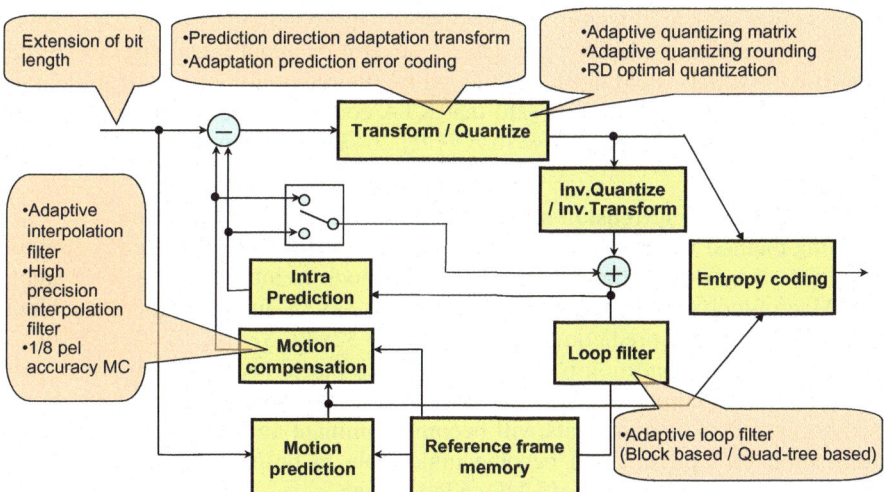

Fig. 55 Improvement points of AVC/H.264

6.3.1.1 Improvement of Coding Performance of KTA

Some new coding tools that have been incorporated into the KTA software that provide improvements in video coding efficiency for motion compensated prediction, transform, loop filter, and Intra prediction, are reviewed below.

(1) Motion Compensated Prediction
When generating the motion compensation signal of sub-pixel accuracy, a suitable filter can be chosen among several candidates and the interpolation

method which increases operation accuracy is used. It is supposed that gains of about 4 to 7% of improvement are realized.

(2) Transform

The size of a macroblock is expanded to maximum of 64x64, and the transform of several block sizes from 4x4 to 16x16 is available. It is effective for high resolution video. It has been claimed that improvements in the range of 10 to 15% for P picture and 15 to 20% for B picture can be achieved.

(3) Loop Filter

The Wiener filter which performs image restoration from a local decoded picture using source picture is designed. The filter can be also turned on and off for each block to improve the quality of a local decoding picture. Gains of about 5 to 10% of improvement have been reported.

(4) Intra Coding

Prediction between pixels is alternatively performed from several directions in the Intra coding of AVC/H.264. In this tool, transform basis is changed according to the directions of the predictions. Improvements of about 7% have been shown.

These tools are the improvements to the existing AVC/H.264 framework and keep the existing coding architecture. They adopt the approach of changing the range of parameters, selecting adaptive case among several candidates and performing the optimization which could not be employed by the restriction of practical memory and operation scale before. It is said that KTA could provide about 20 to 30% of performance improvement relative to AVC/H.264 by adoption of these tools.

6.3.1.2 Block Size Extension and Non-Rectangle Block Application

A coding method which extends AVC/H.264 and performs UHD video coding using block size extension and a non-rectangle block is introduced as an example [31]. Macroblock size is extended from 16x16 to 32x32, and accordingly the block partitions for motion prediction are expanded to 32x32, 32x16, 16x32 and 16x16. When 16x16 block partition is chosen, sub-block partitions of 16x16, 16x8, 8x16 and 8x8 are also employed. Smaller block partitions of 8x4, 4x8 and 4x4 are not used because noise components will become dominant and the essential structure information of a picture will become impossible to be expressed efficiently by such small blocks in the case of UHD video coding.

In addition to extension of macroblock size, non rectangle block partitions are employed for motion prediction. Although non rectangle block partition was proposed for the purpose of expressing more complicated motion with a short motion vector in super low bit rate video coding, it is believed that new partitions will contribute to reducing residual energy by effectively expressing complicated picture structures such as object boundaries in UHD video. On the other hand, there are some problems such as the increase of the motion detection operations, the increase of the shape description information and the necessity of the memory access to complicated shape. Then, the simple diagonal partitions shown in Fig. 57 are adopted, which can be created with the combination of 16x16 blocks.

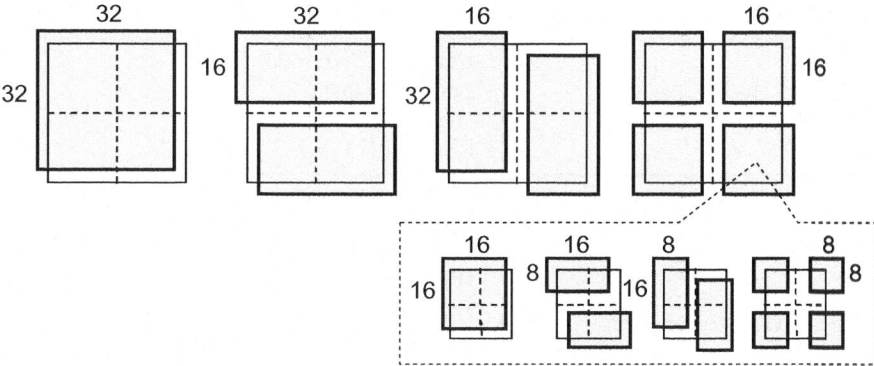

Fig. 56 Hierarchical division of motion compensated prediction blocks

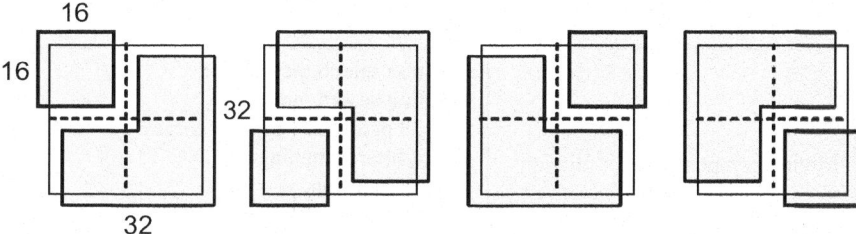

Fig. 57 Diagonal block partition

Moreover, correlation between pixels in wide range can be used if transform size is extended. Then, transform coefficients can be compacted more into the low frequency bands. Since it is preferred to not to include block boundaries in a block partition, it is effective to change the transform block size according to the block partition size for motion prediction. Then, 8x8 and 16x16 block sizes are adopted as transform block sizes and switched adaptively according to chosen motion prediction block partition. The above improvements were applied to P picture, and the computer simulation was performed using a GOP structure of IPPP. The conditions of the experiment are shown in Table 7. Reference software JM15.1 of AVC/H.264 was used as the anchor technique for the quality assessment. Moreover, the test sequences used for the experiment are shown in Fig. 58. Traffic of Class A and Kimono1 of Class B of MPEG test sequence (YUV 4:2:0 format and 8 bpp) were chosen for the experiments. Class A includes sequences with 2560x1600 resolution that have been cropped from 4096x2048 pictures; in this way the quality and compression efficiency of 4Kx2K could be practically evaluated. Class B includes the sequences of Full HD (1920x1080) size.

The results of the experiment show the improvement in efficiency of the motion compensated prediction and the decrease of addition information such as motion vector. The amount of coding bits is successfully reduced by 2 to 30% and PSNR was improved by 0.1-0.9 dB. The improvement of PSNR and the amount of

Table 7 Experimental condition

Test sequence (YUV4:2:0, 8bpp)	- Class A: (2560x1600) Traffic (30 fps) - Class B: (1920x1080) Kimono1 (24 fps)
Number of frames	- Class A: 300 frames - Class B: 240 frames
GOP composition	- IPPP - I picture interval : Class A: 28 frames Class B: 24 frames
MB size	- I picture: 16x16 - P picture Proposed: 32x32 Anchor: 16x16
Motion compensation prediction	- Motion search method: EPZS - Motion search range: ±128 pixels, 1/4-pixel accuracy - Intra/Inter switching ON - Block partition Proposed: rectangle and diagonal Anchor: only rectangle
Qp	- Class A: 25, 29, 33, 37 (fixed) - Class B: 25, 28, 31, 34 (fixed)
R-D optimization	ON
Entropy Coding	CABAC

coding bits are shown also in Fig 58. Moreover, the improvement of coding efficiency is depicted as a comparison of R-D curve with AVC/H.264 is shown in Fig. 59.

6.3.1.3 Other Possibilities of Performance Improvements for UHD Video Coding

In addition to the improvement of KTA tools and the expansion of the block size for UHD video, other possibilities of performance improvement of AVC/H.264 are assumed as follows.

(1) Examination of pre filter processing.
(2) The design of the effective probability table of arithmetic coding in CABAC.
(3) Adaptation of wavelet transform to Intra picture.
(4) Dynamic Vector Quantization.

(5) Super Resolution Plane Prediction.
(6) Fractal Coding

(a) Traffic (Class A)

	BD PSNR [dB]	BD RATE [%]
Traffic	0.42	-12.4
Kimono1	0.89	-29.0

(b) Kimono1 (Class B)   (c) Improvement of PSNR and Bitrate

Fig 58 Test Sequences and Coding performance

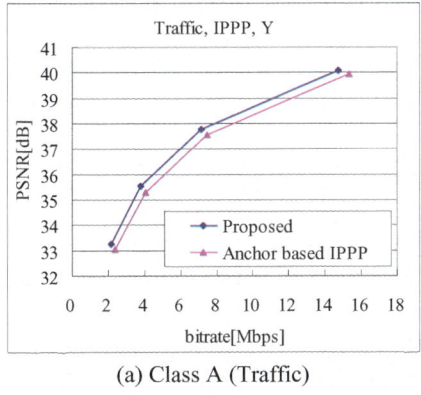

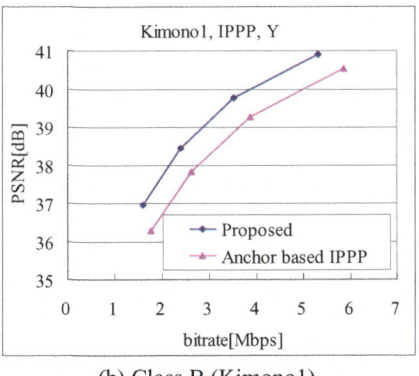

(a) Class A (Traffic)   (b) Class B (Kimono1)

Fig. 59 improvement of coding efficiency

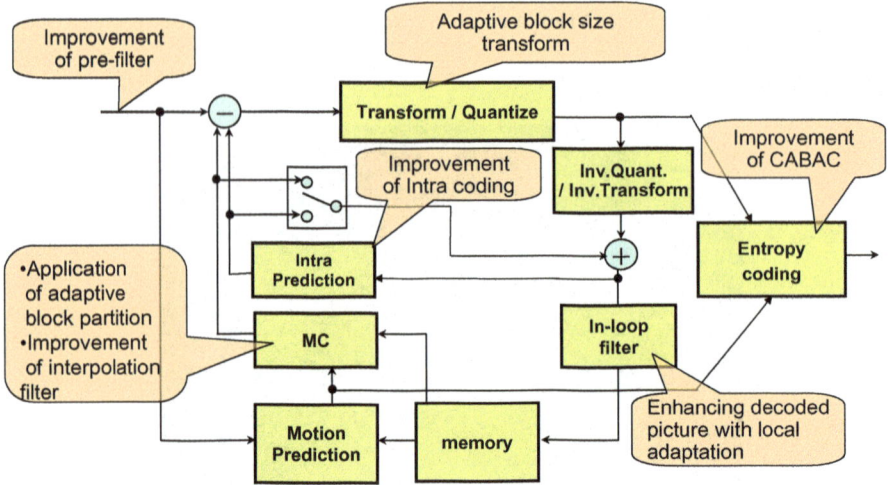

Fig. 60 Trials of coding performance improvement

7 Realization of UHD Service and System

7.1 Realization of UHD Service

UHDTV service will be realized when the next generation video coding technology is accompanied with the progress of the circumference technologies such as advanced sensor, display, storage and transmission infrastructure. UHD video signal compressed by a new video coding standard, which can realize twice as much performance as that of AVC/H.264, will be transmitted and delivered through wired (NGN) or wireless (3.9/4G. wireless LAN, satellite) lines with high speed and ubiquitous. We will be able to enjoy video with high quality and reality by means of large screen TV with over 100 inches diagonal and also to enjoy video with clearness like a gravure picture by means of organic EL display equipped with power saving and flexibility [32].

7.2 Another Possibility of Video Coding Technology in Future

Rather than extensions of the current system, high coding performance and new function for next generation video coding may also be derived from different approaches, e.g., Intelligent Coding as below:

(1) Intelligent Coding advocated by Harashima et al. in the 1980s has recently attracted attention as an alternative path for future video coding.
(2) In this technology, a common model is shared by encoder and decoder, and the model in the decoder is modified synchronously according to the information detected and transmitted by the encoder.

(3) Considering the recent progress of computer, network and image-processing technology, this method could become an interesting candidate and an important video coding technology in the future.

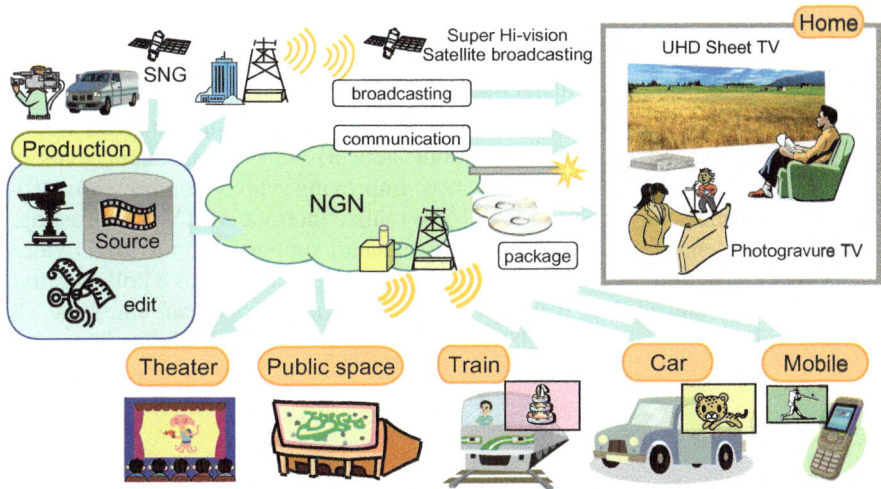

Fig. 61 Next generation video delivery through the Internet and broadcast service

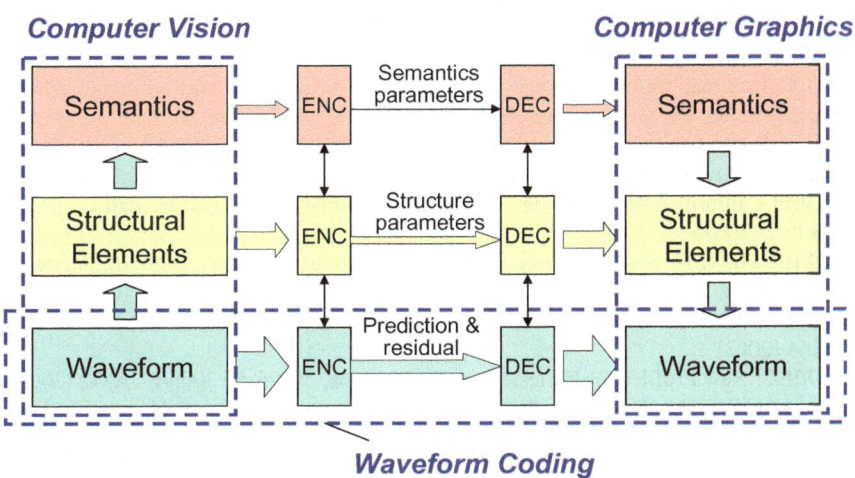

Fig. 62 Integration of graphics and computer vision to video coding [13]

Now, the performance of GPU has been markedly improved and the drawing of 8K picture can be performed on PC in real time. Therefore, a hybrid video coding method is also considered to combine Intelligent Coding and conventional coding to encode, decode and draw pictures using a high speed graphic PC. The future of video technology will spread infinitely if 3-D video technology and Free-viewpoint television technology are combined with Intelligent Coding and conventional coding technologies.

8 Summary

In 1994, MPEG-2 was standardized targeting digital TV services including HDTV by merging MPEG-3 standardization activity, then MPEG-4 standard was developed for video transmission through mobile network. One of the main goals of MPEG-4 AVC/H.264 standardization was improving video compression efficiency by two times compared to MPEG-2 to realize internet HDTV broadcasting.

The development of a new video coding standard supporting up to UHD video such as 4K cinema and 8K Super Hi-Vision is now ready to start as a collaborative work of ISO/IEC and ITU-T. It will not be far off when people can enjoy the 8K resolution video experience through the sheet type organic EL display at home.

References

1. ITU-T: Video Codec for Audiovisual Services at p×64 kbit/s, Recommendation H.261 Version 1 (1990)
2. ISO/IEC: Information technology – Coding of moving pictures and associated audio for digital storage media at up to about 1.5 Mbit/s - Part 2: Video, 11172-2 (1993)
3. ISO/IEC: Generic Coding of Moving Pictures and Associated Audio Information - Part 2: Video, 13818-2 (1994)
4. ITU-T: Information technology - Generic coding of moving pictures and audio information: Video, Recommendation H.262 (1995)
5. ITU-T: Video Coding for Low Bit Rate Communication, Recommendation H.263 version 2 (1998)
6. ISO/IEC: Information technology – Coding of audio-visual objects - Part 2: Visual, 14496-2 (2004)
7. ISO/IEC: Information technology, Coding of audio-visual objects - Part 10: Advanced video coding, 14496-10 (2003)
8. ITU-T: Advanced Video Coding for generic audiovisual services, Recommendation H.264 (2003)
9. ISO/IEC: New Profiles for Professional Applications. 14496-10:2005/FDAM2 (2007)
10. Swartz, C.S. (ed.): Understanding Digital Cinema, A professional Handbook. Focal Press, USA (2005)
11. ISO/IEC: Call for Evidence on High-Performance Video Coding (HVC), JTC1/SC29/WG11 N10553 (2009)
12. Shannon, C.E.: A Mathematical Theory of Communication. Bell System Technical Journal 27, 379–423, 623–656 (1948)
13. Murakami, T.: Future of Video Coding Technology. The MPEG 20th Year Anniversary Commemoration Event, Tokyo (2008),
    http://www.itscj.ipsj.or.jp/forum/forum2008MPEG20.html
14. Oppenheim, A.V., Schafer, R.W.: Digital Signal Processing. Prentice Hall, USA (1975)

15. Shi, Y.Q., Sun, H.: Image and Video Compression for Multimedia Engineering. CRC press, USA (2000)
16. Murakami, T., Itoh, A., Asai, K.: Dynamic Multistage Vector Quantization of Images. Electronics and Communication in Japan 69(3), 93–101 (1986)
17. Gersho, A., Gray, R.: Vector Quantization and Signal Compression. Kluwer Academic Publishers, Dordrecht (1992)
18. Murakami, T.: International Standardization and Future Trend of Video Coding Technology. In: IEEE SSCS Kansai Chapter Technical Seminar (2006)
19. Richardson, I.E.G.: H.264 and MPEG-4 Video Compression. Wiley, England (2003)
20. ITU-R: Parameter values for an expanded hierarchy of LSDI image formats for production and international programme exchange, BT.1769 (2006)
21. Ishida, J., Suzuki, Y.: Study on the design of guardrails at Hoheikyo dam. Hokkaido Civil Engineering Research Institute monthly report Japan 588, 7–18 (2002)
22. Yamada, Y., Sekiguchi, S., Yamagishi, S., Kato, Y.: Standardization Trend of High Quality Video Coding Technology. Mitsubishi Electric technical report 82(12), 7–10 (2008)
23. SMPTE/TC-32NF-30: Proposal for 20 Gb/s Optical Interface (2009)
24. Takahashi, H.: Review of the CMOS Image Sensor Pixel Shrinking Technology. The Journal of the Institute of Image Information and Television Engineers Japan 60(3), 295–298 (2006)
25. Kubota, K.: Toward broadcasting with super-reality. In: URCF symposium, Tokyo, Japan (2009)
26. Numazawa, J.: Status and Future of Image Information Storage. The Journal of the Institute of Image Information and Television Engineers Japan 60(1), 2–5 (2006)
27. SD Association: Bridging Your Digital Lifestyle, SD Association pamphlet (2009)
28. Ishida, O.: 40/100 Gigabit Ethernet Technologies. The Journal of the Institute of Electronics, Information and Communication Engineers Japan 92(9), 782–790 (2009)
29. ISO/IEC JTC 1/SC 29/WG 11/N10361: Vision and Requirements for High-Performance Video Coding (HVC) Codec (2009)
30. ITU-T/SG16: Report of Q6/16 Rapp.Group meeting, Annex Q06.A, VCEG-AL 01 (2009)
31. Sekiguchi, S., Yamagishi, S., Itani, Y., Asai, K., Murakami, T.: On motion block size and partition for 4:4:4 video coding. ITU-T/SG16/Q.6 VCEG-AI31 (2008)
32. Murakami, T., Yoda, F.: Present Status and Future Prospect on Digital Media Technology. Mitsubishi Electric technical report Japan 82(12), 2–6 (2008)

Chapter 5
Compression Formats for HD Recording and Production

Joeri Barbarien, Marc Jacobs, and Adrian Munteanu

Abstract. As a result of the consumer's increasing demand for better quality and higher resolutions, the transit to television broadcasting in HD, with a resolution of at least 1280 x 720 pixels has recently started to take shape. In this respect, an important challenge is the selection of a suitable format for recording and production of HD material. One of the issues is that high definition, unlike standard definition, is quite an ambiguous term. In the first phase of HD television broadcasting deployment, two different formats were put forward: 720p50/59.94/60 or 1080i25/29.97/30. In the first part of this chapter, the benefits and drawbacks of both options will be discussed in detail. Besides the choice between 720p and 1080i, the selection of the video compression format and parameters for HD recording and production is also an important consideration. In this chapter, two state-of-the-art intra-only compression formats will be reviewed and compared: Motion JPEG 2000 and H.264/AVC Intra. First, an in-depth description of both video codecs will be provided. Thereafter, the compression schemes will be evaluated in terms of rate-distortion performance, recompression loss and functionality.

1 Introduction

The quality and viewing experience delivered by standard definition (SD) television have long been considered satisfactory by most consumers. Recently however, the demand for better quality and higher resolutions has dramatically increased. One of the reasons for this is the significant price drop on LCD and plasma televisions capable of displaying high quality, high resolution content. Another is the increasing availability of this type of content. This is the result of the definitive breakthrough of Blu–ray as the format for storage of high-resolution audiovisual material on an optical carrier, at the expense of HD-DVD, and of the

Joeri Barbarien · Marc Jacobs · Adrian Munteanu
Vrije Universiteit Brussel – Interdisciplinary Institute for Broadband Technology (IBBT)
Department of Electronics and Informatics (ETRO)
Pleinlaan 2, B-1050 Brussels, Belgium
e-mail: jbarbari@etro.vub.ac.be, mjacobs@etro.vub.ac.be,
acmuntea@etro.vub.ac.be

success of Sony's PS3 game console which supports Blu-ray playback. As a consequence of the demand for higher quality and higher resolutions, the transit to television broadcasting in high definition (HD), with a resolution of at least 1280 x 720 pixels has started to take shape. This evolution requires fundamental changes to existing file-based television recording, production and distribution processes. Due to the increase in resolution, the bandwidth and storage capacity requirements grow significantly. For the same reason, the computational resources needed for operations such as format conversion and post-processing increase as well.

An entirely different, but equally important challenge is the selection of a suitable format for recording, production and distribution of HD material. While the basic signal properties, such as resolution, frame-rate and scanning format (interlaced or progressive), are fixed for standard definition television, this is not the case for high definition television. A full HD television signal consists of 50, 59.94 or 60 progressive frames per second, each frame having a resolution of 1920 by 1080 pixels; this format is typically denoted as 1080p50/59.94/60. However, the bandwidth and storage requirements associated with full HD production and broadcasting exceed the current capacity of distribution networks and off-the-shelf media production and storage systems. To solve this problem, two different alternatives have been put forward: 720p50/59.94/60 (1280 x 720 pixels per frame, 50, 59.94 or 60 progressive frames per second) or 1080i25/29.97/30 (1920 x 1080 pixels per interlaced frame, 25, 29.97 or 30 interlaced frames per second). In section one of this chapter, the benefits and drawbacks of both formats will be discussed in detail.

For digital production of SD content, relatively simple, intra-only video compression schemes such as DV25 [1, 2] and D-10 [3] were used. Since such standards are exclusively defined for SD material, new compression technology suitable for HD television recording and production must be selected.

For HD acquisition and production, the European Broadcasting Union (EBU) recommends to not subsample the luma component horizontally nor vertically, and to limit chroma subsampling to 4:2:2 [4]. For mainstream production a bit-depth of 8 bits per component is advised and for high-end productions a bit depth of 10 bits per component is suggested [4]. These requirements cannot be met by currently used HD compression formats such as HDCAM and DVCProHD, since both formats apply horizontal sub-sampling of the material prior to compression. Therefore, the transition to more advanced compression techniques is warranted. The final selection of a suitable compression format should take into account several different and sometimes contradictory requirements. First of all, to minimize the impact of the transition to HD on the bandwidth and storage requirements, the selected compression scheme should deliver state-of-the-art compression performance. Secondly, the quality degradation associated with recompression (decoding the compressed material, editing it, and recompressing the modified result), which can occur multiple times in a typical production chain, should be minimized. Moreover, to reduce the number of recompression cycles, the recording format and the production format should ideally be one and the same. Additionally, for optimal edit-friendliness and frame-by-frame random access capability, so-called long-GOP formats, which make use of temporal motion-compensated prediction, should be avoided. Due to the dependency between frames in such formats, a significant delay can occur when the material is randomly accessed. Indeed, decoding an arbitrary frame may require the additional decoding of several other frames which are directly or indirectly needed to complete the decoding process. In other words, an intra-only compression format should be favoured for edit-friendly HD recording and production. Finally, for optimal interoperability, the use of open, international standards is warranted.

In this chapter, two state-of-the-art intra-only compression formats for HD recording and production will be reviewed and compared: Motion JPEG 2000 (ISO/IEC 15444-3:2007/ ITU-T Recommendation T.802: Information technology -- JPEG 2000 image coding system: Motion JPEG 2000,) and H.264/AVC Intra (ISO/IEC 14496-10:2009 Information technology -- Coding of audio-visual objects -- Part 10: Advanced Video Coding / ITU-T Recommendation H.264: Advanced video coding for generic audiovisual services). Section 3 provides a detailed overview of the H.264/AVC standard with an emphasis on Intra-only operation. Similarly, section 4 discusses the Motion JPEG 2000 standard. In section 5 both compression schemes will be evaluated in terms of rate-distortion performance, recompression loss and functionality. Finally, in section 6 the conclusions of this work are presented.

2   720p vs. 1080i

Since full HD production and emission is not yet economically viable, two different alternatives are currently being put forward: 720p50/59.94/60 (1280 x 720 pixels per frame, 50, 59.94 or 60 progressive frames per second) or 1080i25/29.97/30 (1920 x 1080 pixels per interlaced frame, 25, 29.97 or 30 interlaced frames per second). The benefits and drawbacks associated with each alternative largely stem from the different characteristics of interlaced and progressive video transmission.

To avoid flickering and to obtain visually smooth motion, a minimum number of frames per second needs to be displayed. This minimum frame-rate depends on the ambient lighting and on the average luminance of the frames [5]. For television, the commonly accepted minimum lies between 50 and 60 frames per second. In the early days of analogue television, the bandwidth needed for progressive scan transmission, which corresponds to sending 50 or 60 complete frames per second, was deemed to be too high. To solve this problem, interlacing was introduced. Interlacing implies recording, transmitting and presenting only half of the lines for each frame, thereby effectively halving the pixel count. For each pair of consecutive frames, the even-numbered lines of the first frame and the odd-numbered lines of the second frame are retained. The retained lines of each frame form a so-called field. Two consecutive fields are grouped together into one interlaced frame. The resulting interlaced video material consists of 50 to 60 fields per second or 25 to 30 interlaced frames per second. Because of the inherent inertia of the human visual system, the missing lines in a displayed interlaced image are not visible because they coincide with the fading lines of the previously projected image, thus retaining the spatial resolution of the original images. On a CRT screen, the after glowing of the layer of phosphorus increases the effect of the inertia. This advantageous technological side effect makes interlaced video content look better on a classic CRT screen than on a modern LCD or plasma screen. The interlacing technique results in a gross bandwidth reduction of 50%. However, this comes at the expense of visual artefacts associated with interlacing, including interline twitter, line crawling and field aliasing [5].

The higher spatial resolution of 1080i seems to be an advantage but this is relative. First of all, vertical low-pass filtering is typically applied on interlaced video in order to suppress interline twitter [5], effectively reducing the vertical resolution in 1080i. Secondly, it is known that the optical resolution of a human eye limits the ability to distinguish details to an angle of approximately 1 arc minute (1/60th of a degree). Within this angle, details cannot be distinguished. It is typically assumed that in an average living room, the viewing distance for a television set is about 2.7 m [6]. Simple geometry then shows that a screen must be about 85 cm high to allow the distinction of two neighbouring lines in an image of 1920 by 1080 pixels. This corresponds to a screen with a diagonal of 1.75 m or 68". This type of screens are currently too cumbersome, heavy and expensive for the average consumer. In contrast, a vertical resolution of 720 lines for 720p corresponds to a screen with a diagonal of 50", which is more realistic.

Another advantage of 720p is that it offers a better line refresh rate (50-60 frames per second times 720 lines versus 50-60 fields per second times 540 lines for 1080i). This results in a better viewing experience for fast moving content [7]. Motion blur and jitter are less observed with 720p than with 1080i. The fact that LCD and plasma screens are inherently progressive further contributes to the appeal of 720p. When offered interlaced video content, such screens have to convert it to progressive content before displaying. The quality of this conversion depends on the complexity and the processing power, and thus the price of the built-in conversion chips. Professional solutions cost about 10.000 Euro per system. Evidently screen manufacturers are forced to use much cheaper, less efficient devices. For these reasons, existing 1080i video content should be converted to 720p by professional equipment at the broadcast stations prior to distribution, instead of relying on less efficient de-interlacing mechanisms built into current television sets.

Another advantageous factor for 720p is the content representation efficiency. 1080i25 content consists of 1920 x 1080 x 25 = 5184106 pixels per second while 720p50 content consists of 1280 x 720 x 50 = 4608106 pixels per second, corresponding to a difference of more than 10% in favour of 720p. Moreover, subjective tests by EBU indicate that compressed 1080i25 content needs 20% more bit-rate than compressed 720p50 to obtain a comparable visual quality [3].

The above arguments indicate that 720p is the best HD format for recording, production and distribution. However, 1080i has some practical advantages over 720p. A first advantage is the better support for 1080i in current production systems and in (tape-based) cameras, mostly because the format has already been in use for some time in the United States and Japan. For the same reason there is a larger availability of 1080i content. For example, HD recordings of the last Olympic Games were distributed in the 1080i format. A second advantage is the ability to use identical time codes in 1080i as in SD. Time codes are essential for determining the relative place of a frame in a sequence, for editing purposes and for synchronizing video and audio. Usage of identical time codes implies that current production systems can be easily adapted from SD to HD.

3 H.264/AVC

H.264/AVC is a highly efficient video compression standard which was created jointly by ITU-T and ISO/IEC. The initial H.264/AVC standard was finalized in 2003 and focused on video material with 4:2:0 chroma subsampling and a bit-depth of 8 bits [8]. In a later stage, extensions to support higher resolutions and bit-depths and different chroma subsampling options were added under the name FRExt (Fidelity Range Extensions). These extensions include an 8x8 DCT transform and intra prediction, support for adaptive transform sizes, custom quantization matrices, lossless coding, support for 4:2:2 and 4:4:4 chroma subsampling and support for higher bit-depths [9].

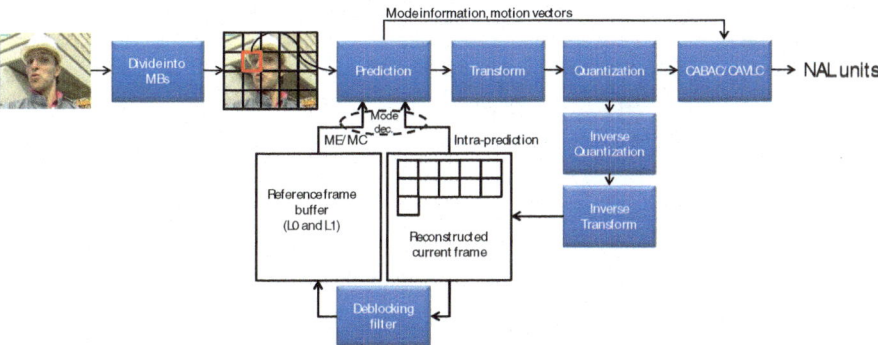

Fig. 1 H.264/AVC encoder architecture

Figure 1 illustrates H.264/AVC's classical block-based hybrid coding architecture. Each picture is divided into one or more slices, each consisting of a number of macroblocks. A macroblock (MB) covers a rectangular picture area of 16x16 luma samples and, corresponding to this, two rectangular areas of chroma samples (when 4:2:0 chroma subsampling is used, two blocks of 8x8 chroma samples). Each MB is either spatially (intra) or temporally (inter) predicted and the prediction residual is transform coded. H.264/AVC supports three basic slice coding types: I slices, P slices and B slices. H.264/AVC Intra, which is used in our experiments, only allows the use of I slices. I slices are coded without any reference to previously coded pictures. Macroblocks in I slices are always spatially predicted using directional intra-prediction. Intra-prediction can be performed on blocks of 4x4, 8x8 and 16x16 samples. Each block is predicted based on neighbouring sample values as illustrated in Figure 2 for 4x4 and 16x16 blocks. The directional 8x8 intra prediction modes introduced in FRExt are similar in design to their 4x4 equivalents [8, 9]. Macroblocks in P and B slices can additionally be temporally predicted using variable block size motion-compensated prediction with multiple reference frames. The macroblock type signals the partitioning of a macroblock into blocks of 16x16, 16x8, 8x16, or 8x8 luma samples. When a macroblock type specifies partitioning into four 8x8 blocks, each of these so-called sub-macroblocks can be further split into 8x4, 4x8, or 4x4 blocks, which is indicated through the sub-macroblock type. For P-slices, one motion vector is transmitted for each block. In addition, the employed reference picture can be independently chosen for each 16x16, 16x8, or 8x16 macroblock partition or 8x8 submacroblock. It is signalled via a reference index parameter, which is an index into a list of reference pictures that is replicated at the decoder [8, 10].

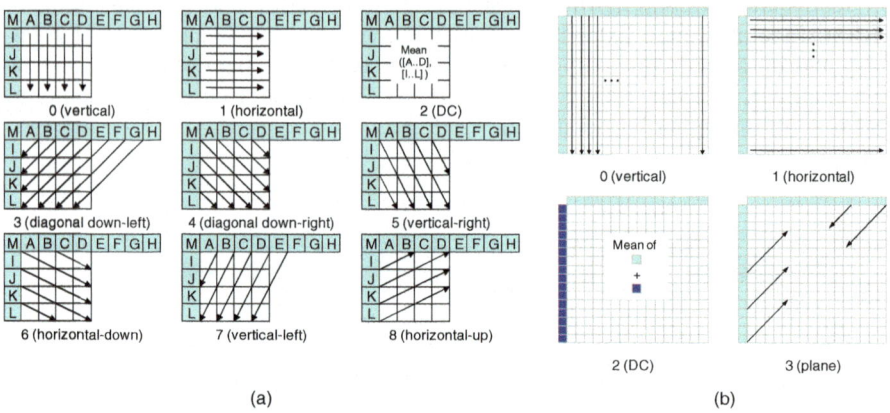

(a)   (b)

Fig. 2 Directional intra-prediction modes. (a) 4x4 prediction modes. (b) 16x16 prediction modes. [10].

In B-slices, two distinct reference picture lists are utilized, and for each 16x16, 16x8, 8x16 macroblock partition or 8x8 sub-macroblock, the prediction method can be selected between list 0, list 1, or bi-prediction. While list 0 and list 1 prediction refer to unidirectional prediction using a reference picture of reference picture list 0 or 1, respectively, in the bi-predictive mode, the prediction signal is formed by a weighted sum of a list 0 and list 1 prediction signal. In addition, special modes such as direct modes in B-slices and skip modes in P- and B-slices are provided, in which motion vectors and reference indexes are derived from previously transmitted information [8]. To reduce the blocking artefacts introduced by the standard's block-based transform and prediction operations, an adaptive deblocking filter is applied in the motion-compensated prediction loop [11]. This filter is only effectively turned on for higher values of the quantization parameter QP [11].

After spatial or temporal prediction, the resulting prediction error is transformed using an approximation of the discrete cosine transform (DCT), in some cases followed by an Hadamard transform on the DC DCT coefficients (Intra 16x16 prediction mode, chroma coefficient blocks). In the initial version of the standard, a 4x4 DCT kernel was used exclusively. Later, during the standardization of FRExt, an 8x8 transform kernel was added, which typically provides superior performance in high resolution scenarios [9]. Additionally, the possibility of adaptively selecting the transform kernel (4x4 or 8x8) on a macroblock basis was also added. The employed transforms consists of two matrix multiplications, which can be executed using integer arithmetic, and a scaling operation which in principle requires floating point operations. For example, the 4x4 transform is defined as:

$$Y = \left(CXC^t \right) \otimes E = \left(\begin{pmatrix} 1 & 1 & 1 & 1 \\ 1 & d & -d & 1 \\ 1 & -1 & -1 & 1 \\ d & -1 & 1 & -d \end{pmatrix} \begin{pmatrix} x_1 & x_2 & x_3 & x_4 \\ x_5 & x_6 & x_7 & x_8 \\ x_9 & x_{10} & x_{11} & x_{12} \\ x_{13} & x_{14} & x_{15} & x_{16} \end{pmatrix} \begin{pmatrix} 1 & 1 & 1 & d \\ 1 & d & -1 & -1 \\ 1 & -d & -1 & 1 \\ 1 & -1 & 1 & -d \end{pmatrix} \right) \otimes \begin{pmatrix} a^2 & ab & a^2 & ab \\ ab & b^2 & ab & b^2 \\ a^2 & ab & a^2 & ab \\ ab & b^2 & ab & b^2 \end{pmatrix}$$

$$a = \frac{1}{2}, b = \sqrt{\frac{2}{5}}, d = \frac{1}{2}$$

However, by absorbing this scaling into the quantization process, floating point arithmetic is avoided and the transform process becomes an integer-only operation. In this way, an efficient implementation is obtained and the drift problem that occurred in previous video coding standards, as a result of differences in accuracy and rounding of the floating-point DCT in the encoder and decoder, is avoided. The standard enforces a link between the prediction mode and the size of the transform kernel used: If prediction on 4x4 blocks is used, the 8x8 transform kernel cannot be employed as it would cross the boundary of the 4x4 blocks used in the prediction, causing high frequency transform coefficients to appear which are expensive to code. For more information concerning the transform part of the standard the reader is referred to [12].

In the next stage, the transform coefficients are quantized. In the initial version of the standard, only uniform scalar quantization was supported. FRExt later introduced support for frequency dependent quantization and rounding, by means of custom quantization and rounding matrices. The quantization strength is determined by the quantization step size which can be defined for each macroblock using the quantization parameter QP which lies in the range [0,51]. The relation between QP and the quantization step size is logarithmic: the quantization step size approximately doubles for each increase of QP by 6. As mentioned earlier, the quantization and the scaling part of the DCT are combined in a single integer-valued operation.

The symbols produced by the encoding process are entropy coded using either context-based adaptive variable length coding (CAVLC) or context-based adaptive binary arithmetic coding (CABAC [13]). While CABAC exhibits a higher computational complexity, it also provides 10 to 15% bit-rate savings compared to CAVLC [13, 14].

Similar to prior video coding standards, H.264/AVC also supports efficient coding of interlaced material. Each interlaced frame can be coded in two ways. The subsequent fields can be coded separately (field coding) or the interlaced frame, i.e. the collection of two successive fields can be coded in the same way as a progressive frame (frame coding). In H.264/AVC, this decision can be made adaptively, on a slice basis (PAFF, picture adaptive frame/field coding). If field coding is used, each field is coded in a manner very similar to a progressive frame, with a few exceptions: motion compensated prediction uses reference fields instead of reference frames, the zig-zag scan of transform coefficients is altered and the deblocking strength is limited when filtering horizontal edges [14]. If frame coding is selected, the entire frame is coded in the same way as a progressive frame or the decision between frame and field coding can be deferred to the macroblock level. The latter is called macroblock adaptive frame/field coding (MBAFF) . In this case, the picture is processed in pairs of macroblocks, i.e. two vertically adjacent macroblocks covering a picture region of 16x32 luma samples are coded together. Each such pair of macroblocks is then either coded as two frame macroblocks or two field macroblocks as illustrated in Figure 3. Since, in this way, each frame will be coded as a mixture of frame and field macroblock pairs, the derivation of the spatially neighbouring pixel values, motion vectors, etc. is significantly complicated. For an in-depth discussion, the reader is referred to [8, 14].

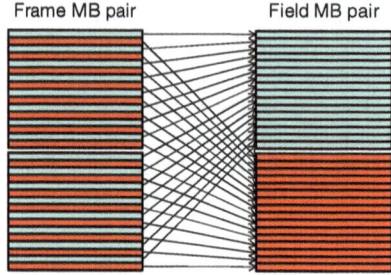

Fig. 3 Frame vs. field macroblock pair in MBAFF [14]

4 Motion JPEG 2000

Motion JPEG 2000 [15] or ITU-T Recommendation T.802 [16] is an intra-only video coding standard which has recently been adopted as the standard format for digital cinema productions. Motion JPEG 2000 is a part of the JPEG 2000 suite of standards, jointly published by ISO/IEC JTC-1 and ITU-T. Essentially, Motion JPEG 2000 applies JPEG 2000 Part 1 [17] still-image compression to each frame in the video sequence and the resulting compressed bit-stream is wrapped in a file format derived from the ISO base media file format [18], which represents timing, structure, and media information for timed sequences of media data. As illustrated in Figure 4, JPEG 2000 Part 1 encoding is performed in two tiers. In Tier 1, the image is first split into rectangular regions called tiles. Typically, when the resolution of the image is relatively low, a single tile is used for the entire picture. Each tile is subsequently transformed to the wavelet-domain using the 2D discrete wavelet transform as shown in Figure 5. To obtain the first decomposition level of the wavelet transform, the image is first horizontally decomposed by performing low-pass and high-pass filtering (filters h and g in Figure 5 respectively) succeeded by dyadic downsampling on the rows of the original image. The resulting image is thereafter vertically decomposed by applying similar operations on its columns. The entire operation yields a set of four subbands, each having half of the resolution of the original image. The subband resulting from low-pass filtering on both rows and columns is called the LL-subband and represents a lower resolution approximation of the original image. The remaining subbands are referred to as the LH, HL and HH subbands (see Figure 5) and respectively contain the high-frequency horizontal, vertical and diagonal details representing the information difference between the original image and the LL-subband. Additional decomposition levels are computed by performing the aforementioned operations on the LL-subband of the previous decomposition level.

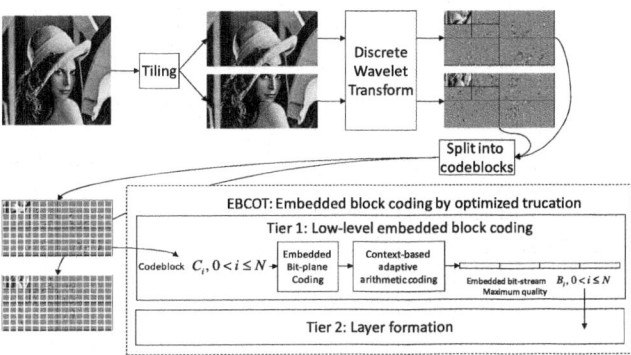

Fig. 4 Overview of a JPEG 2000 Part 1 encoder

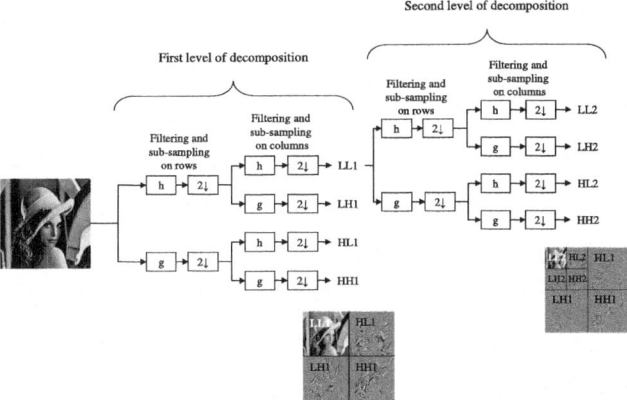

Fig. 5 Two-dimensional wavelet-transform with 2 decomposition levels

Unlike the above description, which defines the wavelet transform as a sequence of filtering and subsampling operations, the JPEG 2000 standard specifies its supported wavelet transforms in terms of their lifting implementation [19]. A lifting implementation of the wavelet transform allows in-place execution, requires less computations and allows for reversible operation, which is needed for lossless compression. Two different transforms are supported, i.e. the irreversible CDF 9/7 transform and the reversible CDF 5/3 transform [20, 21]. Lossless compression requires the use of the reversible CDF 5/3 transform. For lossy compression, the CDF 9/7 transform is typically preferred.

After the transform, the resulting set of coefficients is divided into code-blocks C_i which are typically chosen to be 32 by 32 or 64 by 64 coefficients in size. Each of these code-blocks is thereafter independently coded using embedded bit-plane coding. This means that the wavelet coefficients $c(i), 0 \leq i < N$ are coded in a bit-plane by bit-plane manner by successive comparison to a series of dyadically decreasing thresholds of the form $T_p = 2^p$. A wavelet coefficient $c(j)$ is called significant with respect to a threshold T_p if its absolute value is larger than or equal to this threshold.

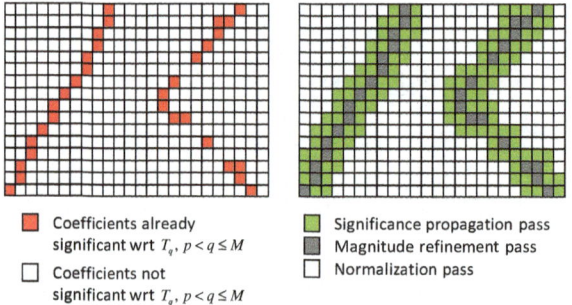

☐ (red) Coefficients already
significant wrt $T_q, p < q \leq M$

☐ Coefficients not
significant wrt $T_q, p < q \leq M$

☐ (green) Significance propagation pass
☐ (gray) Magnitude refinement pass
☐ Normalization pass

Fig. 6 Application of the different coding passes in JPEG 2000

Encoding starts with the most significant bit-plane M and its corresponding threshold $T_M = 2^M$, with $M = \left\lfloor \log_2 \left(\max_{0 \leq i < N} \left(|c(i)| \right) \right) \right\rfloor$. Bit-plane p of the wavelet coefficients in each block is encoded using a succession of three coding passes. First, the significance propagation pass codes the significance of the coefficients which (*i*) were not significant with respect to any of the previous thresholds $T_q, p < q \leq M$ and (*ii*) are adjacent to at least one other coefficient which was already significant with respect to one of the previous thresholds $T_q, p < q \leq M$. Additionally, when a coefficient becomes significant with respect to T_p, its sign is also coded. In the magnitude refinement pass, all coefficients that were significant with respect to one of the previous thresholds $T_q, p < q \leq M$ are refined by encoding the binary value on their p-th bit-plane. Finally, in the normalization pass, the significance of the remaining previously non-significant coefficients is coded. Again, when a coefficient becomes significant with respect to T_p, its sign is also coded. The application of the coding passes is illustrated in Figure 6.

The symbols signaling the significance, the signs and the refinements bits for different coefficients are coded using a combination of run-length coding and context-based binary adaptive arithmetic coding. For further details, the reader is referred to [20].

Since the wavelet coefficients are coded in a bit-plane by bit-plane manner, a quantized version of the original wavelet coefficients in each block C_i can be obtained by simply truncating the resulting compressed bit-stream B_i. This property is exploited in Tier 2 of the codec. In Tier 2, the code-block bit-streams B_i are cut-off at a length l_i, whereby l_i is determined such that the requested bit-rate, resolution and region of interest are met and the decoder-side distortion is minimized. This is achieved using Lagrangian rate-distortion optimization [20, 21].

5 Experiments

In this section, the H.264/AVC Intra and Motion JPEG 2000 coding standards are experimentally evaluated. A first series of experiments reports the compression performance for the two coding systems applied on 1080i25 and 720p50 video material. The employed test material consists of 5 video sequences from the SVT test set [22] in 1080i25 and 720p50 formats: "CrowdRun", "ParkJoy", "DucksTakeOff", "IntoTree" and "OldTownCross". All sequences were converted from the original SGI format to planar $Y'C_bC_r$ 4:2:2 video with 10 bits per component using the commonly employed conversion tools of [23].

For H.264/AVC Intra the JM reference software version 12.2 was used [24]. Although this is not the latest version available, this option was taken due to erroneous operation of more recent versions on 4:2:2, 10 bit video material. For Motion JPEG 2000 the Kakadu software version 6.1 [25] was used. The employed parameter configurations for both codecs are summarized in Table 1 and Table 2 for 720p50 and 1080i25 material respectively.

With respect to Motion JPEG 2000, it must be observed that the standard was not particularly designed to target interlaced material. In the codec's normal configuration, all interlaced frames are coded in the same manner as a progressive frame, which corresponds to the frame coding mode in H.264/AVC. However, field coding can easily be supported by separating the fields of each interlaced frame and coding each field as a separate tile (this corresponds to H.264/AVC's field coding mode).

The results obtained with Motion JPEG 2000 on both frame and field coding modes are reported. The compression performance of the two coding systems is evaluated by measuring the Peak Signal to Noise Ratio (PSNR) at the following bit-rates: 30, 50, 75, 100 Mbit/s. The average PSNR per frame is calculated as

Table 1 Encoder settings for 720p50 material

Motion JPEG 2000		H.264/AVC Intra	
Transform	CDF 9/7	Intra prediction	4x4,8x8,16x16
Transform Levels	6	Transform	Adaptive 4x4/8x8
Codeblock size	64x64	Quantization	Uniform, standard rounding
Tiles	1 Tile per frame	Slices	1 slice per frame
Tier 2 Rate control	Constant slope/lambda per frame, Logarithmic search on Lambda to meet rate.	Rate control	On, RCUpdateMode=1
Perceptual weighting	Off	R-D optimization	High complexity mode
-	-	Entropy coding	CABAC
-	-	Deblocking	On
-	-	Profile/Level	High 4:2:2, Level 5.1

Table 2 Encoder settings for 1080i25 material

Motion JPEG 2000	Frame coding	Field coding	H.264/AVC Intra	
Transform	CDF 9/7	CDF 9/7	Interlaced handling	Frame coding, MBAFF
Transform Levels	7	6	Intra prediction	4x4,8x8,16x16
Codeblock size	64x64	64x64	Transform	Adaptive 4x4/8x8
Tiles	1 Tile per frame	2 Tiles per frame (1 per field)	Quantization	Uniform, standard rounding
Tier 2 Rate control	See Table 1	See Table 1	Slices	1 slice per frame
Tier 2 perceptual weighting	Off	Off	Rate control	On, RCUpdateMode=1
-	-	-	R-D optimization	High complexity mode
-	-	-	Entropy coding	CABAC
-	-	-	Deblocking	On
-	-	-	Profile/Level	High 4:2:2, L 4.1

$\overline{PSNR} = 2\,PSNR_{Y'} + PSNR_{C_b} + PSNR_{C_r}$, as we are compressing 4:2:2 video material. The PSNR results averaged over each sequence are summarized in Figure 7
and Figure 8.

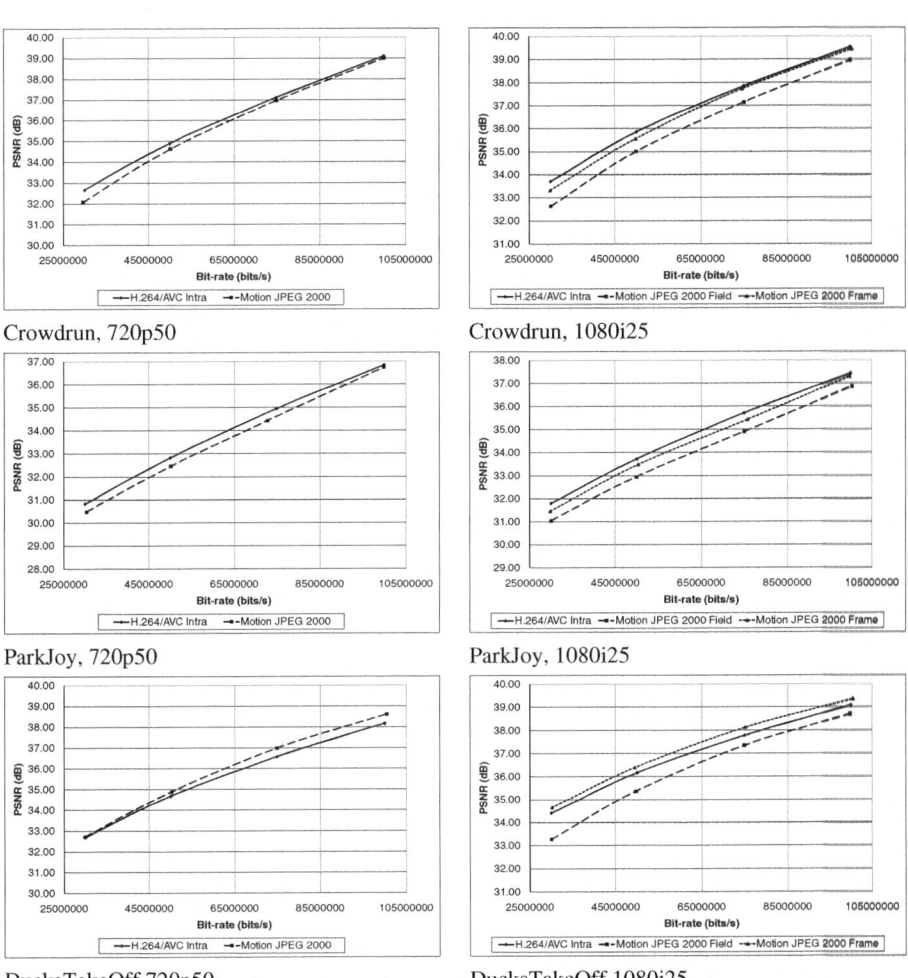

Crowdrun, 720p50   Crowdrun, 1080i25

ParkJoy, 720p50   ParkJoy, 1080i25

DucksTakeOff 720p50   DucksTakeOff 1080i25

Fig. 7 Comparison between the compression performance of H.264/AVC and Motion JPEG
2000 for 720p50 and 1080i25 material (CrowdRun, ParkJoy and DuckTakeOff sequences).

These results show that for progressive material, the compression performance of
H.264/AVC Intra is better than that of Motion JPEG 2000 for CrowdRun, ParkJoy
and OldTownCross. The performance of both coding systems is comparable for
OldTownCross, while Motion JPEG 2000 outperforms H.264/AVC Intra for the

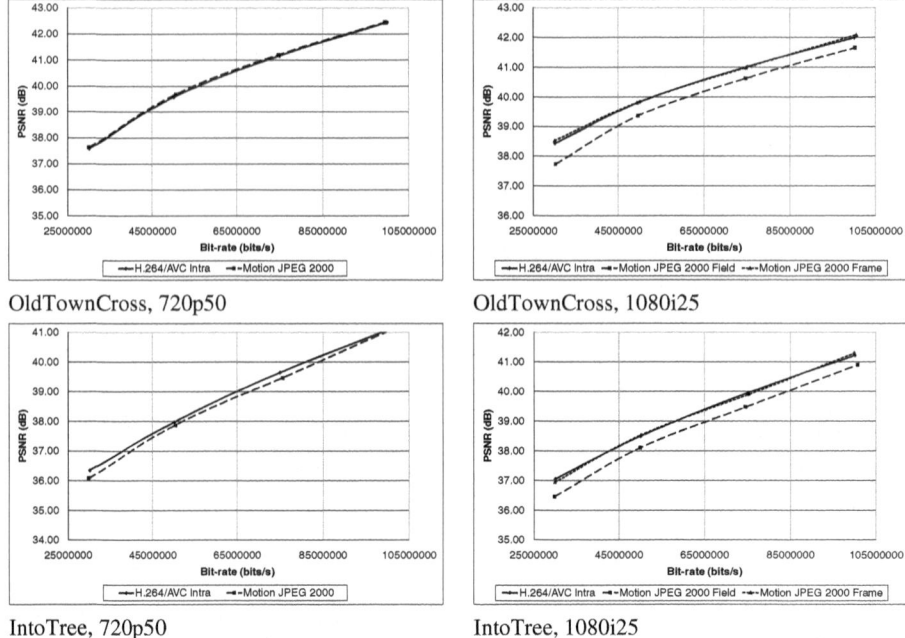

OldTownCross, 720p50   OldTownCross, 1080i25

IntoTree, 720p50   IntoTree, 1080i25

Fig. 8 Comparison between the compression performance of H.264/AVC and Motion JPEG 2000 for 720p50 and 1080i25 material (OldTownCross and IntoTree sequences)

DucksTakeOff sequence. We can conclude that, on average, H.264/AVC Intra offers the best coding performance for 720p50 material.

When looking at the results for 1080i25 material, it becomes clear that the frame coding approach for Motion JPEG 2000 is superior to the field coding approach. This is to be expected since the test material exhibits a relatively high correlation between successive fields, which can only be exploited in the frame coding mode. When comparing Motion JPEG 2000 frame coding with H.264/AVC Intra coding, the performance of H.264/AVC Intra is better for the CrowdRun and ParkJoy sequences. The compression performance of both codecs is similar for IntoTree and OldTownCross. For the DucksTakeOff sequence, Motion JPEG 2000 outperforms H.264/AVC Intra. On average, the same conclusion as for progressive material can be drawn: H.264/AVC Intra offers better coding performance than Motion JPEG 2000.

In a second series of experiments, the quality degradation (or re-compression loss) resulting from successive encoding and decoding of the same video material is assessed. In these experiments four encoding-decoding iterations are performed. No spatial shifting of the reconstructed signal is applied in between successive iterations. The parameter settings from the first experiments are reused. The average PSNR results after the 2nd and the 4th encoding-decoding cycle are reported in Figure 9 - Figure 13.

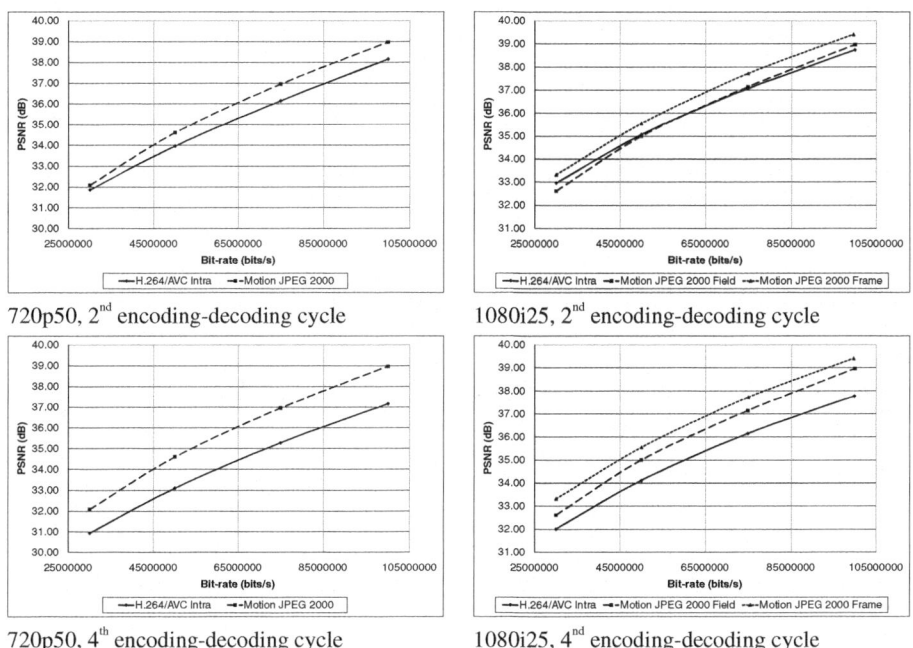

720p50, 2nd encoding-decoding cycle   1080i25, 2nd encoding-decoding cycle

720p50, 4th encoding-decoding cycle   1080i25, 4nd encoding-decoding cycle

Fig. 9 Recompression loss for CrowdRun sequence

These results show that using H.264/AVC Intra, significant quality loss is incurred during successive encoding-decoding cycles, while the quality obtained using Motion JPEG 2000 stays more or less constant. This is likely the result of H.264/AVCs macroblock-based coding using multi-hypothesis intra-prediction, adaptive transform selection and adaptive frame/field coding (interlaced material). The mode decision process, which operates based on actual distortion measurements, will likely make different decisions concerning the intra-prediction mode, transform or frame/field coding of the current macroblock since the reference used in the distortion measurement, i.e. the input frame, is a lower quality version of the corresponding input frame in the first encoding cycle. When the same coefficient is quantized twice with the same quantization step-size, no additional distortion is introduced the second time. However, when different coding options are taken for a macroblock, compared to those used in the previous encoding-decoding cycle, entirely different transform coefficients result, and additional noise is introduced in the quantization stage. Additionally, errors introduced in one macroblock propagate to neighbouring macroblocks through the intra-prediction process, contributing to the overall quality degradation.

The marginal loss incurred during recompression with Motion JPEG 2000 is largely due to rounding errors resulting from the irreversible CDF 9/7 wavelet transform. The quantization of the wavelet coefficients which is applied in each encoding step depends on the estimation of the distortion reduction after each coding pass. This distortion reduction estimate is used to determine the codeblock bitstream truncation in Tier 2 of the codec, which directly translates to the quantization step size. The rounding errors of the transform have a very limited influence on the distortion estimates, especially for the coding passes performed on the higher bit-planes. As a result, quantization of the wavelet coefficients in successive encoding cycles will be almost identical, causing little additional quality loss.

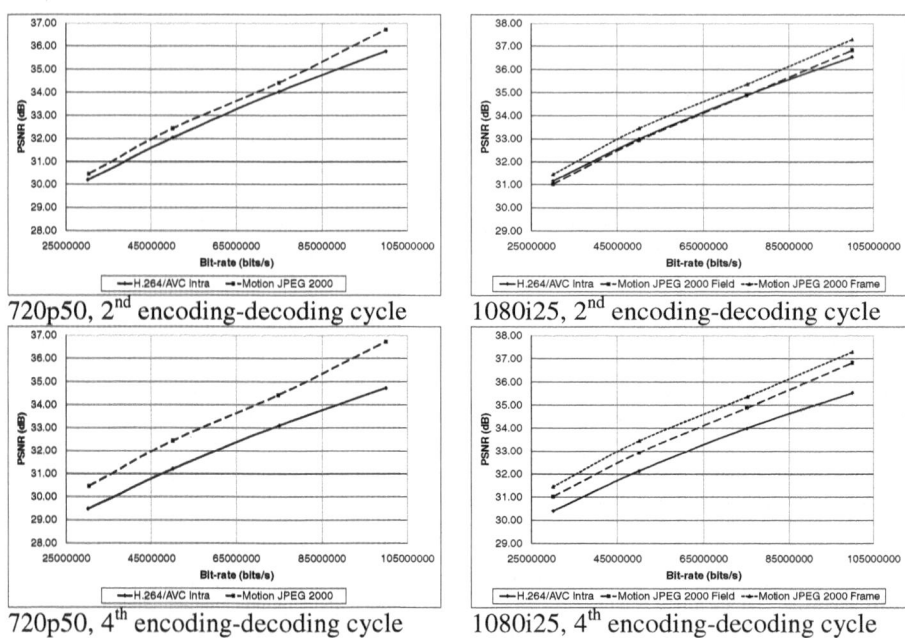

720p50, 2nd encoding-decoding cycle   1080i25, 2nd encoding-decoding cycle

720p50, 4th encoding-decoding cycle   1080i25, 4th encoding-decoding cycle

Fig. 10 Recompression loss for ParkJoy sequence

To conclude this section, a short qualitative evaluation of the functionality and the complexity of both coding standards will be presented. Motion JPEG 2000 inherently supports resolution and quality scalability. To obtain similar functionality, H.264/AVC's scalable extension [8, 26] must be used, which requires a different encoder and decoder. Additionally, Motion JPEG 2000 supports region-of-interest (ROI) coding with near-pixel granularity by using the max-shift method [20, 21]. ROI can also be supported in AVC by using slice groups, but the granularity is limited to a single macroblock. (Motion) JPEG 2000 was also designed to deliver state-of-the-art lossless compression performance. H.264/AVC can also support lossless compression but the performance of this mode is sub-optimal, as it solely relies on crude block-based intra-prediction for decorrelation.

Concerning computational complexity, it is difficult to draw clear conclusions as this is highly implementation and platform dependant. Certain is that Motion JPEG 2000 has similar complexity for encoding and decoding, while for H.264/AVC, the encoder is much more complex than the decoder. In general,

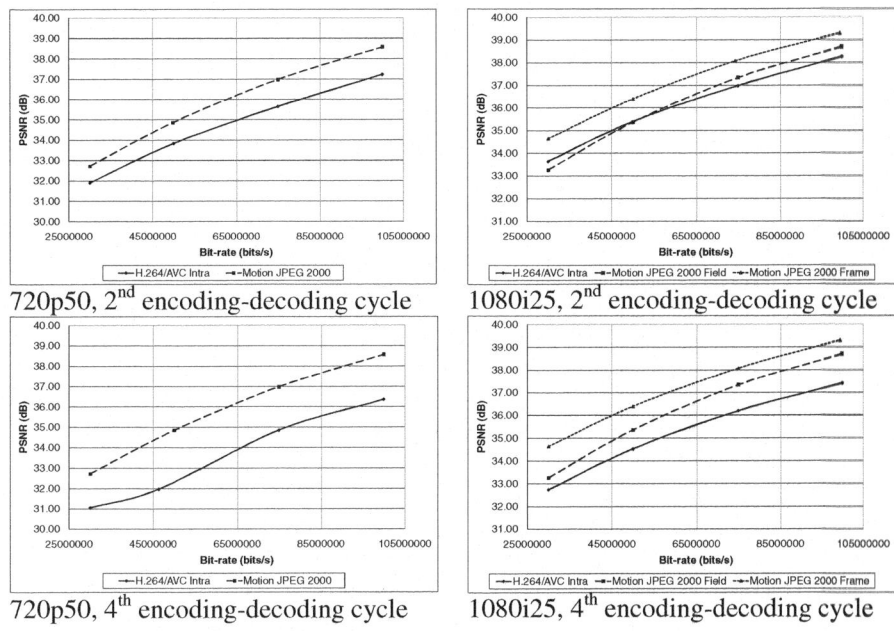

Fig. 11 Recompression loss for DucksTakeOff sequence

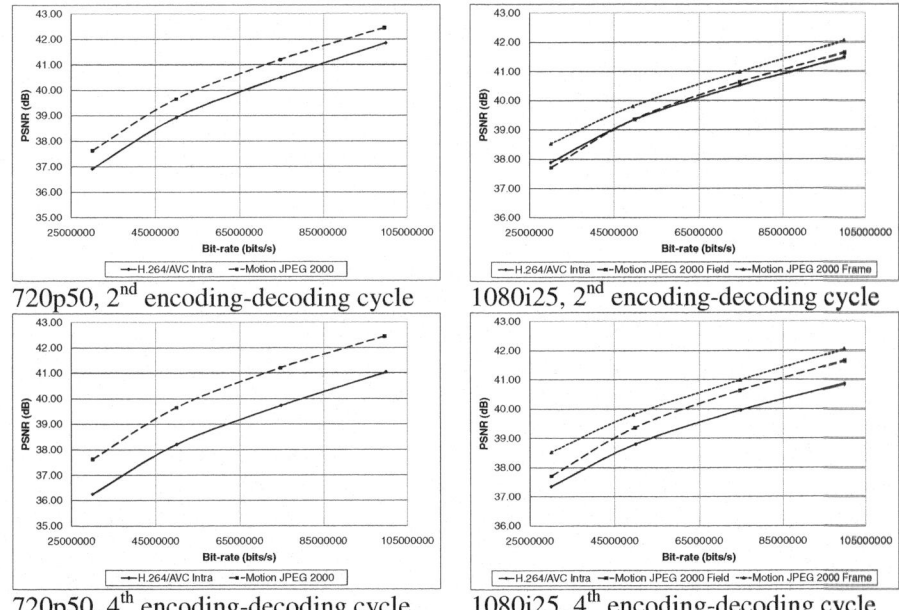

Fig. 12 Recompression loss for OldTownCross sequence

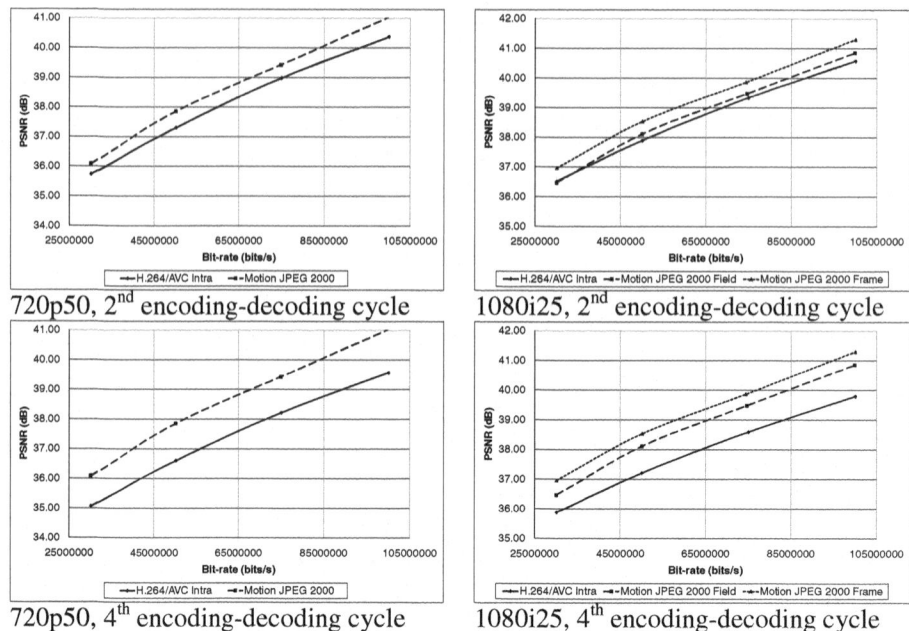

Fig. 13 Recompression loss for IntoTree sequence

H.264/AVC's macroblock-oriented processing is considered to be very beneficial as it allows to preserve locality of reference, which can lead to improved cache usage. Motion JPEG 2000 on the other hand uses a global transform, which makes block-based implementation more difficult. However, solutions for this problem have been proposed in the literature. An overview is given in [21].

6 Conclusions

In this work, we have investigated formats for the digital representation of HD video material. Concerning the choice between 720p and 1080i HD formats, we have shown that, from a technical point of view, 720p is clearly superior to 1080i. However, in practice, other factors, such as the larger availability of material in 1080i format and the better support for 1080i in current production equipment, may lead to the adoption of the latter format. Besides the choice between 720p and 1080i, the selection of the video compression format for HD recording and production was discussed. We have evaluated two state-of-the-art intra-only compression formats, Motion JPEG 2000 and H.264/AVC Intra. In terms of compression performance, H.264/AVC Intra outperforms Motion JPEG 2000 when recompression is not considered. However, Motion JPEG 2000 clearly outperforms H.264/AVC Intra when applying successive encoding-decoding cycles, typically performed in television production. The results have also shown that frame coding clearly offers superior performance over field coding when using Motion JPEG 2000. In terms of functionality, Motion JPEG 2000 has the advantage of inherently supporting resolution and quality scalability, efficient lossless compression and fine-granular region-of-interest coding. H.264/AVC only supports resolution and quality scalability when using the SVC extension; also, region-of-interest coding can be supported through the use of slice groups, but with a limited granularity. H.264/AVC also supports lossless coding, but the compression performance of this mode is limited. From a complexity point of view, little conclusions can be formulated as this is highly platform and implementation dependant.

References

1. Television Digital Recording – 6.35-mm Type D-7 Component Format – Video Compression at 25 Mb/s – 525/60 and 625/50. SMPTE, 306M-2002 (2002)
2. Television – Data Structure for DV-Based Audio, Data and Compressed Video – 25 and 50 Mb/s. SMPTE, 314M-2005 (2005)
3. Type D-10 Stream Specifications — MPEG-2 4:2:2P @ ML for 525/60 and 625/50. SMPTE, 356M-2001 (2001)
4. Choice of HDTV Compression Algorithm and Bitrate for Acquisition, Production & Distribution. EBU, EBU – Recommendation R 124 (2008)
5. Poynton, C.A.: Digital video and HDTV: Algorithms and Interfaces. Morgan Kaufmann/Elsevier Science, San Fransisco (2007)
6. Tanton, N.E.: Results of a survey on television viewing distance. BBC R&D, White Paper WHP 090 (2004)
7. Wood, D.: High Definition for Europe— a progressive approach. EBU Technical Review (2004)
8. ITU-T Recommendation H.264 and ISO/IEC 14496-10 (MPEG-4 AVC): Advanced Video Coding for Generic Audiovisual Services. ITU-T and ISO/IEC JTC 1 (2003-2007)
9. Marpe, D., Wiegand, T., Gordon, S.: H.264/MPEG4-AVC Fidelity Range Extensions: Tools, Profiles, Performance, and Application Areas. In: Proc. IEEE International Conference on Image Processing (ICIP 2005), Genoa, Italy (2005)
10. Richardson, I.E.G.: H.264 and MPEG-4 Video Compression. Wiley, Chichester (2003)
11. List, P., Joch, A., Lainema, J., Bjøntegaard, G., Karczewicz, M.: Adaptive Deblocking Filter. IEEE Transactions on Circuits and Systems for Video Technology 13(7), 614–619 (2003)
12. Wien, M.: Variable Block-Size Transforms for H.264/AVC. IEEE Transactions on Circuits and Systems for Video Technology 13(7), 604–613 (2003)
13. Marpe, D., Schwarz, H., Wiegand, T.: Context-based Adaptive Binary Arithmetic Coding in the H.264/AVC Video Compression Standard. IEEE Transactions on Circuits and Systems for Video Technology 13(7), 620–636 (2003)
14. Wiegand, T., Sullivan, G.J., Bjøntegaard, G., Luthra, A.: Overview of the H.264/AVC Video Coding Standard. IEEE Transactions on Circuits and Systems for Video Technology 13(7), 560–576 (2003)
15. Information Technology - JPEG 2000 Image Coding System: Part 3 - Motion JPEG 2000. ISO/IEC JTC1/SC29/WG1 (2002)

16. ITU-T Recommendation T.802 : Information technology - JPEG 2000 image coding system: Motion JPEG 2000. ITU-T, T.802 (2005)
17. Information Technology - JPEG 2000 Image Coding System: Part 1 - Core Coding System. ISO/IEC JTC1/SC29/WG1, ISO/IEC 15444-1:2000 (2000)
18. Information technology – JPEG 2000 image coding system – Part 12: ISO base media file format. ISO/IEC JTC-1 / SC29, ISO/IEC 15444-12:2005 (2005)
19. Daubechies, I., Sweldens, W.: Factoring Wavelet Transforms into Lifting Steps. Journal of Fourier Analysis and Applications 4(3), 247–269 (1998)
20. Taubman, D., Marcellin, M.W.: JPEG 2000 - Image Compression: Fundamentals, Standards and Practice. Kluwer Academic Publishers, Hingham (2001)
21. Schelkens, P., Skodras, A., Ebrahimi, T.: The JPEG 2000 Suite. Wiley, Chichester (2009)
22. The SVT High Definition Multi Format Test Set, http://www.ebu.ch/CMSimages/en/tec_svt_multiformat_v10_tcm 6-43174.pdf (accessed October 30, 2009)
23. Technische Universität München, Lehrstuhl für Datenverarbeitung, Homepage of Tobias Oelbaum, section Videotools, http://www.ldv.ei.tum.de/Members/tobias (accessed October 30, 2009)
24. H.264/AVC Software Coordination, http://iphome.hhi.de/suehring/tml/index.htm (accessed October 30, 2009)
25. Kakadu software, http://www.kakadusoftware.com/ (accessed October 30, 2009)
26. Schwarz, H., Marpe, D., Wiegand, T.: Overview of the Scalable Video Coding Extension of the H.264/AVC Standard. IEEE Transactions on Circuits and Systems for Video Technology 17(9), 1103–1120 (2007)

Chapter 6
Super Hi-Vision and Its Encoding System

Shinichi Sakaida

Abstract. The Super Hi-Vision (SHV) is an ultra high-definition video system with 4,000 scanning lines. Its video format is 7,680×4,320 pixels, which is 16 times the total number of pixels of high definition television and the frame rate is 60 Hz with progressive scanning. It has been designed to give viewers a strong sense of reality. To make the system suitable for practical use, such as in broadcasting services, a high-efficiency compression coding system is necessary. Therefore, we have developed several Super Hi-Vision codec systems based on MPEG-2 and AVC/H.264 video coding standards. In this chapter, details of these codec systems are described and transmission experiments using the codec systems are introduced.

1 Introduction

Super Hi-Vision (SHV) consists of an extremely high-resolution imagery system and a super surround multi-channel sound system [1]. Its video format consists of 7,680×4,320 pixels (16 times the total number of pixels of high definition television (HDTV)) and a 60-Hz frame rate with progressive scanning. It uses a 22.2 multi-channel sound system (22 audio channels with 2 low frequency effect channels) and has been designed to give viewers a strong sense of reality. The final goal of our research and development of SHV is to deliver highly realistic image and sound to viewers' homes. When SHV becomes applicable as a broadcasting system, we will be able to use it for many purposes, such as archival and medical use.

NHK (Japan Broadcasting Corporation) has developed SHV cameras, projectors, disk recorders, and audio equipment. Several SHV programs have been produced using these devices, and demonstrations of the programs have attracted many visitors at events such as the 2005 World Exposition in Aichi, Japan, as well as NAB 2006 (National Association of Broadcasters) in Las Vegas, USA [2] and IBC 2006 (International Broadcast Conference) in Amsterdam, Netherlands [1].

Shinichi Sakaida
NHK Science and Technology Research Laboratories
1-10-11, Kinuta, Setagaya-ku, Tokyo 157-8510, Japan
e-mail: sakaida.s-gq@nhk.or.jp

To achieve our final goal of broadcasting SHV programs to homes, a new high-efficiency compression coding system is necessary. Therefore, we have developed several SHV codec systems that are based on MPEG-2 (Moving Pictures Experts Group) [3] and AVC (Advanced Video Coding)/H.264 [4] video coding standards. It is necessary to conduct transmission tests and to solicit public opinion to encourage people to recognize SHV as a broadcasting service. However, this cannot be done without an SHV codec. A real-time SHV codec will help people to consider SHV as a functioning broadcasting system. For this reason, we need to develop the codec as soon as possible. In 2006, we successfully conducted the first transmission experiment using the MPEG-2-based SHV codec system via gigabit IP networks [5]. We then continued to carry out several transmission tests in that year. With regard to the AVC/H.264 codec system, we performed the first international satellite transmission tests at IBC2008 [6].

In this chapter, the outline of our SHV system and details of the SHV codec systems are described, and several transmission experiments using the codec systems are explained in detail.

2 Super Hi-Vision Systems

2.1 Video System of SHV

The specifications of the SHV system are listed in Table 1 and compared with those of HDTV. SHV is now the highest resolution TV system available. The basic parameters of the SHV system are designed to enhance viewers' visual experience. There are 7,680 horizontal pixels and 4,320 vertical lines, for approximately 33 million pixels per frame. This is 16 times the total number of pixels per frame of HDTV. The frame rate is 60 Hz progressive, so the total information of SHV is 32 times that of HDTV, i.e. 16 times spatially and 2 times temporally.

International standardization process for SHV is currently in progress. The International Telecommunication Union Radiocommunications sector (ITU-R) has been studying large-screen digital imagery (LSDI) and extremely high-resolution imagery (EHRI) and has produced recommendations for those video systems [7][8]. The Society of Motion Picture and Television Engineers (SMPTE) has produced standard

Table 1 SHV and HDTV specifications

Specifications	SHV	HDTV
Number of pixels	7,680 × 4,320	1,920 × 1,080
Aspect ratio	16 : 9	16 : 9
Standard viewing distance (H: height of screen)	0.75H	3H
Standard viewing angle (horizontal)	100 degrees	30 degrees
Frame / Field rate	60 Hz / 59.94Hz	60 Hz / 59.94 Hz
Interlace ratio	1 : 1	1 : 1/2 : 1

2036-1, which describes the video formats for ultra high-definition TV (UHDTV) system [9]. It corresponds with the ITU-R recommendations.

NHK has developed key equipment for SHV broadcasting systems, such as cameras, display systems, and disk recorders. SHV requires an imaging device and display device with 33 million pixels; however, integrated devices with such high resolution are not yet available. Thus, the SHV camera uses four panels with 8 million pixels each for green 1 (G1), green 2 (G2) (dual-green), red (R), and blue (B) channels, using the pixel-offset method to increase the effective number of pixels both horizontally and vertically. Figure 1 shows the pixel-spatial-sampling arrangement in this method.

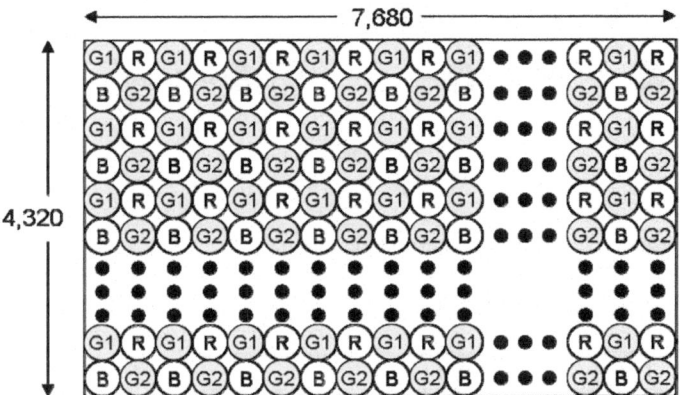

Fig. 1 Pixel-spatial-sampling arrangement in pixel offset method

2.2 Audio System of SHV

The audio format of the SHV system has 22.2 multi-channel sounds as shown in Figure 2 and Table 2. The 22.2 multichannel sound system has 3 loudspeaker layers (top, middle, and bottom), 22 full-bandwidth channels, and 2 low frequency effects (LFE) channels. It is a three-dimensional audio system, whereas the 5.1 multichannel sound system specified in ITU-R BS.775-2 is a two-dimensional audio system without a vertical dimension [10].

The audio sampling frequency is 48 kHz, and 96 kHz can be optionally applied. The bit depth is 16, 20, or 24 bits per audio sample.

International standardization of the 22.2 multichannel sound system is currently underway. ITU-R is studying the system parameters for digital multichannel sound systems. As mentioned earlier, SMPTE has started the standardization process and has already produced standard 2036-2, which describes the audio characteristics and audio channel mapping of 22.2 multichannel sound for production of ultra high-definition television programs [11].

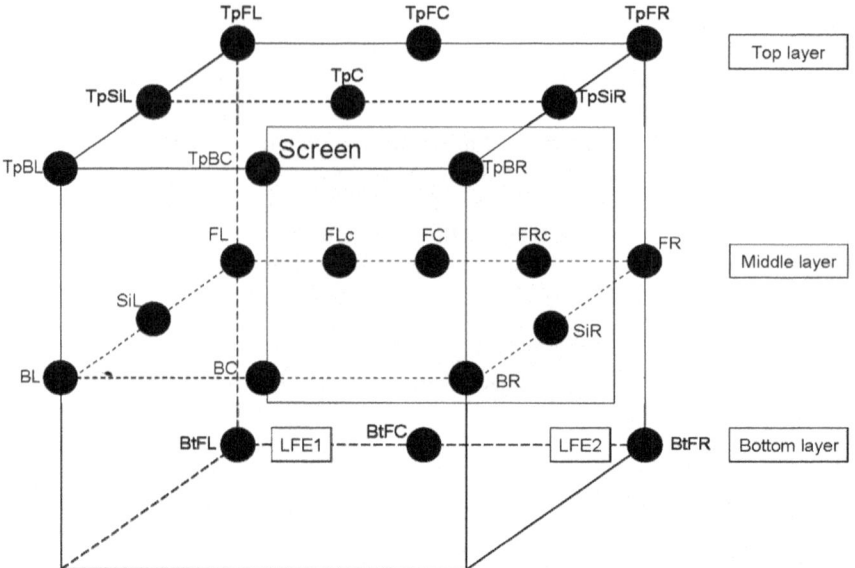

Fig. 2 Loudspeaker layout of 22.2 multichannel sound system

Table 2 Channel maps and labels of 22.2 multichannel sounds

Channel number	Label	Name
1	FL	Front left
2	FR	Front right
3	FC	Front centre
4	LFE1	LFE-1
5	BL	Back left
6	BR	Back right
7	FLc	Front left centre
8	FRc	Front right centre
9	BC	Back centre
10	LFE2	LFE-2
11	SiL	Side left
12	SiR	Side right
13	TpFL	Top front left
14	TpFR	Top front right
15	TpFC	Top front centre
16	TpC	Top centre
17	TpBL	Top back left
18	TpBR	Top back right
19	TpSiL	Top side left

Table 2 *(Cont.)*

20	TpSiR	Top side right
21	TpBC	Top back centre
22	BtFC	Bottom front centre
23	BtFL	Bottom front left
24	BtFR	Bottom front right

2.3 Roadmap of SHV

The provisional roadmap for SHV for the future is shown in Figure 3. NHK Science and Technology Research Laboratories (STRL) started the SHV study in 1995, and in the following decade made steady progress towards its practical use. We expect to launch SHV broadcasting by 2025. The plan is to deliver SHV through a cost-effective network to homes, where it will then be recorded onto a home-use receiver. We anticipate that experimental SHV broadcasts will start in 2020 using a 21-GHz-band satellite, which is a potential delivery media for high-bit-rate transmissions. The display for SHV is another important subject. The widespread use of large, high-resolution flat panel displays is remarkable. We assume that SHV displays for home use will be either a 100–150-inch large screen display or an A3-sized handheld-type paper-thin display with extremely high resolution. The SHV system has the potential for use in various applications in addition to broadcasting, e.g., art, medical use, security, and monitoring. In-theater presentation of sports events, concerts, etc. will be implemented before the broadcasting stage. The SHV systems can also be used in non-theater environments such as for advertisements, image archive materials and background images for program production.

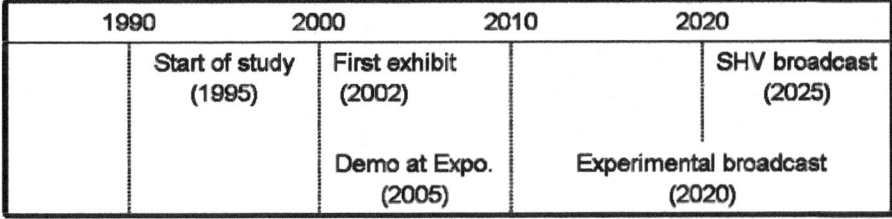

Fig. 3 Roadmap toward achieving SHV broadcasting

3 Codec Systems

The data rate of uncompressed SHV signal reaches about 24 Gbps (bits per second). For achievement of SHV broadcasting services to homes, compression coding is needed to transmit such a huge amount of information. Video codec systems have been developed for the efficient transmission and recording of SHV signals. The developed codec systems consist of a video format converter, video codec, and audio codec. In this section, the video format converter and the video codec are discussed.

3.1 Video Format Converter

To encode SHV signals in the Y/Cb/Cr format, the original SHV signals should first be converted for ease of handling. The video format converter converts the 7,680×4,320 (G1, G2, B and R) format from/into sixteen 1,920×1,080/30 Psf (progressive segmented frame) (Y/Cb/Cr 4:2:2) HDTV size images, where the SHV images are divided spatio-temporally. The color format conversion is shown in formula (3.1), which is based on ITU-R BT. 709. As previously mentioned, the current SHV signals are formatted by the pixel-offset method. Each signal component (G1, G2, B, and R) is a quarter of the size of an SHV signal and is arranged in a pixel shift position as shown in Figure 1. Since there are two luminance signals Y1 and Y2 after the video format conversion, the number of meaningful pixels becomes equal to half the SHV total area of 7,680×4,320. Therefore, in the case of conversion to the 16 HD-SDI (high definition serial digital interface) signals in practice, SHV signals should be spatially divided into 8 parts and temporally into 2 parts. Spatial division should have two modes: (a) four horizontal parts and two vertical parts ("H" division) and (b) two horizontal parts and four vertical parts ("V" division), as shown in Figure 4. The final spatial division is shown in Figure 5 and temporal division is shown in Figure 6. A diagram of the total codec system including the video format converter is shown in Figure 7.

$$
\begin{bmatrix} Y1 \\ Y2 \\ Cb \\ Cr \end{bmatrix} = \begin{bmatrix} 0.7152 & 0 & 0.0722 & 0.2126 \\ 0 & 0.7152 & 0.0722 & 0.2126 \\ -0.1927 & -0.1927 & 0.5000 & -0.1146 \\ -0.2271 & -0.2271 & -0.0458 & 0.5000 \end{bmatrix} \begin{bmatrix} G1 \\ G2 \\ B \\ R \end{bmatrix}
\tag{3.1}
$$

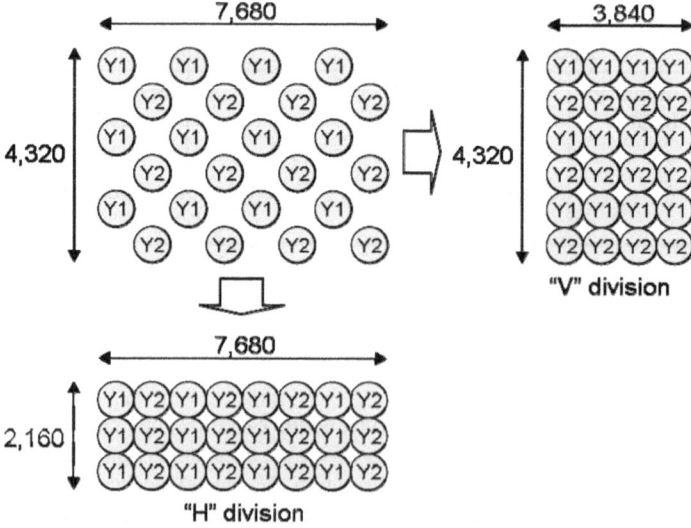

Fig. 4 Spatial division modes

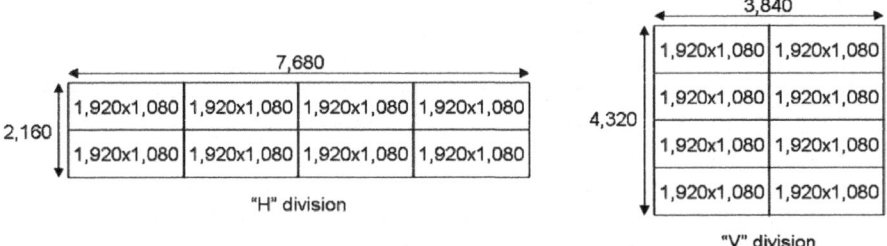

Fig. 5 Spatial division

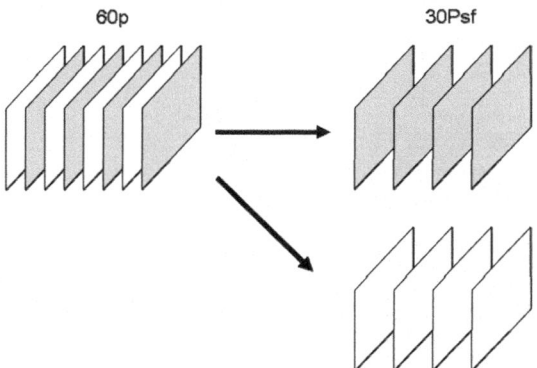

Fig. 6 Temporal division

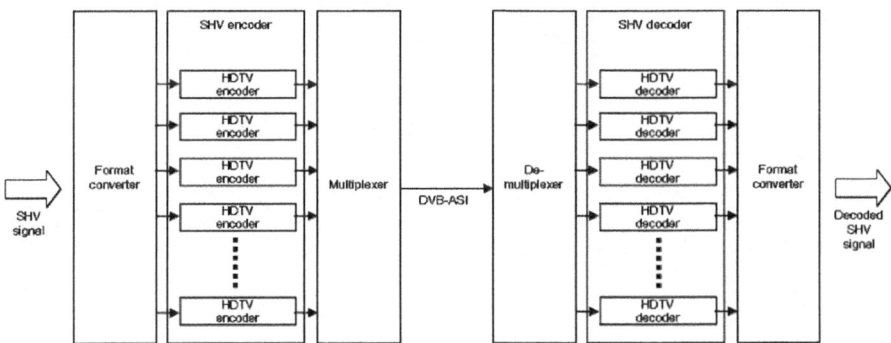

Fig. 7 Diagram of codec system

3.2 *MPEG-2-Based Codec*

For the first version of the SHV codec, we selected the MPEG-2 coding scheme with its proven technology as the base system. The MPEG-2-based video codec consists of 4 sub-codecs for 3,840×2,160 images. A sub-codec contains four single-unit MPEG-2 HDTV codecs and a multichannel frame synchronizer as depicted in Figure 8. To compress large images that exceed the resolution of HDTV by using MPEG-2 video coding, the video format converter divides the SHV image into multiple HDTV units. The HDTV coding conforms to MPEG-2 Main Profile and 4:2:2 Profile @ High Level. Since motion vectors in higher resolution images are often significantly larger than those in lower resolution images, it is necessary to implement motion estimation with a wider search range in the encoder. The search vector range of the developed encoder achieves +/− 211.5 pixels horizontally and +/− 113.5 lines vertically, exceeding existing common MPEG-2 encoders. The total coding delay is about 650 ms, which includes the process by the encoder, the decoder and the video format converters.

Fig. 8 MPEG-2-based codec system

The multichannel sound signals can be transmitted in the form of uncompressed linear pulse code modulation (PCM) (48-kHz sampling, 24 bits per audio sample). In addition to the PCM, Dolby-E codecs with a compression ratio of 1:4 are also equipped. They handle 24 audio signal channels for the 22.2 multichannel sound system. The coded video and audio signals are multiplexed into four MPEG-2 transport stream (TS) signals interfaced via DVB-ASI (Digital Video Broadcasting – Asynchronous Serial Interface).

The MPEG-2 4:2:2 Profile codec system can be used for improving transmission between the broadcasting stations and storage, since the codec achieves a bitrate of 600 Mbps with high image quality. A TS recording device that supports the storage of long program material has also been developed. The storage capacity is 1.2 T bytes, which enables storage of a 4.5-hour-long program that is coded at 600 Mbps.

When public IP (Internet Protocol) networks are used for transmission of the encoded SHV signals, consideration must be given to jitter and time delay depending on the transmission path. To synchronize the four TSs generated by the sub-encoders, the system manages the timing of each video frame by means of a time code and temporal reference in the Group of Pictures (GOP) header of the MPEG-2 video stream. The sub-encoders communicate with each other via Ethernet and their start timing is controlled by a master sub-encoder. At the decoder, the master sub-decoder adjusts the display timing of all the sub-decoders and accounts for transmission delay by referring to the time code and temporal reference. The decoder can cope with the relative delay in the 4 TSs within 15 video frames. All HDTV decoders in sub-decoders work synchronously using black burst as a reference signal.

3.3 AVC/H.264-Based Codec

To achieve lower bit-rate coding for SHV with high-quality images, a codec system based on the AVC/H.264 coding scheme has been developed. AVC/H.264 is currently the most efficient standard video coding scheme and is widely used in various applications, such as broadcasting small images for mobile reception or HDTV services via a satellite network. Since there are no AVC/H.264 codecs for images as large as SHV format ones, 16 HDTV AVC/H.264 codecs are used to construct SHV codecs similar to the MPEG-2-based codecs.

Each HDTV AVC/H.264 codec conforms to Main Profile @ L4 and will be able to handle High Profile in the future. The encoder consists of three field programmable gate array (FPGA) chips and one digital signal processor (DSP); therefore, the encoding process can be modified by replacing the encoder software. One HDTV frame is divided into four slices and each slice is processed in parallel. Motion estimation, which is conducted on FPGA chips, has two phases: pre-motion estimation, which is a rough prediction on one whole HDTV frame with two-pixel precision, followed by precise estimation on each of the four slices with quarter-pixel precision. The DSP chip, used mainly for rate control, administers the entire HDTV encoding processing module. The HDTV encoder is 1 rack unit (RU) in size and has DVB-ASI output.

Frame synchronization of the 16 output images is the most important issue of the system: therefore, a new synchronization mechanism was developed in which one of the 16 encoders becomes a master and the other 15 encoders become slaves. To synchronize the presentation time stamp (PTS) / decoding time stamp DTS) and program clock reference (PCR) for MPEG-2 TS of the output streams, all encoders share the same system date. The master encoder sends to all the other encoders a "start" hardware signal and 27-MHz clock so that all the encoders' date counters increment at the same rate. The signal for synchronization is transmitted with a daisy-chained connection, and the master encoder automatically detects the number of slave encoders using information on the signal. GOP synchronization is also achieved. All encoders generate an intraframe when more than N encoders detect a scene cut change. The value N is programmable and is usually set to nine. When an encoder generates an intraframe independently of the other encoders, it will generate the next intraframe at the beginning of GOP to maintain synchronization with the others encoders. The structure of the synchronization of the encoder units is depicted in Figure 9, and the codec systems themselves are shown in Figure 10.

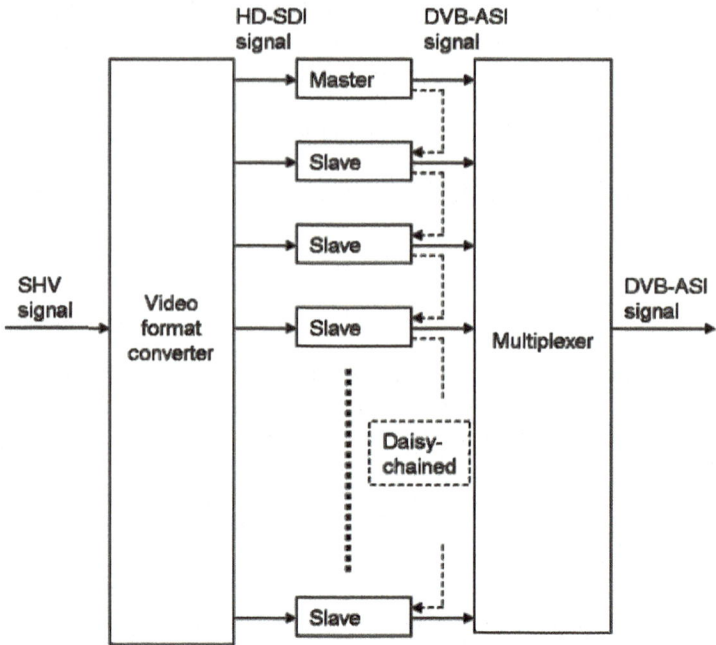

Fig. 9 Daisy-chained structure of encoder units

Fig. 10 AVC/H.264-based codec system

An SHV decoder also consists of 16 HDTV decoders. Each HDTV decoder is 1 RU in size and has the specifications for professional use. One decoder generates a signal for synchronization and supplies it to all the equipment including the 15 HDTV decoders at the decoder side. This construction enables precise synchronization among the decoded 16 HDTV images.

The statistical properties of the SHV source signals are not effectively utilized in the current HDTV encoders. A future goal is to share the SHV source information between the encoders to improve compression performance further.

4 Demonstrations and Experiments

4.1 IP Transmission

On March 14, 2006, NHK carried out an experimental transmission of SHV using the MPEG-2 coding system at bit-rates of 300 and 640 Mbps via gigabit IP networks over a distance of 17.6 km. The IP transmission system was also successfully demonstrated at NAB 2006 in the US from April 24 to 27 [1] and IBC 2006 in the Netherlands from September 8 to 12 [2].

(a)

(b)

Fig. 11 (a) and (b) Setup and appearance of live transmission experiment

Every New Year's Eve, a popular music show is broadcast from the NHK Hall in Tokyo. On December 31, 2006, NHK conducted an experiment on the live relay broadcast from Tokyo to Osaka for public viewing of the music program. This experimental transmission was carried out over 500 km and for about 5 hours. The transmitted program was screened at a special theatre in Osaka. The 22.2 multichannel sound was uncompressed. The video and audio signals were transmitted as multiple IP streams, and the difference of the delay between the IP streams was negligible enough to be ignored. The total TS bit-rate was 640 Mbps, and the total system delay was approximately 700 ms and was dominated by the codec delay. Forward error correction (FEC) was not used, but there was no packet loss for the duration of the live event. This experiment verified that long distance SHV transmission is feasible. The setup and appearance of the experiment are shown in Figure 11.

In September 2008, the first international SHV live transmission was held from London to Amsterdam where IBC 2008 was opened [12]. A camera head and the microphones of the SHV system were placed on top of London City Hall. The coded SHV signal as an IP packet stream was carried via an international fiber-optic undersea cable link to Amsterdam. A diagram of the transmission from London to Amsterdam is shown in Figure 12. To demonstrate the live nature of the link, the scenario was to emulate live news reports from London to Amsterdam with two-way interaction between a reporter at City Hall and a presenter in the SHV theatre in Amsterdam. The appearances of the demonstration are shown in Figure 13. The received picture quality was excellent, enabling the fine details of the scene to be visible, and the surround sound quality was also extremely high. No bit errors were detected on the link over the five days of the experiment. This demonstration showed the possibility of live SHV content being relayed from virtually anywhere in the future.

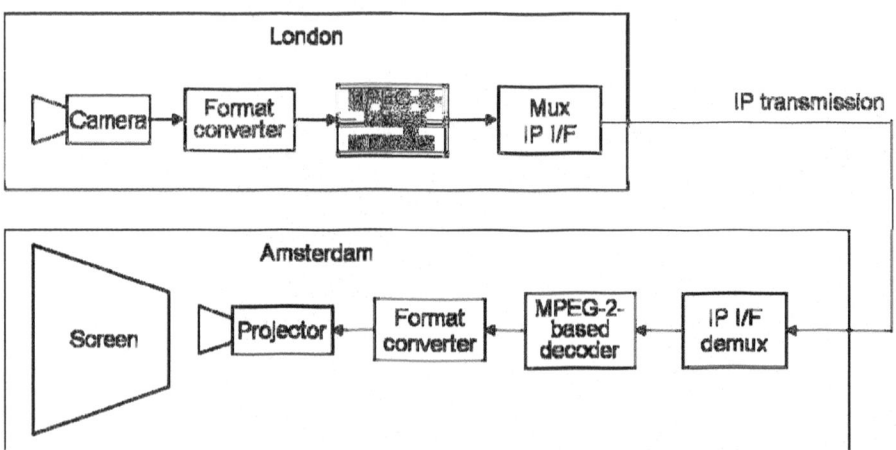

Fig. 12 IP transmission from London to Amsterdam

(a)

(b)

(c)

Fig. 13 (a), (b) and (c) Appearances of international live transmission demonstration

4.2   Satellite Transmission

The AVC/H.264-based SHV codec system developed by NHK was introduced and demonstrated at the NHK STRL open house in May 2007. The demonstration showed that the codec system has the potential to encode the SHV signal at a bit-rate of about 128 Mbps. In combination with satellite technologies, it will be possible to deliver the SHV programs to the home.

The effectiveness of the AVC/H.264 codec and the satellite technologies were demonstrated at IBC 2008 in Amsterdam, where the SHV signals were delivered from the uplink station in Turin, Italy over a Ku-band satellite capacity [13]. For that first public demonstration of SHV transmission by satellite, DVB-S2 modulation technology was indispensable. The SHV video and audio signals were encoded at about 140 Mbps, and the resultant stream was split into two 70-Mbps TS streams, transmitted over two 36-MHz satellite transponders using 8PSK (8 Phase Shift Keying) 5/6 modulation, and re-combined at the receiver using the synchronization and de-jittering features of DVB-S2. A diagram of the transmission from Turin to Amsterdam is shown in Figure 14. The difference of the delay between 2 received streams was 0.256 ms, and splitting of the streams did not affect the decoded signals. The setup and appearance of the transmission experiments are shown in Figure 15.

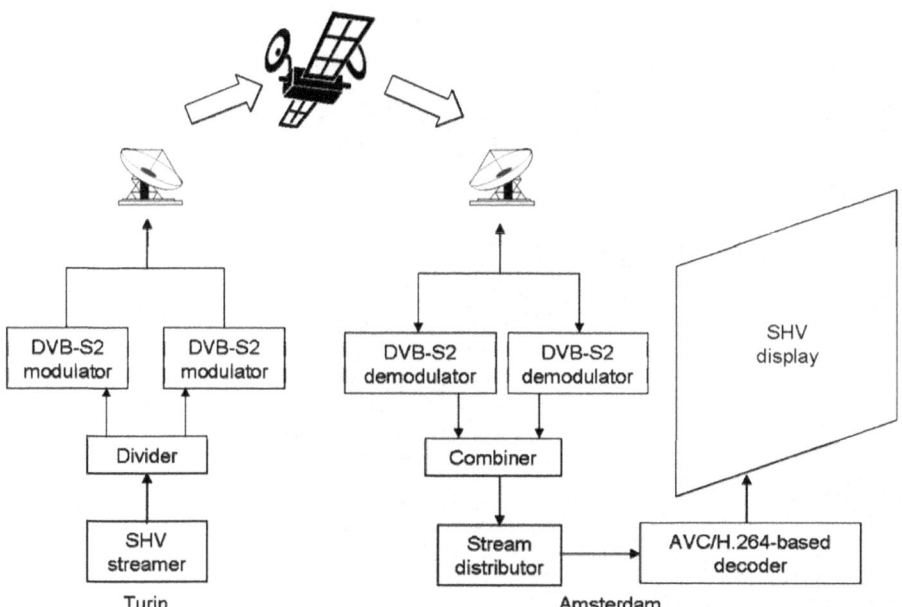

Fig. 14 Satellite transmission diagram

(a)

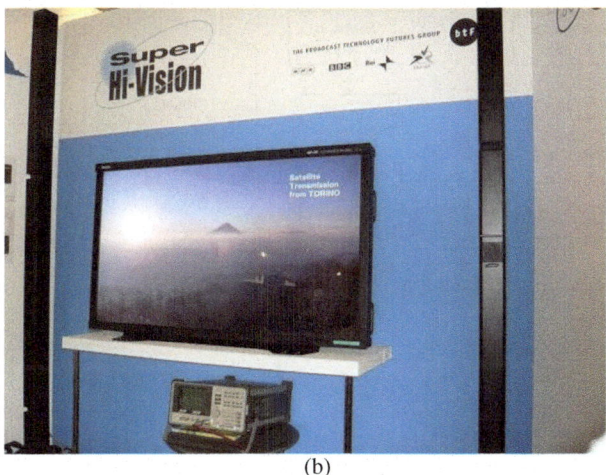

(b)

Fig. 15 (a) and (b) Setup and appearance of the international transmission experiment

In May 2009, NHK conducted a live transmission experiment of multichannel SHV programs using the WINDS, the Wideband InterNetworking engineering test and Demonstration Satellite that was launched in Japan in February 2008. A wideband modulator and demodulator that support a bit-rate of 500 Mbps were used, and a low-density parity check (LDPC) forward error correction code was applied. The transmission was from Sapporo to Tokyo by way of Kashima, as shown in Figure 16. The live camera images and multichannel sound were trans-mitted from Sapporo using gigabit IP to Kashima, then the other two SHV programs were multiplexed in Kashima and relayed via the wideband satellite WINDS. The stream was received at NHK STRL in Tokyo, and the decoded images and sound were presented to the audiences. The appearances of the demonstration are shown in Figure 17.

In the future, a SHV signal may be delivered to the home by Ku- or Ka-band satellites using a single high-power 36–72-MHz transponder and high-order modulation schemes.

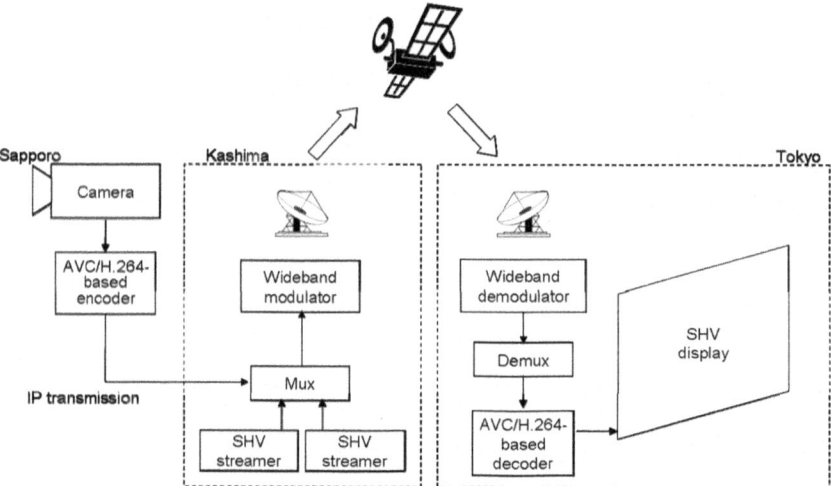

Fig. 16 Transmission via IP and satellite from Sapporo to Tokyo

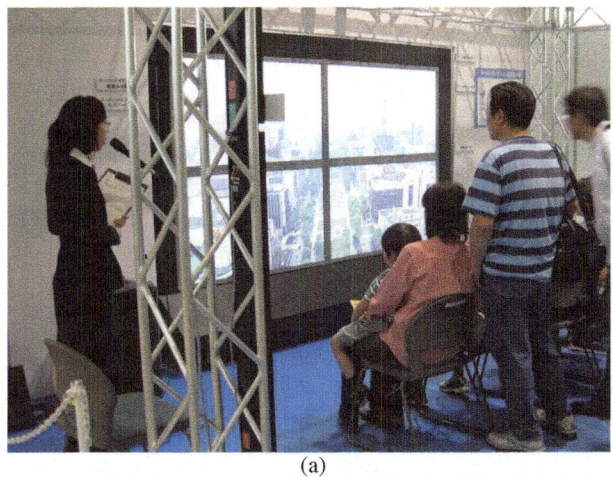

(a)

(b)

Fig. 17 (a) and (b) Appearances of transmission via IP and satellite

5 Conclusion

Significant progress has been made toward implementing the practical use of SHV, particularly in the development of SHV codecs and transmission systems for contribution and distribution purposes. However, current cameras and displays do not have the full resolution of the SHV system because dual-green technology is used.

Imaging devices with 33 million pixels were not available when the cameras and displays were developed. Another important point is that current display devices are projectors, not direct-view displays. Thus, our next research target is the development of a full-resolution camera and a direct-view display. We will develop a full-resolution camera with three 33-million-pixel image sensors. We expect that display manufactures will develop a flat-screen display with the full pixel count. By developing such cameras and direct-view displays, we can provide people with their first experience of seeing SHV images with full resolution in a home viewing environment. We believe this advance will be instrumental for determining the signal parameters of future broadcasting. Besides developing the cameras and displays, we are developing compression techniques. These systems and technologies can be used for various events, such as cooperative museum exhibitions, live relay of sports events on a global scale, and public viewing of SHV programs.

NHK will continue its tireless research and development efforts in accordance with the roadmap aiming to launch SHV broadcasting in 2020 – 2025.

References

1. Nakasu, E., Nishida, Y., Maeda, M., Kanazawa, M., Yano, S., Sugawara, M., Mitani, K., Hamasaki, K., Nojiri, Y.: Technical development towards implementation of extremely high resolution imagery system with more than 4000 scanning lines. In: IBC 2006 Conference Publication, pp. 345–352 (2006)
2. Maeda, M., Shishikui, Y., Suginoshita, F., Takiguchi, Y., Nakatogawa, T., Kanazawa, M., Mitani, K., Hamasaki, K., Iwaki, M., Nojiri, Y.: Steps Toward the Practical Use of Super Hi-Vision. In: 2006 NAB Proceedings, pp. 450–455 (2006)
3. ISO/IEC 13818-2 Information technology – Generic coding of moving pictures and associated audio information: Video (2000)
4. ISO/IEC 14496-10 Information technology – Coding of audio-visual objects – Part 10: Advanced Video Coding | ITU-T Recommendation H.264: Advanced video coding for generic audiovisual services (2009)
5. Nakayama, Y., Nishiguchi, T., Sugimoto, T., Okumura, R., Imai, A., Iwaki, M., Hamasaki, K., Ando, A., Nishida, Y., Mitani, K., Kanazawa, M., Kitajima, S.: Live production and transmission of large-scale musical TV program using 22.2 multichannel sound with ultra-high definition video. In: IBC 2007 Conference Publication, pp. 253–259 (2007)
6. Shishikui, Y., Fujita, Y., Kubota, K.: Super Hi-Vision – the star of the show! EBU Technical review, 4–16 (January 2009)
7. ITU-R Recommendation BT.1769 Parameter values for an expanded hierarchy of LSDI image formats for production and international programme exchange (2006)
8. ITU-R Recommendation BT.1201-1 Extremely high resolution imagery (2004)
9. SMPTE 2036-1 Ultra High Definition Television – Image Parameter Values For Program Production (2007)
10. ITU-R Recommendation BS.775-2 Multichannel stereophonic sound system with and without accompanying picture (2006)

11. SMPTE 2036-2 Ultra High Definition Television – Audio Characteristics and Audio Channel Mapping for Program Production (2008)
12. Zubrzycki, J., Smith, P., Styles, P., Whiston, B., Nishida, Y., Kanazawa, M.: Super Hi-Vision – the London-Amsterdam live contribution link. EBU Technical review, 17–34 (January 2009)
13. Morello, A., Mignone, V., Shogen, K., Sujikai, H.: Super Hi-Vision – delivery perspectives. EBU Technical review, 35–44 (January 2009)

Chapter 7
A Flexible Super High Resolution Video CODEC and Its Trial Experiments

Takeshi Yoshitome, Ken Nakamura, and Kazuto Kamikura

Abstract. We propose a flexible video CODEC system for super-high-resolution videos such as those utilizing 4k x 2k pixel. It uses the spatially parallel encoding approach and has sufficient scalability for the target video resolution to be encoded. A video shift and padding function has been introduced to prevent the image quality from being degraded when different active line systems are connected. The switchable cascade multiplexing function of our system enables various super-high-resolutions to be encoded and super-high-resolution video streams to be recorded and played back using a conventional PC. A two-stage encoding method using the complexity of each divided image has been introduced to equalize encoding quality among multiple divided videos. System Time Clock (STC) sharing has also been implemented in this CODEC system to absorb the disparity in the times streams are received between channels. These functions enable highly-efficient, high-quality encoding for super-high-resolution video. The system was used for the 6k x 1k video transmission of a soccer tournament and the 4k x 2k video recoding of SATIO KIKEN orchestral concert.

1 Introduction

The number of video applications for super-high-resolution (SHR) images has been increasing in the past few years. SHR video images are 2 - 16 times larger than HDTV images, and they have 30 - 60 fps. Because of their high quality and the high level of realism they convey to the viewer, SHR systems[1, 2, 3, 4, 5] are expected to be platforms for many video applications, such as digital cinema, virtual museums, and public viewing of sports, concerts and other events. For SHR video applications, it is important to reduce the network bandwidth, because raw SHR video requires

Takeshi Yoshitome · Ken Nakamura · Kazuto kamikura
NTT Cyber Space Laboratories, NTT Corporation, Japan
e-mail: yoshitome.takeshi@lab.ntt.co.jp,
nakamura.ken@lab.ntt.co.jp, kamikura.kazuto@lab.ntt.co.jp

very high-speed transmission lines or high-speed disks that operate at 3 - 24 Giga bit/sec. SHR video compression schemes are thus needed to reduce the transmission and recording costs.

We have already developed a CODEC system[1] for SHR, it consists of multiple conventional MPEG-2 HDTV CODECs[6] and a frame synchronizer. The main benefit of this system is its ability to handle SHR video of various resolutions. This system can adapt to many kinds of SHR images by increasing or decreasing the number of encoders and decoders used. However, it is difficult to equalize the encoding quality among multiple encoders, and it is also difficult to record and playback the SHR stream because it consists of separated multiple HDTV streams.

In this chapter, first, we describe the basic idea behind our CODEC system, which is spatial image division and multiple stream output. Section 3, we explain the video shift and padding function, which solves the boundary distortion problem caused by spatial image division. Section 4, we show switchable cascade multiplexing that enable to increase the flexibility of the combination of SHR video programs in multiplexing mode. Section 5, we show how to synchronize the multiple outputs in multiplexing mode and in non-multiplexing mode. Section 6, we explain the strategy of the adaptive bit-allocation in multiplexing mode. Sections 7 and 8 discuss our evaluations of the CODEC system. A few trials with transmission and recording using our system are described in Section 9. Section 10 is a brief conclusion.

2   Basic Concept

2.1   Basic Concept of SHR CODEC

SHR image transmission using parallel encoding and decoding architectures consists of several HDTV encoders, a transmission network and several HDTV decoders. SHR image is represented using several HDTV images in such architectures. An SHR camera outputs several synchronous HDTV images, and they are input to HDTV encoders. In HDTV encoders, all HDTV images are encoded independently and generated bitstreams are transferred to HDTV decoders through the network. Decoded images decompressed by the decoders are output to an SHR display system.

2.2   Spatially Parallel SHR CODEC System

This CODEC system adopts the spatial image division and multi-stream output approach. In spatial image division, the input image is divided into multiple sub-images and the encoder modules encode them in parallel, as shown in Fig. 1.

This approach is reasonable in terms of cost performance and scalability and has been used in some HDTV CODEC systems that use multiple SDTV CODEC's [6, 8]. We used it when we constructed the SHR CODEC system based on multiple MPEG-2 HDTV CODEC LSIs. Spatial image division can use a one-stream output system, in which the sub-streams generated by the encoder modules are reconstructed into one SHR elementary stream (ES), or it can use a multiple-stream

Fig. 1 Examples of SHR images

SHR images			Num. of Channel	Size	Target
Left	Right		2	2k x 1k	3D
1	2		1	4k x 1k	Wide View
1	2	3	1	6k x 1k	Sport
1 / 3	2 / 4		1	4k x 2k	Digital Cinema
1 / .... / 12	4 / .... / 16		1	8k x 4k	Future BS

output system, where several HDTV streams generated by the encoder modules are output directly in parallel or multiplexed into one transport stream (TS). We used the multiple stream output system for SHR communication applications, because conventional HDTV decoders can decode its output stream, whereas a one-stream SHR output system needs dedicated SHR decoders. There is an overview of our CODEC system in Fig. 2. The CODEC has two output modes, i.e., a multiplexing and a non-multiplexing mode. In the non-multiplexing mode, each encoder module outputs a constant bit rate (CBR) ES, and this is converted into CBR-TSs which are

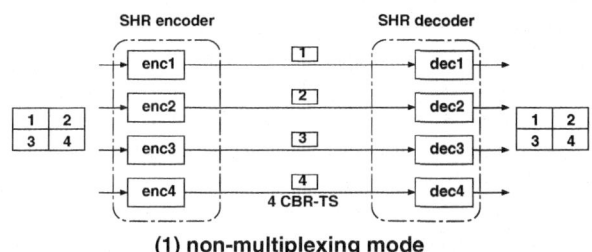

(1) non-multiplexing mode

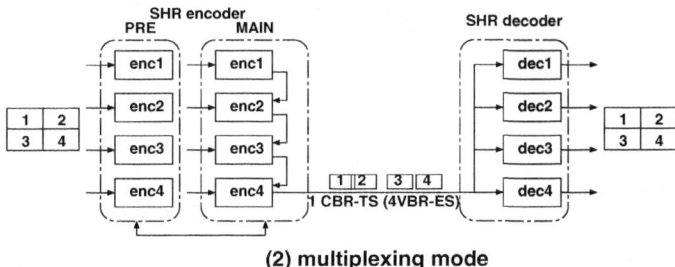

(2) multiplexing mode

Fig. 2 Overview of SHR CODEC system

transmitted in parallel. In the multiplexing mode, the pre-encoder analyzes the input images and each encoder module in the main-encoder outputs a variable bitrate (VBR) ES, and these are multiplexed into one CBR TS. The multiplexing mode has the advantages of available transmitters and efficient coding. The other advantage of the multiplexing mode is that it is easy to record and playback the TS because the TS to be handled is single. However, it is difficult to record and playback in the non-multiplexing mode due to synchronization problems among multiple TSs.

3 Video Shift and Padding Function

Generally, SHR or over-HDTV image camera systems output divided images by using conventional HDTV video-signal interfaces, such as HD-SDI's. There are 1080 active lines in a conventional HDTV system[10]. Here, this system is called a 1080-HD in this chapter. However, some SHR systems use the old HDTV signal system[11], where there are 1035 active lines, we call this the 1035-HD.

There are two problems in connecting SHR equipments that has a different number of active lines. The first is the black line problem. If 1035-HD system signals are received by a conventional 1080-HD system, the received image data do not fill the active line, e.g., 1920 x 1035 image data on 1920 x 1080, and the remaining lines are filled with black or zero data. A 3840 x 2160 projector, which uses the 1080-HD system, will display three horizontal black lines on the screen, as shown in Fig. 3, when the projector is connected to a 3840 x 2070 camera that uses a 1035-HD system. It seems that an effective solution is to vertically shift the SHR image data with the SHR projector to overcome this black-line problem; however, this solution is insufficient rectify to the second problem.

The second problem is distortion caused by mismatch between the image boundary and the DCT block boundary. Because MPEG-2 encoders transform 8 x 8 image blocks by using DCT and cut off high-frequency components, if the edge of the image and the DCT block boundary do not match, the border will not be sharp and the image data will be eroded by black data. Thus, coding distortion will be visible at the boundary of the divided images. Although this problem also occurs at the border of ordinary coded images, it is more visible in SHR systems because the boundaries of the divided images are positioned at their center.

The video shift and padding function modules in our encoder system are placed in front of the encoder modules to solve the mismatch problem. These modules can

Fig. 3 Black line problem caused by mismatch between active lines

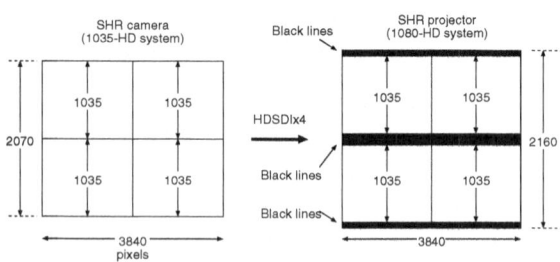

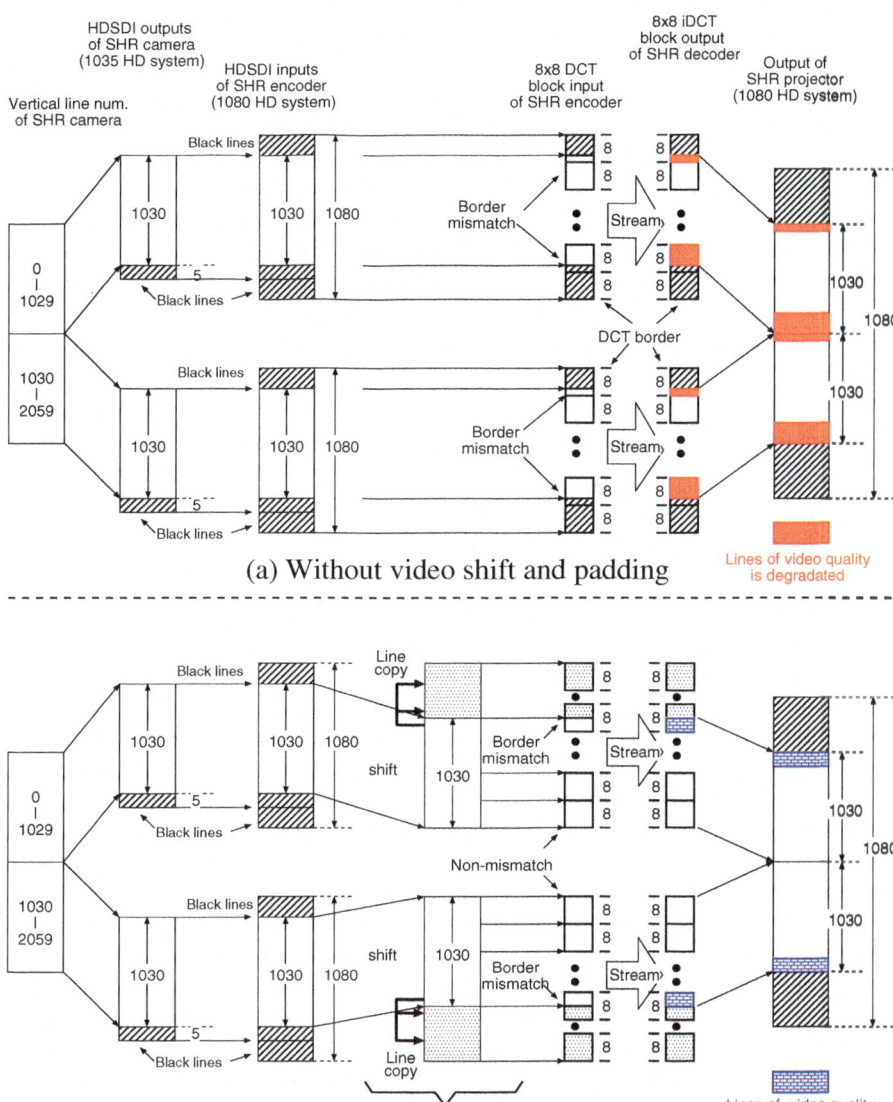

(a) Without video shift and padding

Video shift and padding function

(b) With video shift and padding

Fig. 4 Data flow from SHR camera to SHR projector without and with video shift and padding function

copy the top and bottom lines of image data onto non-image data areas. There is an example of a data flow that includes the video shift padding function in Fig. 4 (b). A data flow without the function is also shown in Fig. 4 (a). In both examples, the 1920 x 2060 pixel images of the SHR camera are output using two 1035-HDSDI

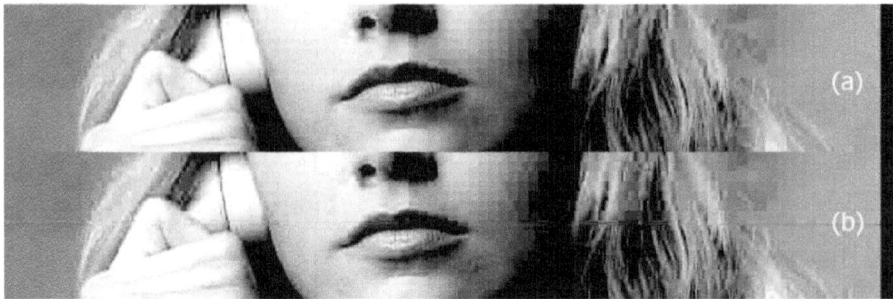

Fig. 5 Output image with and without our shift and padding function

signals that have five blank lines. In Fig. 4 (a), such 1080-HDSDI signals has been input to the encoding module directly without the video shift and padding function, causing boundary mismatch to occur between the edge of the image and the DCT block. These mismatches appear not only at the top of the active image and also at its bottom. In Fig. 4 (a), the hatched areas representing the output of the SHR projector mean that the lines degraded in the video quality. However, in Fig. 4 (b), our encoder has received a 1035-HD signal using the conventional 1080-HDSDI system, and has shifted and padded the input image to prevent the coding distortion mentioned above. In the example in Fig. 4 (b), the shift and padding function cannot prevent the two mismatches from occurring, which appear at the top and bottom of the active image at the same time. This is because "1030" is not a multiple of eight. To make the distortion noise caused by mismatch inconspicuous, the video shift size has been determined to prevent the mismatch border from being allocated to the center of SHR image, and to copy the top and bottom lines of the image data onto the non-image data area. The decoder system does nothing in regard to this operation because the copied data are aborted later. There are examples of output images with and without our function in Figs. 5 (a) and (b). Figure 5 (b) indicates that the video quality is not satisfactory because many viewers may recognize the horizontal center line caused by the mismatch of the image boundary and DCT block boundary as previously mentioned. However, it is difficult to find such a line in Fig. 5 (a). These results demonstrate that video shift and padding prevent image quality from degrading around the center of the frame because the image border matches the DCT block.

4 Switchable Cascade Multiplexing Function

Our system is equipped with a switchable cascade multiplexing function to increase its flexibility in the multiplexing mode. This function of our encoder system is outlined in Fig. 6.

The basic concept behind this function is to mix TS packets from two packet queues, i.e., internal and external queues. The internal queue stores the packets

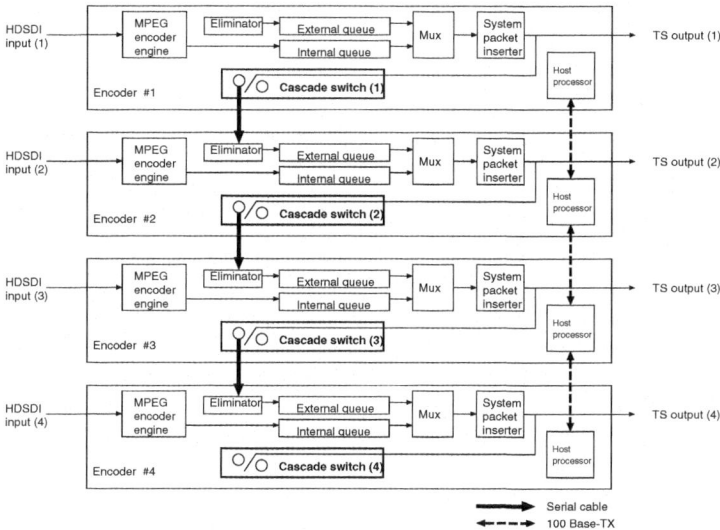

Fig. 6 Switchable cascade multiplexing in multiplexing mode

generated in the i-th encoder itself and the external one stores the packets transferred from the (i-1)th encoder through a high-speed serial cable. The packets in these two queues are mixed at the multiplexing circuit and the system packet inserter adds various system layer packets such as the Program Association Table(PAT), Program Map Table(PMT), Program Clock Reference(PCR), and NULL. The eliminator, which is located at the front of the external queue rejects unnecessary system-layer packets and outputs only video and audio packets to the external queue. Packets from the i-th and (i-1)th encoder's packets are the output of the i-th encoder. The cascade multiplexing function of each encoder is switchable to adapt to many types of SHR images, as shown in Fig. 1. Output of the i-th encoder is the packets from i-th and (i-1)th encoder's packets. Table 1. The left of this table shows the relationship between all encoder inputs and the sub-image of the SHR video to be encoded. The center indicates the cascade multiplexing switch settings of all encoders and the right lists the SHR streams of all encoder outputs. We can see that the 6Kx1K stream is output to encoder #3 and the HD stream is output to encoder #4 when all multiplexing switches except the 3rd encoder's are turned on. The 4Kx2K stream is output to encoder #4 when the multiplexing switches of all encoders turn on. This means many SHR video sizes and many combinations of SHR video programs can be handled using this switchable cascade multiplexing. Also, every stream can be recorded and easily played back using a conventional PC with a DVB-ASI interface card because the stream of each encoder's output is a single TS consisting of several ESs.

There is a block diagram of SHR live transmission for two different sites and local playback using the switchable cascade multiplexing function in Fig. 7. By

Table 1 Examples of cascade multiplexing switch setting for many kinds of SHR video programs

HDSDI input				Cascade sw setting			Stream output			
enc.1	enc.2	enc.3	enc.4	enc.1	enc.2	enc.3	enc.1	enc.2	enc.3	enc.4
\multicolumn{4}{c}{$4k2k$}				on	on	on	-	-	-	$4k2k$
\multicolumn{3}{c}{$6k1k$}	HD_1	on	on	on	-	-	$6k1k$	$6k1k+HD_1$
				on	on	off	-	-	$6k1k$	HD_1
\multicolumn{2}{c}{$4k1k$}	\multicolumn{2}{c}{$2k2k$}	on	on	on	-	$4k1k$	-	$4k1k+2k2k$
				on	off	on	-	$4k1k$	-	$2k2k$
		HD_1	HD_2	on	off	off	-	$4k1k$	HD_1	HD_2
				on	off	on	-	$4k1k$	HD_1	HD_1+HD_2
HD_1	HD_2	HD_3	HD_4	on	on	on	HD_1	HD_1+HD_2	HD_1+HD_2 $+HD_3$	HD_1+HD_2 $+HD_3+HD_4$
				off	off	off	HD_1	HD_2	HD_3	HD_4

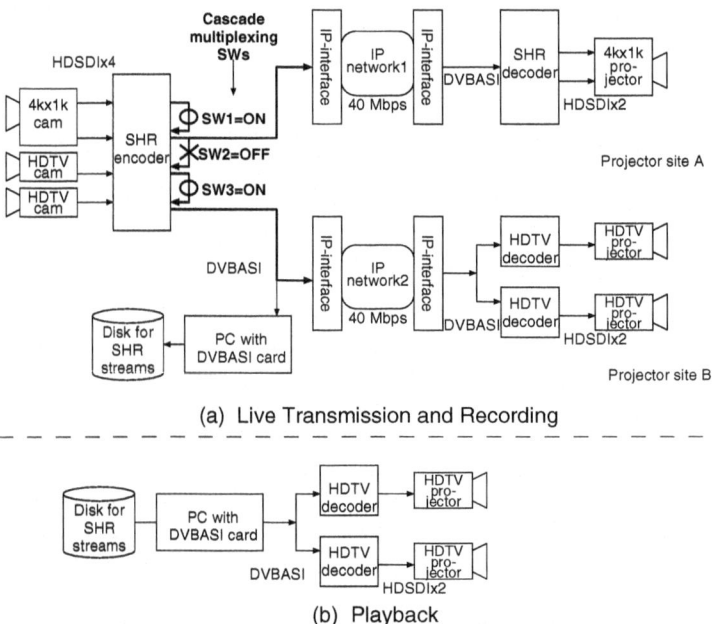

(a) Live Transmission and Recording

(b) Playback

Fig. 7 SHR live transmission and local playback of 4k x 1k program and two HDTV programs

setting three multiplexing switches (Sw1, Sw2, and Sw3) (i.e., On, Off, and On), the 4Kx1K TS can be transmitted from the camera side to the projector at site A, and a single TS that consists of two elementary HDTV streams can be transmitted to the projector at site B. If the setting of the switches is changed to (On, On, and

On) and the bandwidth of IP network 2 can be increased to 80 Mbps, all streams including the two HDTV' streams and the single 4Kx1K stream can be transmitted to the projector at site B. In addition, a conventional PC can easily record and play back the transferred SHR stream at low cost.

5 Two Synchronization Schemes

Even if all decoders can input the same PCR packets using the cascade multiplexing function mentioned above, conventional decoders generate different STCs because each PLL of conventional decoders is made of different crystal. This indicates the possibility of sub-images of SHR decoders being displayed without synchronization. This CODEC system has two schemes to synchronize channels. The first is a multiplexing mode that shares a common STC, as seen in Fig. 8.

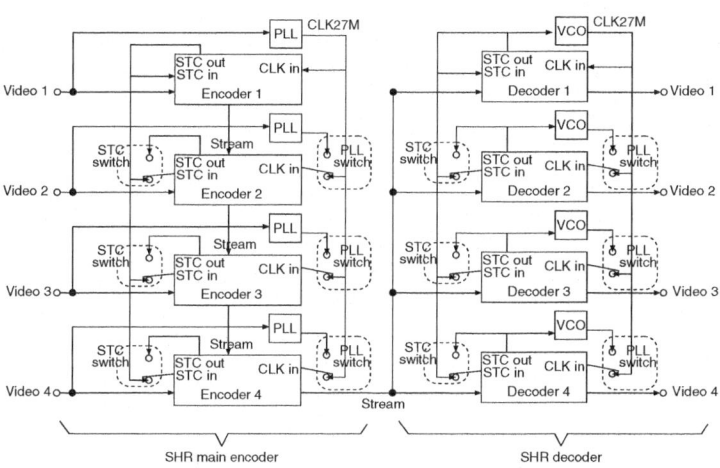

Fig. 8 Synchronization achieved by STC sharing in multiplexing mode

To synchronize all input video signals in the encoder system, a 27-MHz system clock and an STC value are generated from one of the input video signals and distributed to the encoders. Each encoder generates a PCR and a Presenting Time Stamp/Decoding Time Stamp (PTS/DTS) based on the given system clock and STC value. The decoders consist of one master decoder and several slave decoders. The master decoder generates a 27-MHz system clock and an STC from the received PCRs and distributes the system clock and STC value to the slave decoders. To deal with deviations in the received timing and to avoid stream buffer underflow or overflow, the encoder system generates a PTS/DTS with an adequate timing margin.

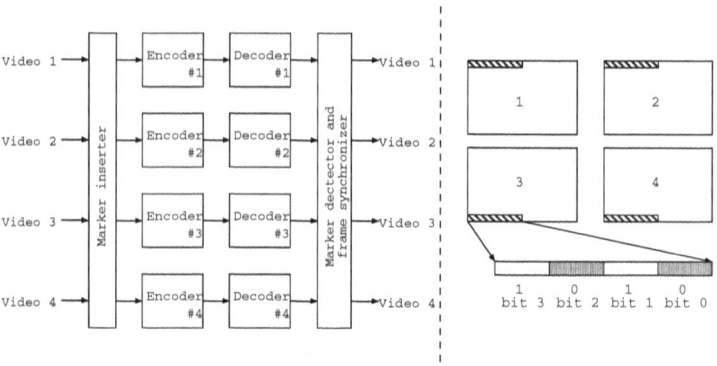

Fig. 9 Synchronization achieved by sync marker in non-multiplexing mode

The second synchronization method is the sync marker scheme, which is mainly useful with multiple conventional HDTV encoders and decoders in the non-multiplexing mode. Many conventional decoders do not have an STC sharing scheme and cannot deal with ancillary data, such as that in time code information. Our encoder places a sync marker on the top or bottom of the active line of each channel of the image, as shown in Fig. 9. The decoded images on the receiving side are synchronized with a multiple frame synchronizer that we have developed [1]. The benefit of this sync marker scheme is that the latest CODEC can be used for SHR video compression. We can replace the MPEG-2 CODEC with the latest H.264 CODEC without changing the construction of the SHR system.

6   Rate Control in Multiplexing Mode

We introduced a two-stage encoding system to equalize and improve the encoding quality of all partial frames. The system consists of pre-encoders, main-encoders, and a controller, as outlined in Fig. 10. The pre-encoder encodes and analyzes the input partial frame, and sends encoding parameters such as image complexity and the number of encoded bits to the controller. All pre- and main-encoders operate in parallel by receiving a common signal from the controller. Because of the time delay generated by the frame buffer in front of the main-encoder, the four main-encoders encode the $(N+1)$-th frame when the four pre-encoders encode the first frame (N is GOP size).

The details of this rate control flow are given in Fig. 11 and it consists of five stages. First, the i-th frame is input to the pre-encoders, and the $(i-N)$-th frame is also input to the main-encoders in Stage 1. The reason the main-encoders input a delayed frame is to improve their own encoding quality. It is well known that scene changes degrade the encoding quality. If the encoders can detect future alterations of the image complexity such as scene changes, they can better distribute target bits to all frames, and minimize the degradation in encoding quality. Large delay

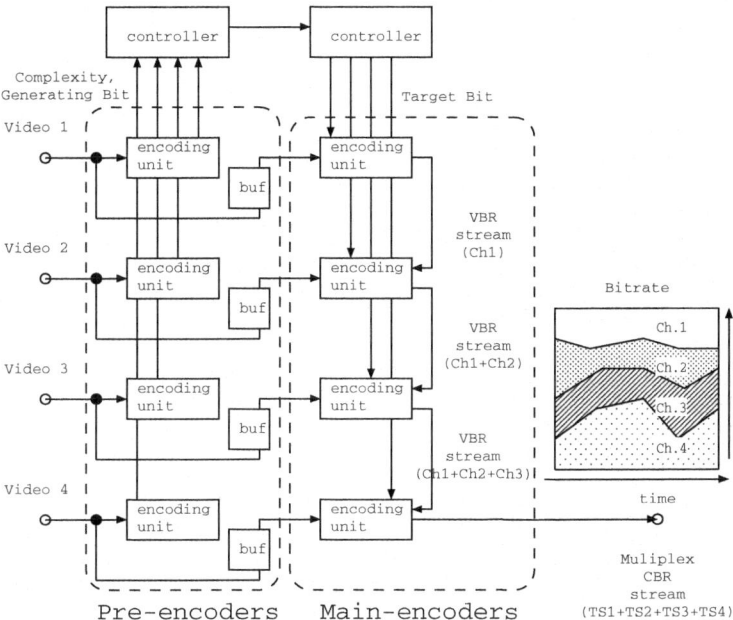

Fig. 10 Two-stage spatially parallel encoding in multiplexing mode

increases encoding quality and the latency of the CODEC. However, small delay decreases them. The value of frame delay is determined by a trade off between the encoding quality and the delay of the CODEC system. We selected one GOP delay as a good trade-off point. In Stage 2, the pre- and main-encoders encode all partial frames. After the i-th frame is encoded by the pre-encoders, the controller obtains complexity $Xp_{i,k}$ and generated bits $S_{i,k}$ of partial frame (i,k) from the k-th pre-encoder in Stage 3. Then, in Stage 4, the function of "Future_X_based_STEP1()" calculates the target bits, TAm_j, of the $(i-N)$-th non-separated frame by taking the complexity of the next N frames into consideration. In other words, this complexity information on the next N frame enables the controller to determine a precise value for the target bits of the main-encoder, and the encoding quality of the main-encoder will thus be superior to that of the pre-encoder. This is why we allocated an N frame buffer in front of the main-encoder.

The output VBR ES of the main encoders are multiplexed into one CBR TS through coordinating the multiplex modules in the encoder LSIs. The output bitrates of the main-encoders are changed at the same timing, while the bitrate of the multiplexed TS is held constant. This bit-allocation process enables the overall picture quality to be kept almost constant, because it updates the rate-distortion models of all frame parts by using the coding information from all encoders in every frame.

```
Two_Stage_Rate_Control ( ) {
 N = GOPSIZE;
 for ( all frame(i) ){

 // Stage.1 : Set parameter for encoder
  i = frame_number;
  for( all encoder(k) ) {
       F(i,k) = Partial_Frame( Frame(i), k );
       set_pre_encode(input:F(i,k),Tp(i,k));
       if ( ( j = i − N ) > 0 )
          set_main_encode(input:F(j,k),Tm(j,k));
 }

 // Stage.2 : Execute encoding
 exec_pre_encode();
 if ( j > 0 ) exec_main_encode();

 /* Stage.3 : Read output parameter from encoder */
 read_pre_encode (output:Sp(i,k),Xp(i,k));
 if ( j > 0 ) read_main_encode(output:Sm(j,k));

 // Stage.4 : Calculate target bits for pre−encoder
 for( all pre_encoder(k) )
       Tp(i+1,k) = tm5_like_STEP1(input: Sp(i,k));

 // Stage.5 : Calculate target bits for main−encoder
 XAp(i) = ∑_k Xp(i,k);
 SAm(i) = ∑_k Sm(i,k);
 if ( j > 0 ) {
   // calculate target bits using next N frame's complexity
   TAm(j+1) = Future_X_based_STEP1(input:
                     XAp(i),XAp(i-1), .., XAp(i-N),
                     SAm(i),SAm(i-1), .., SAm(i-N));
   for( all main_encoder(k) )
      Tm(j+1,k) = TAm(j+1) * Xp(j+1,k) / ∑_k Xp(j+1,k);
 }
}
}
//Frame(i):The i-th frame of SHR video sequence
//Tp(i,k): Target bits for PartialFrame(i,k) in pre-enc(k)
//Tm(i,k): Target bits for PartialFrame(i,k) in main-enc(k)
//Sp(i,k): Generated bits for PartialFrame(i,k) in pre-enc(k)
//Sm(i,k):Generated bits for PartialFrame(i,k) in main-enc(k)
//Xp(i,k): Complexity for PartialFrame(i,k) in main-enc(k)
```

Fig. 11 Rate control flow in multiplexing mode

7   Evaluation

This section discusses the results of several simulations that compared the proposed bit-allocation method with the fixed bit-allocation method. In these simulations, 1920 x 1080 video sequences were divided into four video sequences (960 x 540 pixels each); the total bitrate was 20 Mbps; the coding structure was M=3(IBBP); and 120 frames were encoded.

First, we compared the bitstreams generated by the two bit-allocation methods: fixed allocation and the proposed method. The video sequence "Soccer Action" was used for both simulations. Table. 2 shows the average frame bits are almost the same in number for the fixed allocation method and the proposed method. In the conventional method, the PSNR of the partial frames in the upper parts is 5 dB less than that of lower parts. In the proposed method, the number of frame bits in each partial frame in the upper part are almost double the frame bits of the partial frames in the lower parts, because the proposed method distributes the target bits proportionally according to the complexity of each partial frames. The proposed method reduced the difference in PSNR between partial frames to 1.7 dB.

The disparity in PSNR among the four partial frames for the two bit-allocation methods are depicted in Figs. 12 and 13. The x-axis means the frame number and the y-axis means the PSNR (dB) in these figures. During frames 90-120, the PSNR of partial frame #3 is about 2.5 dB lower than that of #1 with the conventional method. In contrast, the disparity in PSNR among the divided frames decreases to 1.0 dB with the proposed method.

Table 3 lists the average PSNR and disparity in PSNR for the proposed method and the fixed bit-allocation method for five standard video sequences. The PSNR disparity ($DSNR$) is defined by

$$DSNR = max(PSNR_{k=0..3}) - min(PSNR_{k=0..3}), \tag{1}$$

Table 2 Bitstream characteristics with conventional method and proposed method

conventional method			
Partial frame position	Average partial frame bits (kbit)	Average Quantizing parameter	Average PSNR (dB)
upper left	166.2	58.53	25.80
upper right	165.9	57.45	25.87
lower left	165.4	24.13	30.55
lower right	165.3	24.73	30.36
Average	165.7	41.06	28.19
proposed method			
upper left	225.7	38.94	27.14
upper right	227.0	38.86	27.13
lower left	96.4	36.50	29.84
lower right	98.5	36.52	29.75
Average	161.3	37.70	28.48

where $PSNR_k$ is the PSNR of partial frame k. The results for a single encoder using the conventional method have also been listed for reference. Table 3 shows that the average PSNR with the proposed method is 0.24 dB higher than that with the fixed bit-allocation method. The PSNR disparity with the proposed method is 2 dB lower than that with the conventional method and this is nearly that of the single encoder.

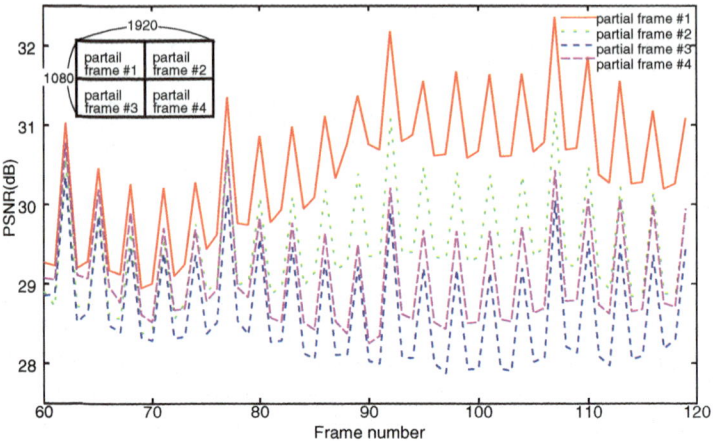

Fig. 12 PSNR of partial frames with conventional method. (sequence : "Church").

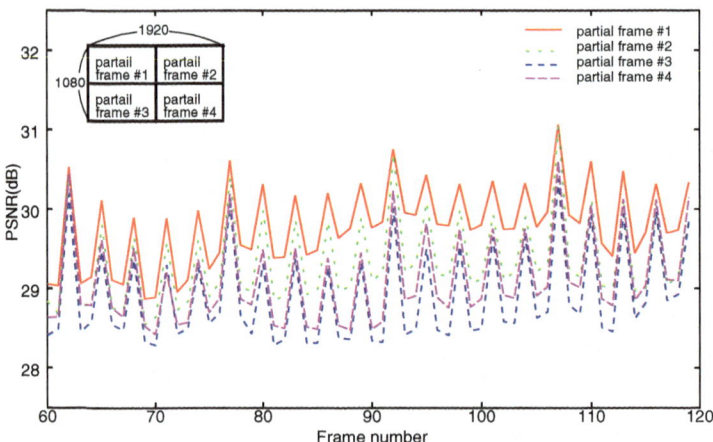

Fig. 13 PSNR of partial frames with proposed method. (sequence : "Church").

Table 3 Average PSNR and disparity PSNR with conventional method and proposed method

Average of PSNR (dB)			
original frame size	1920x1080		
num of encoder	4	4	1
method	proposed	convent.	convent.
Church	29.43	29.47	29.57
Whale Show	28.15	27.85	28.41
Soccer Action	28.48	28.19	28.52
Sprinkling	26.00	25.47	26.53
Horse Race	33.40	33.30	33.41
Average	29.09	28.85	29.28
Disparity of PSNR (dB)			
method	proposed	convent.	convent.
Church	0.74	1.61	0.80
Whale Show	2.07	5.17	1.91
Soccer Action	2.87	4.86	2.77
Sprinkling	4.71	8.66	4.15
Horse Race	0.79	1.52	0.81
Average	2.23	4.26	2.08

8 Experimental System

We developed an experimental SHR CODEC system[9] based on the system we have just described. Its specifications are listed in Table. 4.

The encoder for this CODEC system consists of the pre-encoder unit and the main-encoder unit shown in the photograph in Fig. 14; these units were built with the same hardware, and only the firmware is different. Both units comply with the 1U (44 x 482 x 437 mm) rack size, and includes an audio encoder and a multiplexer, and each has four encoding units that use the new multiple-chip-enhanced 422P@HL

Table 4 Specifications of experimental CODEC system

Video format	1080i(60/59.94i) x 4
	720p(60/59.94p) x 4
Video interface	HDSDI x 4
Stream interface	DVB-ASI x 4
Compression	MPEG-2 422P@HL MP@HL
Bit rate	60 - 160 Mbps
Bitrate control	CBR for TS
	VBR for ES
Power	AC 100 - 240 V
Size(mm)	460(W) x 440(D) x 44(H) x 2

Fig. 14 Photograph of SHR encoding system

Fig. 15 Inside of SHR encoder

MPEG-2 video encoding LSI [7]. There is a photo of the inside of the encoder in Fig. 15.

We used 100 Base-TX for communication between these encoder units. The host processor in the pre-encoder unit performed calculations with the bit-allocation method described in Section. 6, without requiring the use of any external equipment. The SHR decoder consisted of four HDTV decoders, and embodies the STC sharing. The CODEC system can be adapted to many SHR frame sizes by increasing or decreasing the number of CODEC modules instead of having to design and manufacture a new SHR encoder and decoder that can only handle one SHR image resolution.

9 Trial

We carried out live transmission of using a previous version of this system[1] and a 6k x 1k camera[3], we were able to transmit the semifinal game of the 2002 FIFA World Cup Soccer tournament from the Yokohama International Media Center to Yamashita Park as a closed-circuit event.

Fig. 16 Yamashita Park as a closed-circuit event for 2002 FIFA World Cup

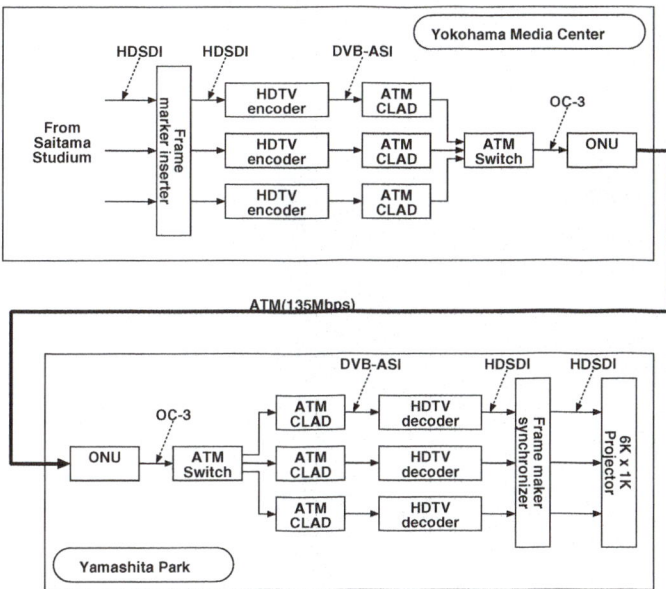

Fig. 17 SHR image transmission system for a 2002 FIFA World Cup

A second example where we used our system was an SHR image recording of an orchestral concert at the Saito Kinen Festival 2004 [12]. The SHR image was captured with a 4K x 2K camera [4], as shown in Fig. 18, and was recorded with our SHR encoder. Four-channel audio streams were recorded without any compression to maintain high audio quality. There is a photograph of the system, which consisted of our SHR encoder and the stream recorder, in Fig. 19. The results obtained demonstrate that the proposed system architecture makes it possible to create high-quality video encoding systems that have scalability in terms of target video resolution.

Fig. 18 4Kx2K camera for SKF2004

Fig. 19 SHR codec and local monitor for SKF2004

10 Conclusion

We propose a multi-channel CODEC system for super-high-resolution video. It uses
the spatially parallel encoding approach and has sufficient scalability for the target
video resolution to be encoded. The switchable cascade multiplexing function of
our system enables various super-high-resolutions to be encoded and super-high-
resolution-video streams to be recorded and played back using a conventional PC.
STC sharing absorbs the disparity in the times streams are received between channels. Two-stage encoding has the ability to equalize the encoding quality of all

partial frames. These functions enable highly-efficient, high-quality encoding for super-high-resolution videos. In the future, we intend to change the MPEG-2 coding LSI of the CODEC system to an H.264 LSI.

References

1. Yoshitome, T., Nakamura, K., Yashima, Y., Endo, M.: A scalable architecture for use in an over-HDTV real-time codec system for multiresolution video. In: Proc. SPIE Visual Communications and Image Processing (VCIP), pp. 1752–1759 (2003)
2. Kanazawa, M.: An Ultra-high-Definition Display Using The Pixel Offset Method. Journal of the Society for Information Display 12(1), 93–103 (2004)
3. MEGA VISION: The Ultra-Widescreen High-Definition Visual System, http://www.megavision.co.jp/pdf/megavision.pdf (accessed October 30, 2009)
4. Octavision: Ultra high-definition digital video camera, http://octavision.olympus-global.com/en (accessed October 30, 2009)
5. RED cameras, http://www.red.com/cameras (accessed October 30, 2009)
6. Yoshitome, T., Nakamura, K., Nitta, K., Ikeda, M., Endo, M.: Development of an HDTV MPEG-2 encoder based on multiple enhanced SDTV encoding LSIs. In: Proc. IEEE International Conference on Consumer Electronics (ICCE), pp. 160–161 (2001)
7. Iwasaki, H., Naganuma, J., Nitta, K., Nakamura, K., Yoshitome, T., Ogura, M., Nakajima, Y., Tashiro, Y., Onishi, T., Ikeda, M., Endo, M.: Single-chip MPEG-2 422P@HL CODEC LSI with Multi-chip Configuration for Large Scale Processing beyond HDTV Level. In: Proc. ACM Design, Automation and Test in Europe Conference (DATE), pp. 2–7 (2003)
8. Ikeda, M., Kondo, T., Nitta, K., Suguri, K., Yoshitome, T., Minami, T., Naganuma, J., Ogura, T.: An MPEG-2 Video Encoder LSI with Scalability for HDTV based on Three-layer Cooperative Architecture. In: Proc. ACM Design, Automation and Test in Europe Conference (DATE), pp. 44–50 (1999)
9. Nakamura, K., Yoshitome, T., Yashima, Y.: Super High Resolution Video CODEC System with Multiple MPEG-2 HDTV CODEC LSIs. In: Proc. IEEE International Symposium Circuits And Systems (ISCAS), vol. III, pp. 793–796 (2004)
10. SMPTE 274-M: 1920x1080 Scanning and Analog and Parallel Digital-Interface for Multiple Picture Rates
11. SMPTE 260-M: 1125/60 High-Definition Production System - Digital Representation and Bit parallel interface
12. SAITO KIKEN FESTIVAL, http://www.saito-kinen.com/e (accessed October 30, 2009)

Chapter 8
Mathematical Modeling for High Frame-Rate Video Signal

Yukihiro Bandoh, Seishi Takamura, Hirohisa Jozawa, and Yoshiyuki Yashima

Abstract. Higher video frame-rates are being considered to achieve more realistic representations. Recent developments in CMOS image sensors have made high frame-rate video signals, over 1000 [Hz], feasible. Efficient coding methods are required for such high frame-rate video signals because of the sheer volume of data generated by such frame rates. Even though it is necessary to understand the statistical properties of these video signals for designing efficient coding methods, these properties have never been clarified, up to now. This chapter establishes, for high frame-rate video, two mathematical models that describe the relationship between frame-rate and bit-rate. The first model corresponds to temporal sub-sampling by frame skip. The second one corresponds to temporal down-sampling by mean filtering, which triggers the integral phenomenon that occurs when the frame-rate is downsampled.

1 Introduction

Highly realistic representations using extremely high quality images are becoming increasingly popular. Realistic representations demand the following four elements:

Yukihiro Bandoh · Seishi Takamura · Hirohisa Jozawa
NTT Cyber Space Laboratories, NTT Corporation, 1-1 Hikari-no-oka, Yokosuka,
Kanagawa, 239-0847 Japan
e-mail: bandou.yukihiro@lab.ntt.co.jp,
takamura.seishi@lab.ntt.co.jp, jozawa.hirohisa@lab.ntt.co.jp

Yoshiyuki Yashima
Faculty of Information and Computer Science, Chiba Institute of Technology,
2-17-1 Tsudanuma, Narashino, Chiba 275-0016, Japan
e-mail: yashima@net.it-chiba.ac.jp

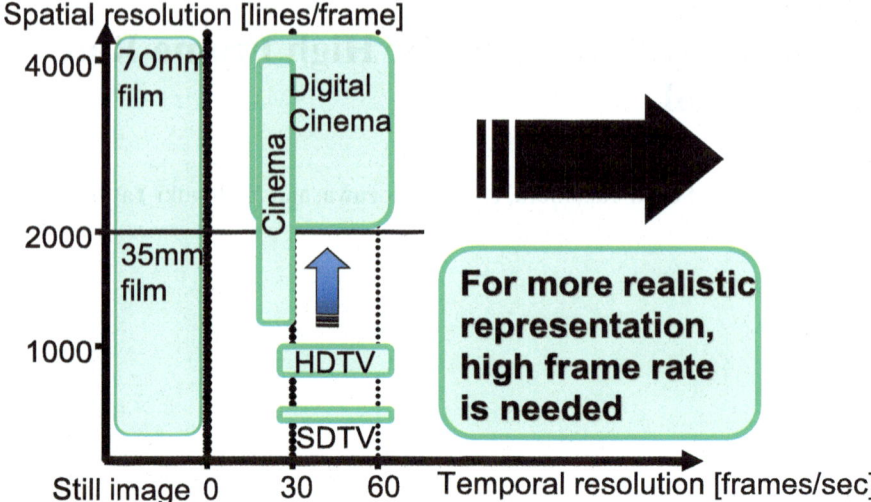

Fig. 1 Spatio-temporal resolution of current video formats

high spatial resolution, large dynamic range, accurate color reproduction, and high temporal resolution. For example, digital cinema[1][2] and Super Hi-Vision TV (SHV)[3][4] offer digital images with high-resolution. As encoders for such high-resolution video formats, H.264 codec for SHV [5] and JPEG2000 codec for digital cinema [6], have been developed. Displays suitable for high-dynamic-range (HDR) images are being developed [7] [8]. JPEG-XR, an HDR image encoding scheme, has been approved as international standard/recommendation [9]. Scalable video coding using tone mapping is one of the approaches being studied for HDR video encoding [10] [11]. Advanced efforts to achieve accurate color reproduction are being made within the Natural Vision Project [12]. H.264/AVC Professional profile supports 4:4:4 color format [13].

In order to create more realistic representations, it is becoming more obvious that the frame-rate is the next factor that will have to be addressed, as shown in Figure 1. The current frame-rate (60 [frames/sec] or [fields/sec]) was simply selected as the lowest rate that well suppressed flickering. Unfortunately, suppressing flicker is not directly connected to the representation of smooth movement. We note that Spillmann found that the gangliocyte of the retina emits up to 300 - 400 [pulses/sec] [14]. Thus, we estimate that the human visual system can perceive light pulses that are 1/150 - 1/200 [sec] long, i.e. the maximum detectable frame-rate is 150 - 200 [frames/sec] from the biochemical viewpoint.

Over the past decade the video acquisition rate has increased drastically. For example, a high-speed HDTV camera that can shoot at 300 [frames/sec] has been developed [15]. Another development is the high speed imaging system that uses large camera arrays [16]. Video technologies that can handle such high frame rates have been opening up a new era in video applications. The Vision Chip architecture [17] realizes a high-speed real-time vision system on a integrated VLSI chip, and has been applied to robotic systems. A real time tracking system that uses a high speed camera has been studied[18]. In addition, an optical flow algorithm for high frame rate sequences has been investigated [19].

Since high frame-rate video requires stronger encoding than low frame-rate video, the statistical properties of high frame-rate video must be elaborated so as to raise encoding efficiency. In particular, it is important to have an accurate grasp of the relationship between frame-rate and bit-rate. When the frame-rate increases, the correlation between successive frames increases. It is easily predicted that increasing the frame-rate decreases the encoded bits of inter-frame prediction error. However, the quantitative effect of frame-rate on bit-rate has not been fully clarified. The modeling of inter-frame prediction error was tackled by [20]. The derivation processes are sophisticated and the results are highly suggestive. Regrettably, however, the model does not consider the effect of frame-rate on prediction error. Modeling the relationship between frame-rate and bit-rate was addressed by the pioneering work of [21] . They assume that some asymptotic characteristics hold, and then inductively generate an interesting model. Unfortunately, however, the model does not consider the effect of motion compensation on the prediction error. In other words, the model covers the bit-rate of the inter-frame prediction error without motion compensation. It is important to consider the inter-frame prediction error with motion compensation, since representative video coding algorithms like H.264/AVC and MPEG-2 adopt inter-frame prediction with motion compensation.

This chapter establishes mathematical models of the relationship between frame-rate and bit-rate in anticipation of encoding high frame-rate video. These models quantify the impact of frame-rate on the bit-rate of inter-frame prediction error with motion compensation. The exact nature of the relationship depends on how the frame-rate is converted. We consider two frame-rate conversion methods. The first one is temporal sub-sampling of the original sequences, that is frame skip, as shown in Figure 2. The shaded rectangles represent frames at each frame-rate and rectangles enclosed by dotted-line represent the down-sampled frames. Figure 2 (b) and (c) illustrate the sequences sub-sampled to 1/2 frame-rate and 1/4 frame-rate, respectively, by subsampling the original sequence shown in Figure 2 (a). The second one is a down-sample filter based on average operator. When the open interval of the shutter in the image pickup apparatus increases, motion blur occurs, which is known as the integral phenomenon. The integral phenomenon changes the statistical properties of the video signal. This integral phenomenon can be formulated as a mean filter. Henceforth, the first model is called temporal sub-sampling and its output sequences are called temporally sub-sampled sequences. The second model is called temporal down-sampling and its output sequences are called temporally down-sampled sequences.

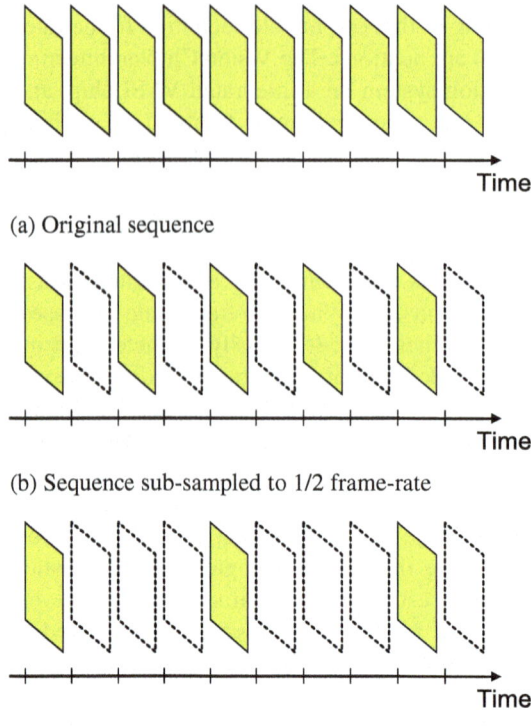

(a) Original sequence

(b) Sequence sub-sampled to 1/2 frame-rate

(c) Sequence sub-sampled to 1/4 frame-rate

Fig. 2 Conversion of frame-rate (Shaded rectangles are frames at each frame-rate. Rectangles enclosed by dotted-lines are down-sampled frames)

2   Relationship between Frame-Rate and Inter-frame Prediction Error

In deriving temporal sub-sampling and temporal down-sampling, this section considers a one-dimensional signal for simplicity. The former is derived in 2.1, and the latter in 2.2. It is trivial matter to extend the following work to cover two-dimensional signals.

2.1   Mathematical Model for Temporal Sub-sampling by Frame Skip

This subsection establishes a mathematical model of the relationship between frame-rate and bit-rate for temporally sub-sampled sequences generated by frame skip. Let $f_t(x)$ denote a one-dimensional signal at position x in the t-th frame with X pixels. Let $B[i]$ $(i = 1, 2, \cdots, X/L)$ be the segment which is a one-dimensional region with L pixels and the i-th segment in $f_t(x)$. When segment $B[i]$ in $f_t(x)$ is predicted from the previous frame by using estimated displacement ($\hat{d}[i]$), the prediction error is given as follows:

$$\sigma[i]^2 = \sum_{x \in B[i]} \{f_t(x) - f_{t-1}(x + \hat{d}[i])\}^2$$

$$= \sum_{x \in B[i]} \{f_{t-1}(x + d[i](x)) - f_{t-1}(x + \hat{d}[i]) + n(x)\}^2$$

$$= \sum_{x \in B[i]} \left\{ \left(\frac{d}{dx} f_{t-1}(x) \right) \zeta[i](x) + \phi(x) + n(x) \right\}^2$$

Where, $\zeta[i](x)$ is displacement estimation error between estimated displacement $\hat{d}[i]$ and true displacement $d[i](x)$ at position x as follows:

$$\zeta[i](x) = d[i](x) - \hat{d}[i]$$

$\phi(x)$ is the second order remainder term of the Taylor expansion, and $n(x)$ is the noise element.

Let us consider the summation of $\sigma[i]^2$ over all segments ($B[i]$ $i(= 1, 2, \cdots, X/L)$). By using the first order approximation of Taylor expansion and the assumption that the noise element is zero-mean white noise and is statistically independent of the video signal, we obtain:

$$\sum_{i=1}^{X/L} \sigma[i]^2 \simeq \sum_{i=1}^{X/L} \sum_{x \in B[i]} \left(\frac{d}{dx} f_{t-1}(x) \right)^2 \zeta[i](x)^2$$

$$+ 2 \sum_{i=1}^{X/L} \sum_{x \in B[i]} \phi(x) \left(\frac{d}{dx} f_{t-1}(x) \right) \zeta[i](x)$$

$$+ \sum_{i=1}^{X/L} \sum_{x \in B[i]} \left(n(x)^2 + \phi(x)^2 \right) \tag{1}$$

In the following, we describe the relationship between displacement and frame-rate. Based on modeling the non-uniform motion of pixels within a block, we have the following approximations about displacement estimation error, as a function of frame-rate F : Let $\hat{d}^F[i]$ and $d^F[i](x)$ be the estimated displacement of segment $B[i]$ and the true displacement, respectively, at position x at frame-rate F.

According to the study by Zhen et al. [22], statistically, block matching based on the sum of squared differences (SSD) criterion will result in displacement that is most likely to be the displacement of block centers. Let $x_c[i]$ be the position of the center of block $B[i]$. Therefore, we have the following approximation about estimated displacement at frame-rate $F = F_0$:

$$\hat{d}^{F_0}[i] \simeq d^{F_0}[i](x_c[i]) \tag{2}$$

Additionally, [22] says that the difference in displacement at position $\hat{x}$ from that at $\hat{x}'$ can be modeled as a zero-mean Gaussian distribution whose variance is proportional to the square of the distance between these positions. Here, position $\hat{x}$ and $\hat{x}'$ are local coordinates in segment $B[i]$ ($i = 0, \ldots, X/L$). According, we have the following statistical model:

$$E[\{(d^{F_0}[i](\hat{x}) - d^{F_0}[i](\hat{x}'))\}^2] \simeq \hat{c}_h^2(\hat{x} - \hat{x}')^2 \tag{3}$$

where $E[\cdot]$ is expectation operator and $\hat{c}_h$ is constant parameter that depends on the original video signal. This model gives a good approximation of the ensemble mean of the difference in displacement $(d^{F_0}[i](\hat{x}) - d^{F_0}[i](\hat{x}'))$ for every segment $B[i]$ ($i = 0, \ldots, X/L$).

From the approximation (2) (3), we create the following approximation :

$$\sum_{i=1}^{X/L} \sum_{x \in B[i]} (d^{F_0}[i](x) - \hat{d}^{F_0}[i])^2$$

$$\simeq \sum_{i=1}^{X/L} \sum_{x \in B[i]} (d^{F_0}[i](x) - d^{F_0}[i](x_c[i]))^2$$

$$\simeq c_h^2 \sum_{\xi} \xi^2 \tag{4}$$

where $c_h^2 = \frac{X}{L}\hat{c}_h^2$ and ξ is $x - x_c[i]$. ξ is a relative coordinate (its origin lies at the center of segment $B[i]$). In other words, ξ is the distance from the center of segment $B[i]$.

We consider the relationship between displacements at different frame-rates. We have the assumption that a moving object exhibits uniform motion across successive frames. This is a highly plausible assumption for high frame rate video signals. In this case, object displacement is proportional to the frame interval. In other words, the displacement is inversely proportional to the frame-rate. It leads to the following equation

$$d^F[i](x) \simeq \frac{F^{-1}}{F_0^{-1}} d^{F_0}[i](x) \tag{5}$$

From the approximation (2) (4) (5), we create the following approximation at frame-rate F :

$$\sum_{i=1}^{X/L} \sum_{x \in B[i]} (d^F[i](x) - \hat{d}^F[i])^2$$

$$\simeq \sum_{i=1}^{X/L} \sum_{x \in B[i]} \left\{ \frac{F^{-1}}{F_0^{-1}} (d^{F_0}[i](x) - \hat{d}^{F_0}[i]) \right\}^2$$

$$= \left(\frac{F^{-1}}{F_0^{-1}}\right)^2 \sum_{i=1}^{X/L} \sum_{x \in B[i]} (d^{F_0}[i](x) - \hat{d}^{F_0}[i])^2$$

$$\simeq F^{-2} F_0^2 c_h^2 \sum_{\xi} \xi^2 \tag{6}$$

By defining $F_0^2 c_h^2 \sum_\xi \xi^2$ as κ_1, we have the following approximation:

$$\sum_{i=1}^{X/L} \sum_{x \in B[i]} \zeta[i](x)^2 \simeq \kappa_1 F^{-2} \tag{7}$$

Next, we consider the relationship between $\sum_{i=1}^{X/L} \sum_{x \in B[i]} \zeta[i](x)$ and frame-rate. Expanding $\left\{ \sum_{i=1}^{X/L} \sum_{x \in B[i]} \zeta^F[i](x) \right\}^2$, we have

$$\left\{ \sum_{i=1}^{X/L} \sum_{x \in B[i]} \zeta^F[i](x) \right\}^2$$
$$= \sum_{i=1}^{X/L} \sum_{x \in B[i]} \left\{ \zeta^F[i](x) \right\}^2 + \sum_{i=1}^{X/L} \sum_{x \in B[i]} \sum_{x' \in B[i], x' \neq x} \zeta^F[i](x) \zeta^F[i](x') \tag{8}$$

where $\zeta^F[i](x)$ is displacement estimation error, defined as follows:

$$\zeta^F[i](x) = d^F[i](x) - \hat{d}^F[i]$$

The first term of equation (8) can be approximated as shown in (6).

About the second term of equation (8), from the Schwarz inequality approach, we have the following inequality:

$$\left\{ \sum_{i=1}^{X/L} \sum_{x \in B[i]} \sum_{x' \in B[i], x' \neq x} \zeta^F[i](x) \zeta^F[i](x') \right\}^2$$
$$\leq \sum_{i=1}^{X/L} \sum_{x \in B[i]} \left\{ \zeta^F[i](x) \right\}^2 \sum_{i=1}^{X/L} \sum_{x' \in B[i], x' \neq x} \left\{ \zeta^F[i](x') \right\}^2$$
$$\simeq \left(F^{-2} F_0^2 c_h^2 \right)^2 \sum_\xi \xi^2 \sum_{\xi'} (\xi')^2$$

where $\xi = x - x_c[i]$ and $\xi' = x' - x_c[i]$. From the above inequalities, we have

$$\sum_{i=1}^{X/L} \sum_{x \in B[i]} \sum_{x' \in B[i], x' \neq x} \zeta^F[i](x) \zeta^F[i](x') \simeq \theta F^{-2} F_0^2 c_h^2 \sqrt{\sum_\xi \xi^2 \sum_{\xi'} (\xi')^2}$$

where θ is a constant in the range -1 to 1.

By inserting the above approximation and approximation (6) into equation (8), we get

$$\left\{ \sum_{i=1}^{X/L} \sum_{x \in B[i]} \zeta^F[i](x) \right\}^2 \simeq F^{-2} F_0^2 c_h^2 \left\{ \sum_\xi \xi^2 + \theta \sqrt{\sum_\xi \xi^2 \sum_{\xi'} (\xi')^2} \right\}$$

Therefore, we have

$$\sum_{i=1}^{X/L} \sum_{x \in B[i]} \zeta^F[i](x) \simeq F^{-1} \gamma F_0 c_h \sqrt{\sum_{\xi} \xi^2 + \theta \sqrt{\sum_{\xi} \xi^2 \sum_{\xi'} (\xi')^2}}$$

where γ is 1 or -1. By defining $\gamma F_0 c_h \sqrt{\sum_{\xi} \xi^2 + \theta \sqrt{\sum_{\xi} \xi^2 \sum_{\xi'} (\xi')^2}}$ as κ_2, we get the following approximation:

$$\sum_{i=1}^{X/L} \sum_{x \in B[i]} \zeta[i](x) \simeq \kappa_2 F^{-1} \tag{9}$$

By assuming that the displacement estimation error $\zeta[i](x)$ is statistically independent of the image intensity derivatives and inserting the above equations into equation (1), we get the following approximation of prediction error per pixel:

$$\frac{1}{X} \sum_{i=1}^{X/L} \sigma[i]^2 \simeq \alpha_1 F^{-2} + \alpha_2 F^{-1} + \alpha_3 \tag{10}$$

where $\alpha_1, \alpha_2, \alpha_3$ are as follows:

$$\alpha_1 = \frac{\kappa_1}{X} \sum_{i=1}^{X/L} \sum_{x \in B[i]} \left(\frac{d}{dx} f_{t-1}(x) \right)^2$$

$$\alpha_2 = \frac{2\kappa_2}{X} \sum_{i=1}^{X/L} \sum_{x \in B[i]} \left(\phi(x) \frac{d}{dx} f_{t-1}(x) \right)$$

$$\alpha_3 = \frac{1}{X} \sum_{i=1}^{X/L} \sum_{x \in B[i]} \left(n(x)^2 + \phi(x)^2 \right)$$

2.2 Mathematical Model of Temporal Down-Sampling by Mean Filter

In this subsection, we establish a mathematical model of the relationship between frame-rate and bit-rate for temporally down-sampled sequences with due consideration of the effect of the integral phenomenon associated with the open interval of the shutter. Let $f_t(x, \delta)$ denote a one-dimensional signal at position x in the t-th frame which was taken with the shutter open in the time interval between t and $t + \delta$. Pixel values in each frame are quantized with 8 [bits] at any interval of shutter open. When the shutter open interval is increased to $m\delta$ (m is a natural number), the corresponding signal $f_{mt}(x, m\delta)$ is given by the following equation:

$$\bar{f}_{mt}(x, m\delta) = \frac{1}{m} \sum_{\tau=mt}^{m(t+1)-1} f_\tau(x, \delta) \tag{11}$$

When segment $B[i]$ in $\bar{f}_{mt}(x, m\delta)$ is predicted from the previous frame by using estimated displacement $(\hat{d}_m[i])$, the prediction error is given as follows:

$$
\begin{aligned}
\sum_{i=1}^{X/L} \sigma_m^2[i] &= \sum_{i=1}^{X/L} \sum_{x \in B[i]} |\bar{f}_{mt}(x, m\delta) - \bar{f}_{m(t-1)}(x + \hat{d}_m, m\delta)|^2 \\
&= \sum_{i=1}^{X/L} \sum_{x \in B[i]} \left\{ \bar{f}_{m(t-1)}(x + d_m(x), m\delta) - \bar{f}_{m(t-1)}(x + \hat{d}_m, m\delta) + n(m) \right\}^2 \\
&= \sum_{i=1}^{X/L} \sum_{x \in B[i]} \left\{ \frac{1}{m} \sum_{j=0}^{m-1} \{ f_{m(t-1)+j}(x + d_m(x), \delta) - f_{m(t-1)+j}(x + \hat{d}_m, \delta) \} + n(m) \right\}^2 \\
&= \sum_{i=1}^{X/L} \sum_{x \in B[i]} \left\{ \frac{\left\{ \sum_{j=0}^{m-1} \frac{d}{dx} f_{m(t-1)+j}(x, \delta) \right\}}{m} \zeta_m[i](x) + \phi(x) + n(m) \right\}^2
\end{aligned}
\tag{12}
$$

where $\phi(x)$ is the second order remainder term of the Taylor expansion, and $n(m)$ is the noise element. $\zeta_m[i](x)$ is displacement estimation error between estimated displacement $\hat{d}_m[i]$ and the true displacement $d_m[i](x)$ at position x as follows:

$$
\zeta_m[i](x) = d_m[i](x) - \hat{d}_m[i]
$$

Henceforth, we substitute $f_t(x)$ for $f_t(x, \delta)$ for simplicity, unless otherwise stated.

By inserting the above equation into equation (7) (9) and using the first order approximation of the Taylor expansion and the assumption that the noise element is statistically independent of the video signal, we obtain:

$$
\sum_{i=1}^{X/L} \sigma_m[i]^2 \simeq \beta_1(m) F^{-2} + \beta_2(m) F^{-1} + \beta_3(m)
\tag{13}
$$

where $\beta_1(m), \beta_2(m), \beta_3(m)$ are as follows:

$$
\beta_1(m) = \kappa_1 \sum_{i=1}^{X/L} \sum_{x \in B[i]} \left\{ \frac{1}{m} \sum_{j=0}^{m-1} \frac{d}{dx} f_{m(t-1)+j}(x) \right\}^2
$$

$$
\beta_2(m) = 2\kappa_2 \sum_{i=1}^{X/L} \sum_{x \in B[i]} \left\{ \frac{1}{m} \sum_{j=0}^{m-1} \frac{d}{dx} f_{m(t-1)+j}(x) \right\} \phi(x)
$$

$$
\beta_3(m) = \sum_{i=1}^{X/L} \sum_{x \in B[i]} \{ \phi(x)^2 + n(m)^2 \}
$$

Henceforth, we set

$$
\mu_{mt}(x) = \frac{1}{m} \sum_{j=0}^{m-1} f_{m(t-1)+j}(x)
$$

$\beta_1(m)$ is expanded as follows:

$$\beta_1(m) = \kappa_1 \sum_{i=1}^{X/L} \sum_{x\in B[i]} \left\{ \frac{d}{dx}\mu_{mt}(x) \right\}^2$$

$$\simeq \kappa_1 \sum_{i=1}^{X/L} \sum_{x\in B[i]} \{\mu_{mt}(x) - \mu_{mt}(x-1)\}^2$$

$$= \kappa_1 \sum_{i=1}^{X/L} \sum_{x\in B[i]} \mu_{mt}(x)^2 + \kappa_1 \sum_{i=1}^{X/L} \sum_{x\in B[i]} \mu_{mt}(x-1)^2 - 2\kappa_1 \sum_{i=1}^{X/L} \sum_{x\in B[i]} \{\mu_{mt}(x)\mu_{mt}(x-1)\}$$

$$\simeq 2\kappa_1 \sum_{i=1}^{X/L} \sum_{x\in B[i]} \mu_{mt}(x)^2 - 2\kappa_1 \sum_{i=1}^{X/L} \sum_{x\in B[i]} \{\mu_{mt}(x)\mu_{mt}(x-1)\}$$

$$\simeq \kappa_1 \frac{2\sigma_s^2(1-\rho)}{m^2} \left\{ m - \frac{1-\rho}{\rho} \sum_{i>j} \eta_{i,j}\rho^{|\bar{d}_i-\bar{d}_j|} \right\}$$

In the above approximation, we assume that κ_1 is statistically independent of $\mu_{mt}(x)$, and we use the following homogeneous model

$$\sum_{i=1}^{X/L} \sum_{x\in B[i]} \{f_t(x)\}^2 = \sigma_s^2$$

$$\sum_{i=1}^{X/L} \sum_{x\in B[i]} \{f_t(x)f_t(x+k)\} = \sigma_s^2\rho^k$$

and the following approximation

$$\sum_{i=1}^{X/L} \sum_{x\in B[i]} \{f_t(x+d_i(x))f_t(x+d_j(x))\}$$

$$\simeq \eta_{i,j} \sum_{i=1}^{X/L} \sum_{x\in B[i]} \{f_t(x+\bar{d}_i)f_t(x+\bar{d}_j)\}$$

$$= \eta_{i,j}\sigma_s^2\rho^{|\bar{d}_i-\bar{d}_j|}$$

where $\bar{d}_i$ and $\bar{d}_j$ are the mean values of $d_m[i](x)$ and $d_m[j](x)$, respectively, and $\eta_{i,j}$ is a parameter to approximate $d_m[i](x)$ and $d_m[j](x)$ using mean displacement ($\bar{d}_i$ and $\bar{d}_j$).

We can assume that ρ is less than but close to one, since ρ is the autocorrelation coefficient of the image signal. Thus, we have

$$\frac{1-\rho}{\rho} \ll 1.$$

Using this inequality, equation (14) can be approximated as follows:

$$\beta_1(m) \simeq \kappa_1 \frac{2\sigma_s^2(1-\rho)}{m}$$

Since m is the ratio of downsampled frame-rate F to maximum frame-rate F_0, we have

$$\beta_1(m) \simeq \kappa_1 \frac{2\sigma_s^2(1-\rho)}{F_0} F$$

In a similar way, we have

$$\beta_2(m) \simeq 2\kappa_2\gamma\phi(x)\sqrt{\frac{2\sigma_s^2(1-\rho)}{F_0}} F$$

where γ is 1 or -1.

Next, let us consider $\beta_3(m)$. Since we assume that the noise element $n(m)$ is statistically independent of the video signal, the averaging procedure denoted by equation (11) reduces $n(m)$ as follows:

$$n(m)^2 = \sum_{i=1}^{X/L} \sum_{x \in B[i]} \frac{n_0^2}{m}$$

$$= X \frac{n_0^2}{F_0} F$$

where, n_0 is the noise signal included in the sequence at frame-rate F_0.

We have the following approximation of prediction error per pixel:

$$\frac{1}{X} \sum_{i=1}^{X/L} \sigma_m^2[i] = \hat{\beta}_1 F^{-1} + \hat{\beta}_2 F^{-1/2} + \hat{\beta}_3 F + \hat{\beta}_4 \tag{14}$$

where, $\hat{\beta}_1$, $\hat{\beta}_2$, $\hat{\beta}_3$, and $\hat{\beta}_4$ are as follows:

$$\hat{\beta}_1 = \frac{2\kappa_1\sigma_s^2(1-\rho)}{XF_0}$$

$$\hat{\beta}_2 = \frac{2\kappa_2\gamma\phi(x)}{X}\sqrt{\frac{2\sigma_s^2(1-\rho)}{F_0}}$$

$$\hat{\beta}_3 = \frac{n_0^2}{F_0}$$

$$\hat{\beta}_4 = \frac{1}{X} \sum_{i=1}^{X/L} \sum_{x \in B[i]} \{\phi(x)\}^2$$

Let us consider the effect of noise components in our models: the third term of equation (10) and the third term of equation (14). The former is constant i.e.

independent of frame-rate. The latter increases with frame-rate, i.e. prediction error may decrease when frame-rate decreases. The reason for this difference is as follows. When the shutter open interval increases, relative noise power decreases because of the increase in light intensity. This relative decrease in noise can lead to a decrease in inter-frame prediction error. Therefore, we understand that the third term of equation (14) decreases as the frame-rate decreases. On the other hand, that of equation (10), the relative noise power is independent of frame-rate, since this model assumes that the shutter open interval is fixed.

3 Evaluation of Proposed Models

3.1 Captured High Frame-Rate Video Signal

The original sequences in the following experiments consisted of 480 frames at 1000 [frames/sec]. Sequences taken with a high speed camera at 1000 [frames/sec] were output in 24 bit RGB format. The frame interval equaled the shutter speed of 1/1000 [sec]. The video signal was not gamma corrected. We converted the color format from RGB data to YCbCr data. Y-data with 8 bit gray scale were used in the rate evaluation experiments. The following three sequences were used; "golf" was the scene of a golf player swinging his club. "tennis" was the scene of a tennis player hitting a ball with her racket. "toy" was the scene where the rings of a toy rotated independently. All sequences were captured without camera work.

In order to identify the relationship between frame-rate and bit-rate, the sequences with different frame rates were generated by two down-sampling methods; temporal sub-sampling and temporal down-sampling. Temporal sub-sampling is frame-rate conversion using frame-skip as shown in Figure 2, and temporal down-sampling means frame-rate conversion using the mean filter described in equation (11). The temporally sub-sampled sequences were utilized for evaluating the model described by equation (10). The temporally down-sampled sequences were utilized for evaluating the model described by equation (14).

3.2 Regression Analyses of the Proposed Models

We performed regression analyses in order to verify the validity of the above-mentioned models. In Figure 3, the dot symbols show the results of rate evaluation experiments on the original sequences and the temporally sub-sampled sequences generated by frame-skip, while the solid lines plot the results of the proposed model given by equation (10). The horizontal axis is the frame-rate [frames/sec] and the vertical axis is the bit-rate [bits / pixel] which is the entropy of inter-frame prediction error. The parameters $(\alpha_1, \alpha_2, \alpha_3)$ were obtained by least-squares estimation. In Figure 4, the dot symbols show the results of rate evaluation experiments on the original sequences and the temporally down-sampled sequences generated by mean-filter, while the solid lines plot the results of the proposed model given by equation (14).

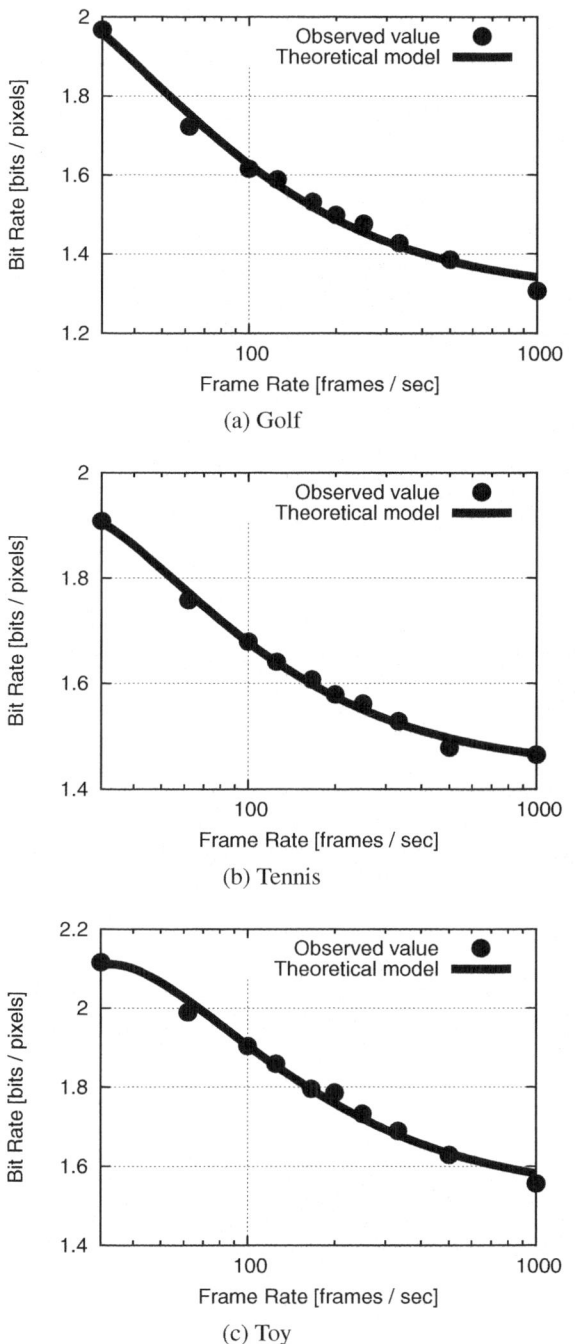

Fig. 3 Relationship between frame-rate and bit-rate of inter-frame prediction error

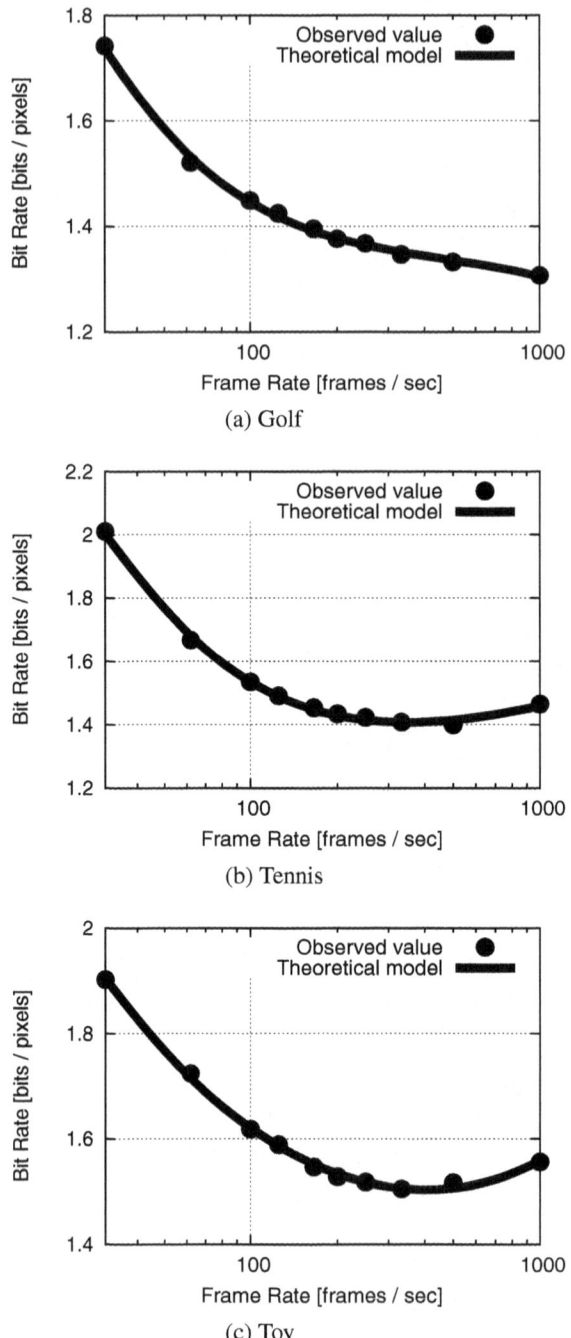

Fig. 4 Relationship between frame-rate and bit-rate of inter-frame prediction error

The horizontal axis and the vertical axis are the same as those of Figure 3, respectively. The parameters $(\hat{\beta}_1, \hat{\beta}_2, \hat{\beta}_3, \hat{\beta}_4)$ were obtained by least-squares estimation.

We used the following inter-frame prediction with motion compensation for rate evaluation, since representative video coding algorithms like H.264/AVC and MPEG-2 adopt inter-frame prediction with motion compensation. Each frame was divided into blocks, and each block was predicted from its previous frame by motion compensation. The number of references was one. The block size used for motion compensation was 16×16 [pixels]. The search range of motion estimation was $\pm$ 8 [pixels] at 1000 [frames/sec]. The search range decreased according as the frame-rate increased. For example, we set $\pm$ 16 [pixels] at 500 [frames/sec]. The motion estimation scheme was full search algorithm. The criterion of motion estimation was the sum of absolute differences (SAD) between current block and reference block. Namely, selected displacement minimized SAD between current block and reference block. Figure 5 shows bit-rate of motion vectors of the original sequences and the temporally sub-sampled sequences, and Figure 6 shows those of the original sequences and the temporally down-sampled sequences. The horizontal axis is the frame-rate [frames/sec] and the vertical axis is the bit-rate [bits/pixel] which is the sum of entropy of the two elements of the motion vector.

As shown in Figure 3, the results of the experiments well agree with the values yielded by the proposed model. In other words, equation (10) and equation (14) well model the relationship between the bit-rate of prediction error and frame-rate. Table 1 and table 2 show residual sum of squares (RSS) between the results of rate evaluation experiments, and the theoretical values from the proposed model, as a measure of the fitness of the proposed model. Table 1 shows the results for the temporally sub-sampled sequences and Table 2 shows those for the temporally down-sampled sequences. In these tables, we compare our model with the conventional model in [21] which is expressed as follows:

$$I(F) = a_1 a_2 (1 - \exp(-1/(a_2 F))) \tag{15}$$

where a_1 and a_2 are constants that depends on the video signal. In this experiment, parameters (a_1 and a_2) were obtained by least-squares estimation. As shown in Table 1 and 2, our model achieved smaller RSS than the conventional model.

Table 1 Residual sum of squares (RSS) of temporally sub-sampled sequences

model	RSS (golf)	RSS (tennis)	RSS (toy)
proposed	3.39e-03	2.96e-04	2.60e-03
conventional	2.24e-01	1.29e-01	1.62e-01

Table 2 Residual sum of squares (RSS) of temporally down-sampled sequences

model	RSS (golf)	RSS (tennis)	RSS (toy)
proposed	2.60e-04	6.33e-04	5.46e-03
conventional	1.27e-01	3.09e-01	1.37e-01

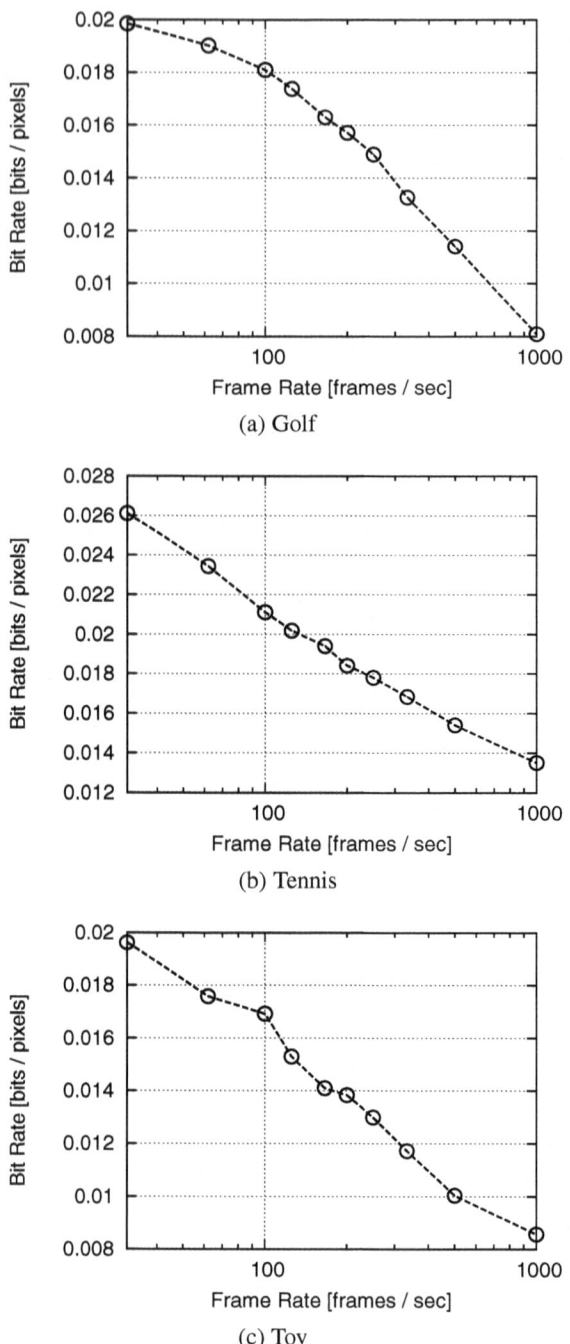

(a) Golf

(b) Tennis

(c) Toy

Fig. 5 Relationship between frame-rate and bit-rate of motion vectors

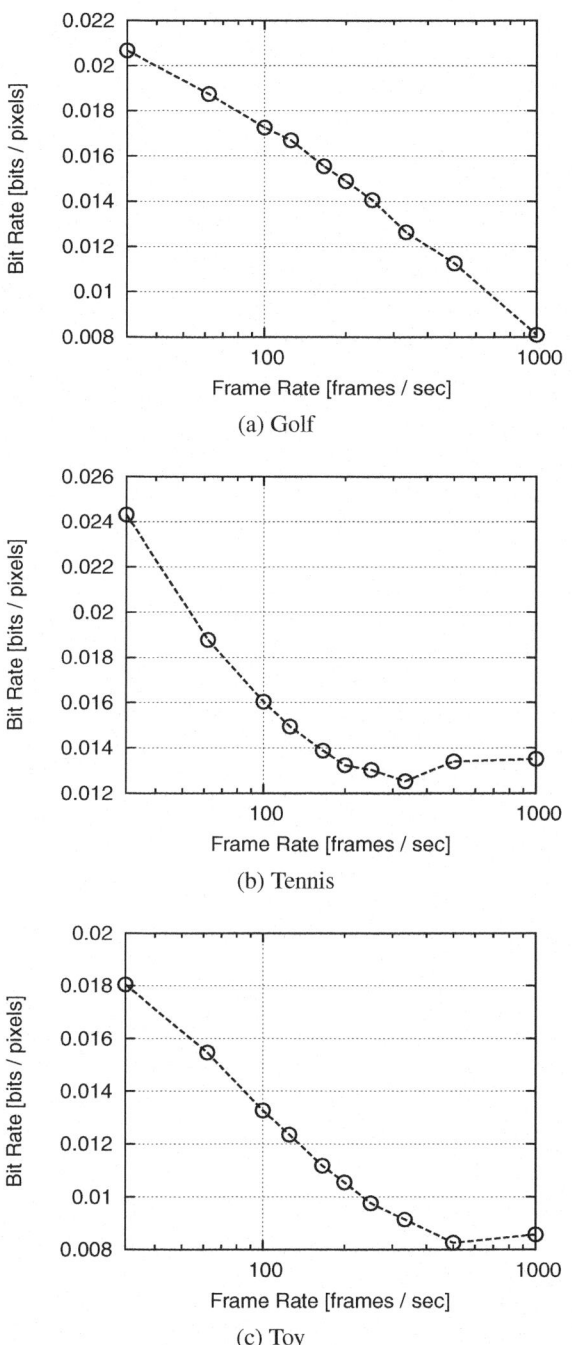

(a) Golf

(b) Tennis

(c) Toy

Fig. 6 Relationship between frame-rate and bit-rate of motion vectors

The superiority of the proposed model over the conventional one is due to the following reasons. The conventional model is constructed based on the assumption that the information of inter-frame prediction error converges to zero at infinitely large frame-rate, i.e. an asymptotic characteristic. The inter-frame prediction is the frame difference and does not consider motion compensation. Furthermore, the conventional model does not consider the above-described noise elements caused by thermal noise in the imaging device. Therefore, the fitness of the model degrades when blocks have large displacements and the effect of the noise elements grows.

From Figure 4(b)(c), we can confirm that the bit-rate of prediction error may decrease as the frame-rate increase to approach 500 [fps]; above this frame-rate, the bit-rate increases. For the case of the temporal down-sampling using the mean filter described in equation (11), the shutter-open interval increases with the decrease in the frame-rate. The increase in the shutter-open interval leads to the suppression of the noise elements caused by the thermal noise in the imaging device and the reduction of the spatio-temporal high frequency components of down-sampled sequences. This is why the bit-rate of the sequences generated by the temporal down-sampling may decrease as the frame-rate increases.

4 Conclusion

In this chapter, we analytically derive two mathematical models that quantify the relationship between frame-rate and bit-rate. The first model supports temporal sub-sampling through the frame skip approach. The second one supports temporal down-sampling realized by a mean filter; it incorporates the integral phenomenon associated with the open interval of the shutter. By using these models, we can describe the properties associated with frame-rate, that have not been clarified in previous studies. We can confirm that the derived models well approximate our experimental results. These evaluation results support the validity of the assumptions used in deriving our models.

Acknowledgements. We would like to thank Dr. T.Nakachi, NTT Network Innovation Laboratories, for his useful suggestions and his valuable advice. We also appreciate the members of video coding group of NTT Cyber Space Laboratories for their cooperation.

References

1. Dettmer, R.: Digital cinema: a slow revolution. IEE Review 49(10), 46–50 (2003)
2. Husak, W.: Economic and other considerations for digital cinema. Signal Processing: Image Communication 19(9), 921–936 (2004)
3. Kanazawa, M., Kondoh, M., Okano, F., Haino, F., Sato, M., Doi, K., Hamada, K.: An ultrahigh-definition display using the pixel offset method. Journal of the SID (2004)
4. Nojiri, Y.: An approach to Ultra High-Definition TV. In: Proc. Int. symposium on universal communication, Kyoto, Japan (2007)

5. Sakaida, S., Nakajima, N., Iguchi, K., Gohshi, S., Kazui, K., Nakagawa, A., Sakai, K.: Development of avc/h.264 codec system for super hi-vision. In: Proc. Int. Workshop on Advanced Image Technology, Bangkok, Thailand (2008)
6. Shirai, D., Yamaguchi, T., Shimizu, T., Murooka, T., Fujii, T.: 4k SHD real-time video streaming system with JPEG 2000 parallel codec. In: IEEE Asia Pacific Conference on Circuits and Systems, Singapore (2006)
7. Durand, F., Dorsey, J.: Fast bilateral filtering for the display of high-dynamic-range images. In: Proc. SIGGRAPH, San Antonio, TX, USA (2002)
8. Meylan, L., Susstrunk, S.: High dynamic range image rendering with a retinex-based adaptive filter. IEEE Trans. on Image Proc. 15(9) (2005)
9. ITU-T Recommendation T.832 and ISO/IEC 29199-2. Information technology - JPEG XR image coding system - Part 2: Image coding specification (2009)
10. Mantiuk, N., Efremov, N., Myszkowski, K., Seidel, H.: Backward compatible high dynamic range MPEG video compression. ACM Trans. Graph. 25(3), 713–723 (2006)
11. Segall, A.: Scalable coding of high dynamic range video. In: Proc. IEEE Int. Conf. on Image Processing, San Antonio, TX, USA (2007)
12. Ohsawa, K., Ajito, K., Fukuda, H., Komiya, Y., Haneishi, H., Yamaguchi, H., Ohyama, N.: Six-band hdtv camera system for spectrum-based color reproduction. Journal of Imaging Science and Technology 48(2), 85–92 (2004)
13. Sullivan, G., Yu, H., Sekiguchi, S., Sun, H., Wedi, T., Wittmann, S., Lee, Y., Segall, A., Suzuki, T.: New standardized extensions of MPEG4-AVC/H.264 for professional-quality video applications. In: Proc. IEEE Int. Conf. on Image Processing, San Antonio, TX, USA (2007)
14. Spillmann L., Werner J.: Visual perception the neurophysiological foundations. Academic Press, San Diego (1990)
15. Ogasawara, T., Yamauchi, M., Tomura, Y., Yamazaki, J., Gotoh, M., Hashimoto, Y., Cho, H., Kochi, E., Kanayama, S.: A 300fps progressive scan HDTV high speed camera. The Journal of ITE 60(3), 358–365 (2007) (in Japanese)
16. Wilburn, B., Joshi, N., Vaish, V., Talvala, E., Antunez, E., Barth, A., Adams, A., Horowitz, M., Levoy, M.: High performance imaging using large camera arrays. ACM Trans. on Graphics 24(3), 765–776 (2005)
17. Komuro, T., Ishii, I., Ishikawa, M., Yoshida, A.: A digital vision chip specialized for high-speed target tracking. IEEE trans. on Electron Devices 50(1), 191–199 (2003)
18. Muehlmann, U., Ribo, M., Lang, P., Pinz, A.: A new high speed CMOS camera for real-time tracking applications. In: Proc. IEEE Int. Conf. on Robotics and Automation, New Orleans, LA, USA (2004)
19. Lim, S., Gamal, A.: Optical flow estimation using high frame rate sequences. In: Proc. IEEE Int. Conf. on Image Processing, Thessaloniki, Greece (2001)
20. Shishikui, Y.: A study on modeling of the motion compensation prediction error signal. IEICE Trans. Communications E75-B(5), 368–376 (1992)
21. Murayama, N.: The law of video data compression. In: Proc. PSCJ, Japan (1988) (in Japanese)
22. Zheng, W., Shishikui, Y., Naemura, M., Kanatsugu, Y., Itho, S.: Analysis of space-dependent characteristics of motion-compensated frame differences based on a statistic motion distribution model. IEEE Trans. on Image Proc. 11(4), 377–386 (2002)

Part III
Visual Content Upscaling

Chapter 9
Next Generation Frame Rate Conversion Algorithms

Osman Serdar Gedik, Engin Türetken, and Abdullah Aydın Alatan

Abstract. There is an increasing trend towards panel displays in consumer electronics, and they are already replacing conventional Cathode Ray Tube (CRT) displays due to their various advantages. However, the main problem of the panel displays, namely motion blur, still remains unsolved. This shortcoming should be overcome efficiently to satisfy increasing demands of viewers such as artifact-free interpolation in dynamic videos. Among many frame-rate up conversion (FRUC) methods that address this problem, motion-compensated frame interpolation (MCFI) algorithms yield superior results with relatively less artifacts. Conventional MCFI techniques utilize block-based translational motion models and, in general, linear interpolation schemes. These methods, however, suffer from blocking artifacts especially at object boundaries despite several attempts to avoid them. Region-based methods tackle this problem by segmenting homogeneous, or smoothly varying, motion regions that are supposed to correspond real objects (or their parts) in the scene. In this chapter, two region-based MCFI methods that adopt 2D homography and 3D rigid body motion models are presented in the order of increasing complexity. As opposed to the conventional MCFI approaches where motion model interpolation is performed in the induced 2D motion parameter space, the common idea behind both methods is to perform the interpolation in the parameter space of the original 3D motion and structure elements of the scene. Experimental results suggest that the proposed algorithms achieve visually pleasing results without halo effects on dynamic scenes with complex motion.

Osman Serdar Gedik · Abdullah Aydın Alatan
Department of Electrical and Electronics Engineering, METU, 06531 Balgat, Ankara, Turkey
e-mail: gedik@eee.metu.edu.tr, alatan@eee.metu.edu.tr

Engin Türetken
Ecole Polytechnique Federale de Lausanne, EPFL, CH-1015 Lausanne, Switzerland
e-mail: engin.turetken@epfl.ch

1 Introduction

Next generation hold-type panel displays emerged into the consumer market and the demand for these items is expected to continue growing. In the near future, due to their compact and aesthetic design, as well as high brightness, the hold type displays are expected to completely replace Cathode Ray Tube (CRT) displays, with scanning type properties. However, one of the main drawbacks of hold-type displays, such as Liquid Crystal Displays (LCD) and Electro-Luminescence Displays (ELD), is the motion blur which is caused by two major reasons [1]:

Long response time of crystal cell: In moving images, pixel luminance changes between two consecutive frames. Due to its long response time, the luminance switch can not be finished within one frame period, which leads to the smearing of object edges.

Constant pixel luminance over the entire frame period (hold-type): Human eye is apt to move and track the moving objects while the successive frames are displayed. In hold-type displays, when eyes move, many pixels are displayed on the same retinal position during one frame period, which causes blurred image observed. The faster the motion is, the more pixels are accumulated at the same retinal position during one frame period and the more motion-blur is observed.

Therefore, for a better perception quality, the frame rate of the original video signal is required to be increased. Frame rate up-conversion (FRUC), or scan/field rate up-conversion, which emerged as a result of this necessity, is a technique of increasing the frame rate of a video signal by inserting interpolated frame(s) in-between the successive frames of the original signal. The two most commonly used FRUC techniques are frame repetition and motion compensation based methods. Frame repetition generally yield inferior results, especially on complex (high depth variation and clutter) or dynamic (moving objects) scenes. Motion compensation based approaches yield relatively more elegant results on these scenes with the cost of a higher computational complexity and provided that accompanying motion estimation and interpolation schemes provide satisfactorily accurate estimates. Although all the methods in this category perform motion estimation, motion model parameter interpolation and motion compensation steps, they exhibit a great variety in the way the three steps are performed. Unlike coding approaches, which simply try to minimize the residual error between adjacent frames, in FRUC algorithms it is crucial to estimate the *correct* motion parameters.

2 Related Work

Most of the existing algorithms in the literature utilize block-based motion estimation, [2] - [18], based on the assumption that all the pixels of an individual block have the same *translational* motion. One of the key elements in the motion estimation for FRUC is the temporal search direction. In addition to the backward directional motion estimation, which is mostly preferred by video coding techniques, FRUC approaches also utilize forward directional motion estimation, [2], bi-directional motion estimation, [8], or an adaptive combination of all three search directions, [10].

As an integral part of the motion estimation, search patterns with fast convergence rates and capability of taking into account *true* object motion should be utilized in order to interpolate more pleasant inter-frames. Among many algorithms, three step[14], spiral [6], hexagonal [15], diamond and modified diamond, [16] - [17], are the most popular search patterns utilized for FRUC. Although, in general, it fails to estimate correct motion vectors, [13], several techniques utilize *full-search* method, [2] - [8].

Mean absolute differences (MAD) ([6, 15, 16]) and *sum of absolute differences (SAD)* ([4, 9, 11, 13]) are widely exploited match error criteria in block-based motion estimation. While SAD avoids division operation, and hence provides computational efficiency over MAD, it does not enforce spatial smoothness. In order to alleviate this problem, Choi et. al, [3], combine SAD measure with side match distortion as a way to estimate relatively smooth motion fields.

The main drawback of block based motion estimation and compensation approaches is that they suffer from blocking artifacts especially occurring at object boundaries. As an attempt to reduce these artifacts, Ha et al., [2], utilize larger blocks for motion estimation, whereas the block size is reduced during motion compensation. In a similar spirit, there are also pixel-based approaches, [20, 21] , which convert block-based motion fields to pixel-based motion fields prior to compensation. Similarly, [4] adaptively changes motion compensation block size depending on the size of the objects. Alternatively, a class of methods cope with blocking artifacts by using overlapping blocks in motion compensation stage [8, 3]. With the same concern, Lee et al. consider neighboring blocks' motion trajectories by computing a weighted average of multiple compensations [19].

Most of the algorithms in the literature employ a two-frame approach for motion estimation and compensation. Although the utilization of two frames may not effectively handle the occlusion problem , such an approach is enforced by existing hardware limitations, caused by the adversity of providing extra storage for multi-frames. However, there exist algorithms that exploit multiple frames at the cost of increased complexity [9, 10].

Techniques that are based on translational motion model (such as the ones mentioned above) generally rely on the idea of linearly interpolating motion vectors between the original successive frames [22] - [26]. Alternatively, a higher order function, such as a polynomial, can be used for the interpolation as proposed in [27]. On the other hand, the interpolation problem becomes more complicated for higher order parametric motion models (e.g., six-parameter affine and eight-parameter homography mappings) due to the need for defining a reasonable model parameter space.

Although blocking artifacts inherent in block-based techniques are intended to be solved by using previously mentioned techniques, they cannot be avoided completely at object boundaries, where multiple-motion exists. In an attempt to segment true object boundaries, region-based FRUC techniques estimate arbitrarily-shaped segments with smoothly varying motion fields that are modeled by higher order parametric models [28, 29].

In this chapter, two region-based FRUC methods with increasing complexity, namely region-based FRUC by homography parameter interpolation (FRUC-RH) by Türetken and Alatan [30], and its multi-view extension, region-based multi-view FRUC by rigid body motion model parameter interpolation (FRUC-RM), are presented. While both methods utilize a motion segmentation step initially, they mainly differ in the order of the parametric model incorporated. The common idea behind both methods is to perform the motion model interpolation step in the parameter space of the original 3D motion and structure elements of the scene, as opposed to the conventional FRUC approaches where the interpolation is performed in the induced 2D motion parameter space. In FRUC-RH, under several practically reasonable assumptions (such as piecewise planar scene model and small angle approximation of 3D rotation), this procedure simplifies to linearly interpolating a set of 2D planar perspective motion parameters corresponding to segments, or layers. Layer support maps at the interpolation time instant(s) are then generated by using these interpolated motion models and the layer maps of the neighboring successive frames. On the other hand, FRUC-RM addresses the more general case of arbitrary scenes by dense reconstructing them prior to motion parameter estimation at the successive frames. Under rigid body motion assumption, motion model parameters are estimated using the reconstructed 3D coordinates of the scene. These parameters are then interpolated in the parameter space of the 3D rigid body motion at the interpolation time instant(s) between the successive frames.

3   Region-Based FRUC by Homography Model Parameter Interpolation (FRUC-RH)

In this section, a region-based motion compensated frame interpolation method, which uses segmented motion layers with planar perspective models, is described. Under several practically reasonable assumptions, it is shown that performing the motion model interpolation in the homography parameter space is equivalent to interpolating the parameters of the real camera motion, which requires decomposition of the homography matrix. Based on this reasoning, backward and forward motion models from the interpolation frame(s) to the successive frames of the original sequence are estimated for each motion layer. The interpolated motion models are then used to warp the layer support maps at the point(s) of interpolation in time. Finally, new interpolation frame(s) are generated from the two successive frames by taking into account layer occlusion relations and local intensity similarities.

The method is comprised of four main steps, which are presented in the following sub-sections: (a) establishing motion layer correspondences, (b) interpolating motion model parameters in both forward and backward directions, (c) interpolating layer maps corresponding to middle and previous frames in time, and (d) generating the interpolated frame.

Fig. 1 Sample segmentation results for *Mobile & Calendar* sequence. From left to right, top row: frames 19, 21 and 23 of the original sequence, bottom row: corresponding motion segmentation maps.

3.1 Motion Layer Correspondence Establishment

The first step of the proposed method is the estimation of a set of spatial regions (hereafter referred to as motion layers), in which motion changes smoothly. Layer segmentation maps for each pair of consecutive frames F^{t-1} and F^t, between which a number of frames $\{F^{t-\Delta t_1}, F^{t-\Delta t_2}, \ldots, F^{t-\Delta t_n}\}$ $(0 < \Delta t_i < 1$, $i = 1, \ldots, n)$ are to be interpolated, are extracted by using a variant of the motion segmentation algorithm by Bleyer et al. [31]. Spatial smoothness in layers is enforced by modeling layer motion by a planar perspective mapping (i.e., eight parameter homography model). An additional cost term for temporal coherence of motion layers is incorporated in the original cost function so as to make the layer extraction process robust to abrupt changes in layer appearances along the temporal axis. Fig. 1 presents sample motion segmentation results for *Mobile & Calendar* sequence. Although temporal coherence of motion layers is enforced explicitly, mismatches between layer boundaries still exist due to several reasons, such as articulated object motion and object-background appearance similarities. In addition, an object entering or exiting the scene naturally results in addition or deletion of a layer. Therefore, as required by the following motion parameter interpolation procedure, motion layers of consecutive time instants have to be linked in time.

The correspondences between the estimated motion layers at time instants $t - 1$ and t are established by mapping the layers and their pair-wise similarity scores to a bipartite graph. In a bipartite graph $G = (U, V, E)$ with two disjoint sets of vertices U and V, and a set of edges E, every edge connects a vertex in U and a vertex in V. The set of layers corresponding to F^{t-1} and F^t are mapped to the disjoint sets U and V with vertices representing the layers and weighted edges representing the pair wise layer similarities. The normalized edge weight between a layer L_i^{t-1} of F^{t-1}, and a backward motion-compensated layer $L_j^{t,w}$ of F^t is defined as

$$E(L_i^{t-1}, L_j^{t,w}) = \frac{|L_i^{t-1} \cap L_j^{t,w}|}{min(|L_i^{t-1}|, |L_j^{t,w}|)}, \qquad (1)$$

where $L_i^{t-1} \cap L_j^{t,w}$ denotes the overlapping region of the layers and $|.|$ denotes the area of a region. The initial graph constructed with the above similarity weights has an edge between every layer in U and every layer in V. This redundancy is eliminated by deleting the links having weights below a predefined threshold. However, the links, whose source or target vertices has only one link left, are retained to ensure that every vertex is connected to the graph.

3.2 Motion Model Parameter Interpolation

Model parameter interpolation refers to estimating forward and backward motion models for a set of layers corresponding to the interpolation frames by using the backward models from F^t to F^{t-1}. Suppose that only a single frame $F^{t-\Delta t}$, corresponding to time instant $t - \Delta t$ $(0 < \Delta t < 1)$, is to be interpolated between the original frames. Given a set of backward layer motion models $\{P_1^t, P_2^t, \ldots, P_n^t\}$, representing the motion from F^t to F^{t-1}, model parameter interpolation problem can be defined as estimating the parameters of forward models $\{P_{1,f}^{t-\Delta t}, P_{2,f}^{t-\Delta t}, \ldots, P_{n,f}^{t-\Delta t}\}$ from $F^{t-\Delta t}$ to F^t and backward models $\{P_{1,b}^{t-\Delta t}, P_{2,b}^{t-\Delta t}, \ldots, P_{n,b}^{t-\Delta t}\}$ from $F^{t-\Delta t}$ to F^{t-1}. The problem is depicted in Fig. 2 for a single layer. In order to compute a meaningful parameter set to define the flow between a layer in the frame to be

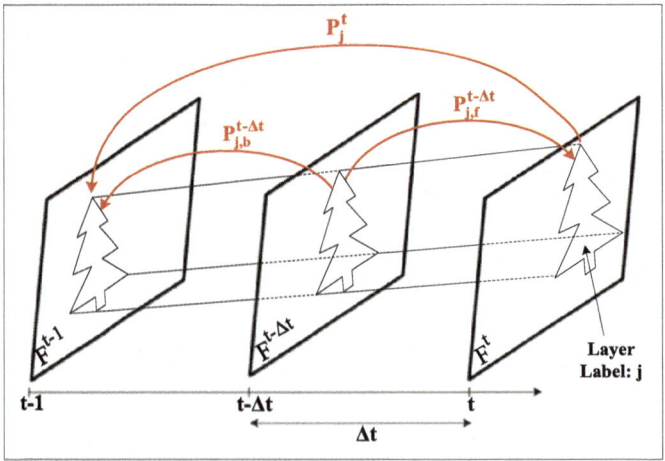

Fig. 2 Motion model parameter interpolation

interpolated and the corresponding layers in the two original frames of an input sequence, it is useful to express the flow in terms of parameters of induced 3D motion. Hence, model parameter interpolation for the eight-parameter homography mapping involves in decomposition of the homography matrix into structure and motion elements of the captured scene.

Homography model is appropriate for compactly expressing the induced 2D motion caused by a moving planar scene or a rotating (and zooming) camera capturing an arbitrary scene. Suppose that the observed world points corresponding to an estimated motion layer lie on a plane. Then, in homogeneous coordinates, two corresponding points $\mathbf{x}^{t-1}$ and $\mathbf{x}^t$ on consecutive frames F^{t-1} and F^t, respectively, are related by

$$\mathbf{x}^{t-1} = sP^t\mathbf{x}^t, \tag{2}$$

where P^t is the projective homography matrix of the backward motion field and s is a scale factor. The projective homography can be expressed in terms of the Euclidean homography matrix H^t and the calibration matrices K^{t-1} and K^t corresponding to time instants $t-1$ and t, respectively as follows:

$$P^t = K^{t-1}H^t(K^t)^{-1}. \tag{3}$$

Suppose that the observed world plane has coordinates $\pi = (\mathbf{n}^T, d)$, where $\mathbf{n}$ is the unit plane normal and d is the orthogonal distance of the plane from the camera center at time t. The Euclidean homography matrix can then be decomposed into structure and motion elements as follows [32]:

$$H^t = R - \mathbf{tn}^T, \tag{4}$$

where R is the rotation matrix, $\mathbf{t}$ is the translation vector of the relative camera motion $\mathbf{t}_d$ normalized with the distance d (See Fig. 3 for an illustration). Although several approaches exist to estimate R, $\mathbf{t}$, and $\mathbf{n}$ from a given homography matrix [33] - [36], the decomposition increases computational complexity and requires the internal calibration matrices to be available. It will be shown in the following paragraphs that the decomposition can be avoided by a series of reasonable assumptions. For the time being, let the decomposed parameters be $\hat{R}$, $\hat{\mathbf{t}}$, and $\hat{\mathbf{n}}$ for the rotation matrix, the translation vector and the surface normal, respectively.

The rotation matrix can be expressed in the angle-axis representation with a rotation angle θ about a unit axis vector $\mathbf{a}$ [32]:

$$\hat{R} = I + sin(\theta)[\,\mathbf{a}\,]_x + (1 - cos(\theta))[\,\mathbf{a}\,]_x^2, \tag{5}$$

where $[\,\mathbf{a}\,]_x$ is the skew-symmetric matrix of $\mathbf{a}$ and I is identity matrix. The unit axis vector $\mathbf{a}$ can be found by solving $(\hat{R} - I)\mathbf{a} = \mathbf{0}$ (i.e., finding null space of $\hat{R} - I$) and the rotation angle can be computed using a two argument (full range) arctangent function [32]:

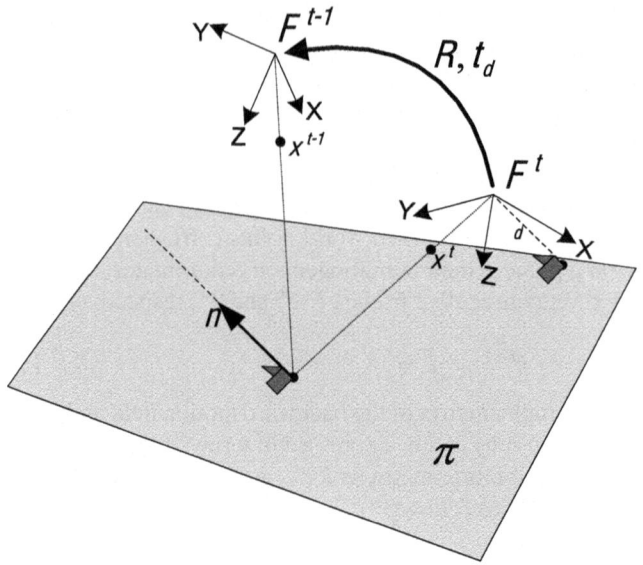

Fig. 3 Relative motion of a camera capturing a planar region

$$cos(\theta) = (trace(\hat{R}) - 1)/2,$$

(6)

$$sin(\theta) = \frac{1}{2}\mathbf{a}^T \begin{bmatrix} \hat{R}_{3,2} - \hat{R}_{2,3} \\ \hat{R}_{1,3} - \hat{R}_{3,1} \\ \hat{R}_{2,1} - \hat{R}_{1,2} \end{bmatrix},$$

where $\hat{R}_{i,j}$ is the element in the i-th row and j-th column of the estimated rotation matrix $\hat{R}$. Under constant velocity assumption, the decomposed motion parameters, that is, the rotation angle θ and the translation vector $\hat{\mathbf{t}}$, can be interpolated at a time instant $t - \Delta t$ by a linear model:

$$\theta^b = \theta\,\Delta t,$$
$$\hat{\mathbf{t}}^b = \hat{\mathbf{t}}\,\Delta t.$$

(7)

In the general case, where the calibration matrices are known or can be estimated, the backward homography matrix $P_b^{t-\Delta t}$ is computed by plugging the interpolated motion parameters θ^b and $\hat{\mathbf{t}}^b$ into Equations 5, 4 and 3, respectively. For the cases where the calibration is not available, it is reasonable to assume that the amount of change in focal length between $t - 1$ and t is negligible (i.e. $K^{t-1} \simeq K^t$). For the sake of further simplicity, we shall use the small angle approximation of rotation:

$$\hat{R} = I + \theta[\,\mathbf{a}\,]_x\,.$$

(8)

Under these assumptions, the backward projective homography matrix $P_b^{t-\Delta t}$ of the interpolation frame is reconstructed as

$$P_b^{t-\Delta t} = K^t (I + \theta_b [\, \mathbf{a} \,]_x - \hat{\mathbf{t}}^b \mathbf{n}^T)(K^t)^{-1} = (1 - \Delta t)I + \Delta t P_b^t , \qquad (9)$$

which reveals that the homography decomposition is not required under the mentioned assumptions. Finally, the forward homography matrix $P_f^{t-\Delta t}$ can be computed from the available models by the following simple linear transformation:

$$P_f^{t-\Delta t} = (P^t)^{-1} P_b^{t-\Delta t}. \qquad (10)$$

3.3 Layer Map Interpolation

The interpolation of the motion models is followed by estimation of the corresponding layer support maps at time instants $t - 1$ and $t - \Delta t$. This is achieved essentially by backward warping the extracted layers (both support maps and intensities) of F^t to the two previous time instants, and updating the overlapping and uncovered regions so as to ensure a single layer assignment for each pixel of $F^{t-\Delta t}$ and F^t. Fig. 4 provides an overview of the layer map interpolation process.

Depth ordering relations of the overlapping layers is extracted by computing a visual similarity measure between each warped layer and the original frame F^{t-1} over the region of overlap. Each pixel of an overlapping region votes for the layer that gives the minimum sum of absolute intensity differences. Visual similarity of

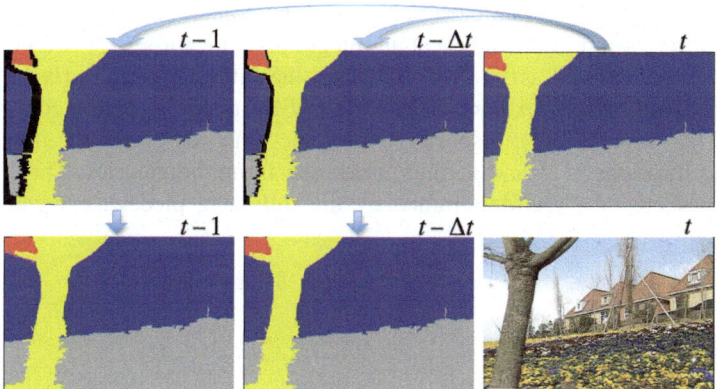

Fig. 4 Overview of the layer map interpolation process. Top row: warping layer maps at time instant t (right) to the two previous time instances $t - 1$ (left) and $t - \Delta t$ (middle). Bottom row: updating the layer assignments on overlapping and uncovered regions of the warped layers at $t - 1$ (left) and $t - \Delta t$ (right). Bottom right: frame $t = 337$ of the original sequence *Flower Garden*.

a warped layer is then modeled as the number of pixels that vote for the layer. The warped layers at $t - 1$ and $t - \Delta t$ are updated by assigning the overlapping regions only to the top layers, which yield the maximum similarity score.

Pixels in the uncovered regions are assigned separately by using the extracted layers of previous time $(t - 1)$. For an uncovered pixel, the set of candidate current time (t) layer labels corresponding to the previous time layer label at that pixel are determined by using the estimated layer correspondences. Finally, the uncovered pixel is assigned to the spatially closest candidate to avoid the creation of disconnected support maps and to ensure a maximum level of compactness.

A similar strategy is followed for uncovered pixels of the interpolation frame. For each pixel of an uncovered region, the candidate layer set is initialized as the neighboring layers of the region. Each pixel of an uncovered region, is then warped to the previous time using the interpolated backward motion models of the candidate layers. If the layer label at the warped location is different for a candidate layer, it is removed from the set. Finally, as before, the spatially closest layer is selected among the candidates for each uncovered pixel.

3.4 Middle Frame Interpolation

In the final stage of the algorithm, pixel intensity vectors at the middle, or interpolation, frame are estimated from the interpolated layer motion models and the interpolated layer visibility images. For each pixel of layer i of the middle frame, corresponding positions in the original frames F^{t-1} and F^t are computed using the previously estimated backward and forward layer motion models $P_{i,b}^{t-\Delta t}$ and $P_{i,f}^{t-\Delta t}$. The interpolated intensity vector is modeled as a function of the intensities at the corresponding sub-pixel locations in the original frames F^{t-1} and F^t.

Let the intensity vector (and layer label) at an integer pixel location $\mathbf{x}^{t-\Delta t}$ of the middle frame $F^{t-\Delta t}$ be denoted as $\mathbf{I}^{t-\Delta t}$ $(L^{t-\Delta t})$, and the intensity vectors (layer labels) at the corresponding sub-pixel locations $\mathbf{x}^{t-1}$ in F^{t-1} and $\mathbf{x}^t$ in F^t are computed to be $\mathbf{I}^{t-1}$ (L^{t-1}) and $\mathbf{I}^t$ (L^t), respectively. For the time being, assume that at least one of the locations $\mathbf{x}^{t-1}$ and $\mathbf{x}^t$ falls inside the frame boundaries. Then, the pixel intensity vector $\mathbf{x}^{t-\Delta t}$ is determined by the following piecewise linear function:

$$
\mathbf{I}^{t-\Delta t} =
\begin{cases}
\mathbf{I}^{t-1} & : L^{t-\Delta t} = L^{t-1}, L^{t-\Delta t} \neq L^t \\[2mm]
\mathbf{I}^t & : L^{t-\Delta t} \neq L^{t-1}, L^{t-\Delta t} = L^t \\[2mm]
(1 - \Delta t)\mathbf{I}^{t-1} + \Delta t \mathbf{I}^t & : \begin{array}{l} L^{t-\Delta t} = L^{t-1}, L^{t-\Delta t} = L^t, \\ \parallel \mathbf{I}^{t-1} - \mathbf{I}^t \parallel < T_{\mathbf{I}} \end{array} \\[3mm]
\underset{\mathbf{I}^j \in \{\mathbf{I}^{t-1}, \mathbf{I}^t\}}{\arg \min} \parallel \mathbf{N}^{t-\Delta t} - \mathbf{I}^j \parallel & : \text{else,}
\end{cases}
\tag{11}
$$

where $\mathbf{N}^{t-\Delta t}$ is the average intensity in the neighborhood of $\mathbf{x}^{t-\Delta t}$ and T_I is an intensity difference threshold for detecting the disagreement between intensities of the corresponding original frame locations, and hence, avoiding the formation of blur and halo effects (i.e. ghosting artifacts around motion boundaries). The above equation states that if the layer label at a transformed location $\mathbf{x}^{t-1}$ or $\mathbf{x}^t$ is different than the layer label at the source location $\mathbf{x}^{t-\Delta t}$, then the intensity at the transformed location should not be taken into account, since the layer is occluded at that point. All the remaining cases that are not covered by the first three pieces of the function are mainly caused by the errors in the estimation process and modeling. In these degenerate cases, smoothness is enforced by using only the intensity that is closer to the average intensity $\mathbf{N}^{t-\Delta t}$ in the neighborhood of the interpolation frame location $\mathbf{x}^{t-\Delta t}$.

Finally, another special degenerate case occurs when both of the transformed locations fall outside the frame boundaries. A similar strategy can be followed in this case in order to enforce smoothness in the results. As an example, such boundary pixels can be interpolated with the intensity values of the closest locations in the interpolation frame.

3.5 Results

Fig. 5 and Fig. 7 present some sample results of the algorithm tested on several well-known video sequences, where bicubic interpolation is used in obtaining intensities at sub-pixel locations. In order to evaluate the performance of the method both qualitatively and quantitatively, only odd frames of the inputted sequences are processed and a single frame corresponding to $\Delta t = 0.5$ is interpolated between each pair of odd frames. The interpolation frames are then compared with the even frames of the original sequences both objectively and subjectively.

The algorithm achieves visually pleasing results without apparent blur or halo effects and with sharpness preserved at object boundaries. For objective evaluation, peak signal-to-noise ratio (PSNR) between the even frames of the original sequences and the interpolated frames are computed. Fig. 6 provides a comparison between the proposed method and a simple frame averaging scheme for the *Flower Garden* sequence. The plot suggests a significant improvement in PSNR as well as an enhanced robustness to changes in motion complexity.

It worths noting that the performance of the proposed technique is closely tied to the quality of the segmentation estimates (both segmentation maps and corresponding motion models). However, the method is observed to be tolerant to changes in the number of layers, provided that the corresponding motion model estimates are reasonably accurate.

Fig. 5 Interpolation results for *Mobile & Calendar* sequence. From left to right, top row: frames 34, 36 and 38 of the original sequence, middle row: corresponding interpolated frames, bottom row: layer support maps corresponding to frames 35, 37 and 39.

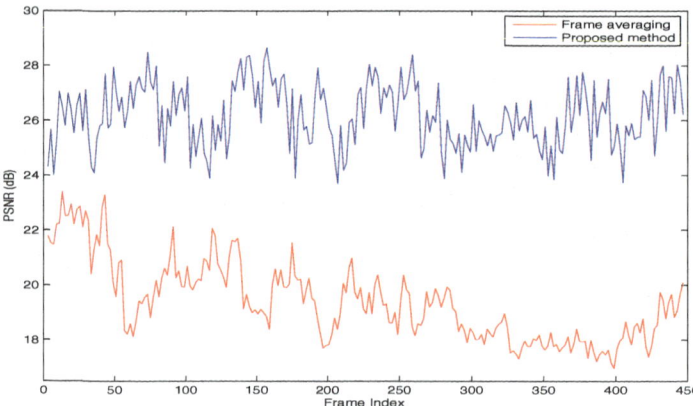

Fig. 6 PSNR curves of the proposed method and frame averaging for *Flower Garden* sequence

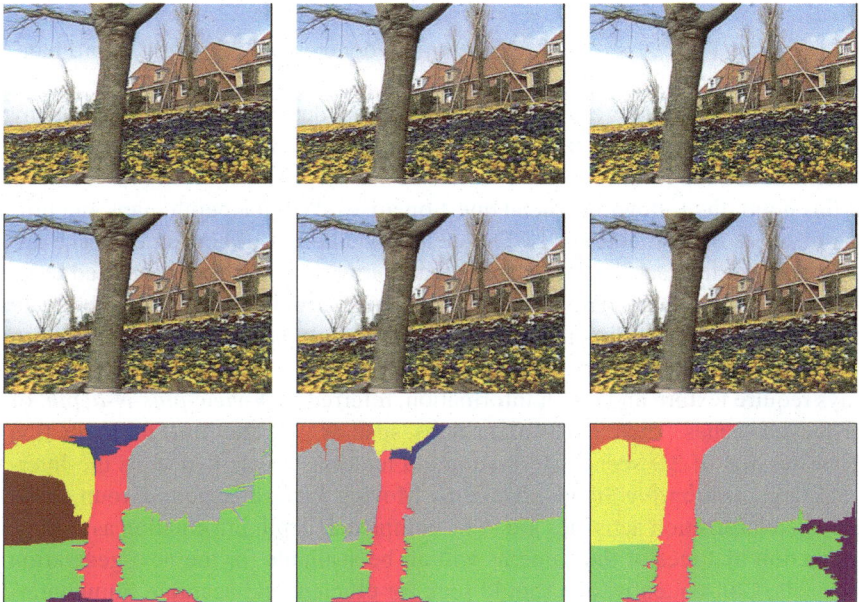

Fig. 7 Sample interpolation results for *Flower Garden* sequence. From left to right, top row: frames 318, 320 and 322 of the original sequence, middle row: corresponding interpolated frames, bottom row: layer support maps corresponding to frames 319, 321 and 323.

4   Region-Based Multi-view FRUC by Rigid Body Motion Model Parameter Interpolation (FRUC-RM)

The rigid object motion can exactly be modeled by using the Euclidean transformation, [32]:

$$\begin{bmatrix} X_t \\ Y_t \\ Z_t \end{bmatrix} = R \begin{bmatrix} X_{t-1} \\ Y_{t-1} \\ Z_{t-1} \end{bmatrix} + t \, , \tag{12}$$

where $[X_t, Y_t, Z_t]^T$ and $[X_{t-1}, Y_{t-1}, Z_{t-1}]^T$ represent 3D coordinates of the moving object between successive frames, R denotes a $3x3$ rotation matrix and t is a $3x1$ translation matrix.

Although only image coordinates of the moving objects are available in video sequences, unfortunately, the utilization of (12) requires 3D coordinates of these moving objects. In order to accomplish such a goal, the calibration of the camera capturing the scene, which defines the projection of 3D world coordinates onto 2D pixel coordinates, as well as the depth map of the scene should be known [32]. The following equation relates pixel coordinates with the world coordinates by the help of a $3x4$ projection matrix, P, as:

$$x = PX ,\tag{13}$$

where x denotes the 2D pixel coordinate and X denotes the 3D world coordinate, respectively.

Once the 3D coordinates of the moving objects are determined, the frames at any desired time instant could be interpolated by the help of the estimated rotation, R, and translation, t, matrices.

At this point, the following crucial question arises: Will the depth maps and projection matrices be available for LCD panels in the homes of consumers? The answer lies behind the fact that 3D display technologies have been progressed drastically in the recent years. The glass-free auto-stereoscopic displays, which create the perception of the 3rd dimension by presenting multiple views, are expected to spread into the consumer market in the very near future. These new generation 3D displays require texture and depth information, referred as *N-view-plus-N-depth*, of the displayed scenes. There exists algorithms, such as [37] and [38], for extraction and transmission of 3D scene information via multiple views; hence, as the International Organization for Standardization - Moving Picture Experts Group (ISO-MPEG) standardization activities are to be completed, depth information as well as the projection matrices of the cameras will all be available for the next generation LCD panels. Consequently, one should seek for efficient and accurate frame interpolation methods specific to multi-view data. The utilization of conventional frame-rate conversion algorithms for multi-view video increases the amount of data to be processed by a factor of 2N. Moreover, conventional techniques mainly exploit 2D motion models, which are only approximations of the 3D rigid body motion model.

In this section, a frame rate conversion system, which estimates the real 3D motion parameters of rigid bodies in video sequences and performs frame interpolation for the desired view(s) of the multi-view set, is presented. The main motivation is firm belief in the utilization of a *true* 3D motion model for development of better FRUC systems, possibly with much higher frame rate increase ratios. Hence, in upcoming sections, a completely novel 3D frame-rate up conversion system is proposed that exploits multi-view video as well as the corresponding dense depth values for all views and every pixel. The system performs moving object segmentation and 3D motion estimation in order to perform MCFI. The overall algorithm is summarized in Fig. 8.

4.1 Depth-Based Moving Object Segmentation

The initial step for the proposed algorithm is the segmentation of the moving rigid objects, which accounts for independently moving object(s) in the scene. For this purpose, a depth-based moving object segmentation scheme is utilized. Fig. 9 illustrates typical color and depth frames for *Rotating Cube* and *Akko-Kayo*, [43], sequences.

The depth maps for *Akko-Kayo* sequence are estimated using the algorithm proposed in [38], whereas those of *Rotating Cube* are generated artificially. The steps of this segmentation algorithm are provided below:

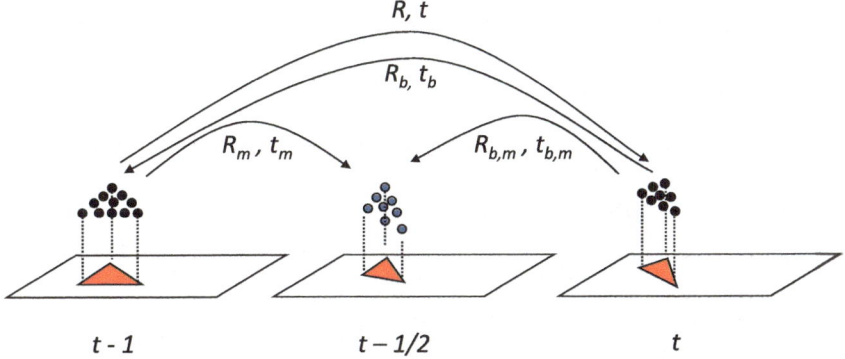

Fig. 8 The overall algorithm FRUC-RM

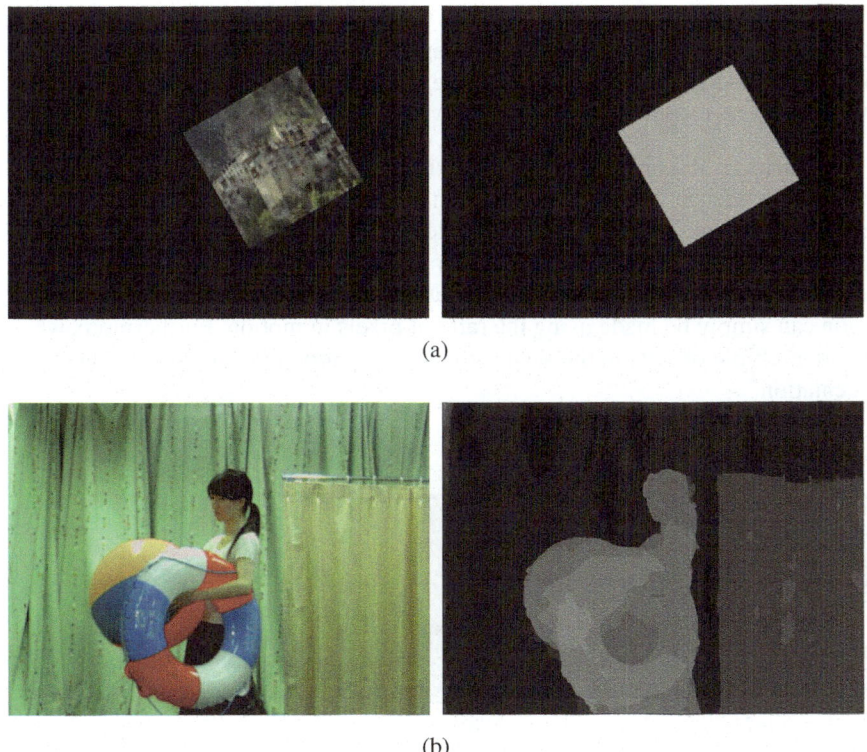

(a)

(b)

Fig. 9 Color and depth frames for (a) *Rotating Cube* sequence and (b) *Akko-Kayo* sequence

1. For the nth view, the differences between consecutive frames are obtained by calculating the pixel-wise absolute differences between depth, D, and texture, C, frames at time instants $t-1$ and t:

$$\Delta C(i,j) = abs(C(i,j,t-1,n) - C(i,j,t,n)),$$
$$\Delta D(i,j) = abs(D(i,j,t-1,n) - D(i,j,t,n)). \quad (14)$$

Then, using these frames, the global segmentation map is calculated as follows: a pixel at location (i,j) is assigned to background, if satisfies the condition given by (15), and assigned to foreground otherwise;

$$\Delta C + \lambda \Delta D < T_{CD}, \quad (15)$$

where, λ (typically 0.5) and the threshold T_{CD} are constants.

2. After the global segmentation map is obtained, the average background depth values of depth maps $D(i,j,t-1,n)$ and $D(i,j,t,n)$ are calculated by using the depth values of background pixels by simple averaging.

3. Finally, the depth values of the pixels at time instants $t-1$ and t, i.e. $D(i,j,t-1,n)$ and $D(i,j,t,n)$ are compared to the average background depth values calculated in Step 2, and the foreground pixels are determined as the pixels having depth values different from the average depth values by a certain threshold.

Fig. 10 illustrates the obtained global, previous and current segmentation maps for *Akko-Kayo* sequence. It should be noted that most of the available multi-view sequences have static camera arrays, which lets this simple algorithm yield satisfactory results. In the case that the camera is not stationary, more complex segmentation algorithms such as [39] should be utilized. The static/dynamic camera distinction can simply be made using the ratio of pixels in motion. Furthermore, when there are multiple objects in the scene, connected component labeling is crucial for segmentation.

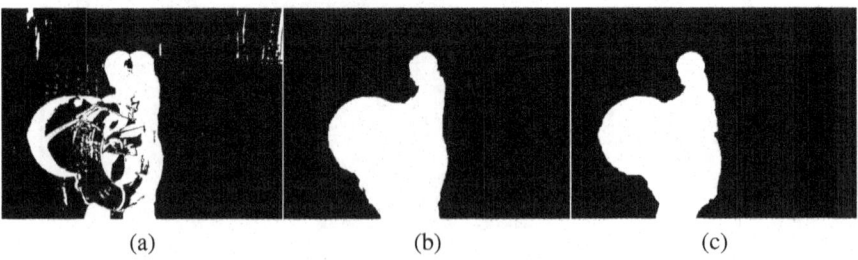

(a)   (b)   (c)

Fig. 10 (a) Global, (b) previous and (c) current segmentation maps for *Akko-Kayo* sequence

Fig. 11 Matched SIFT features

4.2   3D Motion Estimation

Having segmented foreground objects and the static background, the 3D motion parameters of the moving objects, namely rotation and translation matrices, are estimated by using the features matched between the successive frames of the video sequence. Scale Invariant Feature Transform (SIFT) algorithm detailed in [40] is utilized in order to test the proposed multi-view FRUC algorithm. Fig. 11 illustrates SIFT features from the successive frames of *Akko-Kayo* sequence.

Once the features on the moving rigid objects are matched, the next step is the calculation of rotation and translation parameters. Fortunately, the depth values of extracted SIFT features are available for the *N-view-plus-N-depth* content type; hence, 3D motion estimation step is relatively simple. For this purpose, the initial step is the determination of the 3D coordinates of the matched features, which is achieved via the back-projection equation [32]:

$$X(\lambda) = P^+ x + \lambda C,$$ (16)

where λ is a positive number, C denotes camera center, P^+ represents pseudo inverse of the projection matrix. The inherent scale ambiguity is solved by using the known depth of the point, and the exact coordinates are calculated. After determining the 3D coordinates of the matched features, we solve for R and t in (12) using Random Sample Consensus (RANSAC) algorithm, [41], in a robust manner in order to account for outliers in feature matching step:

1. R and t matrices are estimated using quaternion approach, [42]
2. 3D coordinates of the features in the first frame are rotated and translated using the estimated R and t matrices
3. Euclidean distances between available 3D coordinates of the features in the second frame, and the 3D coordinates obtained by rotating and translating those of features in the first frame are calculated,
4. The number of inliers are obtained by comparing these Euclidean distances by a threshold,

5. If the number of inliers is greater than maximum number of inliers, maximum iteration number is updated [41]
6. The number of iterations is increased by one, and if maximum number of iterations is reached process is terminated. Otherwise, Steps 1-6 are repeated.

The maximum number of iterations, N, is selected sufficiently high in order to ensure with a probability, p, that at least one of the random samples of s points is free from outliers. Suppose e is the probability that any selected data point is an outlier (hence, $w = 1 - e$ is the probability that it is an inlier). Then, N can be obtained as [41]:

$$N = \frac{log(1-p)}{log(1-(1-e)^s)}. \tag{17}$$

Note that, prior to 3D motion estimation, the depth maps, which are estimated using [38], are smoothed with a Gaussian filter of 9x9 kernel, in order to compensate for the erroneously estimated depth regions.

4.3 Middle Frame Interpolation

The final step of the algorithm is the interpolation of the middle frame pixel intensities, which is achieved through interpolation of the estimated 3D motion models of foreground objects. Although the methods mentioned in this section can be used to render multiple interpolated frames at any time instant in-between the successive frames, for simplicity and without loss of generality, we assume that a single middle frame is to be generated corresponding to time instant $t - \frac{1}{2}$. Basically, *all* 3D points on the foreground objects are rotated and translated to time instant $t - \frac{1}{2}$ and then projected to the image plane of the desired view(s) of the multi-view set. In Sect. 4.2, the 3D motion, i.e. the rotation and translation, parameters are estimated between the successive frames at the time instants $t - 1$ and t. Let R_m and t_m, respectively, represent the interpolated 3D rotation matrix and translation vector corresponding to rigid motion of a foreground object from $t - 1$ to $t - \frac{1}{2}$. Under constant velocity assumption of objects between successive frames, the interpolated translation vector can be determined easily as follows:

$$t_m = \frac{1}{2}t. \tag{18}$$

Similarly, using the angle-axis representation highlighted in Sect. 3.2, the interpolated rotation R_m is written as:

$$R_m = R\left(\frac{\theta}{2}, \mathbf{a}\right), \tag{19}$$

where, as defined previously, θ is the rotation angle and $\mathbf{a}$ is the rotation axis.

Using R_m and t_m, 3D coordinates of foreground objects at $t - 1$ are rotated and translated to $t - \frac{1}{2}$ by the following relation:

$$\begin{bmatrix} X_m \\ Y_m \\ Z_m \end{bmatrix} = R_m \begin{bmatrix} X_{t-1} \\ Y_{t-1} \\ Z_{t-1} \end{bmatrix} + t_m \, . \tag{20}$$

The interpolation frame for any desired view of the multi-view set can then be re-constructed by re-projecting the 3D points onto the corresponding image plane:

$$\begin{bmatrix} x_m \\ y_m \end{bmatrix} = P_{desired} \begin{bmatrix} X_m \\ Y_m \\ Z_m \end{bmatrix} \, . \tag{21}$$

Intensities corresponding to foreground objects are interpolated by using the Red-Green-Blue (RGB) values of corresponding pixels at time instant $t - 1$. For interpolating the background regions, the segmentation maps, given in Fig. 10 are exploited by assigning average RGB values for common background regions and assigning RGB values from either frame for uncommon background regions. Fig. 12 shows the resulting interpolated frame of *Akko-Kayo* sequence.

Due to the rounding effects, some parts of the foreground object remain unfilled; thus, in order to alleviate such problems, the interpolated frame is post-processed by a 3x3 median filter. Fig. 13 illustrates the resulting frame after post processing.

So far, 3D coordinates of the foreground objects, and hence, the corresponding 2D pixel locations at the interpolation frame of time instant $t - \frac{1}{2}$ are calculated using the frame at $t - 1$ via (20) and (21). In order to increase the quality of the interpolation, bi-directional filling, which utilizes the intensities of frames at time instants $t - 1$ and t, is employed. For bi-directional interpolation, backward rotation matrix R_b and translation vector t_b transforming 3D foreground coordinates from

Fig. 12 Interpolated frame using 3D motion information

the frame at t to the frame at $t - 1$ are calculated. After some algebra, R_b and t_b can be obtained in terms of the forward rotation matrix R and the translation vector t as follows:

$$R_b = R^{-1},$$
$$t_b = -R^{-1}t. \tag{22}$$

As in the case of forward directional interpolation, interpolated motion parameters for the desired temporal location are obtained via linear interpolation:

$$R_{b,m} = R_b\left(\frac{\theta_b}{2}, \mathbf{a}_b\right),$$
$$t_{b,m} = \frac{t_b}{2}. \tag{23}$$

For bi-directional interpolation, the 3D coordinates of the foreground objects at the time instants $t - 1$ and t are rotated and translated to the time instant $t - \frac{1}{2}$ using forward and backward motion parameters and then projected to the image plane of a desired view. Intensities at overlapping 2D locations of the middle frame are found by simply averaging corresponding intensities at the time instants $t - 1$ and t. Figures 14 and 13 reveals that utilizing bi-directional interpolation instead of only forward directional interpolation improves the visual quality of the resulting frames significantly.

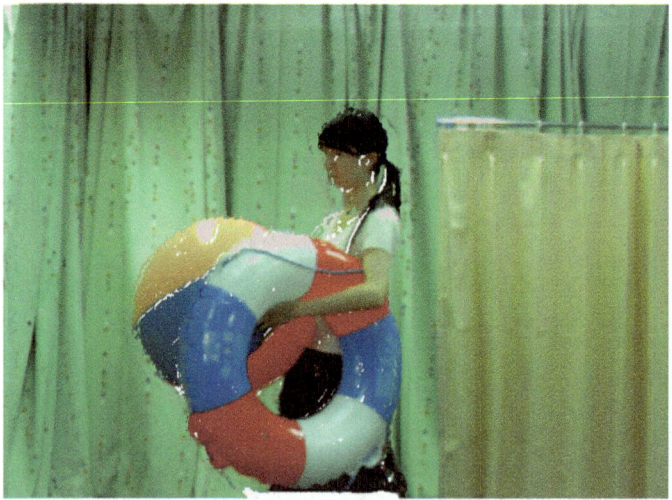

Fig. 13 Interpolated frame after post-processing

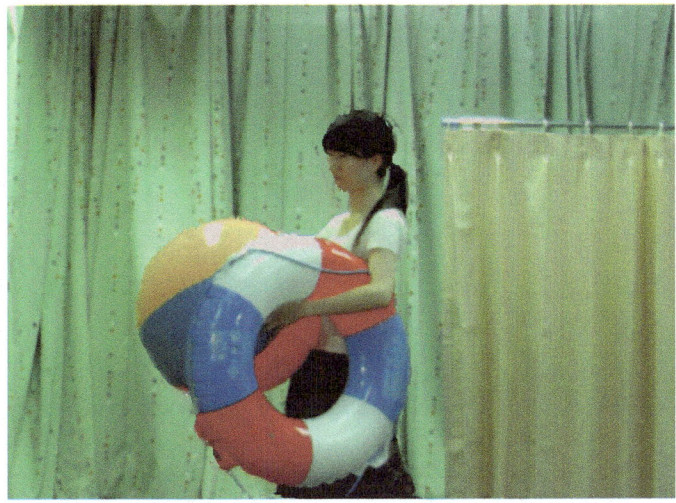

Fig. 14 Bi-directionally interpolated frame

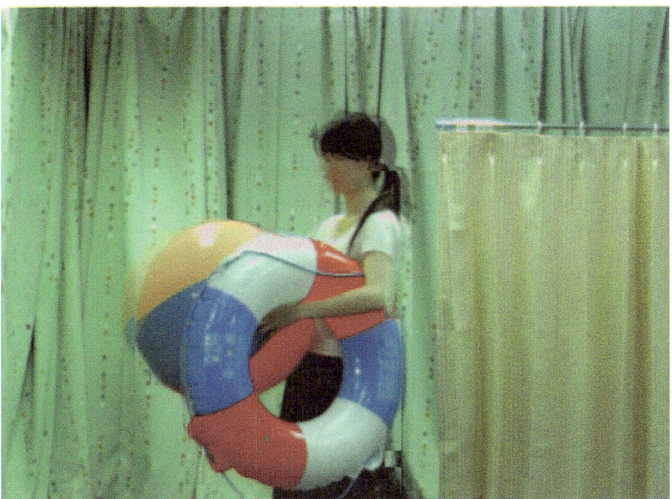

Fig. 15 Bi-directionally interpolated frame using conventional method.

For the sake of comparison to the block-based conventional methods, Fig. 15 shows the resulting middle frame interpolated using a conventional FRUC algorithm that utilizes bi-directional motion estimation, [8], with hexagonal search pattern [15]. It is clear that our approach interpolates visually more pleasant frames without blocking artifacts at object boundaries.

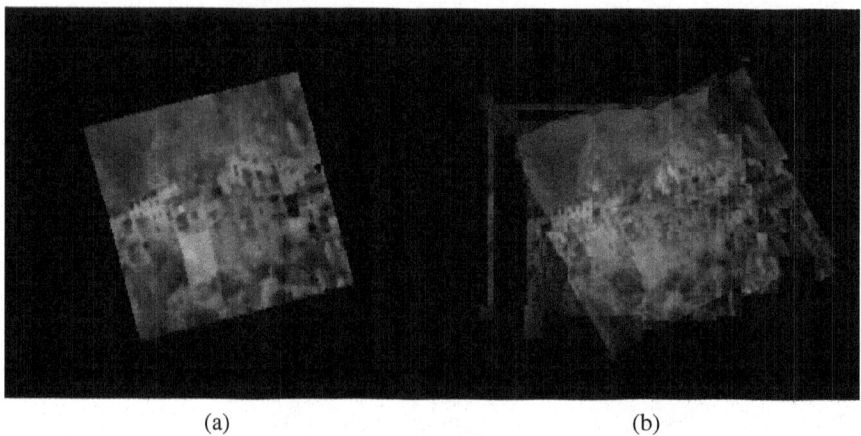

(a)   (b)

Fig. 16 Frames interpolated using (a) the proposed method and (b) the conventional method

Fig. 16 illustrates sample interpolation results for synthetically generated *Rotating Cube* sequence using both the conventional method and the proposed method. The figure reveals that conventional translational motion model is not capable of handling complex motion types, and hence the interpolated frames are significantly distorted. On the other hand, the rigid body motion model utilized by the FRUC-RM algorithm yields qualitatively better frames.

The utilization of 3D motion parameters during frame interpolation enables rendering of any desired number of frames in between the successive frames of the original multi-view sequences, by the following motion parameter interpolation scheme:

$$
\begin{aligned}
R_i &= R\left(\frac{k}{n+1}, \mathbf{a}\right), \\
t_i &= \frac{k}{n+1}t, \\
R_{b,i} &= R_b\left(\left(1 - \frac{k}{n+1}\right)\theta_b, \mathbf{a}_b\right), \\
t_{b,i} &= \left(1 - \frac{k}{n+1}\right)t_b, \quad k = 1, 2, .., n.
\end{aligned}
\tag{24}
$$

where n denotes the number of frames to be interpolated and k represents the index of an interpolation frame. Fig. 17 shows three interpolation frames obtained via the above equations for *Rotating Cube* sequence.

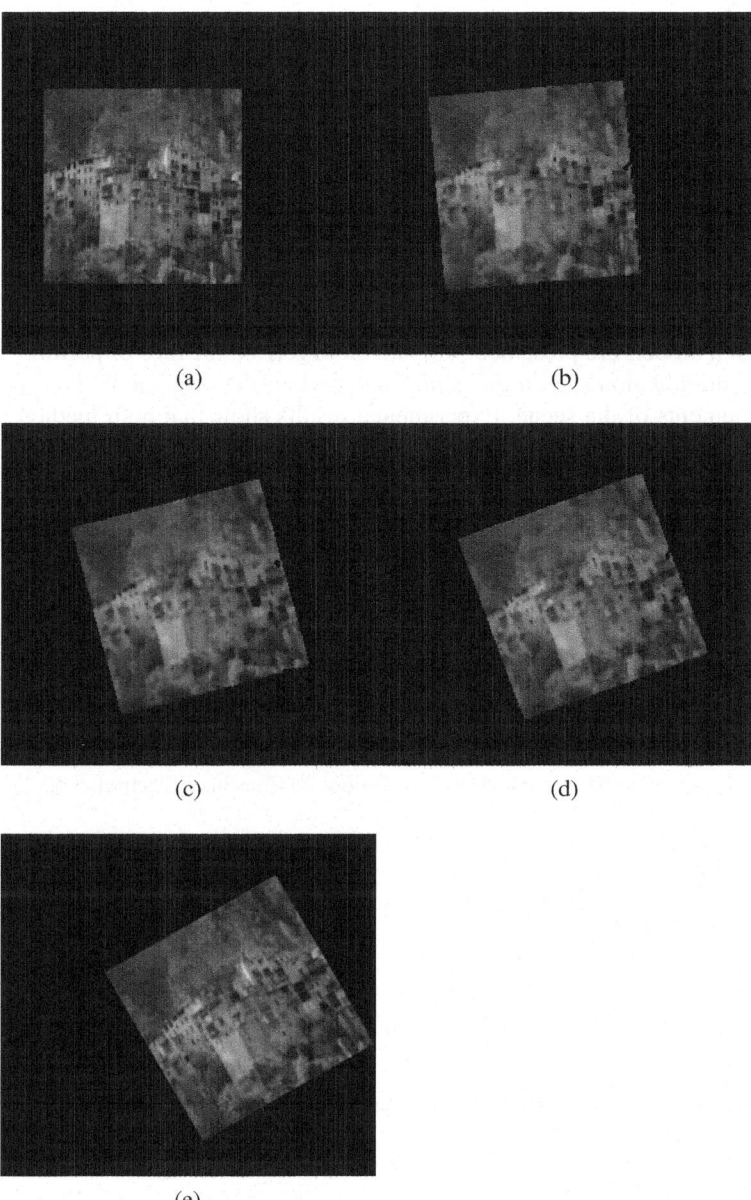

Fig. 17 Original frames (a) $t-1$, (e) t and interpolated frames at time instants (b) $t - \frac{5}{6}$, (c) $t - \frac{3}{6}$ and (d) $t - \frac{1}{6}$

5   Summary

FRUC is an important problem for consumer electronics, especially for improving the performance of hold-type panel displays. Increasing frame-rates on typical media content, which often involves dynamic scenes, requires modeling and utilization of motion field more precisely.

In this chapter, first an overview of the conventional FRUC algorithms that mainly utilize block-based motion estimation and compensation with translational motion model is provided. In order to avoid several shortcomings of conventional methods such as blocking artifacts and failure to robustly model correct object motion, two region-based FRUC techniques utilizing 2D planar perspective and 3D rigid body motion models are presented. Both methods rely on the idea of performing the motion interpolation step in the parameter space of the original 3D motion and structure elements of the scene. Experimental results show that both methods yield visually pleasing results without halo effects on dynamic scenes.

Acknowledgements. This work is funded by Vestek R&D Corp., Turkey.

References

1. Chen, H., Kim, S.S., Lee, S.H., Kwon, O.J., Sung, J.H.: Nonlinearity Compensated Smooth Frame Insertion for Motion-blur Reduction in LCD. In: Proc. IEEE 7th Multimedia Signal Processing Workshop, Shangai, China (2005)
2. Ha, T., Lee, S., Kim, J.: Motion Compensated Frame Interpolation by New Block-based Motion Estimation Algorithm. IEEE Transactions on Consumer Electronics 50(2), 752–759 (2004)
3. Choi, B.D., Han, J.W., Kim, C.S., Ko, S.J.: Motion-Compensated Frame Interpolation Using Bilateral Motion Estimation and Adaptive Overlapped Block Motion Compensation. IEEE Transactions on Circuits and Systems for Video Technology 17(4), 407–416 (2007)
4. Fujiwara, S., Taguchi, A.: Motion-compensated Frame Rate Up-conversion Based on Block Matching Algorithm with Multi-size Blocks. In: Proc. IEEE International Symposium on Intelligent Signal Processing and Communication Systems, Hong Kong, China (2005)
5. Lee, S.H., Yang, S., Jung, Y.Y., Park, R.H.: Adaptive Motion-compensated Interpolation for Frame Rate Up-conversion. In: Proc. IEEE International Conference on Computers in Education, Auckland, New Zeland (2002)
6. Jeon, B.W., Lee, G.I., Lee, S.H., Park, R.H.: Coarse-to-Fine Frame Interpolation for Frame Rate Up-Conversion Using Pyramid Structure. IEEE Transactions on Consumer Electronics 49(3), 499–508 (2003)
7. Biswas, M., Nguyen, T.: A Novel Motion Estimation Algorithm Using Phase Plane Correlation for Frame Rate Conversion. In: Proc. IEEE Conference Record of the Thirty-Sixth Asilomar Conference on Signals, Systems and Computers (2002)
8. Choi, B.T., Lee, S.H., Ko, S.J.: New Frame Rate Up-Conversion Using Bi-Directional Motion Estimation. IEEE Transactions on Consumer Electronics 46(3), 603–609 (2000)
9. Mishima, N., ltoh, G.: Novel Frame Interpolation Method For Hold-Type Displays. In: Proc. IEEE International Conference on Image Processing, Singapore (2004)

10. Sugiyama, K., Aoki, T., Hangai, S.: Motion Compensated Frame Rate Conversion Using Normalized Motion Estimation. In: Proc. IEEE Workshop on Signal Processing Systems Design and Implementation, Athens, Greece (2005)
11. Chen, T.: Adaptive Temporal Interpolation Using Bi-directional Motion Estimation and Compensation. In: Proc. IEEE International Conference on Image Processing, Rochester, NY, USA (2002)
12. Haan, G., Biezen, P.W.A.C.: An Efficient True-Motion Estimator Using Candidate Vectors from a Parametric Motion Model. IEEE Transactions on Circuits and Systems for Video Technology 8(1), 85–91 (1998)
13. Hilman, K., Park, H.W., Kim, Y.: Using Motion-Compensated Frame-Rate Conversion for the Correction of 3: 2 Pulldown Artifacts in Video Sequences. IEEE Transactions On Circuits and Systems for Video Technology 10(6), 869–877 (2000)
14. Koga, T., Iinuma, K., Hirano, A., Iijima, Y., Ishigora, T.: Motion Compensated Interframe Coding for Video Conferencing. In: Proc. IEEE National Telecommunications Conference, New Orleans, LA, USA (1981)
15. Zhu, C., Lin, X., Chau, L.P.: Hexagon-Based Search Pattern for Fast Block Motion Estimation. IEEE Transactions on Circuits and Systems for Video Technology 12(5), 349–355 (2002)
16. Zhu, S., Ma, K.K.: A New Diamond Search Algorithm for Fast Block-Matching Motion Estimation. IEEE Transactions on Image Processing 9(2), 287–290 (2000)
17. Cheng, Y., Wang, Z., Dai, K., Guo, J.: A Fast Motion Estimation Algorithm Based on Diamond and Triangle Search Patterns. In: Marques, J.S., Pérez de la Blanca, N., Pina, P. (eds.) IbPRIA 2005. LNCS, vol. 3522, pp. 419–426. Springer, Heidelberg (2005)
18. Beric, A., Haan, G., Meerbergen, J., Sethuraman, R.: Towards An Efficient High Quality Picture-Rate Up-Converter. In: Proc. IEEE International Conference on Image Processing, Barcelona, Spain (2003)
19. Lee, S., Kwon, O., Park, R.: Weighted-adaptive Motion-compensated Frame Rate Upconversion. IEEE Transactions on Consumer Electronics 49(3), 485–491 (2003)
20. Al-Mualla, M.E.: Motion Field Interpolation for Frame Rate Conversion. In: Proc. IEEE International Symposium on Circuit and Systems, Bankok, Thailand (2003)
21. Kuo, T., Kim, J., Jay Kuo, C.C.: Motion-Compensated Frame Interpolation Scheme For H.263 Codec. In: Proc. IEEE International Symposium on Circuit and Systems, Orlando, Florida, USA (1999)
22. Tiehan, L.U.: Motion-Compensated Frame Rate Conversion with Protection Against Compensation Artifacts. WIPO, Patent No. 2007123759 (2007)
23. Ohwaki, K., Takeyama, Y., Itoh, G., Mishima, N.: Apparatus, Method, and Computer Program Product for Detecting Motion Vector and for Creating Interpolation Frame. US Patent Office (US PTO), Patent No. 2008069221 (2008)
24. Sato, K., Yamasaki, M., Hirayama, K., Yoshimura, H., Hamakawa, Y., Douniwa, K., Ogawa, Y.: Interpolation Frame Generating Method and Interpolation Frame Generating Apparatus. US PTO, Patent No. 2008031338 (2008)
25. Chen, H.F., Kim, S.S., Sung, J.H.: Frame Interpolator, Frame Interpolation Method and Motion Reliability Evaluator. US PTO, Patent No. 2007140346 (2007)
26. Ohwaki, K., Itoh, G., Mishima, N.: Method, Apparatus and Computer Program Product for Generating Interpolation Frame. US PTO, Patent No. 2006222077 (2006)
27. Bugwadia, K., Petajan, E.D., Puri, N.N.: Motion Compensation Image Interpolation and Frame Rate Conversion for HDTV. US PTO, Patent No. 6229570 (2001)
28. Benois-Pineau, J., Nicolas, H.: A New Method for Region-based Depth Ordering in a Video Sequence: Application to Frame Interpolation. Journal of Visual Communication and Image Representation 13(3), 363–385 (2002)

29. Wang, J.: Video Temporal Reconstruction and Frame Rate Conversion. Wayne State University, Detroit (2006)
30. Turetken, E., Alatan, A.A.: Region-based Motion-compensated Frame Rate Upconversion by Homography Parameter Interpolation. In: Proc. IEEE International Conference on Image Processing, Cairo, Egypt (2009)
31. Bleyer, M., Gelautz, M., Rhemann, C.: Region-based Optical Flow Estimation with Treatment of Occlusions. In: Proc. Joint Hungarian-Austrian Conference on Image Processing and Pattern Recognition, Veszprem, Hungary (2005)
32. Hartley, R.I., Zisserman, A.: Multiple View Geometry in Computer Vision. Cambridge University Press, Cambridge (2004)
33. Tsai, R.Y., Huang, T.S.: Estimating Three-dimensional Motion Parameters of a Rigid Planar Patch. IEEE Transactions on Acoustics, Speech and Signal Processing 29(6), 1147–1152 (1981)
34. Faugeras, O., Lustman, F.: Motion and Structure from Motion in a Piecewise Planar Environment. International Journal of Pattern Recognition and Artifcial Intelligence 2(3), 485–508 (1988)
35. Zhang, Z., Hanson, A.R.: 3D Reconstruction Based on Homography Mapping. In: Proc. ARPA Image Understanding Workshop, Palm Springs, CA, USA (1996)
36. Malis, E., Vargas, M.: Deeper Understanding of the Homography Decomposition for Vision-based Control. Research Report INRIA (2007)
37. Ozkalayci, B., Gedik, S., Alatan, A.: 3-D Structure Assisted Reference View Generation for H.264 Based Multi-View Video Coding. In: Proc. IEEE Picture Coding Symposium, Lisbon, Portugal (2007)
38. Cigla, C., Zabulis, X., Alatan, A.: Region-Based Dense Depth Extraction From Multi-View Video. In: Proc. IEEE International Conference on Image Processing, San Antonio Texas, USA (2007)
39. Cigla, C., Alatan, A.: Object Segmentation in Multi-view Video via Color, Depth and Motion Cues. In: Proc. IEEE International Conference on Image Processing, San Diego, California, USA (2008)
40. Lowe, D.G.: Distinctive Image Features from Scale-Invariant Keypoints. International Journal of Computer Vision 60(2), 91–110 (2004)
41. Fischler, M.A., Bolles, R.C.: Random Sample Consensus: A Paradigm for Model Fitting with Applications to Image Analysis and Automated Cartography. Communications of the ACM 24(6), 381–395 (1981)
42. Paragios, N., Chen, Y., Faugeras, O.: Hand Book of Mathematical Models in Computer Vision. Springer, USA (2006)
43. Akko-Kayo: Multi-view Sequence by Department of Information Electronics at Nagoya University,
http://www.tanimoto.nuee.nagoya-u.ac.jp/ fukushima/ mpegftv/Akko.htm (accessed September 5, 2009)

Chapter 10
Spatiotemporal Video Upscaling Using Motion-Assisted Steering Kernel (MASK) Regression

Hiroyuki Takeda, Peter van Beek, and Peyman Milanfar

Abstract. In this chapter, we present *Motion Assisted Steering Kernel* (MASK) regression, a novel multi-frame approach for interpolating video data spatially, temporally, or spatiotemporally, and for video noise reduction, including compression artifact removal. The MASK method takes both local spatial orientations and local motion vectors into account and adaptively constructs a suitable filter at every position of interest. Moreover, we present a practical algorithm based on MASK that is both robust and computationally efficient. In order to reduce the computational and memory requirements, we process each frame in a block-by-block manner, utilizing a block-based motion model. Instead of estimating the local dominant orientation by singular value decomposition, we estimate the orientations based on a technique similar to vector quantization. We develop a technique to locally adapt the regression order, which allows enhancing the denoising effect in flat areas, while effectively preserving major edges and detail in texture areas. Comparisons between MASK and other state-of-the-art video upscaling methods demonstrate the effectiveness of our approach.

1 Introduction

Advances in video display technology have increased the need for high-quality and robust video interpolation and artifact removal methods. In particular, LCD flat-panel displays are currently being developed with very high spatial resolution and very high frame rates. For example, so-called "4K" resolution panels are capable of displaying 2160×4096 full color pixels. Also, LCD panels with frame rates of

Hiroyuki Takeda · Peyman Milanfar
Electrical Engineering Dept., University of California,
1156 High St. Santa Cruz, CA, 95064, USA
e-mail: htakeda@soe.ucsc.edu, milanfar@soe.ucsc.edu

Peter van Beek
Sharp Laboratories of America, 5750 NW Pacific Rim Blvd., Camas, WA, 98607, USA
e-mail: pvanbeek@sharplabs.com

120[Hz] and 240[Hz] are becoming available. Such displays may exceed the highest spatial resolution and frame rate of video content commonly available, namely 1080 × 1920, 60[Hz] progression High Definition (HD) video, in consumer applications such as HD broadcast TV and Blu-ray Disc. In such (and other) applications, the goal for spatial and temporal video interpolation reconstruction is to enhance the resolution of the input video in a manner that is visually pleasing and artifact-free. Common visual artifacts that may occur in spatial and temporal interpolation are: edge jaggedness, ringing, blurring of edges and texture detail, as well as motion blur and judder. In addition, the input video usually contains noise and other artifacts, e.g. caused by compression. Due to increasing sizes of modern video displays, as well as incorporation of new display technologies (e.g. higher brightness, wider color gamut), artifacts in the input video and those introduced by scaling are amplified, and become more visible than with past display technologies. High quality video upscaling requires resolution enhancement and sharpness enhancement as well as noise and compression artifact reduction.

A common approach for spatial image and video upscaling is to use linear filters with compact support, such as from the family of cubic filters [1]. In this chapter, our focus is on multi-frame methods, which enable resolution enhancement in spatial upscaling, and allow temporal frame interpolation (frame rate upconversion). Although many algorithms have been proposed for image and video interpolation, spatial upscaling and frame interpolation (temporal upscaling) are generally treated separately. The conventional super-resolution technique for spatial upscaling consists of image reconstruction from irregularly sampled pixels, provided by registering multiple low resolution frames onto a high resolution grid using motion estimation, see [2, 3] for overviews. A recent work by Narayanan et al. ([4]) proposed a video-to-video super resolution algorithm using a partition filtering technique, in which local image structures are classified into vertical, horizontal, and diagonal edges, textures, and flat areas by vector quantization [5] (involving off-line learning), and prepare a suitable filter for each structure class beforehand. Then, with the partition filter, they interpolate the missing pixels and recover a high resolution video frame. Another recent approach in [6] uses an adaptive Wiener filter and has a low computational complexity when using a global translational motion model. This is typical for many conventional super-resolution methods, which as a result often don't consider more complex motion.

For temporal upscaling, a technique called *motion compensated frame interpolation* is popular. In [7], Fujiwara et al. extract motion vectors from a compressed video stream for motion compensation. However, these motion vectors are often unreliable; hence they refine the motion vectors by the block matching approach with variable-size blocks. Similar to Fujiwara's work, in [8], Huang et al. proposed another refinement approach for motion vectors. Using the motion reliability computed from prediction errors of neighboring frames, they smooth the motion vector field by employing a vector median filter with weights decided based on the local motion reliability. In [9, 10], instead of refining the motion vector field, Kang et al. and Choi et al. proposed block matching motion estimation with overlapped and variable-size block technique in order to estimate motion as accurately as possible.

However, the difficulty of the motion-based approach is that, even though the motion vector field may be refined and/or smoothed, more complex transitions (e.g. occlusions, transparency, and reflection) are not accurately treated. That is, motion errors are inevitable even after smoothing/refining motion vector fields, and, hence, an appropriate mechanism that takes care of the errors is necessary for producing artifact-free outputs.

Unlike video processing algorithms which depend directly on motion vectors, in a recent work, Protter *et al.* [11] proposed a video-to-video super-resolution method without explicit motion estimation or compensation based on the idea of Non-Local Means [12]. Although the method produces impressive spatial upscaling results even without motion estimation, the computational load is very high due to the exhaustive search (across space and time) for blocks similar to the block of interest. In a related work [13], we presented a space-time video upscaling method, called *3-D iterative steering kernel regression* (3-D ISKR), in which explicit subpixel motion estimation is again avoided. 3-D ISKR is an extension of 2-D *steering kernel regression* (SKR) proposed in [14, 15]. SKR is closely related to bilateral filtering [16, 17] and normalized convolution [18]. These methods can achieve accurate and robust image reconstruction results, due to their use of robust error norms and locally adaptive weighting functions. 2-D SKR has been applied to spatial interpolation, denoising and deblurring [15, 18, 19]. In 3-D ISKR, instead of relying on motion vectors, the 3-D kernel captures local spatial and temporal orientations based on local covariance matrices of gradients of video data. With the adaptive kernel, the method is capable of upscaling video with complex motion both in space and time.

In this chapter, we build upon the 2-D steering kernel regression framework proposed in [14], and develop a spatiotemporal (3-D) framework for processing video. Specifically, we propose an approach we call *motion-assisted steering kernel* (MASK) regression. The MASK function is a 3-D kernel, however, unlike as in 3-D ISKR, the kernel function takes spatial (2-D) orientation and the local motion trajectory into account separately, and it utilizes an analysis of the local orientation and local motion vector to steer spatiotemporal regression kernels. Subsequently, local kernel regression is applied to compute weighted least-squares optimal pixel estimates. Although 2-D kernel regression has been applied to achieve super-resolution reconstruction through fusion of multiple pre-registered frames on to a 2-D plane [14, 18], the proposed method is different in that it does not require explicit motion compensation of the video frames. Instead, we use 3-D weighting kernels that are "warped" according to estimated motion vectors, such that the regression process acts directly upon the video data. Although we consider local motion vectors in MASK, we propose an algorithm that is robust against errors in the estimated motion field. Prior multi-frame resolution-enhanced or super-resolution (SR) reconstruction methods ([2, 3]) often consider only global translational or affine motions; local motion and object occlusions are often not addressed. Many SR methods require explicit motion compensation, which may involve interpolation or rounding of displacements to grid locations. These issues can have a negative impact on accuracy and robustness. Our proposed method is capable of handling local motions, avoids explicit motion compensation, and is more robust. The proposed MASK approach is capable of simultaneous spatial interpolation with resolution enhancement, temporal video interpolation, noise reduction, and preserving high frequency components. Initial results using MASK were presented in [20].

An overview of this chapter is as follows. Firstly, we provide a review of 2-D SKR in Section 2. Then, we extend 2-D SKR to 3-D SKR and describe the MASK approach in Section 3. Subsequently, we propose a practical video upscaling algorithm based on MASK in Section 4, proposing further novel techniques to reduce computational complexity and improve robustness. We present several example results of our algorithm in Section 5 and conclude in Section 6.

2   Review of Steering Kernel Regression

This section gives an overview of SKR, which is the basis of MASK. We begin with describing the fundamental framework of SKR, called *kernel regression* (KR), in which we estimate a pixel value of interest from neighboring pixels using a weighted least-square formulation. We propose an effective weighting function for the weighted least-square estimator, called *steering kernel function*, that takes not only spatial distances between the samples of interest into account, but also the radiometric values of those samples.

2.1   Kernel Regression in 2-D

The KR framework defines its data model as

$$y_i = z(\mathbf{x}_i) + \varepsilon_i, \quad i = 1, \cdots, P, \quad \mathbf{x}_i = [x_{1i}, x_{2i}]^T, \tag{1}$$

where y_i is a noisy sample at $\mathbf{x}_i$ (Note: x_{1i} and x_{2i} are spatial coordinates), $z(\cdot)$ is the (hitherto unspecified) *regression function* to be estimated, ε_i is an i.i.d. zero mean noise, and P is the total number of samples in an arbitrary "window" around a position $\mathbf{x}$ of interest as shown in Fig. 1. As such, the kernel regression framework provides a rich mechanism for computing point-wise estimates of the regression function with minimal assumptions about global signal or noise models.

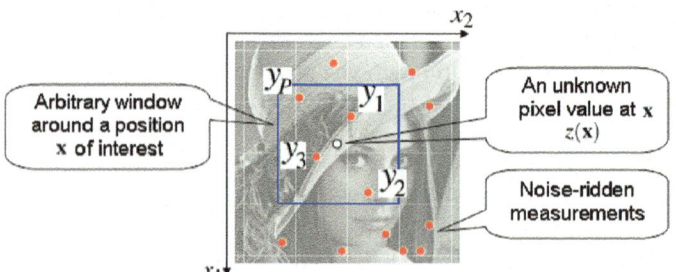

Fig. 1 The data model for the kernel regression framework

While the particular form of $z(\cdot)$ may remain unspecified, we can develop a generic local expansion of the function about a sampling point $\mathbf{x}_i$. Specifically, if the position of interest $\mathbf{x}$ is near the sample at $\mathbf{x}_i$, we have the N-th order Taylor series

$$z(\mathbf{x}_i) \approx z(\mathbf{x}) + \{\nabla z(\mathbf{x})\}^T (\mathbf{x}_i - \mathbf{x}) + \frac{1}{2}(\mathbf{x}_i - \mathbf{x})^T \{Hz(\mathbf{x})\} (\mathbf{x}_i - \mathbf{x}) + \cdots$$

$$= \beta_0 + \boldsymbol{\beta}_1^T (\mathbf{x}_i - \mathbf{x}) + \boldsymbol{\beta}_2^T \text{vech}\left\{(\mathbf{x}_i - \mathbf{x})(\mathbf{x}_i - \mathbf{x})^T\right\} + \cdots \qquad (2)$$

where ∇ and H are the gradient (2×1) and Hessian (2×2) operators, respectively, and $\text{vech}(\cdot)$ is the half-vectorization operator that lexicographically orders the lower triangular portion of a symmetric matrix into a column-stacked vector. Furthermore, β_0 is $z(\mathbf{x})$, which is the signal (or pixel) value of interest, and the vectors $\boldsymbol{\beta}_1$ and $\boldsymbol{\beta}_2$ are

$$\boldsymbol{\beta}_1 = \left[\frac{\partial z(\mathbf{x})}{\partial x_1}, \quad \frac{\partial z(\mathbf{x})}{\partial x_2}\right]^T,$$

$$\boldsymbol{\beta}_2 = \frac{1}{2}\left[\frac{\partial^2 z(\mathbf{x})}{\partial x_1^2}, \quad 2\frac{\partial^2 z(\mathbf{x})}{\partial x_1 \partial x_2}, \quad \frac{\partial^2 z(\mathbf{x})}{\partial x_2^2}\right]^T. \qquad (3)$$

Since this approach is based on *local* signal representations, a logical step to take is to estimate the parameters $\{\boldsymbol{\beta}_n\}_{n=0}^{N}$ from all the neighboring samples $\{y_i\}_{i=1}^{P}$ while giving the nearby samples higher weights than samples farther away. A (weighted) least-square formulation of the fitting problem capturing this idea is

$$\min_{\{\boldsymbol{\beta}_n\}_{n=0}^{N}} \sum_{i=1}^{P}\left[y_i - \beta_0 - \boldsymbol{\beta}_1^T (\mathbf{x}_i - \mathbf{x}) - \boldsymbol{\beta}_2^T \text{vech}\left\{(\mathbf{x}_i - \mathbf{x})(\mathbf{x}_i - \mathbf{x})^T\right\} - \cdots\right]^2 K_{\mathbf{H}}(\mathbf{x}_i - \mathbf{x})$$

$$(4)$$

with

$$K_{\mathbf{H}}(\mathbf{x}_i - \mathbf{x}) = \frac{1}{\det(\mathbf{H})} K\left(\mathbf{H}^{-1}(\mathbf{x}_i - \mathbf{x})\right), \qquad (5)$$

where N is the regression order, $K(\cdot)$ is the kernel function (a radially symmetric function such as a Gaussian), and $\mathbf{H}$ is the smoothing (2×2) matrix which dictates the "footprint" of the kernel function. In the classical approach, when the pixels (y_i) are equally spaced, the smoothing matrix is defined as

$$\mathbf{H} = h\mathbf{I} \qquad (6)$$

for every sample, where h is called the *global smoothing parameter*. The shape of the kernel footprint is perhaps the most important factor in determining the quality of estimated signals. For example, it is desirable to use kernels with large footprints in the smooth local regions to reduce the noise effects, while relatively smaller footprints are suitable in the edge and textured regions to preserve the signal discontinuity. Furthermore, it is desirable to have kernels that adapt themselves to the local structure of the measured signal, providing, for instance, strong filtering along an edge rather than across it. This last point is indeed the motivation behind the *steering* KR framework [14] which we will review in Section 2.2.

Returning to the optimization problem (4), regardless of the regression order and the dimensionality of the regression function, we can rewrite it as a weighted least squares problem:

$$\widehat{\mathbf{b}} = \arg\min_{\mathbf{b}} \left[(\mathbf{y} - \mathbf{X}\mathbf{b})^T \mathbf{K} (\mathbf{y} - \mathbf{X}\mathbf{b}) \right], \tag{7}$$

where

$$\mathbf{y} = [y_1, \, y_2, \, \cdots, \, y_P]^T, \quad \mathbf{b} = \left[\beta_0, \, \boldsymbol{\beta}_2^T, \, \cdots, \, \boldsymbol{\beta}_N^T \right]^T, \tag{8}$$

$$\mathbf{K} = \mathrm{diag}\left[K_{\mathbf{H}}(\mathbf{x}_1 - \mathbf{x}), \, K_{\mathbf{H}}(\mathbf{x}_2 - \mathbf{x}), \cdots, \, K_{\mathbf{H}}(\mathbf{x}_P - \mathbf{x}) \right], \tag{9}$$

and

$$\mathbf{X} = \begin{bmatrix} 1, \, (\mathbf{x}_1 - \mathbf{x})^T, \, \mathrm{vech}^T\{ (\mathbf{x}_1 - \mathbf{x})(\mathbf{x}_1 - \mathbf{x})^T \}, \, \cdots \\ 1, \, (\mathbf{x}_2 - \mathbf{x})^T, \, \mathrm{vech}^T\{ (\mathbf{x}_2 - \mathbf{x})(\mathbf{x}_2 - \mathbf{x})^T \}, \, \cdots \\ \vdots \quad \vdots \qquad\qquad \vdots \qquad\qquad \vdots \\ 1, \, (\mathbf{x}_P - \mathbf{x})^T, \, \mathrm{vech}^T\{ (\mathbf{x}_P - \mathbf{x})(\mathbf{x}_P - \mathbf{x})^T \}, \, \cdots \end{bmatrix} \tag{10}$$

with "diag" defining a diagonal matrix. Using the notation above, the optimization (4) provides the weighted least square estimator

$$\widehat{\mathbf{b}} = \left(\mathbf{X}^T \mathbf{K} \mathbf{X} \right)^{-1} \mathbf{X}^T \mathbf{K} \, \mathbf{y} = \begin{bmatrix} \mathbf{W}_N \\ \mathbf{W}_{N,x_1} \\ \mathbf{W}_{N,x_2} \\ \vdots \end{bmatrix} \mathbf{y}, \tag{11}$$

where $\mathbf{W}_N$ is a $1 \times P$ vector that contains filter coefficients, which we call the *equivalent kernel* weights, and $\mathbf{W}_{N,x_1}$ and $\mathbf{W}_{N,x_2}$ are also $1 \times P$ vectors that compute the gradients along the x_1- and x_2-directions at the position of interest $\mathbf{x}$. The estimate of the signal (i.e. pixel) value of interest β_0 is given by a weighted *linear* combination of the nearby samples:

$$\hat{z}(\mathbf{x}) = \widehat{\beta}_0 = \mathbf{e}_1^T \widehat{\mathbf{b}} = \mathbf{W}_N \mathbf{y} = \sum_{i=1}^{P} W_i(K, \mathbf{H}, N, \mathbf{x}_i - \mathbf{x}) \, y_i, \quad \sum_{i=1}^{P} W_i(\cdot) = 1, \tag{12}$$

where $\mathbf{e}_1$ is a column vector with the first element equal to one and the rest equal to zero, and we call W_i the equivalent kernel weight function for y_i (q.v. [14] or [21] for more detail). For example, for zero-th order regression (i.e. $N = 0$), the estimator (12) becomes

$$\hat{z}(\mathbf{x}) = \hat{\beta}_0 = \frac{\sum_{i=1}^{P} K_{\mathbf{H}}(\mathbf{x}_i - \mathbf{x}) \, y_i}{\sum_{i=1}^{P} K_{\mathbf{H}}(\mathbf{x}_i - \mathbf{x})}, \tag{13}$$

which is the so-called *Nadaraya-Watson* estimator (NWE) [22].

What we described above is the "classic" kernel regression framework, which, as we just mentioned, yields a pointwise estimator that is always a local *linear* combination of the neighboring samples. As such, it suffers from an inherent limitation. In the next sections, we describe the framework of *steering* KR in two and three dimensions, in which the kernel weights themselves are computed from the local window, and therefore we arrive at filters with more complex (nonlinear) action on the data.

2.2   Steering Kernel Function

The steering kernel framework is based on the idea of robustly obtaining local signal structures (e.g. discontinuities in 2-D and planes in 3-D) by analyzing the radiometric (pixel value) variations locally, and feeding this structure information to the kernel function in order to affect its shape and size.

Consider the (2×2) smoothing matrix $\mathbf{H}$ in (5). As explained in the previous section, in the generic "classical" case, this matrix is a scalar multiple of the identity. This results in kernel weights which have equal effect along the x_1- and x_2-directions. However, if we properly choose this matrix locally (i.e. $\mathbf{H} \to \mathbf{H}_i$ for each y_i), the kernel function can capture local structures. More precisely, we define the smoothing matrix as a symmetric matrix

$$\mathbf{H}_i = h\mathbf{C}_i^{-\frac{1}{2}}, \tag{14}$$

which we call the *steering* matrix and where, for each given sample y_i, the matrix $\mathbf{C}_i$ is estimated as the local covariance matrix of the neighborhood spatial gradient vectors. A naive estimate of this covariance matrix may be obtained by

$$\widehat{\mathbf{C}}_i^{\text{naive}} = \mathbf{J}_i^T \mathbf{J}_i, \tag{15}$$

with

$$\mathbf{J}_i = \begin{bmatrix} z_{x_1}(\mathbf{x}_1) & z_{x_2}(\mathbf{x}_1) \\ \vdots & \vdots \\ z_{x_1}(\mathbf{x}_P) & z_{x_2}(\mathbf{x}_P) \end{bmatrix}, \tag{16}$$

where $z_{x_1}(\cdot)$ and $z_{x_2}(\cdot)$ are the first derivatives along x_1- and x_2-axes, and P is the number of samples in the local analysis window around a sampling position $\mathbf{x}_i$. However, the naive estimate may in general be rank deficient or unstable. Therefore, instead of using the naive estimate, we can obtain the covariance matrix by using the (compact) singular value decomposition (SVD) of $\mathbf{J}_i$:

$$\mathbf{J}_i = \mathbf{U}_i \mathbf{S}_i \mathbf{V}_i^T, \tag{17}$$

where $\mathbf{S}_i = \text{diag}[s_1, s_2]$, and $\mathbf{V}_i = [\mathbf{v}_1, \mathbf{v}_2]$. The singular vectors contain direct information about the local orientation structure, and the corresponding singular values represent the energy (strength) in these respective orientation directions. Using the singular vectors and values, we compute a more stable estimate of our covariance matrix as:

$$\widehat{\mathbf{C}}_i = \gamma_i \mathbf{V}_i \begin{bmatrix} \rho_i & \\ & \frac{1}{\rho_i} \end{bmatrix} \mathbf{V}_i^T = \gamma_i \left(\rho_i \mathbf{v}_1 \mathbf{v}_1^T + \frac{1}{\rho_i} \mathbf{v}_2 \mathbf{v}_2^T \right), \qquad (18)$$

where

$$\rho_i = \frac{s_1 + \lambda'}{s_2 + \lambda'}, \quad \gamma_i = \left(\frac{s_1 s_2 + \lambda''}{P} \right)^{\alpha}. \qquad (19)$$

The parameters ρ_i and γ_i are the *elongation* and *scaling* parameter, respectively, and λ' and λ'' are "regularization" parameters, respectively, which dampen the effect of the noise and restrict γ_i and the denominator of ρ_i from becoming zero. The parameter α is called the *structure sensitivity* parameter. We fix $\lambda' = 0.1$, $\lambda'' = 0.1$, and $\alpha = 0.2$ in this work. More details about the effectiveness and the choice of the parameters can be found in [14]. With the above choice of the smoothing matrix and a Gaussian kernel, we now have the steering kernel function as

$$K_{\mathbf{H}_i}(\mathbf{x}_i - \mathbf{x}) = \frac{\sqrt{\det(\mathbf{C}_i)}}{2\pi h^2} \exp \left\{ -\frac{(\mathbf{x}_i - \mathbf{x})^T \mathbf{C}_i (\mathbf{x}_i - \mathbf{x})}{2h^2} \right\}. \qquad (20)$$

Fig. 2 illustrates a schematic representation of the estimate of local covariance matrices and the computation of steering kernel weights. First we estimate the gradients and compute the local covariance matrix $\mathbf{C}_i$ by (18) for each pixel. Then, for example, when denoising y_{13}, we compute the steering kernel weights for each neighboring pixel with its $\mathbf{C}_i$. In this case, even though the spatial distances from y_{13} to y_1 and y_{21} are equal, the steering kernel weight for y_{21} (i.e. $K_{\mathbf{H}_{21}}(\mathbf{x}_{21} - \mathbf{x}_{13})$) is larger

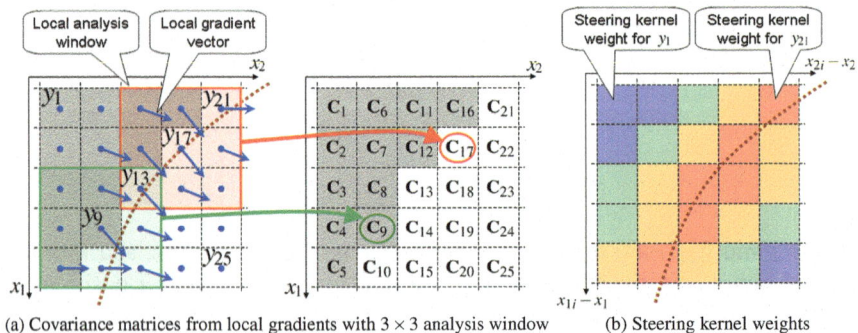

(a) Covariance matrices from local gradients with 3 × 3 analysis window   (b) Steering kernel weights

Fig. 2 A schematic representation of the estimates of local covariance metrics and the steering kernel weights at a local region with one dominant orientation: (a) First, we estimate the gradients and compute the local covariance matrix $\mathbf{C}_i$ by (18) for each pixel, and (b) Next, when denoising y_{13}, we compute the steering kernel weights with $\mathbf{C}_i$ for neighboring pixels. Even though, in this case, the spatial distances between y_{13} and y_1 and between y_{13} y_{21} are equal, the steering kernel weight for y_{21} (i.e. $K_{\mathbf{H}_{21}}(\mathbf{x}_{21} - \mathbf{x}_{13})$) is larger than the one for y_1 (i.e. $K_{\mathbf{H}_1}(\mathbf{x}_1 - \mathbf{x}_{13})$). This is because y_{13} and y_{21} are located along the same edge.

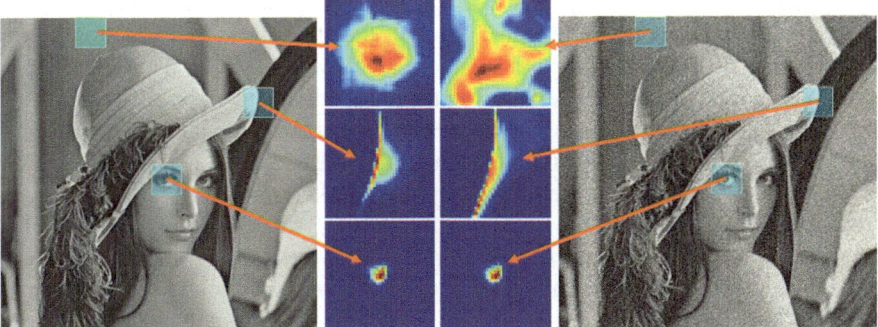

Fig. 3 Steering kernel weights for Lena image without/with noise (white Gaussian noise with standard deviation $\sigma = 25$) at flat, edge, and texture areas

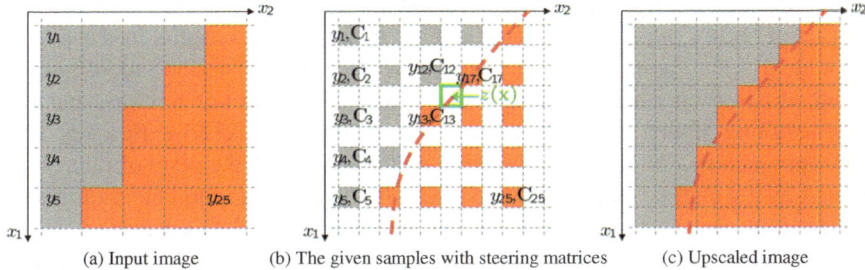

(a) Input image   (b) The given samples with steering matrices   (c) Upscaled image

Fig. 4 Steering kernel regression for image upscaling: (a)Input image. (b)We compute steering matrices for each pixel and then estimate. Then, estimate the missing position $\mathbf{z}(\mathbf{x})$ and denoise the given pixels y_i. The red dashed line is a speculative local orientation. (c)Upscaled image by steering kernel regression.

than the one for y_1 (i.e. $K_{\mathbf{H}_1}(\mathbf{x}_1 - \mathbf{x}_{13})$). Moreover, Fig. 3 shows visualizations of the 2-D steering kernel function for noise-free Lena image and a low PSNR[1] case (we added white Gaussian noise with standard deviation 25, the corresponding PSNR being 20.16[dB]). As shown in Fig. 3, the steering kernel weights (which are the normalized $K_{\mathbf{H}_i}(\mathbf{x}_i - \mathbf{x})$ as a function of $\mathbf{x}_i$ with $\mathbf{x}$ held fixed) illustrate the relative size of the actual weights applied to compute the estimate as in (12). We note that even for the highly noisy case, we can obtain stable estimates of local structure.

At this point, the reader may be curious to know how the above formulation would work for the case where we are interested not only in denoising, but also upscaling the images. Fig. 4 illustrates a summary of image upscaling by steering kernel regression. Similar to the denoising case, we begin with computing steering (covariance) matrices, $\mathbf{C}_i$ for all the pixels, y_i, from the input image shown in Fig. 4(a) by (18) as depicted in Fig. 2(a). Once $\mathbf{C}_i$'s are available, we compute

---

[1] Peak Signal to Noise Ratio $= 10\log_{10}\left(\frac{255^2}{\text{Mean Square Error}}\right)$ [dB].

steering kernel weights by (20). For example, when we estimate the missing pixel $z(\mathbf{x})$ at $\mathbf{x}$ shown as the green box in Fig. 4(b), the steering kernel function gives high weights to the samples y_{13} and y_{17} and a small weight to y_{12}. This is because the missing pixel, $z(\mathbf{x})$, most likely lies on the same edge (shown by the red dashed curve) as y_{13} and y_{17}. Next, plugging the steering kernel weights into (11), we compute the equivalent kernel $\mathbf{W}_N$ and the estimator (12) gives the estimated pixel $\hat{z}(\mathbf{x})$ at $\mathbf{x}$. Fig. 4(c) shows the upscaled image by steering kernel regression. In [14], we introduced an iterative scheme where we recompute $\mathbf{C}_i$ from the upscaled image one more time, and, using the new covariance matrices, we estimate the missing pixels and denoise the given samples again. However, in this work, to keep the computational load low, we compute the steering matrices only once from the given samples.

3   Motion Assisted Steering Kernel Regression

SKR estimates an unknown pixel value in a single image by a weighted combination of neighboring pixels in the same image, giving larger weights to the pixels along a local orientation. In this section, we develop a multi-frame video upscaling method based on SKR by additionally utilizing local motion vectors, and we call the resulting method *motion-assisted steering kernel* (MASK) regression. The MASK approach is a 3-D kernel regression method in which the pixel of interest is estimated by a weighted combination of pixels in its spatiotemporal neighborhood, involving multiple video frames. Hence, we first extend the 2-D kernel regression framework into a 3-D framework. Then, we present our 3-D data-adaptive kernel, the MASK function, which relies not only on local spatial orientation but also local motions. Finally, we describe the process of spatial upscaling and temporal frame interpolation based on MASK. While we focus on the principles of our approach in this section, we present a specific algorithm for video processing based on the MASK method in the next section.

3.1   Spatiotemporal Kernel Regression

For video processing, we define a spatiotemporal data model as

$$y_i = z(\mathbf{x}_i) + \varepsilon_i, \quad i = 1, \cdots, P, \quad \mathbf{x}_i = [x_{1i}, x_{2i}, t_i]^T, \tag{21}$$

where y_i is a given sample (pixel) at location $\mathbf{x}_i$, x_{1i} and x_{2i} are the spatial coordinates, t_i is the temporal coordinate, $z(\cdot)$ is the *regression function*, and ε_i is i.i.d zero mean noise. P is the number of samples in a spatiotemporal neighborhood of interest, which spans multiple video frames.

Similar to the 2-D case, in order to estimate the value of $z(\cdot)$ at point $\mathbf{x}$, given the above data samples y_i, we can rely on a local N^{th} order Taylor expansion about $\mathbf{x}$. We denote the pixel value of interest $z(\mathbf{x})$ by β_0, while $\boldsymbol{\beta}_1, \boldsymbol{\beta}_2, \cdots, \boldsymbol{\beta}_N$ denote vectors containing the first-order, second-order, $\cdots$, N^{th} order partial derivatives of

$z(\cdot)$ at $\mathbf{x}$, resulting from the Taylor expansion. For example, $\beta_0 = z(\mathbf{x})$ and $\beta_1 = [z_{x_1}(\mathbf{x}), z_{x_2}(\mathbf{x}), z_t(\mathbf{x})]^T$.

The unknowns, $\{\beta_n\}_{n=0}^N$, can be estimated from $\{y_i\}_{i=1}^P$ using the following weighted least-squares optimization procedure:

$$\min_{\{\beta_n\}_{n=0}^N} \sum_{i=1}^P \left[y_i - \beta_0 - \beta_1^T(\mathbf{x}_i - \mathbf{x}) - \beta_2^T \text{vech}\left\{ (\mathbf{x}_i - \mathbf{x})(\mathbf{x}_i - \mathbf{x})^T \right\} - \cdots \right]^2 K_{\mathbf{H}_i^{3D}}(\mathbf{x}_i - \mathbf{x})$$

$$(22)$$

where N is the regression order and $K(\cdot)$ is a *kernel function* that weights the influence of each sample. Typically, samples near $\mathbf{x}$ are given higher weights than samples farther away.

A *3-D steering kernel* is a direct extension of the 2-D steering kernel defined in [14]. The 3×3 data-dependent steering matrix $\mathbf{H}_i^{3D}$ can be defined as

$$\mathbf{H}_i^{3D} = h\left(\mathbf{C}_i^{3D}\right)^{-\frac{1}{2}} \qquad (23)$$

where h is a global smoothing parameter and $\mathbf{C}_i^{3D}$ is a 3×3 covariance matrix based on the sample variations in a local (3-D) neighborhood around sample $\mathbf{x}_i$. We can construct the matrix $\mathbf{C}_i^{3D}$ parametrically as $\mathbf{C}_i^{3D} = \gamma_i \mathbf{R}_i \Lambda_i \mathbf{R}_i^T$, where $\mathbf{R}_i$ is a 3-D rotation matrix, Λ_i is a 3-D elongation matrix, and γ_i is a scaling parameter. We have found that such an approach performs quite well for spatial upscaling of video [13]. However, this 3-D kernel does not consider the specific spatiotemporal characteristics of video data. In particular, problems may occur in the presence of large object displacements (fast motion). This may result in either shrinking of the kernel in the temporal direction, or spatial blurring (as the kernel weights spread across unrelated data samples), both undesirable effects.

3.2 Motion Assisted Steering Kernel Function

A good choice for steering spatiotemporally is to consider local motion or optical flow vectors caused by object motion in the scene, in conjunction with spatial steering along local edges and isophotes. Spatial steering should consider the locally dominant orientation of the pixel data and should allow elongation of the kernel in this direction. Spatiotemporal steering should allow alignment of the kernel weights with the local optical flow or motion trajectory, as well as overall temporal scaling. Hence, we construct our spatiotemporal kernel as a product of a spatial- and motion-steering kernel, and a kernel that acts temporally:

$$K_{\text{MASK}} \equiv \frac{1}{\det(\mathbf{H}_i^s)} K\left((\mathbf{H}_i^s)^{-1} \mathbf{H}_i^m (\mathbf{x}_i - \mathbf{x})\right) K_{h_t}(t_i - t), \qquad (24)$$

where $\mathbf{H}_i^s$ is a 3×3 *spatial steering* matrix, $\mathbf{H}_i^m$ is a 3×3 *motion steering* matrix, $K_{h_t}(\cdot)$ is a temporal kernel function, and h_t is the temporal smoothing parameter which controls the temporal penalization. These data-dependent kernel components determine the steering action at sample $\mathbf{x}_i$, and are described next.

Following [14], the spatial steering matrix $\mathbf{H}_i^s$ is defined by:

$$\mathbf{H}_i^s = h_s \begin{bmatrix} \mathbf{C}_i \\ & 1 \end{bmatrix}^{-\frac{1}{2}}, \tag{25}$$

where h_s is a global spatial smoothing parameter, and $\mathbf{C}_i$ is a 2×2 covariance matrix given by (18), which captures the sample variations in a local spatial neighborhood around $\mathbf{x}_i$. $\mathbf{C}_i$ is constructed in a parametric manner, as shown in (18).

The motion steering matrix $\mathbf{H}_i^m$ is constructed on the basis of a local estimate of the motion (or optical flow vector) $\mathbf{m}_i = [m_{1i}, m_{2i}]^T$ at $\mathbf{x}_i$. Namely, we warp the kernel along the local motion trajectory using the following shearing transformation:

$$\begin{cases} (x_{1i} - x_1) \leftarrow (x_{1i} - x_1) - m_{1i} \cdot (t_i - t) \\ (x_{2i} - x_2) \leftarrow (x_{2i} - x_2) - m_{2i} \cdot (t_i - t) \end{cases}.$$

Hence,

$$\mathbf{H}_i^m = \begin{bmatrix} 1 & 0 & -m_{1i} \\ 0 & 1 & -m_{2i} \\ 0 & 0 & 0 \end{bmatrix}. \tag{26}$$

Assuming a spatial prototype kernel was used with elliptical footprint, this results in a spatiotemporal kernel with the shape of a tube or cylinder with elliptical cross-sections at any time instance t. Most importantly, the center point of each such cross-section moves along the motion path.

The final component of (24) is a temporal kernel that provides temporal penalization. A natural approach is to give higher weights to samples in frames closer to t. An example of such a kernel is the following:

$$K_{h_t}(t_i - t) = \frac{1}{h_t} \exp\left(-\frac{|t_i - t|^2}{2h_t^2}\right), \tag{27}$$

where a temporal smoothing parameter h_t controls the relative temporal extent of the kernel. We use the temporal kernel (27) in this section to illustrate the MASK approach. However, we will introduce a more powerful adaptive temporal weighting kernel in Section 4.2, which acts to compensate for unreliable local motion vector estimates.

3.3 Spatial Upscaling and Temporal Frame Interpolation

Having introduced our choice of 3-D smoothing matrix, $\mathbf{H}_i^{3D}$, using Gaussian kernel for K, we have the MASK function as

$$K_{\text{MASK}}(\mathbf{x}_i - \mathbf{x}) = \frac{1}{\det(\mathbf{H}_i^s)} K\left((\mathbf{H}_i^s)^{-1} \mathbf{H}_i^m(\mathbf{x}_i - \mathbf{x})\right) \cdot K_{h_t}(t_i - t)$$

$$= \frac{1}{\det(\mathbf{H}_i^s)} K\left((\mathbf{H}_i^s)^{-1} \left(\mathbf{x}_i - \mathbf{x} - \begin{bmatrix} \mathbf{m}_i \\ 1 \end{bmatrix}(t_i - t)\right)\right) \cdot K_{h_t}(t_i - t)$$

$$= \frac{\sqrt{\det(\mathbf{C}_i)}}{h_s^2 h_t^2} \exp\left(-\frac{1}{2h_s^2} \left\| \mathbf{x}_i - \mathbf{x} - \begin{bmatrix} \mathbf{m}_i \\ 1 \end{bmatrix}(t_i - t) \right\|_{\mathbf{C}_i}^2\right)$$

$$\cdot \exp\left(-\frac{|t_i - t|^2}{2h_t^2}\right) \quad (28)$$

where $\| \cdot \|_{\mathbf{C}_i^s}^2$ is weighted squared L_2-norm. Figs. 5(a-i)-(a-iii) graphically describe how the proposed MASK function constructs its weights for spatial upscaling. For ease of explanation, suppose there are 5 frames at times from t_1 to t_5, and we upscale

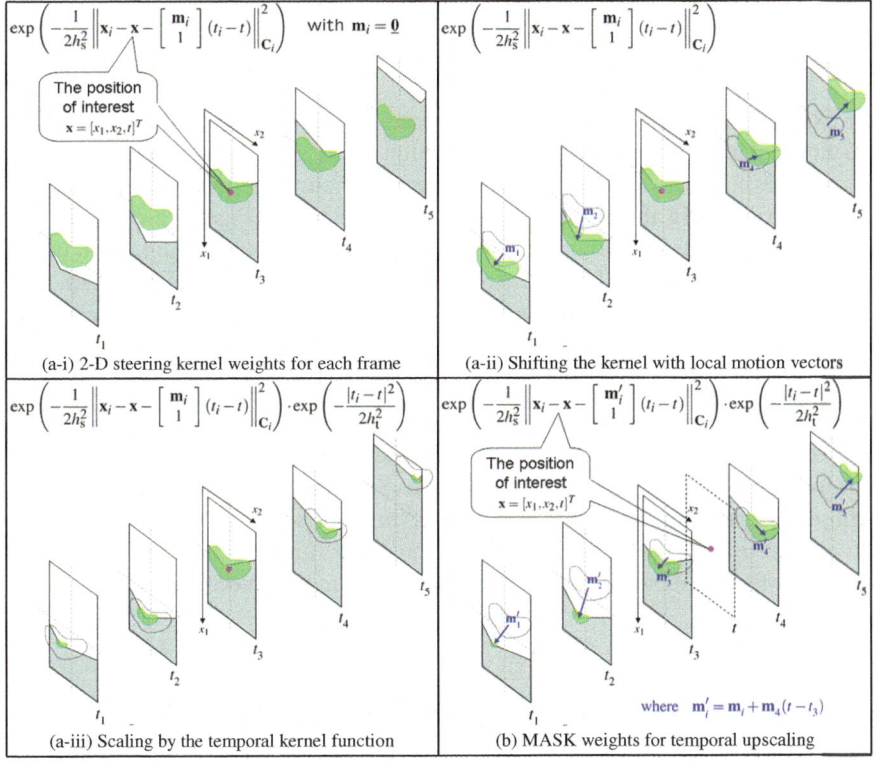

Fig. 5 Schematic representations of the construction of MASK weights: the proposed MASK weights are constructed by the following procedure (a-i) compute 2-D steering kernel weights for each frame (with $\mathbf{m}_i = \underline{\mathbf{0}}$ at this moment), (a-ii) shift the steering kernels by the local motion vectors, and (a-iii) scale the shifted steering kernels by the temporal kernel function. Fig.(b) shows the weight construction for the estimation of an intermediate frame at time t.

the third frame (spatial upscaling). When estimating the pixel value at $\mathbf{x} = [x_1, x_2, t]$, where $t = t_3$, first we compute 2-D steering kernel weights for each frame, as illustrated in Fig. 5(a-i), using the first Gaussian kernel function in (28). Motions are not taken into account at this stage. Second, having motion vectors, $\mathbf{m}_i$, which we estimate using the optical flow technique with the translational motion model and the frame at $t_{i=3}$ as the anchor frame, we shift the steering kernels for each frame by $\mathbf{m}_i$ as illustrated in Fig. 5(a-ii). Finally, as in Fig. 5(a-iii), the temporal kernel function penalizes the shifted steering kernels so that we give high weights to closer neighboring frames.

Local steering parameters and spatio-temporal weights are estimated at each pixel location $\mathbf{x}_i$ in a small region of support for the final regression step. Once the MASK weights are available, similar to the 2-D case, we plug them into (11), compute the equivalent kernel $\mathbf{W}_N$, and then estimate the missing pixels and denoise the given samples from the local input samples (y_i) around the position of interest $\mathbf{x}$. Similar to (12), the final spatio-temporal regression step can be expressed as follows:

$$\hat{z}(\mathbf{x}) = \sum_{i=1}^{P} W_i(\mathbf{x}; \mathbf{H}_i^s, \mathbf{H}_i^m, h_t, K, N) \, y_i. \tag{29}$$

The MASK approach is also capable of upscaling video temporally (also called frame interpolation or frame rate upconversion). Fig. 5(b) illustrates the MASK weights for estimating an intermediate frame at sometime between t_3 and t_4. Fundamentally, following the same procedure as described in Figs. 5(a-i)-(a-iii), we generate MASK weights. However, for the motion vector with the unknown intermediate frame as the anchor frame, we assume that the motion between the frames at t_3 and t_4 is constant, and using the motion vectors, $\mathbf{m}_{i=1,\cdots,5}$, we linearly interpolate motion vectors $\mathbf{m}_i'$ as

$$\mathbf{m}_i' = \mathbf{m}_i + \mathbf{m}_4 (t - t_3). \tag{30}$$

Note that when $\mathbf{m}_4$ is inaccurate, the interpolated motion vectors for other frames in the temporal window ($\mathbf{m}_i'$) are also inaccurate. In that case, we would shift the kernel toward the wrong direction, and the MASK weights would be less effective for temporal upscaling. Therefore, one should incorporate a test of the reliability of $\mathbf{m}_4$ into the process, and use vectors $\mathbf{m}_i$ instead of $\mathbf{m}_i'$ if it is found to be unreliable. Our specific technique to compute the reliability of motion vectors is described in Section 4.2.

4   A Practical Video Upscaling Algorithm Based on MASK

In this section, we describe a complete algorithm for spatial upscaling, denoising and enhancement, as well as temporal frame interpolation, based on the MASK approach. We introduce several techniques that enable a practical implementation of the MASK principles explained in the previous section. In particular, we develop an algorithm with reduced computational complexity and reduced memory requirements, that is suitable for both software and hardware implementation.

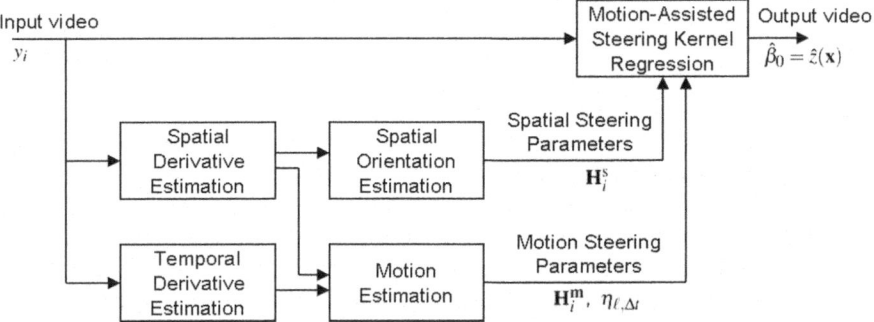

Fig. 6 Illustration of video processing based on motion-assisted spatiotemporal steering kernel (MASK) regression

An overview of the proposed video interpolation and denoising algorithm based on motion-assisted spatiotemporal steering kernel regression is provided in Fig. 6. The algorithm estimates spatial and motion steering parameters using gradient-based techniques. Hence, we first compute initial estimates of the spatial and temporal derivatives, e.g. based on classic kernel regression. In this work, we obtain a quick and robust estimate of the spatial orientation angle (θ_i), elongation (ρ_i) and scaling (γ_i) parameters at $\mathbf{x}_i$ by applying a vector quantization technique to the covariance matrix obtained from the spatial gradient data. This will be described in Section 4.3. Motion vectors are estimated using the well-known Lucas and Kanade method, based on both spatial and temporal gradients in a local region. This is followed by computing estimates of the temporal motion reliability (η), and is described further in Section 4.2. Given spatial and motion steering parameters, final MASK regression is applied directly on the input video samples; further details of this step are provided in Section 4.4.

The following are further salient points for our algorithm based on MASK. We first summarize them, and then provide details in subsequent subsections.

▷ **Block-by-Block Processing**
Since the kernel-based estimator is a pointwise process, it is unnecessary to store the orientations and motion vectors of all the pixels in a video frame ($\mathbf{H}_i^s$ and $\mathbf{H}_i^m$ for all i) in memory. However, strict pixel-by-pixel processing would result in a large number of redundant computations due to the overlapping neighborhoods of nearby pixels. In order to reduce the computational load while keeping the required memory space small, we break the video data into small blocks (e.g. 8×8 pixels), and process the blocks one-by-one.

▷ **Adaptive Temporal Penalization**
MASK relies on motion vectors, and the visual quality of output video frames is strongly associated with the accuracy of motion estimation. Even though our motion estimation approach is able to estimate motion vectors quite accurately, the estimated vectors become unreliable when the underlying scene motion and

camera projection violate the motion model. In practice, errors in motion vectors are inevitable and it is important to provide a fall-back mechanism in order to avoid visual artifacts.

▷ **Quantization of Orientation Map**
The estimation of spatial orientations or steering covariance matrices $\mathbf{C}_i^s$ in (18) involves singular value decomposition (SVD), which represents significant computational complexity. Instead of using the SVD, we use a pre-defined lookup table containing a set of candidate covariance matrices, and locally select an appropriate matrix from the table. Since the lookup table contains only stable (invertible) covariance matrices, the estimation process remains robust.

▷ **Adaptive Regression Order**
A higher regression order (e.g. $N = 2$ in this chapter) preserves high frequency components in filtered images, although it requires more computation (11). On the other hand, zeroth regression order ($N = 0$) has lower computational cost, but it has a stronger smoothing effect. Although second order regression is preferable, it is only needed at pixel locations in texture and edge regions. Moreover, in terms of noise reduction, zeroth order regression is more suitable in flat regions. We propose to adjust the order N locally, based on the scaling parameter (γ_i). Consequently, this adaptive approach keeps the total computational cost low while it preserves, and even enhances, high frequency components.

4.1 Block-by-Block Processing

The overall MASK algorithm consists of several operations (i.e. estimating spatial and temporal gradients, spatial orientations, and motions as shown in Fig. 6 and finally applying kernel regression), and it is possible to implement these in, e.g., a pixel-by-pixel process or a batch process. In a pixel-by-pixel process, we estimate gradients, orientations, and motions one-by-one, and then finally estimate a pixel value. Note that most of these operations require calculations involving other pixels in a neighborhood around the pixel of interest. Since the neighborhoods of nearby pixels may overlap significantly, frequently the same calculation would be performed multiple times. Hence, a pixel-by-pixel implementation suffers from a large computational load. On the other hand, this implementation requires very little memory. In a batch process, we estimate gradients for all pixels in an entire frame and store the results in memory, then estimate orientations of all pixels and store those results, etc. In the batch implementation, we need a large memory space to store intermediate results for all pixels in a frame; however, it avoids repeated calculations. This type of process is impractical for a hardware implementation.

As a compromise, in order to limit both the computational load and the use of memory, we process a video frame in a block-by-block manner, where each block contains, e.g., 8×8 or 16×16 pixels. Further reduction of the computational load is achieved by using a block-based motion model: we assume that, within a block, the motion of all the pixels follow a parametric model, e.g, translational or affine. In this chapter, we fix the block size to 8×8 pixels and we use the translational motion model. A variable block size and the use of other motion models are also possible, and are the subject of ongoing research.

4.2 Motion Estimation and Adaptive Temporal Penalization

As mentioned, motion estimation is based on the well-known Lucas and Kanade method [23, 24], applied in a block-by-block manner as follows. Assume we computed initial estimates of the local spatial and temporal derivatives. For example, spatial derivatives may be computed using classic kernel regression or existing derivative filtering techniques. Temporal derivatives are computed by taking the temporal difference between pixels of the current frame and one of the neighboring frames. Let $\hat{\mathbf{z}}_{x_1}$, $\hat{\mathbf{z}}_{x_2}$ and $\hat{\mathbf{z}}_t$ denote vectors containing (in lexicographical order) derivative estimates from the pixels in a local analysis window w_l associated with the ℓ-th block in the frame. This window contains and is typically centered on the block of pixels of interest, but may include additional pixels beyond the block (i.e. analysis windows from neighboring blocks may overlap). A motion vector $\mathbf{m}_l$ for block ℓ is estimated by solving the optical flow equation $[\hat{\mathbf{z}}_{x_1}, \hat{\mathbf{z}}_{x_2}]\mathbf{m}_\ell + \hat{\mathbf{z}}_t = \underline{\mathbf{0}}$ in the least-squares sense. The basic Lucas and Kanade method is applied iteratively for improved performance. As explained before, MASK uses multiple frames in a temporal window around the current frame. For every block in the current frame, a motion vector is computed to each of the neighboring frames in the temporal window. Hence, if the temporal window contains 4 neighboring frames in addition to the current frame, we compute 4 motion vectors for each block in the current frame.

In practice, a wide variety of transitions/activies will occur in natural video. Some of them are so complex that no parametric motion model matches them exactly, and motion errors are unavoidable. When there are errors in the estimated motion vectors, visually unacceptable artefacts may be introduced in the reconstructed frames due to the motion-based processing. One way to avoid such visible artifacts in upscaled frames is to adapt the temporal weighting based on the correlation between the current block and the corresponding blocks in other frames determined by the motion vectors. That is to say, before constructing MASK weights, we compute the reliability (η_ℓ) of each estimated motion vector. A simple way to define η_ℓ is to use the mean square error or mean absolute error between the block of interest and the corresponding block in the neighboring frame towards which the motion vector is pointing. Once the reliability of the estimated motion vector is available, we penalize the steering kernels by a temporal kernel K_t, a kernel function of η. Fig. 7 illustrates the temporal weighting, incorporating motion reliability. Suppose we upscale the ℓ-th block in the frame at time t using 2 previous and 2 forward frames, and there are 4 motion vectors, $\mathbf{m}_{\ell,i}$, between a block in the frame at t and the 4 neighboring frames. First, we find the blocks that the motion vectors indicate from the neighboring frames shown as $\mathbf{y}_{\ell,i}$ in Fig. 7. Then, we compute the motion reliability based on the difference between the ℓ-th block at t and other blocks and decide the temporal penalization for each neighboring block.

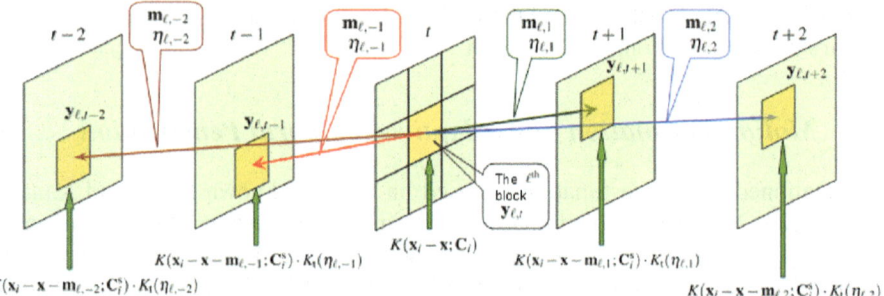

Fig. 7 A schematic representation of temporal weighting in MASK for upscaling the ℓ-th block ($\mathbf{y}_{\ell,t}$) of the frame at time t. First, we locate the neighboring blocks ($\mathbf{y}_{\ell,i}$ for $i = -2, -1, 1, 2$) indicated by the motion vectors ($\mathbf{m}_{\ell,i}$). Then, we compute the motion reliability ($\eta_{\ell,i}$) based on the difference between the ℓ-th block at t and the neighboring blocks, and combine the temporal penalization by K_t with the spatial kernel function K.

More specifically, we define $\eta_{\ell,\Delta t}$ and K_t as

$$\eta_{\ell,\Delta t} = \frac{\left\| \mathbf{y}_{\ell,t} - \mathbf{y}_{\ell,t+\Delta t} \right\|_{\mathrm{F}}}{M}, \tag{31}$$

$$K_{h_t}(\eta_{\ell,\Delta t}) = \frac{1}{1 + \dfrac{\eta_{\ell,\Delta t}}{h_t}} \tag{32}$$

where h_t is the (global) smoothing parameter, which controls the strength of temporal penalization, $\mathbf{y}_{\ell,t}$ is the ℓth block of the frame at time t, $t + \Delta t$ is a neighboring frame, M is the total number of pixels in a block, and $\| \cdot \|_{\mathrm{F}}$ is Frobenius norm. We replace the temporal kernel in (28) by (32). This temporal weighting technique is similar to the Adaptive Weighted Averaging (AWA) approach proposed in [25]; however, the weights in AWA are computed pixel-wise. In MASK, the temporal kernel weights are a function of radiometric distances between small pixel blocks and are computed block-wise.

4.3 Quantization of Orientation Map

The computational cost of estimating local spatial steering (covariance) matrices is high due to the SVD. In this section, using the well-known technique of *vector quantization* [5], we describe a way to obtain stable (invertible) steering matrices without using the SVD. Briefly speaking, first, we construct a look-up table which has a certain number of stable (invertible) steering matrices. Second, instead of computing the steering matrix by (18), we compute the naive covariance matrix (15), and then find the most similar steering matrix from the look-up table. The advantages of using the look-up table are that (i) we can lower the computational complexity by avoiding singular value decomposition, (ii) we can control and trade-off accuracy and computational load by designing an appropriate vector quantization scheme with almost any desired number of steering matrices in the look-up table, and (iii) we can pre-calculate kernel weights to lower the computational load further (since the steering matrices are fixed).

From (18), the elements of the spatial covariance matrix $\mathbf{C}_i$ are given by the steering parameters with the following equations:

$$\mathbf{C}_j(\gamma_j, \rho_j, \theta_j) = \begin{bmatrix} c_{11} & c_{12} \\ c_{12} & c_{22} \end{bmatrix}, \tag{33}$$

with

$$c_{11} = \gamma_j \left(\rho_j \cos^2 \theta_j + \rho_j^{-1} \sin^2 \theta_j \right) \tag{34}$$

$$c_{12} = -\gamma_j \left(\rho_j \cos \theta_j \sin \theta_j + \rho_j^{-1} \cos \theta_j \sin \theta_j \right) \tag{35}$$

$$c_{22} = \gamma_j \left(\rho_j \sin^2 \theta_j + \rho_j^{-1} \cos^2 \theta_j \right) \tag{36}$$

where γ_j is the scaling parameter, ρ_j is the elongation parameter, and θ_j is the orientation angle parameter. Fig. 8 visualizes the relationship between the steering parameters and the values of the covariance matrix. Based on the above formulae, using a pre-defined set of the scaling, elongation, and angle parameters, we can generate a lookup table for covariance matrices, during an off-line stage.

During the on-line processing stage, we compute a naive covariance matrix $\mathbf{C}_i^{\text{naive}}$ (15) and then normalize $\mathbf{C}_i^{\text{naive}}$ so that the determinant of the normalized naive covariance matrix $\det(\widetilde{\mathbf{C}}_i^{\text{naive}})$ equals 1.0:

$$\widetilde{\mathbf{C}}_i^{\text{naive}} = \frac{\mathbf{C}_i^{\text{naive}}}{\sqrt{\det\left(\mathbf{C}_i^{\text{naive}}\right)}} = \frac{1}{\gamma_i} \mathbf{C}_i^{\text{naive}}, \tag{37}$$

where again γ_i is the scaling parameter. This normalization eliminates the scaling parameter from the look-up table and simplifies the relationship between the elements of covariance matrices and the steering parameters, and allows us to reduce the size of the table. Table 1 shows an example of a compact lookup table. When the elongation parameter ρ_i of $\widetilde{\mathbf{C}}_i$ is smaller than 2.5, $\widetilde{\mathbf{C}}_i$ is quantized as an identity matrix (i.e. the kernel spreads equally every direction). On the other hand, when $\rho_i \geq 2.5$, we quantize $\widetilde{\mathbf{C}}_i$ with 8 angles. Using $\widetilde{\mathbf{C}}_i^{\text{naive}}$, we obtain the closest covariance matrix $\widetilde{\mathbf{C}}_i$ from the table. I.e.,

$$\widetilde{\mathbf{C}}_i = \arg\min_{\rho_j, \theta_j} \left\| \mathbf{C}(\rho_j, \theta_j) - \widetilde{\mathbf{C}}_i^{\text{naive}} \right\|_{\text{F}}, \tag{38}$$

where $\| \cdot \|_{\text{F}}$ is the Frobenius norm. The final matrix $\widehat{\mathbf{C}}_i$ is given by:

$$\widehat{\mathbf{C}}_i = \gamma_i \widetilde{\mathbf{C}}_i. \tag{39}$$

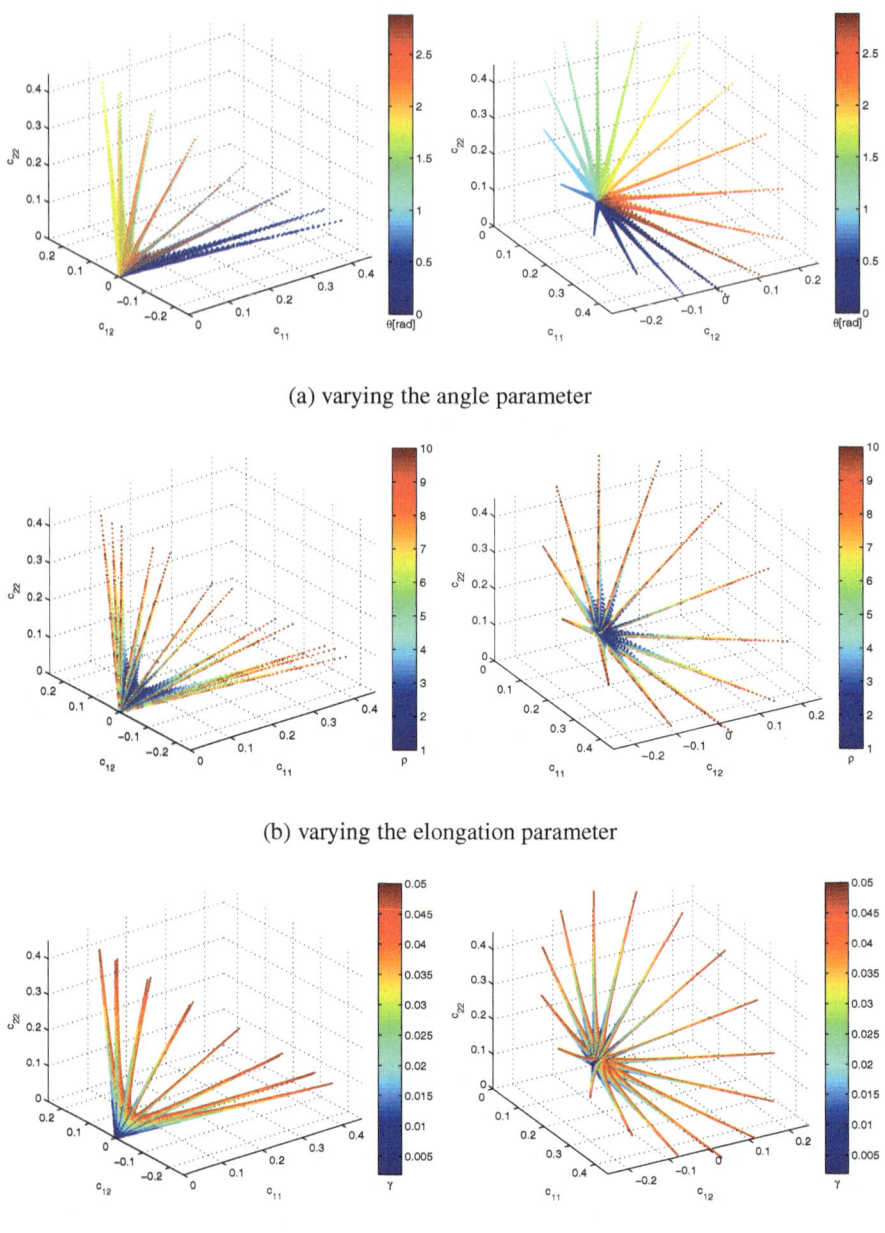

(a) varying the angle parameter

(b) varying the elongation parameter

(c) varying the scaling parameter

Fig. 8 The graphical relationship between the steering kernel parameters and the values of covariance matrix

Table 1 A compact lookup table for covariance matrices

c_{11}	c_{12}	c_{22}	ρ_j	θ_j
1.0000	0	1.0000	1.0	0
2.5000	0	0.4000	2.5	0
2.1925	1.0253	0.7075	2.5	$\frac{1}{8}\pi$
1.4500	1.4500	1.4500	2.5	$\frac{2}{8}\pi$
0.7075	1.0253	2.1925	2.5	$\frac{3}{8}\pi$
0.4000	0	2.5000	2.5	$\frac{4}{8}\pi$
0.7075	-1.0253	2.1925	2.5	$\frac{5}{8}\pi$
1.4500	-1.4500	2.1925	2.5	$\frac{6}{8}\pi$
2.1925	-1.0253	0.7075	2.5	$\frac{7}{8}\pi$

4.4 Adaptive Regression Order

As mentioned earlier, although the kernel estimator with a higher regression order preserves high frequency components, the higher order requires more computation. In this section, we discuss how we can reduce the computational complexity, while enabling adaptation of the regression order. According to [26], the second order equivalent kernel, $\mathbf{W}_2$, can be obtained approximately from the zeroth order one, $\mathbf{W}_0$, as follows. First, we know that the general kernel estimator (12) can be expressed as:

$$\hat{z}(\mathbf{x}) = \mathbf{e}_1^T \left(\mathbf{X}^T \mathbf{K} \mathbf{X} \right)^{-1} \mathbf{X}^T \mathbf{K} \, \mathbf{y} = \mathbf{W}_N \mathbf{y} \tag{40}$$

where again $\mathbf{W}_N$ is a $1 \times P$ vector containing the filter coefficients and which we call the equivalent kernel. The zeroth order equivalent kernel can be modified into $\mathbf{W}_2$ by

$$\widetilde{\mathbf{W}}_2^T = \mathbf{W}_0^T - \kappa \mathbf{L} \mathbf{W}_0^T, \tag{41}$$

where $\mathbf{L}$ is Laplacian kernel in matrix form (we use $[1,1,1;1,-8,1;1,1,1]$ as a discrete Laplacian kernel) and κ is a regression order adaptation parameter. This operation can be seen to "sharpen" the equivalent kernel, and is equivalent to sharpening the reconstructed image. Fig. 9 shows the comparison between the actual second order equivalent kernel, $\mathbf{W}_2$, and the equivalent kernel, $\widetilde{\mathbf{W}}_2$, given by (41). In the comparison, we use the Gaussian function for K, and compute the zeroth order and the second order equivalent kernels shown in Fig. 9(a) and (b) respectively. The equivalent kernel, $\widetilde{\mathbf{W}}_2$, is shown in Fig. 9(c), and Fig. 9(d) shows the horizontal cross section of $\mathbf{W}_0$, $\mathbf{W}_2$, and $\widetilde{\mathbf{W}}_2$. As seen in Fig. 9(d), $\widetilde{\mathbf{W}}_2$ is close to the exact second order kernel $\mathbf{W}_2$.

There are two advantages brought by (41): (i) The formula simplifies the computation of the second order equivalent kernels, i.e. there is no need to generate the basis matrix, $\mathbf{X}$, or take inversion of matrices. (ii) Since the effect of the second

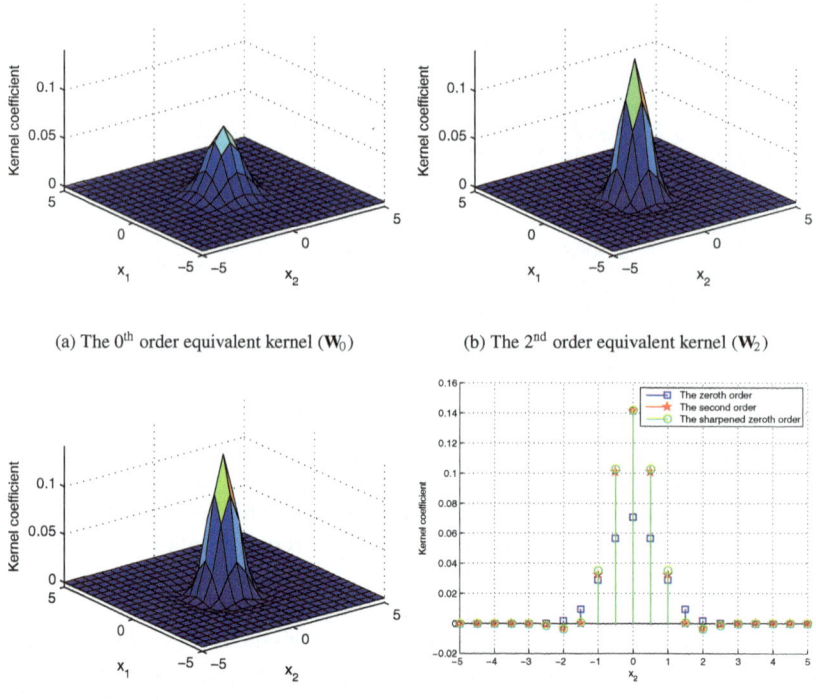

(a) The 0th order equivalent kernel ($\mathbf{W}_0$)   (b) The 2nd order equivalent kernel ($\mathbf{W}_2$)

(c) A sharpened 0th order equivalent kernel ($\widetilde{\mathbf{W}}_2$)   (d) Horizontal cross sections of the equivalent kernels

Fig. 9 Equivalent kernels given by classic kernel regression: (a) the 0th order equivalent kernel with the global smoothing parameter $h = 0.75$, (b) the 2nd order equivalent kernel ($\mathbf{W}_2$) with $h = 0.75$, (c) a sharpened 0th order equivalent kernel ($\widetilde{\mathbf{W}}_2$) with a 3×3 Laplacian kernel ($\mathbf{L} = [1, 1, 1; 1, -8, 1; 1, 1, 1]$) and $\kappa = 0.045$, and (d) Horizontal cross sections of the equivalent kernels $\mathbf{W}_0$, $\mathbf{W}_2$, and $\widetilde{\mathbf{W}}_2$. For this example, we used a Gaussian function for $K(\cdot)$.

order regression is now explicitly expressed by $\kappa \mathbf{L} \mathbf{W}_0$ in (41), the formulation allows for adjustment of the regression order across the image, but also it allows for "fractional" regression orders, providing fine control over the amount of sharpening applied locally.

We propose a technique to automatically select the regression order parameter (κ) adaptively as follows. By setting κ near zero in flat regions and to a large value in edge and texture regions, we can expect a reduction of computational complexity, prevent amplifying noise component in flat regions, and preserve or even enhance texture regions and edges. In order to select spatially adapted regression factors, we can make use of the scaling parameter γ_i, which we earlier used to normalize the covariance matrix in (37). This makes practical sense since γ_i is high in texture and edge areas and low in flat area as shown in Fig. 10. Because γ_i is already computed

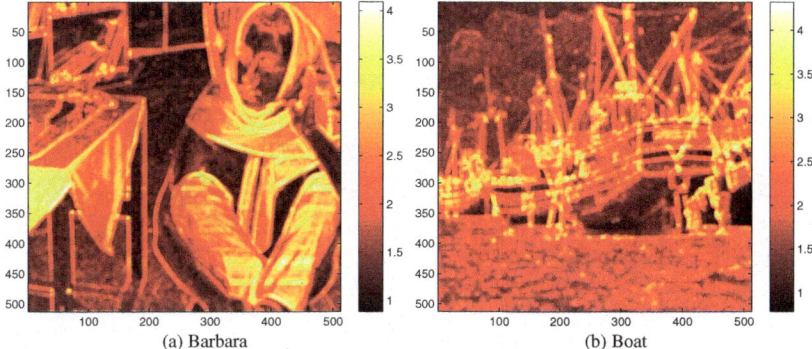

(a) Barbara   (b) Boat

Fig. 10 Local scaling parameters (γ_i) for (a) Barbara image and (b) Boat image. With the choice of the adaptive regression order $\kappa_i = 0.01\gamma_i$ (42), the regression order becomes nearly zero in the areas where γ_i is close to zero, while in areas where γ_i is around 5, the resulting equivalent kernel given by (41) approximately becomes second order.

when computing the steering matrices, no extra computation is required. A good way to choose the regression factor (κ) locally is to make it a simple function of γ_i. Specifically, we choose our adaptive regression factor by

$$\kappa_i = 0.01\gamma_i, \tag{42}$$

where 0.01 is a global parameter controlling the overall sharpening amount. E.g. it is possible to choose a larger number if a stronger sharpening effect is desired globally. As shown in Fig. 10, with the choice of the adaptive regression order $\kappa_i = 0.01\gamma_i$ (42), the regression order becomes close to zero in the area where γ_i is close to zero, while the resulting equivalent kernel given by (41) approximately becomes a second order kernel in the area where γ_i is around 5. Setting κ too large results in overshoot of pixel values around texture and edges. We process color video in the YCbCr domain and estimate spatial orientations in the luminance component only, since the human visual system is most sensitive to orientations in the luminance component.

5   Example Video Upscaling and Denoising Results

In this section, we provide video frames generated by the proposed MASK algorithm as visual illustrations of its performance. We will provide examples of spatial upscaling, temporal frame interpolation, and denoising. We compare MASK to two other state-of-the-art multi-frame video upscaling methods: Non Local-mean based super resolution [11] and 3-D iterative spatial steering kernel regression (3-D ISKR) [13]. The algorithm proposed in [11] consists of multi-frame fusion with Non Local-mean based weighting, as well as explicit deblurring. 3-D ISKR is an

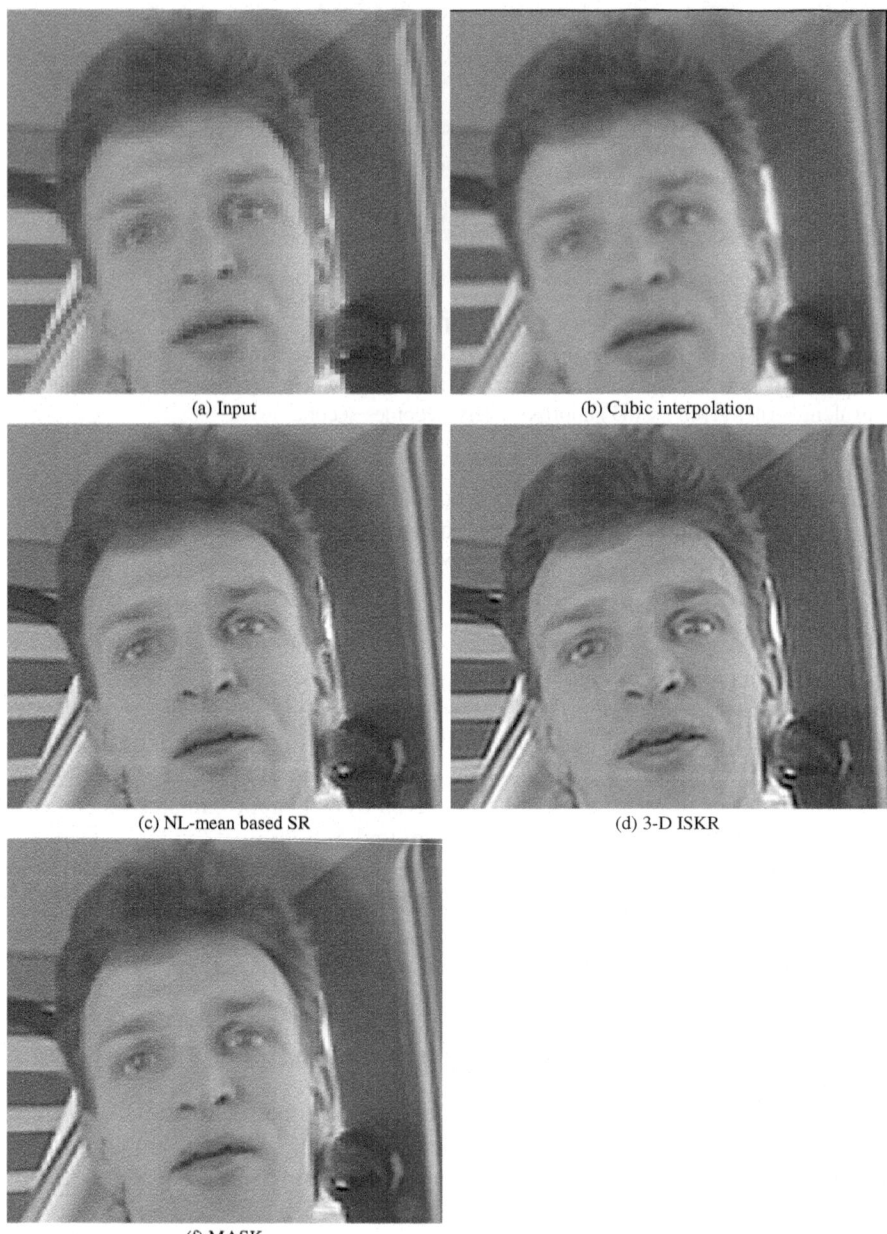

Fig. 11 Spatial upscaling of Car-phone video sequence: (a) input video frames at time $t = 25$ (144×176, 30 frames) and (b)-(f) the upscaled frames by single frame bicubic interpolation, NL-mean based SR [11], 3-D iterative SKR [13], and MASK, respectively.

algorithm closely related to MASK involving iterative 3-D steering kernel regression; however, it does not require accurate (subpixel) motion estimation. For 3-D ISKR and MASK, we set the temporal window of support 5, and NL-based SR approach searches similar local patches across all the frames in time and the window of support 21×21 in space.

The first example shown in Fig. 11 is a visual comparison of spatial upscaling and temporal frame interpolation results, using MASK, NL-mean based SR, and 3-D ISKR. For this example, we used the Car-phone video sequence in QCIF format (144×176 pixels, 30 frames) as input, and spatially upscaled the video with an upscaling factor of $1:3$. Fig. 11(a) shows the input frame at time $t = 25$ (upscaled by pixel-replication). The upscaled results by single frame bicubic interpolation, NL-mean based SR, 3-D ISKR, and MASK are shown in Figs. 11(b)-(f), respectively. In addition, Fig. 12 shows a spatiotemporal upscaling example (both spatial upscaling and temporal frame interpolation) of the Car-phone sequence by 3-D ISKR and MASK. For this example, we estimated an intermediate frame at time $t = 25.5$ as well as spatially upscaling the intermediate frames with the upscaling factor of $1:3$. Comparing to the result by bicubic interpolation, all the adaptive methods, NL-mean based SR, 3-D ISKR, and MASK, reconstruct high-quality upscaled frames, although each has a few artifacts: jaggy artifacts on edge regions for NL-mean based SR and MASK, and overshooting artifact for 3-D ISKR.

The second example is spatio-temporal video upscaling using two color real video sequences: Spin-Calendar (504×576 pixels, 30 frames) and Texas (504×576 pixels, 30 frames). Fig. 13(a) and 14(a) show an input frame of each sequence at time $t = 5$, respectively. Spin-Calendar has relatively simple motions, namely rotations. Texas sequence contains more complicated motions, i.e., occlusion, 3-D rotation of human heads, and reflection on the helmet. Furthermore, Spin-Calendar

(a) 3-D ISKR   (b) MASK

Fig. 12 Spatiotemporal upscaling of Car-phone video sequence: (a) upscaled frames by 3-D iterative SKR [13] at $t = 25.5$, and (b) upscaled frames by MASK at $t = 25.5$. In this example, we upscale Car phone sequence shown in Fig. 11(a) with the spatial upscaling factor $1:3$ and the temporal upscaling factor $1:2$.

contains camera noise, while Texas contains significant compression artifacts (e.g. blocking). Video frames that were spatially upscaled by a factor of 1 : 2 using single frame bicubic interpolation, 3-D ISKR, and MASK are shown in Figs. 13(b)-(d) and 14(b)-(d), respectively. Also, Figs. 13(e)-(h) and 14(e)-(h) show selected portions of

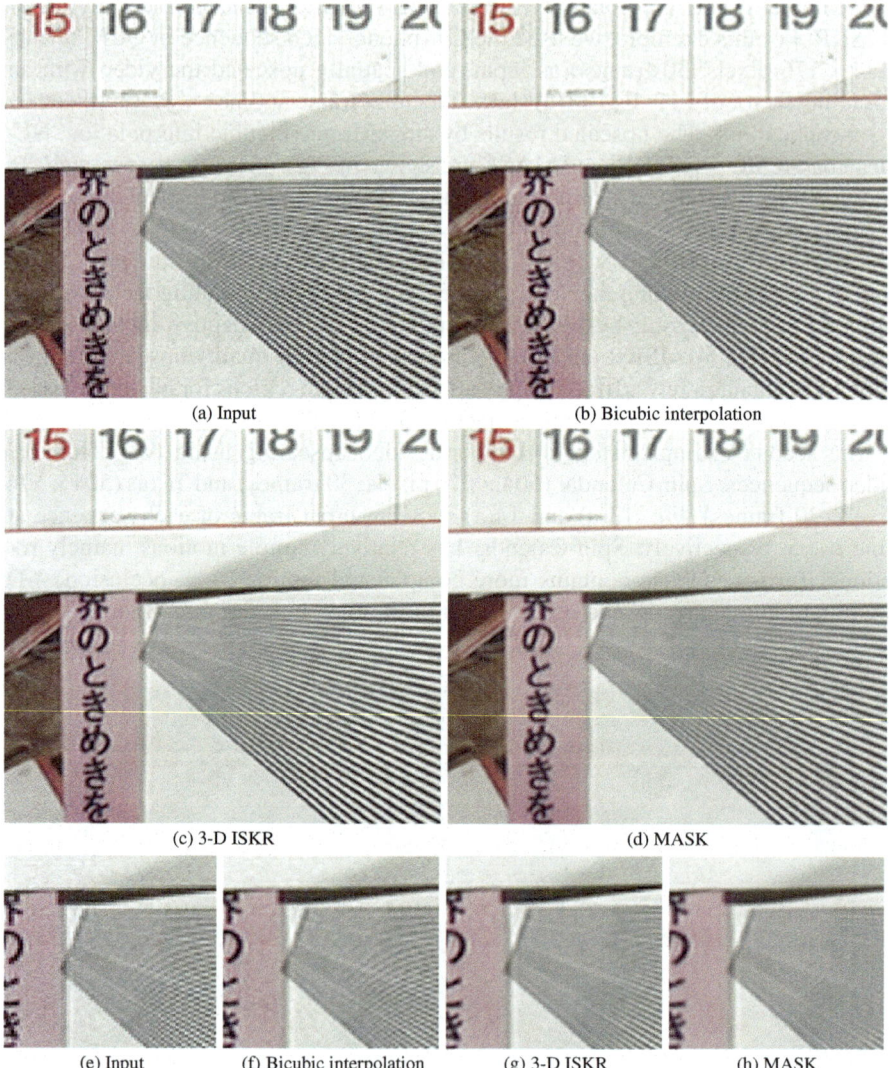

Fig. 13 Spatial upscaling of Spin-Calendar video sequence: (a) the input frame at $t = 5$, (b)-(d) the upscaled video frames by bicubic interpolation, 3-D ISKR, and MASK, respectively. (e)-(h) Enlarged images of the input frame and the upscaled frames by cubic interpolation, 3-D ISKR, and MASK, respectively.

the input frame, the upscaled frame using single frame bicubic interpolation, 3-D ISKR, and MASK at a large scale. Next, we estimated an intermediate frame at time $t = 5.5$ for both Spin-Calendar and Texas sequences by 3-D ISKR and MASK, and the results are shown in Fig. 15. The intermediate frames are also spatially upscaled

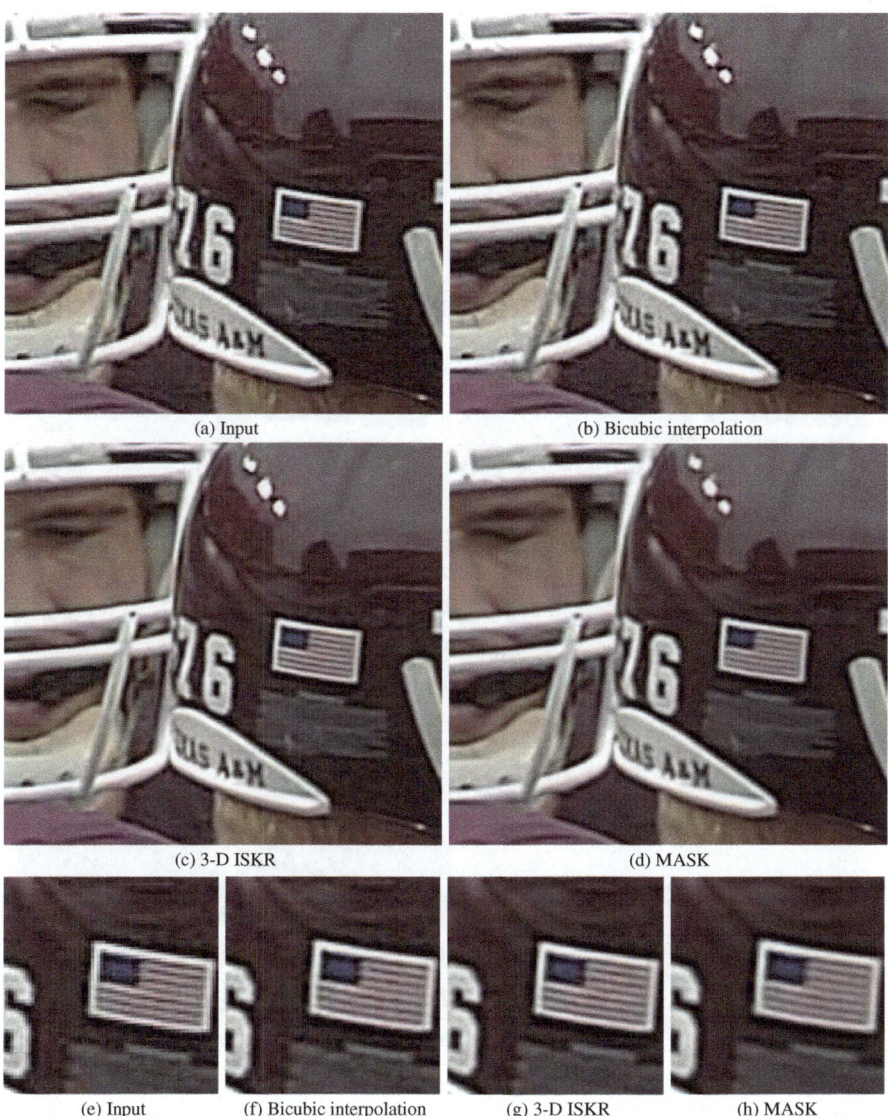

(a) Input   (b) Bicubic interpolation

(c) 3-D ISKR   (d) MASK

(e) Input   (f) Bicubic interpolation   (g) 3-D ISKR   (h) MASK

Fig. 14 Spatial upscaling of Texas video sequence: (a) the input frame at $t = 5$, (b)-(d) the upscaled video frames by bicubic interpolation, 3-D ISKR, and MASK, respectively. (e)-(h) Enlarged images of the input frame and the upscaled frames by cubic interpolation, 3-D ISKR, and MASK, respectively.

by the same factor (1 : 2). Again, both 3-D ISKR and MASK produce high quality frames in which camera noise and blocking artifacts are almost invisible while the important contents are preserved.

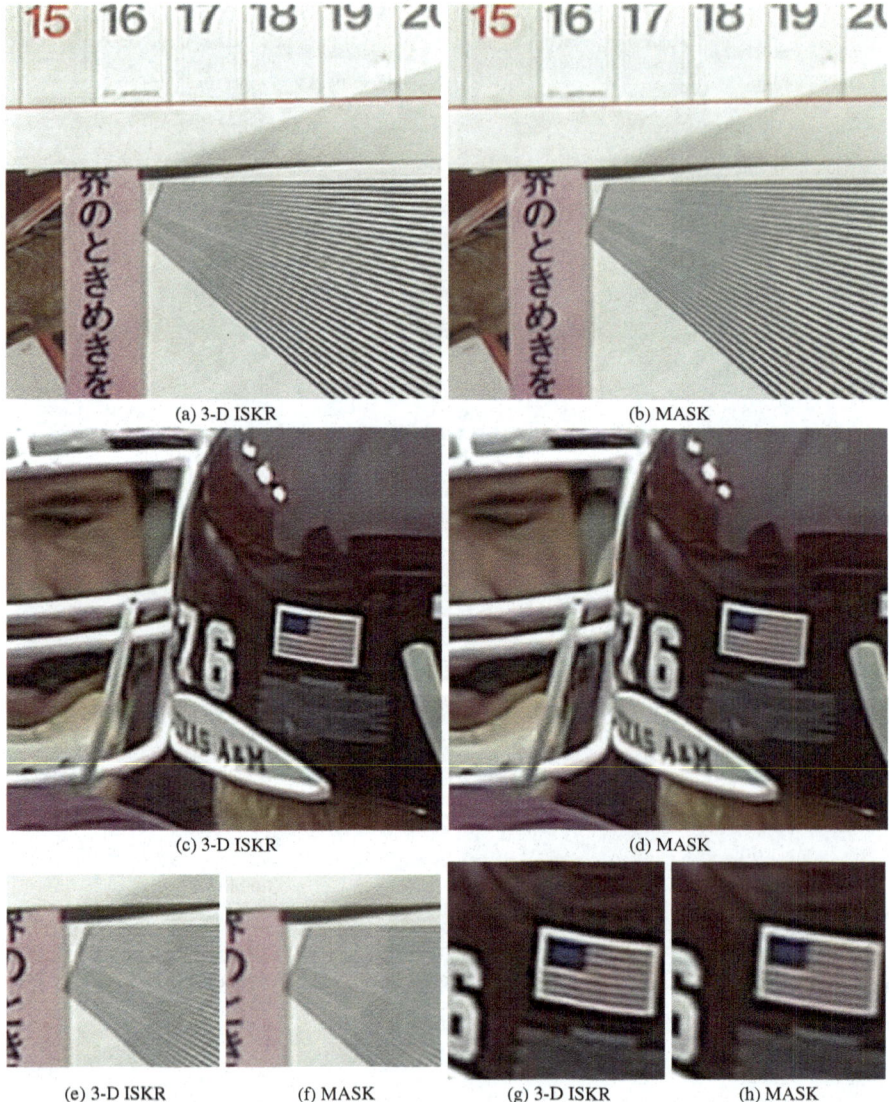

Fig. 15 Spatiotemporal upscaling of Spin-Calender and Texas video sequences: (a),(c) the estimated intermediate frames at time $t = 5.5$ by 3-D ISKR, (b),(d) the estimated intermediate frames by MASK. The frames are also spatially upscaled with the upscaling factor of 1 : 2. The images in (e)(f) and (g)(h) are the enlarged images of the upscaled frames by 3-D ISKR and MASK, respectively.

6   Conclusion

In this chapter, we presented an extension of steering kernel regression for video upscaling. Our proposed algorithm is capable of spatial upscaling with resolution enhancement, temporal frame interpolation, noise reduction, as well as sharpening. In the proposed algorithm, we construct 3-D kernels based on local motion vectors, unlike our previous work [11, 13]. The algorithm includes motion estimation, but doesn't use explicit motion compensation. Instead, the spatio-temporal kernel is oriented along the local motion trajectory, and subsequent kernel regression acts directly on the pixel data. In order to avoid introducing artifacts due to motion estimation errors, we examine the motion vectors for their reliability. We apply a temporal weighting scheme, which allows us to suppress data from neighboring frames in the case of a motion error. Also, we reduce the computational cost of MASK by using a block-based motion model, using a quantized set of local orientations, and adapting the regression order. The adaptive regression order technique not only reduces the computational cost, but also provides sharpening while avoiding noise amplification.

We have presented several video upscaling examples showing that the MASK approach recovers resolution, suppresses noise and compression artifacts, and is capable of temporal frame interpolation with very few artifacts. The visual quality of the upscaled video is comparable to that of other state-of-the-art multi-frame upscaling methods, such as the Non-Local-Means based super-resolution method [11] and 3-D ISKR [13]. However, the computational complexity of MASK in terms of processing and memory requirements is significantly lower than these alternative methods. In order to improve the visual quality of MASK further, it may be necessary to include more accurate motion estimation, for example by using smaller block sizes (currently 8×8), or extending the motion model, e.g. to an affine model.

References

1. Mitchell, D.P., Netravali, A.N.: Reconstruction Filters in Computer Graphics. Computer Graphics 22(4), 221–228 (1988)
2. Park, S.C., Park, M.K., Kang, M.G.: Super-Resolution Image Reconstruction: A Technical Overview. IEEE Signal Processing Magazine 20(3), 21–36 (2003)
3. Farsiu, S., Robinson, D., Elad, M., Milanfar, P.: Advances and Challenges in Super-Resolution. International Journal of Imaging Systems and Technology, Special Issue on High Resolution Image Reconstruction (invited paper) 14(2), 47–57 (2004)
4. Narayanan, B., Hardie, R.C., Barner, K.E., Shao, M.: A Computationally Efficient Super-Resolution Algorithm for Video Processing Using Partition Filters. IEEE Transactions on Circuits and Systems for Video Technology 17(5), 621–634 (2007)
5. Gersho, A., Gray, R.M.: Vector Quantization and Signal Compression. Kluwer Academic Publishers, Boston (1992)
6. Hardie, R.: A Fast Image Super-Resolution Algorithm Using an Adaptive Wiener Filter. IEEE Transactions on Image Processing 16(12), 2953–2964 (2007)

7. Fujiwara, S., Taguchi, A.: Motion-Compensated Frame Rate Up-Conversion Based on Block Matching Algorithm with Multi-Size Blocks. In: Proc. International Symposium on Intelligent Signal Processing and Communication Systems, Hong Kong, China (2005)
8. Huang, A., Nguyen, T.Q.: A Multistage Motion Vector Processing Method for Motion-Compensated Frame Interpolation. IEEE Transactions on Image Processing 17(5), 694–708 (2008)
9. Kang, S., Cho, K., Kim, Y.: Motion Compensated Frame Rate Up-Conversion Using Extended Bilateral Motion Estimation. IEEE Transactions on Consumer Electronics 53, 1759–1767 (2007)
10. Choi, B., Han, J., Kim, C., Ko, S.: Motion-Compensated Frame Interpolation Using Bilateral Motion Estimation and Adaptive Overlapped Block Motion Compensation. IEEE Transactions on Circuits and Systems for Video Technology 17(4), 407–416 (2007)
11. Protter, M., Elad, M., Takeda, H., Milanfar, P.: Generalizing the Non-Local-Means to Super-resolution Reconstruction. IEEE Transactions on Image Processing 16(2), 36–51 (2009)
12. Buades, A., Coll, B., Morel, J.M.: A Review of Image Denoising Algorithms, with a New One. In: Proc. Multiscale Modeling and Simulation, Society for Industrial and Applied Mathematics (SIAM) Interdisciplinary Journal, New Orleans, LA, USA (2005)
13. Takeda, H., Milanfar, P., Protter, M., Elad, M.: Superresolution without Explicit Subpixel Motion Estimation. IEEE Transactions on Image Processing 18(9), 1958–1975 (1958)
14. Takeda, H., Farsiu, S., Milanfar, P.: Kernel Regression for Image Processing and Reconstruction. IEEE Transactions on Image Processing 16(2), 349–366 (2007)
15. Takeda, H., Farsiu, S., Milanfar, P.: Robust Kernel Regression for Restoration and Reconstruction of Images from Sparse Noisy Data. In: Proc. International Conference on Image Processing (ICIP), Atlanta, GA, USA (2006)
16. Tomasi, C., Manduchi, R.: Bilateral Filtering for Gray and Color Images. In: Proc. IEEE International Conference of Compute Vision, Bombay, India (1998)
17. Elad, M.: On the Origin of the Bilateral Filter and Ways to Improve it. IEEE Transactions on Image Processing 11(10), 1141–1150 (2002)
18. Pham, T.Q., van Vliet, L.J., Schutte, K.: Robust Fusion of Irregularly Sampled Data Using Adaptive Normalized Convolution. EURASIP Journal on Applied Signal Processing, 1–12 (2006)
19. Takeda, H., Farsiu, S., Milanfar, P.: Deblurring Using Regularized Locally-Adaptive Kernel Regression. IEEE Transactions on Image Processing 17(4), 550–563 (2008)
20. Takeda, H., van Beek, P., Milanfar, P.: Spatio-Temporal Video Interpolation and Denoising Using Motion-Assisted Steering Kernel (MASK) Regression. In: Proc. IEEE International Conference on Image Processing (ICIP), San Diego, CA, USA (2008)
21. Wand, M.P., Jones, M.C.: Kernel Smoothing. Chapman and Hall, London (1995)
22. Nadaraya, E.A.: On Estimating Regression. Theory of Probability and its Applications, 141–142 (1964)
23. Lucas, B., Kanade, T.: An Iterative Image Registration Technique with an Application to Stereo Vision. In: Proc. DARPA Image Understanding Workshop (1981)
24. Stiller, C., Konrad, J.: Estimating Motion in Image Sequences - A Tutorial on Modeling and Computation of 2D Motion. IEEE Signal Processing Magazine 16(4), 70–91 (1999)
25. Ozkan, M., Sezan, M.I., Tekalp, A.M.: Adaptive Motion-Compensated Filtering of Noisy Image Sequences. IEEE Transactions on Circuits and Systems for Video Technology 3(4), 277–290 (2003)
26. Haralick, R.M.: Edge and Region Analysis for Digital Image Data. Computer Graphic and Image Processing (CGIP) 1(12), 60–73 (1980)

Chapter 11
Temporal Super Resolution Using Variational Methods

Sune Høgild Keller, François Lauze, and Mads Nielsen

Abstract. Temporal super resolution (TSR) is the ability to convert video from one frame rate to another and is as such a key functionality in modern video processing systems. A higher frame rate than what is recorded is desired for high frame rate displays, for super slow-motion, and for video/film format conversion (where also lower frame rates than recorded is sometimes required). We discuss and detail the requirements imposed by the human visual system (HVS) on TSR algorithms, of which the need for (apparent) fluid motion, also known as the phi-effect, is the principal one. This problem is typically observed when watching video on large and bright displays where the motion of high contrast edges often seem jerky and unnatural. A novel motion compensated (MC) TSR algorithm using variational methods for both optic flow calculation and the actual new frame interpolation is presented. The flow and intensities are calculated simultaneously in a multiresolution setting. A frame doubling version of our algorithm is implemented and in testing it, we focus on making the motion of high contrast edges to seem smooth and thus reestablish the illusion of motion pictures.

1 Background

TSR is most asked for in displaying low frame rate recordings on high frame rate displays, but is also needed both for super slow-motion (super $\equiv$ high quality) and for combining different frame rate recordings into one common frame rate program.

Sune Høgild Keller
PET and Cyclotron Unit, Copenhagen University Hospital, Blegdamsvej 9,
DK-2100 Copenhagen, Denmark
e-mail: sune@pet.rh.dk

François Lauze · Mads Nielsen
Department of Computer Science, University of Copenhagen, Universitetsparken 1,
DK-2100 Copenhagen, Denmark
e-mail: francois@diku.dk, madsn@diku.dk

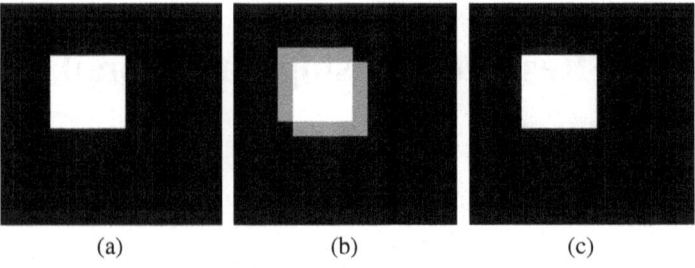

(a)   (b)   (c)

Fig. 1 a True frame from a sequence where the square moves diagonally, **b** bad TSR by frame interpolation, and **c** good motion compensated TSR (variational in this case).

In terms of number of frames per second (fps) created from a given input, TSR is not just an upscaling, but also a downscaling. This means we will have to create entirely new frames even if we downscale in time, and thus we use the term TSR also for downscaling (although the term 'super resolution' implies a higher resolution). Both spatial super resolution and TSR are ill-posed *sampling* problems.

Upscaling in time is widely needed as most displays (projectors, plasma, LCDs and CRTs) have higher frame refresh rates than the frame rate used when recording the displayed material. All cinematographic movies are for instance shot at 24 fps, while practically all displays today have refresh rates of at least 50 Hz. The higher display frequencies are necessary to stop the perceived image sequence from flickering. In the human visual system (HVS), any part of a scene projected onto the retina away from the fovea (center of a focus) is subject to flicker as these regions of the retina are highly sensitive to flickering.

To increase the frame rate, many flat panel TV-sets and most PC displays do not use real TSR but just repeat the same frame once or twice, and in cinemas every frame is shown again two or three times to avoid flicker. At small viewing angles (the part of the field of view covered by the screen) frame repetition works fine most of the time, but at larger viewing angles motion will start to appear jerky. The *phi-effect*—the effect of perceiving a sequences of still images as motion pictures—is halted [1]. The problem is typically seen around high contrast edges in motion as edges are the major information perceived and processed in lower level vision. The archetype example of jerky motion is a horizontal camera pan, e.g. in well lit interior scenes or exterior shots of houses and cities.

There are three different methods for frame rate conversion: a) frame repetition, b) frame averaging and c) motion compensated interpolation. Frame repetition is the simplest and does not create any artifacts from bad temporal interpolation, but the motion portrayal stays unnatural. When conversion ratio it not integer, e.g. 24 to 60 fps, frames will be repeated a different number of time adding some nonlinear jumps to the motion, possibly increasing the unnaturalness of the motion. Frame averaging, where the two nearest known frames are weighed by the inverse of their distance to the new frame, yields double exposure-like images in case of motion as shown in Fig. 1(b). Frame averaging will to a certain extent smooth motion as compared to frame repetition, but the blending is an undesired artifact.

Better results are obtained when one computes the 2D optic flow between known frames and then compensate for the motion in the new frame: Only when knowing the flow one can truly create the data in moving regions of the frame. This is motion compensated frame rate conversion, which we denote temporal super resolution, TSR. Variational frameworks offer methods for computing both dense and accurate flow fields and high quality motion compensated intensity interpolation to get optimal TSR results as seen in Fig. 1(c).

1.1 Frame Rate Requirements

We will focus on the frame rate requirements of humans viewers as our TSR algorithm is aimed at application in video processors in broadcast or home entertainment systems where pleasing human viewers is the final goal. The properties of the human visual system guides what minimum frame rates should be used to keep the viewing experience pleasing. The two main requirements are:

- The *phi-effect* should be obtained to create apparent living pictures with natural motion portrayal.
- Flickering should be avoided when displaying image sequences.

The phi-effect is the effect of recording and showing a set of still images so fast after each other that any motion in the depicted scene appears real and natural as the HVS will interpolate the simplest (linear) motion between the frames [1]. To create the phi-effect, the frame rate has to high enough for the HVS to *perceive* all motion as natural in spite of this interpolation.

Flickering occurs on when an image is not updated often enough, that is the update frequency (frame refresh rate) is so low that the eye senses flicker.

Determining the exact minimum required frame rate of image sequences is a difficult, multi-parameter problem (in HVS properties and viewing conditions) but the consensus is that more than 50 fps is needed to fulfill the requirements above. However, it ultimately depends on the tracking done in the eye of the viewer. The rise of 100 Hz TV in Europe and no 120Hz in the US and Asia indicates that 60 Hz might suffice for TVs. Flicker can be avoided by frame repetition, but to get the phi-effect—perceived smooth and natural motion—motion compensated frame rate conversion is necessary.

1.2 Blur Acceptance in Human Vision

Blur is a very common artifact in image sequences, but depth of focus blur and motion blur is accepted by viewers and in de-noising of images blur is often the side effect, but is preferred over local, high contrast artifacts like noise, block effect (JPEG and MPEG material) as edges (high contrast) is the key input to the HVS. Doing TSR in a wrong or incomplete way will most likely create artifacts in the new, interpolated frames. Using a motion compensated variational method, blur will be the most likely artifact. Thus we need to know (if possible) how much blur is acceptable to the human visual system.

Judging generally and objectively how unsharp we can allow parts (say every other frame) of an image sequence to be, is still an open question. Vision research does not offer an answer as the boundary between sharp and blurred in human perception is still sought for in more limited sub-problems, e.g. in [2] where the authors try to find out when blur becomes bothersome on simple stationary text/characters. In [3] (Burr and Morgan) and [4] (Morgan and Benton) it is shown experimentally that moving objects often appear sharp to the HVS, not because some mechanism removes blur, but because the HVS is unable to decide whether the object is really sharp or not. Even though we do not get an answer to our question off blur acceptance from vision research, we do get helping pointers: It seems we can allow for some blur when doing temporal super resolution and still get subjectively good results (evaluation by the HVS of the viewers). In [5] Chen et al. shows that motion blur in LCD displays can be reduced by inserting blurred frames between frames that are enhanced correspondingly in the high frequencies. The safest way towards optimal TSR is, however, to make the new frames in the output as sharp as possible.

1.3 Related Work

Temporal interpolation of signals is not new, it has been done for a long time for 1D signals in signal processing, but these methods cannot be applied to frame rate conversion due to the presence of motion.

In medical imaging interpolation of new frames or volumes of a time sequence of 2D or 3D scans are of interest, mainly in lung (respiratory gated) and heart (heart gated) imaging. The work by Ehrhardt et al. in [6] is a typical and recent example, where temporal super resolution in heart gated imaging is performed using an accurate flow algorithm, but with simple motion compensated interpolation of intensities along the flow lines to get the new frames. In the field of video processing there are several TSR patents, e.g. [6, 7, 8], mostly doing flow calculation (good or bad) followed by some simple, non-iterative averaging along the flow. TSR is also done in integrated circuits (ICs) as described by de Haan in [9] using 8×8 block matching flow with a median filter for motion compensated interpolation (see [10] for details). In a recent paper [11] by Dane and Nguyen motion compensated interpolation with adaptive weighing to minimize the error from imprecise or unreliable flow is presented. This elaborate scheme is surely needed as the flow used in [11] is the MPEG coding vectors, typically prediction error minimizing vectors, which can be very different from the optical flow.

In [12] Karim et al. focus on improving block matching flow estimation for motion compensated interpolation in low frame rate video and no less then 16 references to other TSR algorithms are given. An overview of early work on motion compensated temporal interpolation in general (TSR, coding, deinterlacing etc.) is given by Dubois and Konrad in [13] where they state that even though motion tra-jectories are often nonlinear, accelerated and complex, a simple linear flow model will suffice in many cases. In [14] Chahine and Konrad state that complex motion modeling can improve objective results (PSNR). The work by Brox et al. [15] shows how variational optic flow algorithms can model and compute complex flows.

A problem, somewhat more complex than the interpolation of new frames, is trying to create a new arbitrary viewpoint 2D sequence from a multi-camera record-ing of a scene as done by Vedula et al. in [16]. The TSR used in that work is flow computations using the method from [17] followed by simple intensity interpola-tion. Shechtman et al. [18] use a multiple camera approach to TSR (and spatial SR) where all the cameras are assumed to be close spatially, or the scene assumed planar, allowing simple registration to replace flow computation. This technique can not be used on standard single camera recordings of film/television/video.

Using patches to represent salient image information is well-known [19, 20] and an extension to spatiotemporal image sequences as video epitomes is presented and used for TSR by Cheung et al. in [21]. It is unclear from [21] if video epitome TSR can handle more than simple and small motion, and its learning strategy is (still) computationally very costly.

1.4   Motion Compensated Frame Rate Conversion with Simultaneous Flow and Intensity Calculations

The traditional approach to temporal super resolution is to first compute the flow of the sequence and then interpolate the intensity values in the new frames. The sim-plest TSR methods use linear interpolation along the flow trajectories; they weigh each of the two original input frame contribution inversely by their distance to the new frame being interpolated. Simple TSR gives perfect results if the computed flow field is always reliable and precise, but this is rarely the case. Thus a fall back option is needed, often to simple temporal averaging with no use of motion information. Dane and Nguyen e.g. reports 4–42% fall back in [11].

When interpolating or warping the flow computed between two known frame into any new frame(s) positioned between them, it will not always be so that there is a flow vector in all pixel positions of the new frame(s). A fall back as above could be used, but one could also fill in neighboring flow vectors hoping they will be correct. This is a very complex strategy as seen in [8]. Without knowing the intensities of the new frame(s), it is impossible to know if the guessed flow is correct, but to get the intensities we need to know the flow! This case of two unknowns each depending on the other is truly a hen–egg problem.

In order to work around this problem, we use an approach that aims at recovering both the image sequence and the motion field simultaneously. In actual computa-tions, a pure simultaneous approach might become very complex and instead we use an iterative procedure: Given an estimate of the image sequence, we can update the estimate of the motion field, and given an estimate of the motion field, we can produce a new estimate of the image sequence. This procedure is embedded into a multiresolution framework, which is common approach in motion calculations (to enable computation of large flows).

1.5 Benchmarking in Testing

When presenting a new solution to a problem, any claims of improved results or performance should be supported by a reasonable validation of the claims. In most cases the other methods used to compare ones own method to are rather simple, as authors often have to balance the effort between improving one's own method and implementing other advanced methods for comparison. A common benchmark for testing TSR algorithms would make it easy to compare performances and results, but none exist. Even with a data set for benchmarking generally agreed upon, the question of how to evaluate the results remains—should it be done objectively and/or subjectively and what exact method(s) of evaluation should be used.

1.6 Outline

The rest of this chapter is organized as follows. In the next section, a generic energy formulation for TSR is proposed, and we study in more details the case of frame doubling proposing two algorithms. Then in Section 3 we evaluate our two frame doubling methods through a series of experiments before drawing our conclusions in section 4.

2 Energy Minimization Formulation

We use a probabilistic formulation to get to our energy formulation. This formulation was first proposed in [22] for image sequence inpainting and then used for deinterlacing in [23] and well as video (spatial) super resolution in [24].

2.1 Variational Temporal Super Resolution

We assume we are given an "degradation process" D which produces a "low-resolution" observed output from an hypothetical high-resolution input, as well as such a low resolution observation u_0. Following [22], the probability that a high-resolution input u and motion field $\mathbf{v}$ produces the output u_0 via D, $p(u, \mathbf{v}|u_0, D)$, is factored as

$$p(u, \mathbf{v}|u_0, D) \propto \underbrace{p(u_0|u, D)}_{P_0} \underbrace{p(u_s)}_{P_1} \underbrace{p(u_t|u_s, \mathbf{v})}_{P_2} \underbrace{p(\mathbf{v})}_{P_3} . \tag{1}$$

where u_s and u_t are the spatial and temporal distribution of intensities respectively. On the left hand side we have the a posteriori distribution from which we wish to extract a maximum a posteriori (MAP) estimate. The right side terms are: P_0, the image sequence likelihood, P_1 the spatial prior on image sequences, P_3 the prior on motion fields, and P_2 a term that acts both as likelihood term for the motion field and as spatiotemporal prior on the image sequence. In the case of frame doubling, D consists in "forgetting every second frame". It becomes a linear map of a bit more complex form for different frame rate ratios, typically a projection and inverting it is generally ill-posed.

In this work we do not consider noise contamination between the "ideal" u and the observed u_0, i.e. we do not try to denoise u_0. The likelihood term P_0 is then a Dirac distribution

$$p(u_0|u,D) = \delta_{Du-u_0}.$$

We use a Bayesian to variational rationale à la Mumford [25], $E(x) = -\log p(x)$, to transform our MAP estimation into a continuous variational energy minimization formulation, taking into account the form of the likelihood term

$$\underset{(u,\mathbf{v}),Du=u_0}{\text{arg.min}}\ E(u,\mathbf{v}) = E_1(u_s) + E_2(u_s,u_t,\mathbf{v}) + E_3(\mathbf{v}) \tag{2}$$

(using the same notation for discrete and continuous formulations). Then assuming some mild regularity assumptions, a minimizing pair $(u,\mathbf{v})$ must satisfy the condition $\nabla E(u,\mathbf{v}) = 0$ where ∇ is the gradient, and the solution expressed by the coupled system of equations

$$\begin{cases} \nabla_u E(u,\mathbf{v}) &= 0\,, \quad Du = u_0 \\ \nabla_\mathbf{v} E(u,\mathbf{v}) &= 0\,. \end{cases} \tag{3}$$

This system can be considered simultaneous when alternatingly updating the guesses on solutions to $\nabla_u E = 0$ and $\nabla_\mathbf{v} E = 0$ down through the multiresolution pyramid as discussed in Sect. 1.4. We thus minimize both the flow and intensity energy on each level of the pyramid as we iterate down through it.

We now discuss the choice of the actual terms in the energy (2). The term E_0 has already been described above: its function is to preserve input frames unaltered whenever they are at the position of an output frame.

The term E_2 is important as it models the consistent temporal transport of information into the new frames along the flows (forward and backward). It acts both as a prior on the intensities and as the likelihood of the motion field. We derive it from the classical brightness constancy assumption (BCA) which assume intensity preservation along the motion trajectories. We in fact use its linearized version, the optic flow constraint (OFC) $\nabla u \cdot \mathbf{v} + u_t = 0$, where ∇ denotes the spatial gradient $(\partial_x, \partial_y)^t$ used first in [26], but we punish a regularized 1-norm form of it, not the (original) quadratic ones. In the sequel, we write

$$\nabla u \cdot \mathbf{v} + u_t = \mathscr{L}_\mathbf{v} u$$

often referred to as the *Lie derivative* of u along $\mathbf{v}$ (although, it should, more correctly be along the spatiotemporal extension $(\mathbf{v}^t, 1)^t$ of $\mathbf{v}$).

The term E_3 is the prior on the flow. It serves the purpose of filling in good estimates of flow vectors in smooth regions from accurate values calculated where salient image data is available (edges, corners etc. giving nonzero image gradients).

To insure that flow discontinuities are preserved, we use a regularized form of the Frobenius norm of the Jacobian of $\mathbf{v}$ [27].

The term E_1 ensures spatial regularity of the recreated frames. We use a smooth form of the classical total variation ROF model [28]. It is especially useful when the motion estimates are unreliable. Assuming nevertheless that we can reliably estimate motion most of the time, this term should have a limited influence, by means of giving it a small weight.

The detailed energy formulation we use thus is

$$E(u,\mathbf{v}) = \underbrace{\lambda_1 \int_\Omega \psi(|\nabla u|^2)dx}_{E_1} + \underbrace{\lambda_2 \int_\Omega \psi(|\mathscr{L}_\mathbf{v} u|^2)dx}_{E_2}$$
$$+\underbrace{\lambda_3 \int_\Omega \left(\psi(|\nabla_3 v_1|^2 + |\nabla_3 v_2|^2)\right)dx}_{E_3}, \qquad Du = u_0 \qquad (4)$$

where Ω is the entire image sequence domain, $\nabla_3 = (\partial_x, \partial_y, \partial_t)^T$ is the spatiotemporal gradient, and the λ's are positive constants weighing the terms with respect to each other. v_1 and v_2 are the x- and y-components of the flow field, i.e. $\mathbf{v} = (v_1, v_2)^T$. (In the implementation we use a double representation of the flow field in the forward and backward directions respectively. In theory and in the continuous domain they are one and the same, but is split in practice—mainly due to discretization.) $\psi(s^2) = \sqrt{s^2 + \varepsilon^2}$ is an approximation of the $|\cdot|$ function as the latter is non-differentiable at the origin. ε is a small positive constant (10^{-8} in our implementation).

Splitting the energy (4) accordingly in an intensity and a flow part, we get this energy to be minimized for the intensities

$$E^i(u) = \underbrace{\lambda_s \int_\Omega \psi(|\nabla u|^2)dx}_{E_1} + \underbrace{\lambda_t \int_\Omega \psi(|\mathscr{L}_\mathbf{v} u|^2)dx}_{E_2}, \qquad Du = u_0 \qquad (5)$$

where $\lambda_s = \lambda_1$ and $\lambda_t = \lambda_2$ in (4). For the flow we need to minimize

$$E^f(\mathbf{v}) = \underbrace{\lambda_2 \int_\Omega \psi(|\mathscr{L}_\mathbf{v} u|^2)dx}_{E_2} + \underbrace{\lambda_3 \int_\Omega \left(\psi(|\nabla v_1|^2) + \psi(|\nabla v_2|^2)\right)dx}_{E_3}. \qquad (6)$$

In order to improve quality, the BCA in E_2 could be supplemented with the gradient constancy assumption (GCA) proposed first by Brox et al. in [15] for optical flows only. The GCA assumes that the spatial gradients remain constant along trajectories, and can be written as

$$\begin{pmatrix} u_{xx} & u_{xy} \\ u_{xy} & u_{yy} \end{pmatrix} \mathbf{v} + \nabla u_t = 0.$$

We use the more compact form $\mathcal{L}_{\mathbf{v}}\nabla u = 0$. It will improve the quality of the flow to add it to $E^f(\mathbf{v})$ in (6), but as shown in [29], the added complexity only pays off in a minimal quality improvement if it is added to the intensity energy $E^i(u)$ in (5). Adding the GCA to E_2 for the flow in $E^f(\mathbf{v})$, gives

$$E_2^f(\mathbf{v}) = \int_\Omega \left(\lambda_2 \psi(|\mathcal{L}_{\mathbf{v}}u|^2 + \gamma |\mathcal{L}_{\mathbf{v}}\nabla u|^2) \right) dx \tag{7}$$

where γ is a positive constant weight. If we set $\gamma = 0$ in our implementation, we are back at (6) and thus we are able to test both with and without GCA for the flow energy in one joint implementation.

2.2 Implementation of TSR: Frame Doubling

To test our ideas we have chosen to implement a frame rate doubler, but implementing solutions for other conversion rates would be easy. Frame doubling here means that the projection D forgets every second frame. We may decompose the domain Ω as the domain of *known frames* K and its complement $\Omega \backslash K$. The constraint $Du = u_0$ becomes

$$u|_K = u_0.$$

2.2.1 Euler-Lagrange Equations and Their Solvers

To minimize the intensity and flow energies given in (5), (6) and (7) we derive and solve the associated Euler-Lagrange equations. Let us start with the flow energy minimization: After exchanging the E_2-term of the flow energy in (6) with the E_2-term from (7) to incorporate to option of using GCA, the flow Euler-Lagrange equation is derived. It is implemented numerically along the lines given by Brox et al. in [15] and by Lauze in [22, 30] and minimized iteratively by repeated linearizations of it, each linear equation being solved by a Gauss-Seidel solver.

Details on the computation of the gradient of the intensity energy (5) can be found in [22] (details on discretization in [24]), and we here recall the final result:

$$\nabla_u E^i = -\lambda_s \nabla_2 \cdot (A(u)\nabla u) - \lambda_t \nabla_3 \cdot (B(u)(\mathcal{L}_{\mathbf{v}}u)\mathbf{V}) \tag{8}$$

where $\mathbf{V} = (\mathbf{v}^T, 1)^T$ is the spatiotemporal extension of $\mathbf{v}$, $\nabla_2 \cdot$ is the 2-dimensional divergence operator, while $\nabla_3 \cdot$ is the 3-dimensional one and the coefficients $A(u)$ and $B(u)$ are, respectively

$$A(u) = \psi'(|\nabla u|^2), \qquad B(u) = \psi'(|\mathcal{L}_{\mathbf{v}}u|^2).$$

In order to solve (8) numerically, we again use a fixed point approach: At each fixed point iteration, $A(u)$ and $B(u)$ are computed from the estimated values of u and $\mathbf{v}$ and thus frozen. Equation (8) then becomes linear. It is discretized and solved here too by Gauss-Seidel relaxation. Only the discretization of $\mathscr{L}_\mathbf{v}u$ is somewhat non standard. It uses the OFC/BCA approximation

$$\mathscr{L}_\mathbf{v}u \approx u(x+\mathbf{v}, t+1) - u(x,t)$$

which gives rise to a numerical intensity diffusion along trajectories with correction for the potential trajectories divergence. Details can be found in [29].

2.2.2 Algorithm

Before detailing our frame doubling algorithm, here it is in overview (leaving out the special initialization at the top level):

At each level from the top, coarse to fine, for $k = levels$ until $k = 1$

1. Calculate the forward and backward flows, $\mathbf{v}_0^f$ and $\mathbf{v}_0^b$, of the resized original input sequence $u_{0,k}$ minimizing (6) with/without E_2 from (7).
2. Initialize new frames: $u(\mathbf{x},t,k) = resize[u(\mathbf{x},t,k-1)]$ in the domain D.
3. Initialize forward and backward flows of new frames: $v(\mathbf{x},t,k) = resize[\mathbf{v}(\mathbf{x},t,k-1)]$ in the domain D.
4. Calculate the flows $\mathbf{v}^f$ and $\mathbf{v}^b$ of the output sequence u minimizing (6) with/without E_2 from (7).
5. Calculate new frames in $u|_D$ by minimizing (5).

In our multiresolution settings, on each level k of the pyramid, we first compute the forward and backward flows, $\mathbf{v}_0^f$ and $\mathbf{v}_0^b$, of the *original input sequence* u_0 (resized to the size of the current level), minimizing (6) (including E_2 from (7) to give the option of using GCA or not) with the resized input sequence u_0 simply replacing u. ($E^f(\mathbf{v})$ is minimized over the domain K instead of over Ω.) This is to have a highly reliable anchor point flow when calculating the flows $\mathbf{v}^f$ and $\mathbf{v}^b$ of the full output sequence. At the given level of the pyramid, k, we then initialize intensities and the flows of the new frames by resizing the intensities and flows calculated at the above coarser level $k+1$. Then we calculate the flows from these initializations by minimizing either (6) or (7) (w/o GCA). Next we calculate u at level k by minimizing the energy (5) knowing $\mathbf{v}^f$ and $\mathbf{v}^b$ and using the resized intensities from level $k+1$ as initialization of u in the new frames, just as when calculating $\mathbf{v}^f$ and $\mathbf{v}^b$. The resizing function (*resize*) used is given in [31].

The use of a multiresolution schemes is considered essential when doing variational flow calculations. In TSR, calculating both flow and intensities at each level solves the hen–egg problem of what comes first in a new frame: The flow or the intensities. Thus we iteratively improve first one and then the other to get simultaneous computations and optimize our solution using a small scale factor between levels to get optimal initializations.

2.2.3 Initialization

At the coarsest level at the top of the pyramid we do not have a $k+1$ level to initialize our data from and thus have to use temporal initialization (inferior to $k+1$ initialization). For the flow calculation we have chosen to do frame averaging of both flow and intensities. If the new frame is located at time n and the two know frames are at times $n \pm 1/2$ then $\mathbf{v}(\mathbf{x},n) = \big(\mathbf{v}_0(\mathbf{x},n-1/2) + \mathbf{v}_0(\mathbf{x},n+1/2)\big)/2$ and $u(\mathbf{x},n) = \big(u_0(\mathbf{x},n-1/2) + u_0(\mathbf{x},n+1/2)\big)/2$. Even though the flow we compute at the top level is (almost) of subpixel size due to the downscaling, we still use it to re-initialize the intensities by simple interpolation along the flow $u(\mathbf{x},n) = \big(u_0(\mathbf{x}+\mathbf{v}^b,n-1/2) + u_0(\mathbf{x}+\mathbf{v}^f,n+1/2)\big)/2$ before we minimize $E^i(u)$.

3 Experiments

We have implemented our frame doubling algorithm in such a way that we can test it in two versions: With and without the gradient constancy assumption on the flow. With GCA on the flow, we expect the most correct results as both flow and intensities are subject to minimal blurring. Without GCA on the flow, a more blurred flow is expected and thus also a more blurred intensity output.

The tests conducted have focused on the major problem of having too low a frame rate in image sequences: Unnatural, jerky motion, which is typically most prominent when the camera pans on scenes containing high contrast (vertical) edges. By doubling the frame rate we will aim at reestablishing the phi-effect. The images sequences chosen for testing all have the problem of perceived jerky, unnatural motion. The sequences are a mix of homemade and cutouts of real world motions pictures on standard PAL 25 fps DVDs originating from film. All inputs and results discussed are also given as video files (*.avi) online at: http://image.diku. dk/sunebio/TSR/TSR.zip [32]. The shareware AVI video viewer/editor VirtualDub is included in the material, and we would like to stress the importance of viewing the results as video: The effects, artifacts and improvements discussed are mainly temporal and not seen in stills.

3.1 Parameters

There are eleven parameters to tune in our algorithm and we have focused on optimizing the output quality, not speed (yet). Through extensive empirical parameter testing we have optimized two sets of parameters for variational frame doubling TSR; with and without GCA in the flow. The settings found to be optimal are given in Table 1 for both versions of our algorithm. We see that settings for the intensity energy minimization is the same for both algorithm versions, but the given values proved optimal with both flow energy minimizations. The temporal to spatial diffusion weight ratio, $\lambda_t{:}\lambda_s$, is high, favoring temporal diffusion, which ensures that spatial diffusion is only used when temporal information is highly unreliable. Lowering the ratio from 50:1 to 20:1 gave similar results when evaluating on video, but

Table 1 Optimal parameter settings for variational TSR. The eleventh parameter of the algorithms is the convergence threshold set to 10^{-7} in all tests.

		Flow without GCA	Flow with GCA
Multiresolution	Scale factor	1.04	1.04
	Levels	55 or 75	55 or 75
Flow	Fixed point iterations	10	5
	Relaxation iterations	40	20
	λ_3 in (6)	30	100
	λ_2 in (7)	1	1
	γ in (7)	0	100
Intensities	Fixed point iterations	5	5
	Relaxation iterations	10	10
	λ_s in (5)	1	1
	λ_t in (5)	50	50

judging from stills, there where minor degradations, thus we recommend $\lambda_t{:}\lambda_s = 50{:}1$.

The number of flow iterations needed are higher than the number of intensity iterations needed, which illustrates the larger complexity of flow calculations.

Table 1 also shows that without GCA in the flow we need more iterations to get optimal flows. This is because fewer point give reliable flow information when only evaluating brightness constancy, which increases the need for flow diffusion by the regularization term on the flow, E_3 in (6). The E_3-term is weighed 30 times over BCA when we do not use GCA, but is given the same weight as GCA when GCA is used (with the BCA nearly neglected due to the $\lambda_2{:}\gamma = 1{:}100$ ratio).

The number of levels in the multiresolution pyramid is set to either 55 or 75 depending on the frame size of the given image sequence with a 1.04 (coarse to fine) scale factor between the levels. The low scaling factor ensures good information transfer down through the pyramid, but increasing it would give a speedup (the effect on quality have not been thoroughly investigated).

We also conducted experiments using zero as initial values for both flows and intensities in the new frames at the top level. Given the many levels we use, the error introduced was corrected down through pyramid, showing great robustness against bad initializations.

3.2 Evaluation Methodology

As the human visual system is the final judge when evaluating the enhancement achieved, we will focus on subjective results (although we have not used a standardized and complex evaluation as described in [33]). To give a broader validation, we have also given objective results. Still images are an borderline acceptable and easy way to evaluate the quality of frame doubling, but it is imperative to evaluate on video to judge if the motion portrayal has become natural during real-time playback.

In our tests we have doubled frame rates from 25 to 50 fps, which should enable viewing the video examples in the online material [32] on any modern PC screen at refresh rates of 50 Hz or 100 Hz, whereas viewing at other rates above 50 Hz might add some jerkiness from frame repetition (by the graphics/video card and/or playback software). Comparing the 25 fps input sequences in [32] with the 50 fps results should however clearly illustrate the difference in quality (e.g. using Virtual-Dub included in [32]).

As objective measures we have used the mean square error (MSE) and the peak signal to noise ratio (PSNR). Using the notations given in Sect. 2, the MSE and PSNR are

$$MSE = \frac{1}{N} \sum_{\Omega} (u - u_{gt})^2 \qquad PSNR = 10 \log_{10} \left(\frac{255^2}{MSE} \right) \qquad (9)$$

where u is the frame doubled output and u_{gt} is the ground truth. We sum over all pixels of the sequence (also the old frames from the input that are not changed in the output). PSNR is measured relative to the maximum grey value, 255.

3.3 Frame Doubling Results

We generally do not discuss frame repetition results as they are identical to the input. Thus any description of the input also fits on the corresponding frame doubling output.

In Fig. 2 results for the sequence Square is given. Square has 50×50 frame size, is 5 frames long in the input and 9 frames in the output. The 10×10 square moves diagonally down to the right. The speed of the square is 2 pixels/frame in the output.

Frame averaging creates a double, semitransparent square as seen in Fig. 2(b). Variational TSR perfectly recreate the square with GCA on the flow as seen in Fig. 2(d), but not without GCA (shown in Fig. 2(c)). When watched as video [32], the square is not perceived as unsharp in the result without GCA and the motion has become fluent as compared to the input and the result looks identical to TSR with GCA. The motion in the frame averaging output is jerky and has a clear trail of the square.

The flows computed by the variational TSR algorithm on Square are shown in Figs. 2(e)–(h). A (dis)occlusion trail can be seen in the flows, which in the non-GCA version gives some artifacts (Fig. 2(c)). We also see an nice filling in of the flow (by the E_3-term) in the center of the completely uniform square (image gradient zero, which gives no flow locally from the BCA and GCA). Even though it is very hard to detect visually in the flow fields, the flow field of the GCA version is closer to the correct magnitude and direction at the corners of the square, yielding better intensity result as seen in Fig. 2(d).

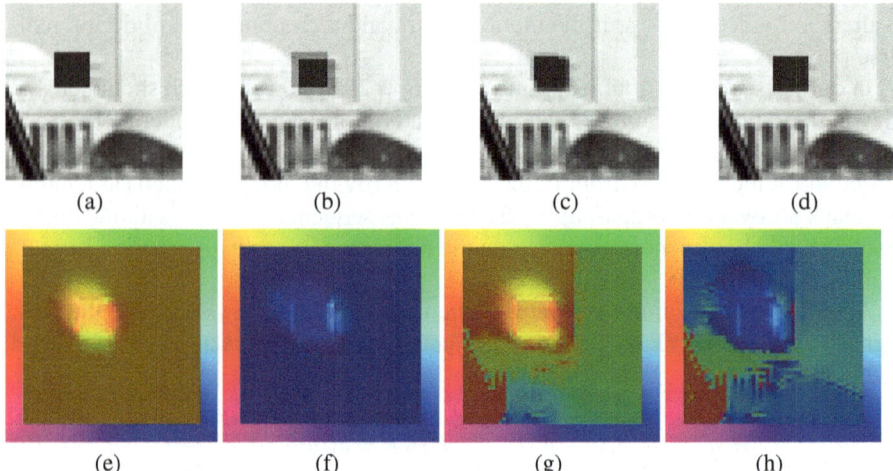

Fig. 2 Frame doubling of the 50×50 sequence Square. **a** Original frame 3 (5 in the output). The new frame 6 of the output created by **b** frame averaging, **c** variational TSR without GCA, and **d** variational TSR with GCA. Optic flows computed by variational TSR: **e** backward from frame 6 to 5 without GCA, **f** forward from frame 6 to 7 without GCA, **g** backward from frame 6 to 5 with GCA, and **h** forward from frame 6 to 7 with GCA. In this color representation, the hue value gives the flow direction as coded on the boundary and the intensity gives the flow magnitude (normalized in [0.5–1]).

The sequence Cameraman Pan is a pan (10 pixel/frame in the input) across the image Cameraman in a 130×100 window. On the sequence Square, frame averaging was performing bad but not unacceptable. On Cameraman Pan the performance of frame averaging is unacceptably bad as Fig. 3(b) illustrates. Artifacts this bad are clearly visible when viewing the result as video [32] and the motion seems, if possible, even more jerky than the motion in the 25 fps input. The motion of the two variational TSR outputs are much smoother and appears very natural. In Figs. 3(c) and 3(d) it is also seen how the new frames produced with variational TSR are very similar to the original frame shown in Fig. 3(a). The only difference is a slight smoothing, which is only seen in the stills.

Some minor (dis)occlusion errors occur at the frame boundaries of Cameraman Pan when details leave or enter the frame, which are only spotted during video playback if one looks for them, or happens to focus on that particular part of the frame. The cause of this problem is our use of Neumann boundary condition: When a flow vector points out of the frame, we use Neumann BC to spatially find a replacement value in side the frame. This creates a high magnitude temporal gradient if the replacement pixel (in the neighboring frame) is very different from the pixel currently being processed, resulting in increased spatial diffusion (which is often unreliable).

The sequence Building is a 284×236 cutout of a PAL DVD (telecined from film). The scene chosen has a camera tilt down a building with discrepancies from

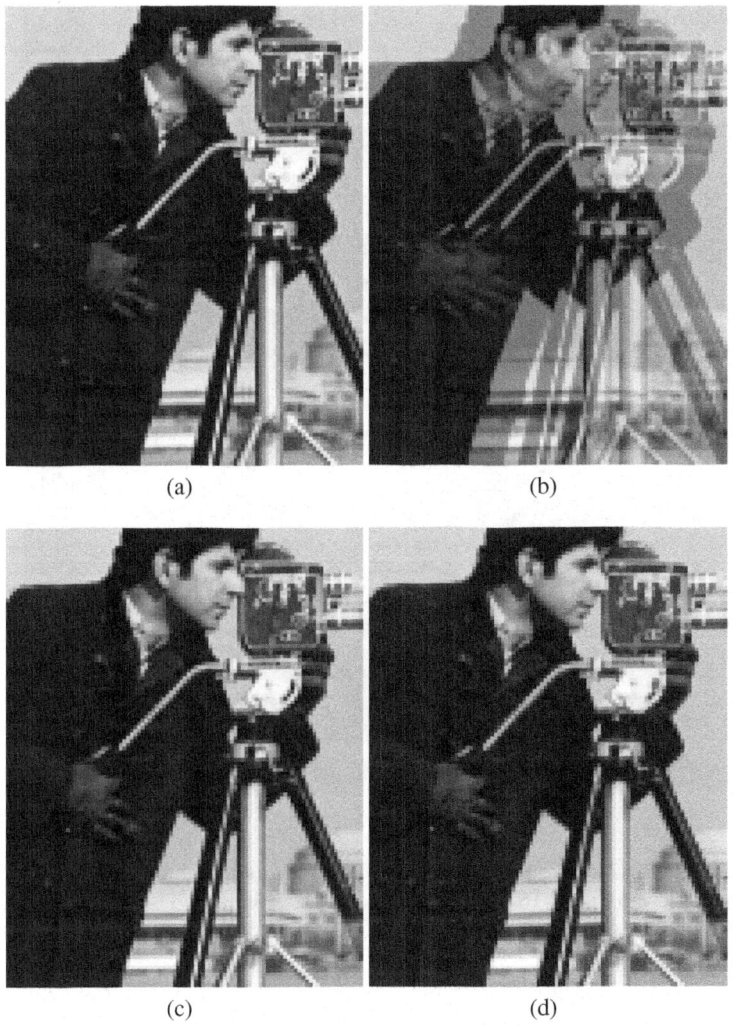

(a)   (b)

(c)   (d)

Fig. 3 Frame doubling of the 130×100 sequence Cameraman Pan. **a** original frame 5 (9 in the output). New frame 10 by **b** frame averaging, **c** variational TSR without GCA, and **d** variational TSR with GCA.

uniform global translational motion due to the depth of the scene and variations in the motion of the camera. Frame averaging on Building blurs the new frames quite a lot as can be seen in Fig. 4(b). During video playback [32] this blurring is not sensed, but the motion is as jerky as in the input. The lamps seen inside the building seems to flicker as they are blurred to middle grey in the new frames. These artifacts make the frame averaging result very annoying to watch. The motion portrayal in the two variational TSR results is natural and no flickering occurs. As seen in Figs. 4(c) and 4(d) the new frames are a bit smoothed when compared to

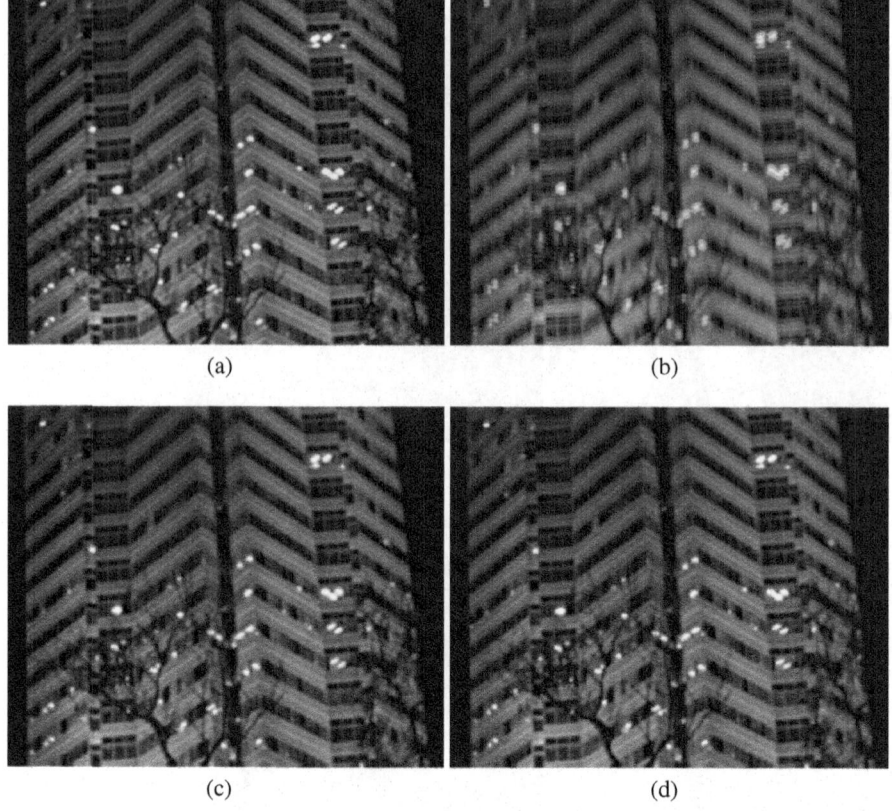

Fig. 4 Frame doubling of the 284×236 sequence Building. **a** original frame 1. New frame 2 by **b** frame averaging, **c** variational TSR without GCA, and **d** variational TSR with GCA.

the original frames, but this is not noticeable in the videos (which in some sense supports the results in [5] on alternating between sharp and blurred frames without loss of overall perceived sharpness).

The sequence Control Panel is taken from the same movie as Building and has a fast camera pan and complex motion (the person walking behind the control panel and being tracked by the camera pan). Results are given in Fig. 5 and as videos in [32]. As with Building the frame averaging result still has jerky motion and flickering and the new frames are also blurry, while TSR w/o GCA produce only slightly blurred frames and have overall natural motion. When looking at single frames as stills, it becomes clear that the complexity of the motion with many and fast (dis)occlusion is too much for our variational optic flow scheme in its current version (both with and without GCA). As with the boundary problems in Cameraman Pan the problems might be spotted during video playback if the attention of the viewer happens to get focussed on just that region. Whether that will happen in a very dynamic scene like Control Panel is hard to say. We discuss

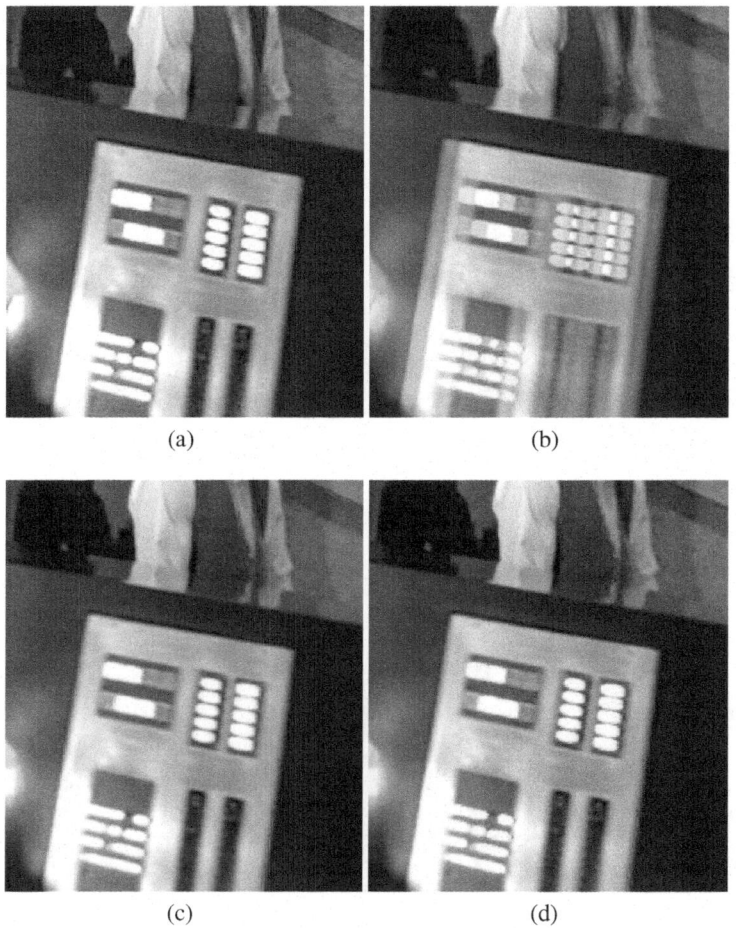

(a)   (b)

(c)   (d)

Fig. 5 Frame doubling of the 256×220 sequence Control Panel. **a** original frame 4 (7 in the output). New frame 8 by **b** frame averaging, **c** variational TSR without GCA, and **d** variational TSR with GCA.

how our flow scheme can be improved to possibly handle more complex motion in Sec. 3.4.

The sequence Boat (320×306 cutout) taken from another PAL DVD has an even faster pan than Control Panel and object motion as well. The motion is very stuttering when the 25 fps input sequence is played back (Boat25fps.avi from [32]) and Boat has the most unnatural motion of all the sequences we have run tests on. Again the frame averaging result is of poor quality as seen in Fig 6(b) and the video (BoatFrameAv50fps.avi), and again the two variational TSR schemes produce high quality results, only slightly smoothed (Figs. 6(c) and 6(d)) but with natural motion portrayal as seen in the videos. Repeated watching of the

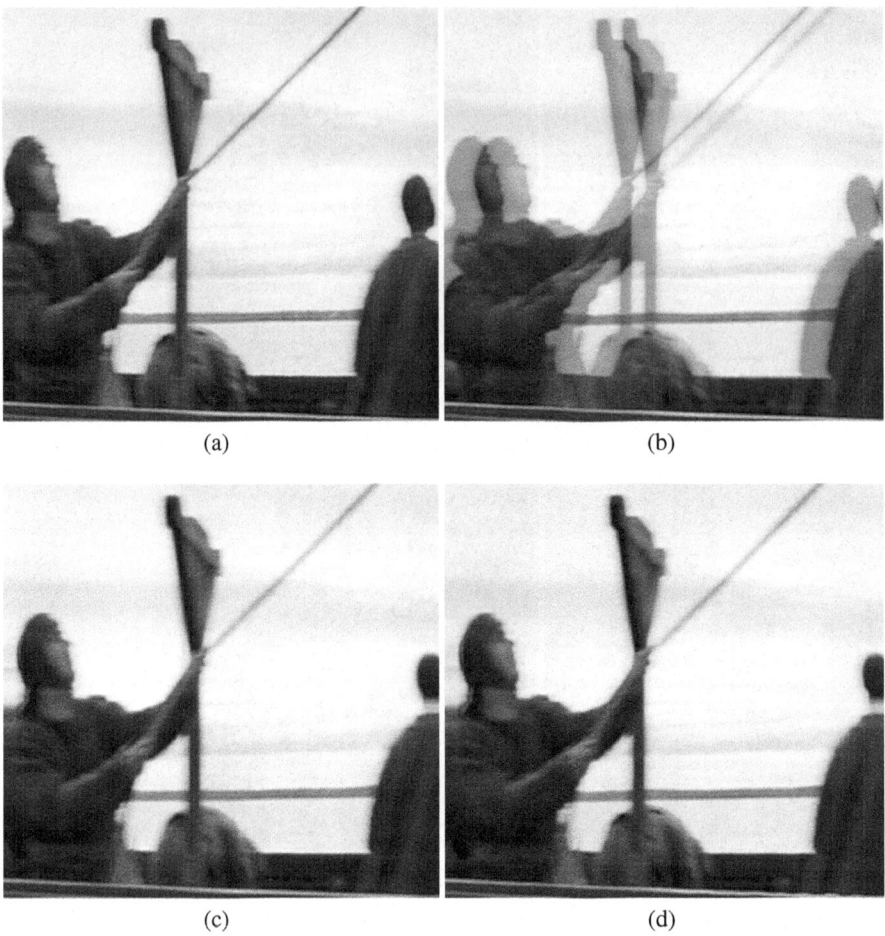

Fig. 6 Frame Doubling on the 320 × 306 sequence Boat. **a** Original frame 2, now frame 3 of the double frame rate output, new frame 4 by **b** frame averaging, **c** variational TSR without GCA, and **d** variational TSR with GCA.

variational TSR results on Boat gives a sense of a slight stutter in the motion, indicating that 50 fps is not enough on this and other sequences with similar motion.

3.3.1   Objective Evaluation

Table 2 gives objective results for four of the test sequences evaluated subjectively in the previous section. For the real sequences Building and Control Panel we have created ground truth sequences by taking out every other frame of the inputs and used these shortened sequences as frame doubling inputs, and for the artificial ones we have simply created the ground truth frames. On the real sequences this

Table 2 Objective evaluation of the four frame doubling methods in test: Frame repetition, frame averaging and variational TSR without/with GCA. MSE and PSNR scores are given for the four test sequences Square, Cameraman Pan, Building and Control Panel.

		Square	Cameraman Pan	Building	Control Panel
Frame repetition	MSE	158.4	1887.9	462.5	1369.8
	PSNR	*26.13*	*15.37*	*21.48*	*16.76*
Frame averaging	MSE	82.54	1208.3	287.8	849.9
	PSNR	*28.96*	*17.31*	*23.54*	*18.84*
TSR without GCA	MSE	13.05	**39.47**	**13.82**	**76.59**
	PSNR	*36.97*	***32.17***	***36.73***	***29.29***
TSR with GCA	MSE	**8.97**	107.9	16.44	96.87
	PSNR	***38.60***	*27.80*	*35.97*	*28.27*

means larger panning/tilting motions from frame to frame as we now do 12.5 to 25 fps frame doubling. Since frame repetition is not really worth comparing with other results in stills, and since its motion portrayal is the same as in the input, we left it out of the subjective evaluation but have included it here as it is the most widely use method for frame rate up-conversion.

As the results in Table 2 show, our variational TSR algorithms outperforms frame averaging as it was also the case in the subjective evaluation. It is also no surprise that in the presence of motion, frame repetition is clearly the objectively worst performing frame doubling algorithm. Whether it is subjectively worse than frame averaging is however a question up for debate because of the double exposure in new frame in frame averaging, which introduces additional artifacts.

Returning to the far better variational TSR frame doublers, the use of GCA helps in the case of *object* motion. Variational TSR with GCA gives the best objective result on Square, which corresponds well with the subjective results. For the two sequences Cameraman Pan and Building dominated by global motions, the non-GCA version is objectively better than the GCA version, which can be explained by the GCA version tending to overfit the flows. The boundary problems in Cameraman Pan are judged from the objective result worse in the GCA version. On Control Panel the non-GCA version produces a smoother flow field and thus the intensity output is also somewhat smoother, which helps dampen the problems with wrong flow estimations of the complex flow in Control Panel.

From our combined tests results, we can conclude that variational TSR without GCA performs slightly better or the same as TSR with GCA in cases where the sequences are dominated by global flow (camera motion). It is clear that our motion compensated variational TSR frame doublers are producing outputs far superior to the outputs from the simple methods frame averaging and frame repetition.

Variational TSR needs to be benchmarked against other motion compensated TSR algorithms to show its full potential.

3.4 Discussion: Improving Variational Optic Flow for Motion Compensated Methods

Scenes with complex motion can cause problems even to advanced optic flow algorithms. It is important to have a robust motion compensated algorithm that switches off temporal input when flows are unreliable, but the limitations of the human visual system will also help in optimizing the algorithm: In scenes with complex motion the HVS will not be able to track all the motions and thus we might get away with producing suboptimal outputs. Still, optimal results require precise and reliable flows, but the modeling of e.g. accelerations, whirls and transparent motions is still complex. Variational methods are getting better and better at this, while e.g. block matching motion estimation will fail by its basic motion modeling assumptions.

The problems we see with changes motion magnitude and directions (acceleration) e.g. in Control Panel is most likely due to the fact that we use a 3D local spatiotemporal prior on the flow, E_3 in (6), reported to give better flow results than a purely spatial 2D prior on sequences with slow temporal changes (e.g. Yosemite) in [15, 34]. In [24] we showed that the problem might be solved by processing the sequence in very short bites (2–4 frames) but a more robust solution would be to consider spatiotemporal sequences as 2D+1D instead of the unnatural 3D, separating space and time, but still linking them together (naturally) along the optic flow field. Alternatively, a flow acceleration prior could be added to the variational formulation of our problem.

4 Conclusion

In this chapter we have discussed the requirements put on the design of temporal super resolution algorithms by the human visual system, and have presented a novel idea of simultaneous flow and intensity calculation in new frames of an image sequence. A novel variational temporal super resolution method has been introduced, and it has been implemented and tested for the subproblem of frame rate doubling. Our results showed that the use of the gradient constancy assumption gives no major improvements on the image sequence output, but as indicated in our discussion, it might do so as variational optic flow modeling improves.

Even though the new variational TSR algorithms do not always create perfect new frames, they do provide high quality 50 fps video from 25 fps video without noticeable artifacts during video playback, thus reestablishing the phi-effect in the troublesome case of high contrast edges in motion. The framework presented also has the potential to be used for other frame rate conversions than frame rate doubling, the problems of implementation being mainly of practical character.

References

1. Matlin, M.W., Foley, H.J.: Sensation and Perception, 4th edn. Allyn and Bacon, Boston (1997)
2. Ciuffreda, K.J., Selenow, A., Wang, B., Vasudevan, B., Zikos, G., Ali, S.R.: Bothersome blur: A functional unit of blur perception. Vision Research 46(6-7), 895–901 (2006); E-published December 6 2005
3. Burr, D.C., Morgan, M.J.: Motion deblurring in human vision. Proceedings of the Royal Society B: Biological Sciences 264(1380), 431–436 (1997)
4. Morgan, M.J., Benton, S.: Motion-deblurring in human vision. Nature 340, 385–386 (1989)
5. Chen, H.F., Lee, S.H., Kwon, O.J., Kim, S.S., Sung, J.H., Park, Y.J.: Smooth frame insertion method for motion-blur reduction in LCDs. In: IEEE 7th Workshop on Multimedia Signal Processing, pp. 581–584 (2005), ieeexplore.ieee.org
6. Ehrhardt, J., Säring, D., Handels, H.: Optical flow based interpolation of temporal image sequences (2006)
7. Cornog, K.H., Dickie, G.A., Fasciano, P.J., Fayan, R.M.: Interpolation of a sequence of images using motion analysis (2003)
8. Robert, P.: Method for the temporal interpolation of images and device for implementing this method (1993)
9. de Haan, G.: IC for motion-compensated de-interlacing, noise reduction, and picture-rate conversion. IEEE Transactions on Consumer Electronics, 617–624 (1999)
10. Ojo, O.A., de Haan, G.: Robust motion-compensated video up-conversion. IEEE Transactions on Consumer Electronics 43(4), 1045–1055 (1997)
11. Dane, G., Nguyen, T.Q.: Optimal temporal interpolation filter for motion-compensated frame rate up conversion. IEEE Transactions on Image Processing 15(4), 978–991 (2006)
12. Karim, H.A., Bister, M., Siddiqi, M.U.: Multiresolution motion estimation for low-rate video frame interpolation. EURASIP Journal on Applied Signal Processing (11), 1708–1720 (2004)
13. Dubois, E., Konrad, J.: Estimation of 2-d motion fields from image sequences with application to motion-compensated processing. In: Sezan, M., Lagendijk, R. (eds.) Motion Analysis and Image Sequence Processing, pp. 53–87. Kluwer Academic Publishers, Dordrecht (1993)
14. Chahine, M., Konrad, J.: Motion-compensated interpolation using trajectories with acceleration. In: Proc. IS&T/SPIE Symp. Electronic Imaging Science and Technology, Digital Video Compression: Algorithms and Technologies 1995, vol. 2419, pp. 152–163 (1995)
15. Brox, T., Bruhn, A., Papenberg, N., Weickert, J.: High Accuracy Optical Flow Estimation Based on a Theory for Warping. In: Pajdla, T., Matas, J(G.) (eds.) ECCV 2004. LNCS, vol. 3024, pp. 25–36. Springer, Heidelberg (2004)
16. Vedula, S., Baker, S., Kanade, T.: Image-based spatio-temporal modeling and view interpolation of dynamic events. ACM Transactions on Graphics 24(2), 240–261 (2005)
17. Lucas, B., Kanade, T.: An iterative image registration technique with an application to stereo vision. In: International Joint Conference on Artificial Intelligence, pp. 674–679 (1981)
18. Shechtman, E., Caspi, Y., Irani, M.: Space-Time Super-Resolution. IEEE Trans. on Pattern Analysis and Machine Intelligence 27(4), 531–545 (2005)
19. Freeman, W.T., Pasztor, E.C., Carmichael, O.T.: Learning low-level vision. International Journal of Computer Vision 40(1), 25–47 (2000)

20. Griffin, L.D., Lillholm, M.: Hypotheses for image features, icons and textons. International Journal of Computer Vision 70(3), 213–230 (2006)
21. Cheung, V., Frey, B.J., Jojic, N.: Video epitomes. In: Proceedings of Computer Vision and Pattern Recognition, CVPR 2005, pp. 42–49 (2005), ieeexplore.ieee.org
22. Lauze, F., Nielsen, M.: A Variational Algorithm for Motion Compensated Inpainting. In: Hoppe, S.B.A., Ellis, T. (eds.) British Machine Vision Conference, BMVA, vol. 2, pp. 777–787 (2004)
23. Keller, S.H., Lauze, F., Nielsen, M.: Deinterlacing using variational methods. IEEE Transactions on Image Processing 17(11), 2015–2028 (2008)
24. Keller, S.H.: Video Upscaling Using Variational Methods. Ph.D. thesis, Faculty of Science, University of Copenhagen (2007), http://image.diku.dk/sunebio/Afh/SuneKeller.pdf (accessed November 16, 2009)
25. Mumford, D.: Bayesian rationale for the variational formulation. In: ter Haar Romeny, B.M. (ed.) Geometry-Driven Diffusion In Computer Vision, pp. 135–146. Kluwer Academic Publishers, Dordrecht (1994)
26. Horn, B., Schunck, B.: Determining Optical Flow. Artificial Intelligence 17, 185–203 (1981)
27. Papenberg, N., Bruhn, A., Brox, T., Didas, S., Weickert, J.: Highly Accurate Optic Flow Computation With Theoretically Justified Warping. International Journal of Computer Vision 67(2), 141–158 (2006)
28. Rudin, L., Osher, S., Fatemi, E.: Nonlinear total variation based noise removal algorithms. Physica D 60, 259–268 (1992)
29. Lauze, F., Nielsen, M.: On Variational Methods for Motion Compensated Inpainting. Tech. rep., Dept. of Computer Science, Copenhagen University, DIKU (2009), http://image.diku.dk/francois/seqinp (accessed November 16, 2009)
30. Lauze, F.: Computational methods for motion recovery, motion compensated inpainting and applications. Ph.D. thesis, IT University of Copenhagen (2004)
31. Bruhn, A., Weickert, J., Feddern, C., Kohlberger, T., Schnörr, C.: Variational Optic Flow Computation in Real-Time. IEEE Trans. on Image Processing 14(5), 608–615 (2005)
32. Keller, S.H.: Selected electronic results of TSR experiments (2007), http://image.diku.dk/sunebio/TSR/TSR.zip (accessed November 16, 2009)
33. ITU: ITU-R recommendation BT.500-11: Methodology for the subjective assessment of the quality of television pictures (2002)
34. Bruhn, A., Weickert, J., Schnörr, C.: Lucas/Kanade Meets Horn/Schunck: Combining Local and Global Optic Flow Methods. International Journal of Computer Vision 61(3), 211–231 (2005)

Chapter 12
Synthesizing Natural Images Using Spatial Layout Information

Ling Shao and Ruoyun Gao

Abstract. We propose an algorithm for synthesizing natural images consisting of composite textures in this chapter. Existing texture synthesis techniques usually fail to preserve the structural layout of the input image. To overcome this drawback, a target image which contains the layout information of the input is used to control the synthesis process. With the guidance of this target image, the synthesized output resembles the input globally and is composed of the original textures. For images composed of textured background and a non-textured foreground object, segmentation is first applied on the main object and texture synthesis is used on the background texture. In comparison to other synthesis methods, the proposed solution yields significantly better results for natural images.

1 Introduction

Texture synthesis has experienced great development in the past two decades, and manifested its significance in rendering synthetic images. By reproducing visual realism of the physical world, it is used to create texture of any size. The goal of the texture synthesis can be stated as follows: given a texture sample, synthesize a new texture that, when perceived by a human observer, appears to be generated by the same underlying stochastic process [4].

Texture synthesis techniques are mainly divided into two groups: parametric and non-parametric. Parametric texture synthesis usually only works for simple and regular textures. Non-parametric or example-based texture synthesis can deal with a wide variety of texture types, from regular to stochastic. Among example-based

Ling Shao
Department of Electronic & Electrical Engineering, The University of Sheffield, UK
e-mail: ling.shao@sheffield.ac.uk

Ruoyun Gao
Department of Automation, Xiamen University, China
e-mail: ruoyun.gao@gmail.com

texture synthesis methods, the works of Efros and Leung [1] and Wei and Levoy [2] have been influential. Both methods are pixel-based, and Wei and Levoy's method is basically an improved and more efficient version of [1]. The algorithm synthesizes an image in a raster scan order and for each unfilled pixel it does a full search in the sample image to find a pixel whose neighbourhood matches the best to the filled neighbourhood of the unfilled pixel. Tree-structured Vector Quantisation is used to significantly speed up the search process. Ashikhmin [3] presented a constrained search scheme in which only shifted locations of neighbourhood pixels of the current unfilled pixel are used for matching. Patch-based texture synthesis is proposed by Efros and Freeman [4] and Kwatra et al. [5]. These techniques are usually more effective for structured textures and are faster than pixel-based methods.

The above mentioned synthesis methods do not consider the structural layout in the input sample. Therefore, the output image would look very different to the input globally though is composed of the same textures. For the purpose of image scaling, the proposed algorithm can capture both the texture elements and the global layout of the input. For an image composed entirely of texture, we use a target image which has the same size as the output to guide the synthesis process. For an image composed of both a foreground object and the background texture, foreground object is segmented and treated differently to the background texture.

In the remainder of this chapter, we will first describe in Section 2 the Image Quilting algorithm introduced by Efros and Freeman [4] and use it as a benchmark for our proposed method. In Section 3, a novel synthesis algorithm particularly for natural composite textures is introduced and compared with state-of-the-art techniques. The texture synthesis technique based on foreground object segmentation is presented in Section 4. Finally, we conclude this chapter in Section 5.

2 Image Quilting

The Image Quilting (IQ) algorithm [4] is an efficient method to synthesize textures, taking the place of pixel based algorithms which are slow and unreliable. The IQ algorithm finds the best matched patches within certain error tolerance, called *candidate blocks*, from the input texture, and randomly selects one of the candidates to stitch with the existing patches of the output image. The *stitching* is implemented by finding the minimum error boundary cut in the overlap region. In this case, the IQ algorithm guarantees the smoothness of the output image. Generally, the image quilting algorithm works as follows:

- Go through the image to be synthesized in the raster scan order in steps of one block (minus the overlap).
- For every location, search the input texture for a set of blocks that satisfy the overlap constraints (above and left, see Fig. 1) within certain error tolerance. Randomly pick one such block.

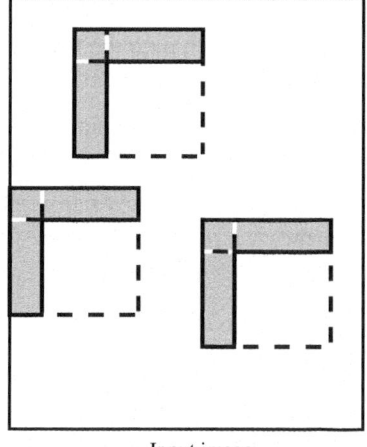

Input image

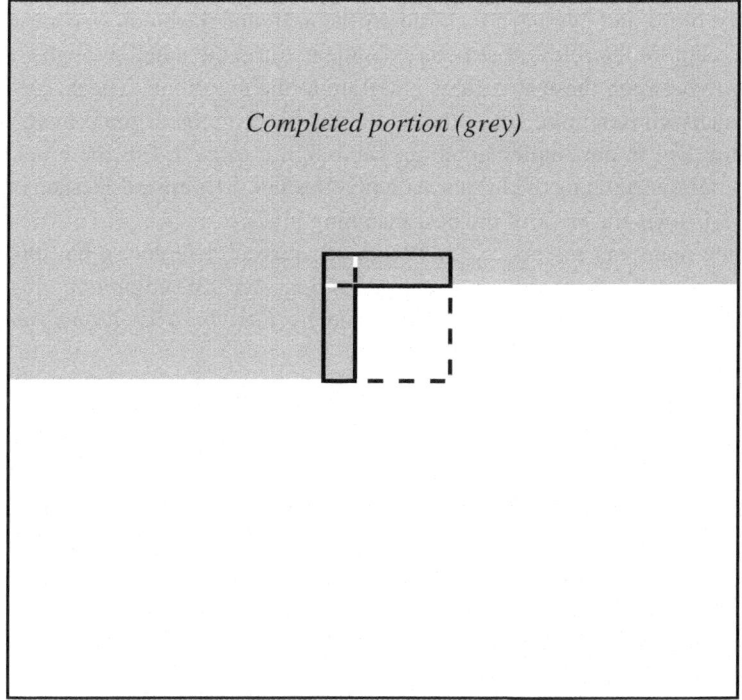

Output image

Fig. 1 The three blocks indicated in the input texture are the candidate blocks. The block in the output image is under processing.

- Compute the error surface between the newly chosen block and the old blocks at the overlap region. Find the minimum cost path along this surface and make that the boundary of the new block. Paste the block onto the texture.
- Repeat the above steps until the whole image is completed.

The minimal cost path through error surface of two overlapping blocks $B1$ and $B2$, is calculated as follows. Taking the vertical overlaps for example, the error surface is defined as $(B^{ov}_1 - B^{ov}_2)^2$ (B^{ov}_1 and B^{ov}_2 indicate the overlap regions). To find the minimal vertical cut *through* this surface we traverse the error function $e(i,j)$ and compute the cumulative minimum error E for all paths:

$$E(i,j) = e(i,j) + \min(E(i-1,j-1), E(i-1),j, E(i-1,j+1)) \tag{1}$$

In the end, the minimum value of the last row in E will indicate the end of the minimal vertical path though the surface and one can trace back and find the path of the best cut. Similar procedure can be applied to horizontal overlaps. Fig. 1 gives an intuitive outline of the IQ algorithm.

The sizes of the block and overlap are chosen by the user, and the block size must be big enough to capture the relevant structures in the texture, but small enough so that the interaction between these structures is left up to the algorithm. The size of the overlap is usually chosen to be 1/6 of the size of the block in the original image quilting algorithm. But in our implementation, we assign it to be 1/4 of the block size. The error is computed using the L2 norm on pixel values. The error tolerance is set to be within 1.1 times the error of the best matching block.

In IQ, the patch matching method is SSD (sum of squared difference), but this method is computationally expensive (on the order of hours) if SSD computation is carried out literally. However, it can be accelerated by *Fast Fourier Transform (FFT)* [6].

2.1 Discussion

The IQ algorithm is a milestone of the development of texture synthesis. With such a simple idea, it works extremely well for a wide range of textures. For regular textures, see Fig. 2(a), 2(b), the patterns of the input textures are monotonously repetitive, and the output images naturally preserve the repeatability of the input images and give satisfying results. For irregular textures, see Fig. 2(c), since the input texture is stochastic and no specific layout to preserve, this algorithm also works well.

However, IQ has limitations in synthesizing textures consisting of an arrangement of small objects. These textures are commonly available in the real world. Fig. 2(d) depicts the result of IQ for a natural image composing two

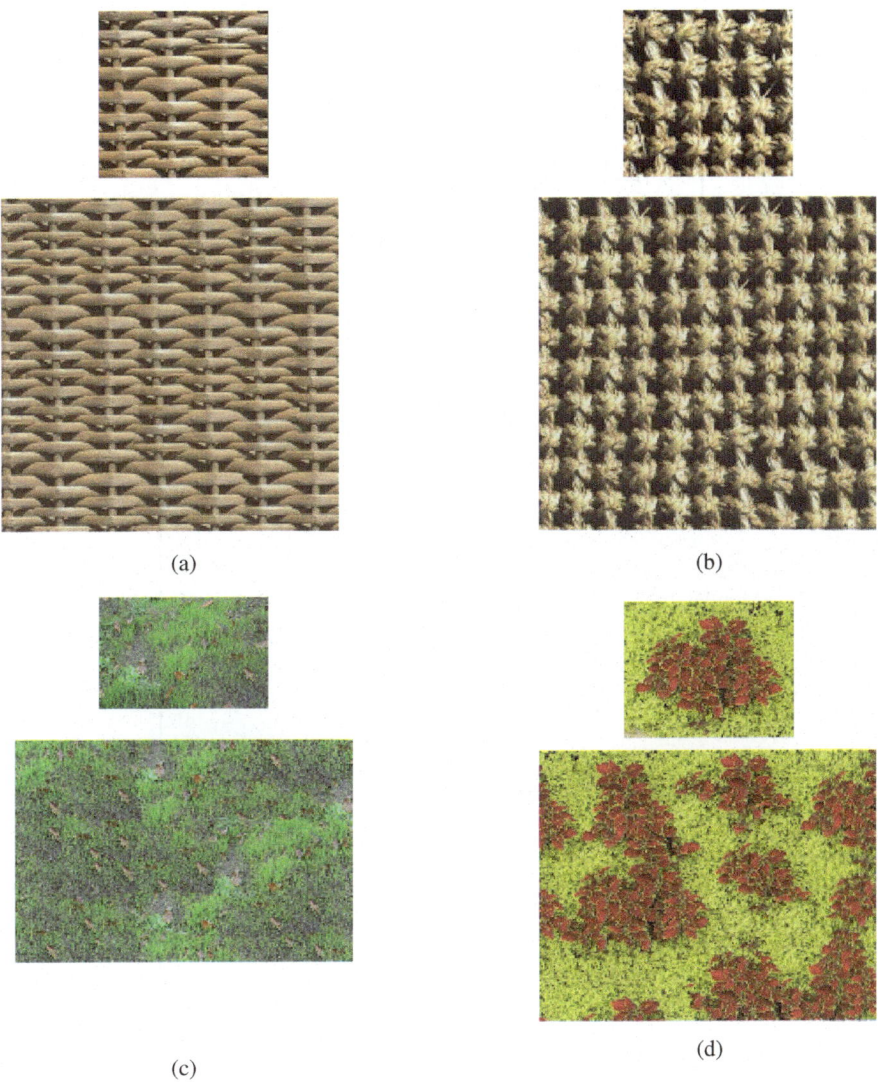

(a)   (b)

(c)   (d)

Fig. 2 Results of the IQ algorithm. Top ones are inputs and bottom ones are outputs.

distinctive textures. Although this output is smooth, yet it is not what we desire. We want the output image not only to have a larger size but also to have the similar layout as the input image. For the image quilting algorithm, it does not take the layout information into account. To solve this problem, we propose a new algorithm in Section 3.

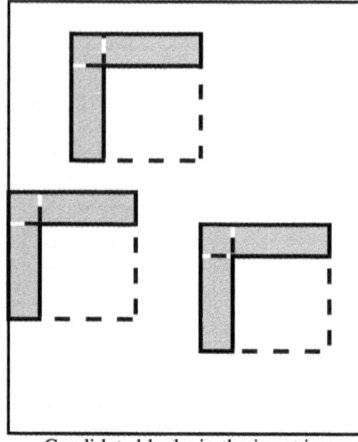

Candidate blocks in the input image

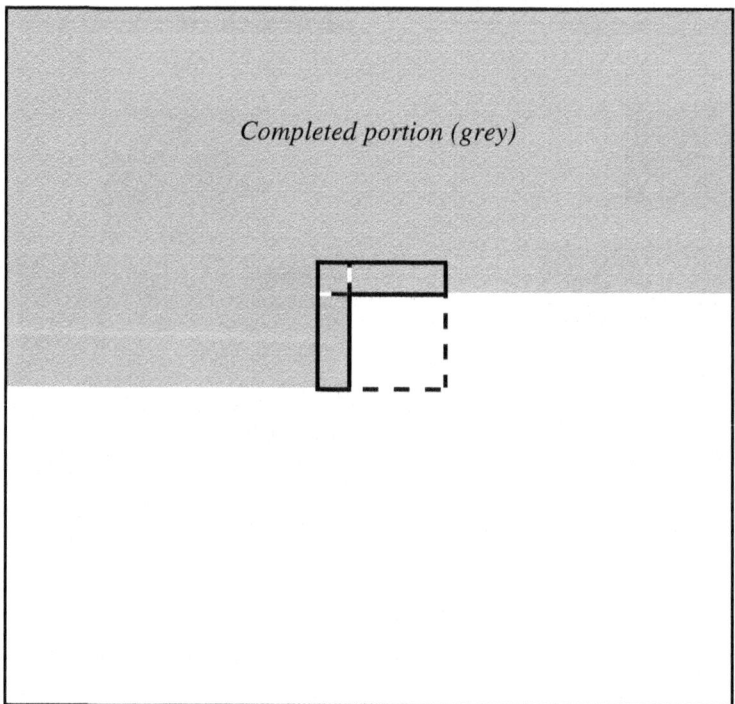

Output image (initialized by the target image)

Fig. 3 Illustration of one patch in the new algorithm. Comparing to image quilting, our method uses the whole block instead of just the overlap region to match all the possible blocks from the input.

3 Texture Synthesis for Natural Images

Based on the image quilting algorithm, our idea is to utilize a target image which has the same size as the desired output image and contains the layout information as the input. In the IQ algorithm, the error surface of searching for the best match patch is calculated in the region of the overlap, which has the size of 40 x 10 pixels in our implementation. For the sake of the layout information, we simply upscale the input image to be the same size as the output image, and use it as the target image for controlling the synthesis process.

In the original IQ algorithm, only the overlap region as indicated in Fig. 1 is used to find good matches in the input texture. Since we want our new algorithm to be constrained by the target image, the match window is a whole block consisting of the overlap region from the completed portion and the remaining region from the target image, as illustrated in Fig. 3. The synthesis process starts with the target image as the initialization of the output. For the first patch in the top-left corner, the match window in the output image contains all the pixels from the target image, because no completed portion is available yet. Similar as IQ, a specified error tolerance is set to find several good matches from which a randomly selected patch is copied into the corresponding location in the output. Minimum error boundary cut is also adopted on the overlap regions to make the transitions between patches smooth. Fig. 4 depicts an example of finding of the minimum error boundary cuts. The process for finding good matches from the input texture is repeated until all the pixels in the output image are updated.

(a) Input image   (b) Target image   (c) Zoom-in synthesis process

Fig. 4 Illustration of updating patches in the proposed algorithm. From left to right shows (a) the input image, (b) the target image and (c) a zoom-in synthesis process of the top-left part of (b), and the white curves illustrate the found minimum error boundary cuts.

Since the target image is the layout constraint of the input, the output should follow its feature information, such as color, structure and luminance etc. In Fig. 5, we compare synthesis results generated by different methods. Clearly, the proposed method captures much better structural layout for composite textures.

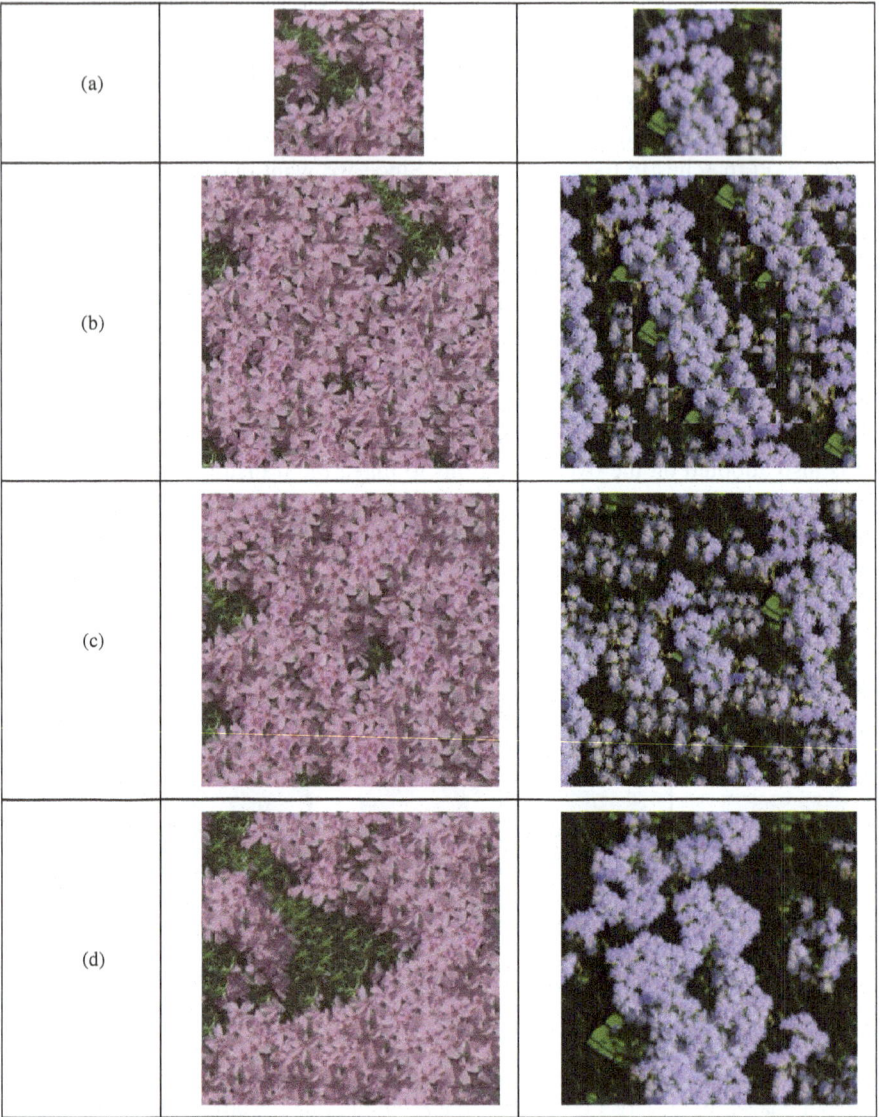

Fig. 5 Each column from top to bottom shows (a) the input texture, (b) result of image quilting without finding the minimum boundary cut, (c) result of image quilting and (d) result of the proposed algorithm, respectively.

4 Texture Synthesis Based on Segmentation

The method described in the previous section can only be applied on images totally composed of textures. For objects, it does not make sense to use texture synthesis techniques for image up-scaling. Therefore, for images composed of both texture and object, we first use Intelligent Scissors [7] to outline the main object, and then obtain a mask image to indicate where the object is. In the second step, we input both the original image and the mask image to do the texture synthesis using modified IQ. After this step we obtain the upscaled background. At last, we enlarge the main object twice indicated by the mask image and stick it to the synthesized background. Refer to Fig. 6 for a schematic diagram of the method.

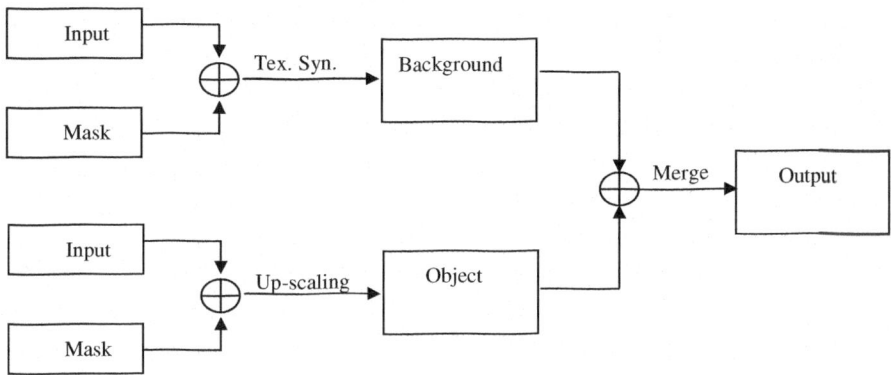

Fig. 6 Diagram of the proposed approach based on object segmentation

In our approach we want the synthesized output image to be twice as large as the input. During the synthesis procedure, we use the mask image to differentiate the area of object and the area of the textured background. Fig. 7 shows an example of the algorithm. Firstly, texture synthesis as described in Section 3 is used to synthesize the background texture (Fig.2 (c)). Then, an image up-scaling technique, such as Bicubic interpolation, is adopted to resize the foreground object. The upscaled object is finally merged with the synthesized background and the joint boundary is smoothed to remove noticeable artifacts.

In the blending step, we use the seamless cloning method proposed in [8] to merge the object with the background smoothly and naturally. Different from the original seamless cloning [8], which replaces certain features of one object by alternative features, we dilate the object area in the mask map and apply the seamless cloning only in this dilated area. Therefore, the sharpness of the main object can be preserved better.

Fig. 8 illustrates the results of the proposed method with and without boundary blending. The output after using the seamless cloning (Fig.3 (b)) has a smoother boundary and looks more natural. More synthesized results using the algorithm

(a).Input   (b).Mask

(c).Background   (d).Output

Fig. 7 Example of the synthesis process

(a).Without boundary blending   (b).With boundary blending

Fig. 8 Comparison of output with and without boundary blending

based on object segmentation are depicted in Fig. 9. We can see that the foreground objects are up-scaled and the background textures are synthesized from the original textures in the input images.

Fig. 9 Other experimental results

5 Conclusion

In this chapter, a texture synthesis algorithm particularly for composite textures is proposed. The algorithm takes a sample texture image as input and outputs a larger image filled with the same texture from the sample. The synthesis process is controlled by a target image which has the same size as the output image and yields a similar layout as the input. Due to the guidance of the target image, the output image looks similar as the input globally and is composed of the same texture. The proposed texture synthesis method can produce much more pleasant-looking results for natural images than state-of-the-art texture synthesis techniques. For images composed of both a non-texture foreground object and a textured background, an additional object segmentation step is applied to differentiate the foreground from background. Controlled texture synthesis is used for the background texture and normal image up-scaling techniques are utilized on the foreground object to preserve the scale of the object.

References

1. Efros, A., Leung, T.: Texture Synthesis by Non-parametric Sampling. In: Proc. IEEE International Conference on Computer Vision, Corfu, Greece (1999)
2. Wei, L., Levoy, M.: Fast Texture Synthesis using Tree-structured Vector Quantization. In: Proc. ACM 27th International Conference on Computer Graphics and Interactive Techniques, New Orleans, Louisiana, USA (2000)
3. Ashikhmin, M.: Synthesizing Natural Textures. In: Proc. ACM Symposium on Interactive 3D Graphics, Research Triangle Park, North Carolina, USA (2001)
4. Efros, A., Freeman, W.: Image Quilting for Texture Synthesis and Transfer. In: Proc. ACM 28th International Conference on Computer Graphics and Interactive Techniques, Los Angeles, CA, USA (2001)
5. Kwatra, V., Schodl, A., Essa, I., Turk, G., Bobick, A.: Graphcut Textures: Image and Video Synthesis Using Graph Cuts. In: Proc. ACM 30th International Conference on Computer Graphics and Interactive Techniques, San Diego, CA, USA (2003)
6. Kilshau, S., Drew, M., Moller, T.: Full search content independent block matching based on the fast fourier transform. In: Proc. IEEE International Conference on Image Processing, Rochester, NY, USA (2002)
7. Mortensen, E., Barrett, W.: Intelligent Scissors for Image Composition. In: Proc. ACM 22nd International Conference on Computer Graphics and Interactive Techniques, Los Angeles, CA, USA (1995)
8. Perez, P., Gangnet, M., Blake, A.: Poisson Image editing. ACM Transactions on Graphics 22(3), 313–318 (2003)

Part IV
3D Visual Content Processing and Displaying

Chapter 13
The Use of Color Information in Stereo Vision Processing

Wided Miled and Béatrice Pesquet-Popescu

Abstract. Binocular stereovision is the process of recovering the depth information of a visual scene, which makes it attractive for many applications like 3-D reconstruction, multiview video coding, safe navigation, 3-D television and free-viewpoint applications. The stereo correspondence problem, which is to identify the corresponding points in two or more images of the same scene, is the most important and difficult issue of stereo vision. In the literature, most of the stereo matching methods have been limited to gray level images. One information that has been largely neglected in computational stereo algorithms, although typically available in the stereo images, is color information. Color provides much more distinguishable information than intensity values and can therefore be used to significantly reduce the ambiguity between potential matches, while increasing the accuracy of the resulting matches. This would largely profit to stereo and multiview video coding, since efficient coding schemes exploit the cross-view redundancies based on a disparity estimation/compensation process. This chapter investigates the role of color information in solving the stereo correspondence problem. We test and compare different color spaces in order to evaluate their efficiency and suitability for stereo matching.

1 Introduction

Research on multiview technologies has been widely enhanced recently, covering the whole media processing chain, from capture to display. A multiview video system consists in generating multiple views by capturing from different viewpoints the same scene via a set of multiple cameras. By presenting the corresponding image

Wided Miled · Béatrice Pesquet-Popescu
TELECOM ParisTech, TSI Department, 46 rue Barrault, 75634 Paris Cédex 13, France
e-mail: miled@telecom-paristech.fr, pesquet@telecom-paristech.fr

of two slightly different views to the left/right eye, the viewer perceives the scene in three dimensions. Such a 3-D scene representation enables functionalities like 3-D television (3DTV) [1] and free viewpoint video (FVV) [2]. While 3DTV offers depth perception of program entertainments without wearing special additional glasses, FVV allows the user to freely change his viewpoint position and viewpoint direction around a 3-D reconstructed scene. Efficient coding of stereo and multi-view video data, as well as their associated depth or disparity maps, is crucial for the success of 3DTV and FVV systems. Powerful algorithms and international coding standards achieve high compression efficiency by exploiting the statistical dependencies from both temporal and inter-view reference pictures via motion and disparity compensated prediction [3]. When processing stereo or multiview video sequences, the most important and difficult step is to find an accurate correspondence between points in images taken from different viewpoints. This procedure, often referred to as the stereo correspondence problem, leads to the computation of the disparity map, which represents the displacement between the considered scene views. Several approaches have been developed for solving the stereo correspondence problem. A good survey of the various strategies is addressed in [4]. Methods that produce dense disparity maps are preferred in computer vision to those which yield sparse displacement results based on the matching of extracted salient features from both images, such as edges, corners or segments. Indeed, methods computing only sparse matches cannot be considered in many applications of stereo, such as view synthesis and 3-D reconstruction. Stereo algorithms that produce dense disparity maps can be classified in local or global optimization methods. Local methods, in which the disparity at each pixel only depends on the intensity values within a local window, perform well in highly textured regions, however they often produce noisy disparities in textureless regions and fail in occluded areas. Global methods formulate stereo matching in terms of an energy functional, which is typically the sum of a data term and a smoothness term, and solve the problem through various minimization techniques. In recent years, global optimization approaches have attracted much attention in the stereo vision community due to their excellent experimental results [4]. Many global stereo algorithms have, therefore, been developed dealing with ambiguities in stereo such as occlusions, depth discontinuities, lack of texture and photometric variations. These methods exploit various constraints on disparity such as smoothness, visibility, view consistency etc., while using efficient and powerful optimization algorithms. However, they are mostly limited to gray level images, although color information is typically available in the stereo images.

Color images provide more useful information than gray value images. Therefore, using color information in the matching process helps reducing the ambiguity between potential matches and improves the accuracy of disparity maps. Recently, a number of approaches dealing with color stereo matching have been proposed in the literature, showing that the matching results have been considerably improved when using the color information instead of gray value information. Furthermore, it has been recognized from prior color evaluation studies conducted in [5, 6, 7] that the accuracy of color stereo matching highly depends on the selection of the appropriate color space. In particular, luminance-chrominance systems, such as YC_rC_b and Luv spaces, have been shown to produce more favorable results than the commonly used RGB color space since they are very close to human perception [7].

The organization of this chapter is as follows. Section 2 provides necessary background on stereo vision. Gray level based stereo methods are also presented. The most used color spaces are addressed in Section 3, along with a survey of color based stereo methods. Finally, some comparisons are drawn in Section 4 before concluding the chapter by Section 5.

2   Stereo Vision Basics

Stereo vision is the process of extracting three dimensional information from two or multiple cameras, which is the same process that works in human visual systems to achieve depth perception. By identifying in two images pixels locations that correspond to the same 3-D position, we compute the disparity map which can be used to recover the 3-D positions of the scene elements via a simple triangulation process, given knowledge about the camera configurations. So, the primary task of stereo vision algorithms is to obtain pixel matches, which is known as the correspondence or stereo matching problem. This section is devoted to presenting stereo vision basic concepts, necessary to understand the stereo correspondence problem.

2.1   The Pinhole Camera Model

The image formation process can be modeled using a pinhole camera system, as illustrated in Fig. 1. This model, which transforms a spatial 3-D point in a 2-D image point under perspective projection, is the most commonly used camera model. All of the rays departing from a scene object pass through a small hole, called *center of projection* or *optical center*, and form the inverted image on the screen. The pinhole

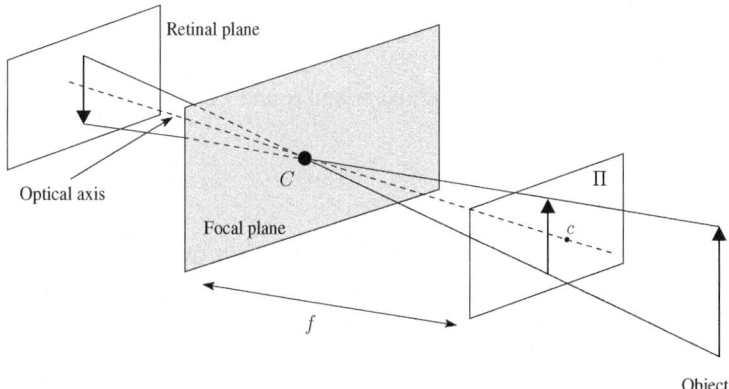

Fig. 1 The pinhole camera model

camera model is defined by its center of projection C, which corresponds to the position of the camera, and the *retinal plane* where points in three-dimensional space are projected onto. The distance from the center of projection to the image plane is called *focal length*. A point M with coordinates (X, Y, Z) in three-dimensional space is projected onto a point m with coordinates (x, y) on the retinal plane by the intersection of this latter with the ray (CM). This projection is defined by the following linear equation

$$\lambda (x\ y\ 1)^T = P\ (X\ Y\ Z\ 1)^T\ , \quad \lambda \in \mathbb{R}^\star \tag{1}$$

where P is the perspective transformation matrix which depends upon the camera parameters.

2.2 The Stereo Camera Setup

We consider now the case of binocular stereovision, where two images, commonly referred as left and right images, are acquired by two cameras observing the same scene from two different positions. Each camera is characterized by its optical center and a perspective transformation matrix. The stereo camera system is assumed to be fully calibrated, i.e. the camera parameters as well as the positions and orientations of the cameras are known. Both cameras capture the scene point M whose projections, m and m', onto the left and right images are given by the intersection of the lines (CM) and $(C'M)$ with the corresponding image planes. Using the perspective transformation (1), we can derive

$$m = P\,M\ , \tag{2}$$

$$m' = P'M\ , \tag{3}$$

where P and P' are the left and right camera projection matrices, respectively. Suppose that the origin is located at the left camera, projection matrices are given by:

$$P = A\,(I_3 \quad 0) \quad \text{and} \quad P' = A'\,(R \quad t)\ , \tag{4}$$

where A and A' are the internal camera parameters and R and t are the rotation and translation operating between both cameras.

2.2.1 Epipolar Geometry

Epipolar geometry describes the geometrical relation between two images of the same scene, taken from two different viewpoints. It is independent of scene structure, and only depends on the internal cameras parameters [8]. Epipolar geometry establishes a geometric constraint between a point in one image and its corresponding in the other image, resulting in a very powerful restriction in correspondence estimation. Let two images be taken by two left and right cameras with optical centers C and C', respectively. The point C projects to the point e' in the right image,

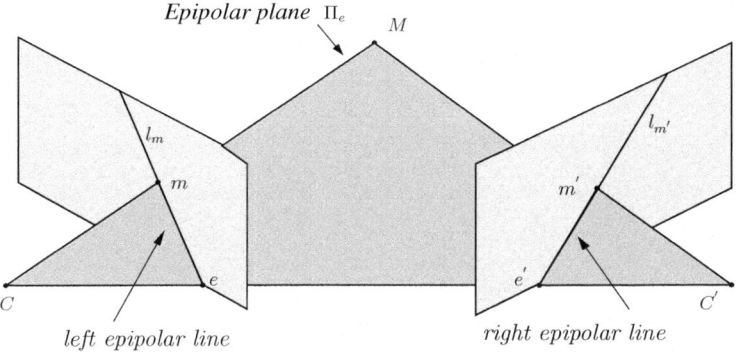

Fig. 2 Epipolar geometry

and the point C' projects to the point e in the left image (see Fig. 2). The two points e and e' are the *epipoles* and the lines through e and e' are the *epipolar lines*. Let a space point M be projected on m and m' respectively in the left and the right image planes. The camera centers, the space point M and its image projections are coplanar and form the plane Π_e, called *epipolar plane*. The projections of this plane into the left and the right image are respectively the epipolar lines l_m and $l_{m'}$. The epipolar constraint states that the optical ray passing through m and M is mapped into the corresponding epipolar line $l_{m'}$ in the right image plane and therefore that m' must lie on $l_{m'}$. Reciprocally, m necessarily lies on the homologous epipolar line l_m which represents the projection of the optical ray of m' onto the left image plane. In terms of a stereo correspondence algorithm, due to this epipolar constraint, the search of corresponding points m and m' does not need to cover the entire image plane but can be reduced to a 1D search along the epipolar lines.

2.2.2 Parallel Cameras Geometry

The parallel camera configuration uses two cameras with parallel optical axes. In this configuration, the epipoles move to infinity and the epipolar lines coincide with horizontal scanlines. The matching point of a pixel in one view can then be found on the same scanline in the other view. In a general camera setup, a technique called rectification [9] is used to adjust images so that they are re-projected onto a plane parallel to the baseline, as in the case of a parallel camera setup.

Consider a point M in the 3-D world with coordinates (X, Y, Z), Z being the distance between the point M and the common cameras plane. Let the coordinates of points m and m', projections of M on the left and right image planes, be (x, y) and (x', y'), respectively. By applying Thales theorem in similar triangles of Fig. 3, we can derive

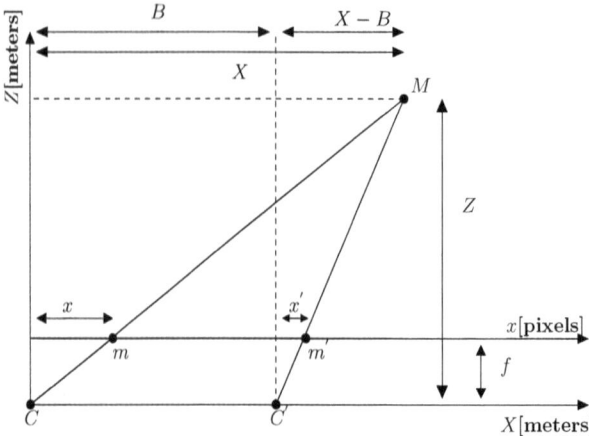

Fig. 3 Reconstruction of the depth coordinate via triangulation

$$\frac{x}{f} = \frac{X}{Z}, \qquad \text{and} \qquad \frac{y}{f} = \frac{Y}{Z}, \tag{5}$$

$$\frac{x'}{f} = \frac{X-B}{Z}, \qquad \text{and} \qquad \frac{y'}{f} = \frac{Y}{Z}, \tag{6}$$

where B is the horizontal distance between the two cameras' optical centers, called the *baseline* distance and f is the focal length of the cameras. It is obvious that y and y' are identical, which constitutes the advantage of the parallel camera setup. The disparity d corresponding to the horizontal displacement between corresponding pixels is defined as

$$d = x - x'.$$

Once the correspondence problem is solved, the reconstruction of a point's depth can then be accomplished via triangulation, as shown in Fig. 3. Indeed, the depth Z is simply derived by combining Eqs. (5) and (6)

$$Z = \frac{B\,f}{x - x'} = \frac{B\,f}{d}. \tag{7}$$

From the above equation, we conclude that disparity is inversely proportional to depth. A disparity map that records the disparity for each image point is therefore sufficient for a complete three-dimensional reconstruction of the scene.

2.3   The Stereo Correspondence Problem

Although epipolar geometry helps reducing the computationnal load of searching corresponding points, the stereo correspondence problem still remains a difficult task due to several factors. This includes

Photometric variations: it is quite reasonable to assume that homologous points in the left and right views have identical intensity values. This assumption, commonly referred to as the brightness constancy assumption, is sometimes violated in practice due to varying shadows, illumination changes and specular reflections. Thus, although commonly used, this hypothesis may lead to incorrect disparity estimates and consequently may reduce the efficiency of depth recovery.

Untextured regions: image areas which contain little or repetitive textures result in ambiguities in the matching process caused by the presence of multiple possible matches of very similar intensity patterns.

Depth discontinuities: the presence of discontinuities causes occlusions, which are points only visible in one image of the stereo pair, making the disparity assignment very difficult at object boundaries.

In order to overcome these ambiguities and make the problem more tractable, a variety of constraints and assumptions are typically made. The most commonly used constraints are related to the following factors:

Smoothness: this assumption imposes a continuous and smooth variation in the uniform areas of the disparity field. It is motivated by the observation that natural scenes consist of objects with smooth surfaces. It holds true almost everywhere except at depth boundaries.

Uniqueness: states that a pixel of one view can have, at most, one corresponding pixel in the other view. This constraint is often used to identify occlusions by enforcing one-to-one correspondences for visible pixels across images.

Ordering: constrains the order of points along epipolar lines to remain the same. The advantage of using this assumption is that its application allows for the explicit detection of occlusions. However, it does not always hold true, especially for scenes containing thin foreground objects.

2.3.1 A Survey of Stereo Vision Methods

There is a considerable amount of literature on the stereo correspondence problem. An extensive review is addressed by Scharstein and Szeliski in [4]. The authors identify four steps that characterize stereo algorithms. These are:

(i) matching cost computation,

(ii) cost aggregation,

(iii) optimization,

(iv) disparity refinement.

In the analysis below, we focus on the optimization component and classify stereo algorithms as local, progressive, cooperative or global optimization methods.

Local approaches

Local (or window-based) approaches are based on the similarity between two sets of pixels. These methods are very popular for their simplicity and have been widely used in computer vision for applications such as image registration, motion estimation, video compression etc. In stereo correspondence, matching pixels consists in comparing the neighborhood of the point for which a correspondence needs to be established with the neighborhood of potential corresponding points located on the associated epipolar line in the other image. Using a predefined search range, the matching score for a pixel (x,y) at each allowed disparity d is derived by comparing the intensity values of the window centered at (x,y) of the first view against the window centered at the position $(x+d,y)$ of the second image. Commonly used matching measures include sum of absolute differences (SAD), sum-of-squared-differences (SSD) and normalized cross-correlation (NCC). This latter measure is insensitive to affine transformations between intensity images, which makes it robust to illumination inconsistencies that may occur between both views.

The choice of an appropriate window size and shape is crucial for window-based local methods. Indeed, the use of windows of fixed size and shape may lead to erroneous matches in the most challenging image regions. In less-textured regions, small windows do not capture enough intensity variations to make reliable matching, whereas large windows tend to blur the depth boundaries and do not capture well small details and thin objects. The different approaches for adaptive/ shiftable windows [10, 11, 12] attempt to solve these problems by varying the size and shape of the window according to the intensity variation. The work in [13] uses a multiple window method where a number of distinct windows are tried and the one providing the highest correlation is retained.

Progressive approaches

Progressive methods first establish correspondences between points that can be matched unambiguously and then iteratively propagate the results of these matched pixels on neighboring pixels [14, 15, 16]. The advantage of these approaches is their computational cost, since they generally avoid a computationally expensive global optimization. However, they propagate errors if an early matching point was not well matched. The method of Wei and Quan [14] attempts to overcome this problem by matching regions derived from color segmentation. Since regions contain richer information than individual pixels, the likelihood of early wrong matches is reduced.

Cooperative approaches

Cooperative approaches make use of both local and global methods. They first calculate a three dimensional space (x,y,d) where each element corresponds to the pixel (x,y) in one reference image and all possible disparities d. A cooperative algorithm is initialized by computing, for all possible matches, a matching score using a local method. These initial matching scores are then refined by an iterative update function based on the uniqueness and smoothness assumptions. Cooperative methods have shown to give strong results in various publications. However, limitations include the higher computational effort compared to local methods and the dependence on a good initialization. Furthermore, depth boundaries tend to be blurred, since rectangular support regions are employed. Zhang and Kambhamettu [17] try to overcome these problems by using image segmentation to estimate an appropriate support window and ensure that the initial matches are correct.

Global approaches

Global approaches formulate stereo matching in terms of a global energy function, which consists of two terms and takes typically the following form:

$$E(d) = E_{\text{data}}(d) + \lambda \, E_{\text{smooth}}(d) . \tag{8}$$

The first term measures the distance between corresponding pixels, while the second one enforces the smoothness of the disparity field and λ is a positive constant weighting the two terms. Several different energy minimization algorithms have been proposed to solve Eq. (8). The most common approach is dynamic programming, which uses the ordering and smoothness constraints to optimize correspondences in each scanline. The matching costs of all points on a scanline describe the disparity search space. Finding the correct disparities is akin to finding the minimum cost path through this space. The most significant limitation of dynamic programming is its inability to enforce smoothness in both horizontal and vertical directions. The work of [15] proposes a way to cope with this problem while maintaining the dynamic programming framework. Recently, powerful algorithms have been developed based on graph cuts [18, 12] and belief propagation [19] for minimizing the full 2-D global energy function. The idea is to cast the stereo matching problem as a pixel labelling problem to find the minimum cut through a certain graph. Variational approaches have also been very effective for minimizing Eq. (8) via an iterative scheme derived from the associated Euler-Lagrange differential equations [20, 21]. However, these techniques often are computationally demanding and they require a careful study for the discretization of the associated partial differential equations. Besides, they require the determination of the Lagrange parameter λ which may be a difficult task. The latter problem becomes even more involved when a sum of regularization terms has to be considered to address multiple constraints, which may arise in the problem. The work in [22] attempts to overcome these difficulties by formulating the stereo matching problem in a set theoretic framework. Each available constraint is then represented by a convex set in the solution space and the intersection of these sets constitutes the set of admissible solutions. An appropriate convex quadratic objective function is finally minimized on the feasibility set using an efficient block-iterative algorithm which offers great flexibility in the incorporation of a wide range of complex constraints.

3 The Use of Color Information in Stereo Vision

Nowadays, most digital cameras enable the acquisition of color images. The use of color images becomes, therefore, more and more common in image processing and computer vision. In this section, an investigation of the efficiency of color information as an aid in solving the stereo correspondence problem is described.

3.1 Color Spaces

Color is an important attribute of visual information. Human color perception relies upon three distinct types of photoreceptor cells in the retina, called cones. Similarly, color images require three numbers per pixel position to represent color accurately. The model chosen to represent color in a 3-D coordinate system is described as a color space. The selection of an appropriate color space is of specific importance in many applications, including stereo matching. The color spaces that are most used can be distinguished into four categories [23]:

- Primary systems: RGB and XYZ;
- luminance-chrominance systems: Luv, Lab, YC_rC_b;
- Perceptual system: HSV;
- Statistical independent component systems: $I_1I_2I_3$ and $H_1H_2H_3$.

3.1.1 Primary Color Systems

Primary systems rely on the use of the three primary colors : Red, Green and Blue. The most common system in this category is the RGB space, as its three coordinates are the reference colors in almost all the image acquisition processes. In this additive color space, a color image is represented with three numbers that indicate the relative proportions of Red, Green and Blue. Any color can be created by combining Red, Green and Blue values in varying proportions. The three components of this primary system are highly correlated and dependent on the luminance information. However, two colors may have the same chrominance, but a different luminance. To consider only the chrominance information, coordinates can be normalized as follows:

$$r = \frac{R}{R+G+B}, \quad g = \frac{G}{R+G+B}, \quad b = \frac{B}{R+G+B}.$$

The advantage of this new color representation is that it is invariant to affine changes in illumination intensity. In 1931, the international lighting commission (CIE) [24], recommended the XYZ color system. While the RGB space is not able to represent all the colors in the visible spectrum, any perceived color can be described mathematically by the amounts of the three color primaries X, Y and Z. These coordinates can be computed by using the following transformation matrix [5]

$$\begin{pmatrix} X \\ Y \\ Z \end{pmatrix} = \begin{pmatrix} 0.607 & 0.174 & 0.201 \\ 0.299 & 0.587 & 0.114 \\ 0.000 & 0.066 & 1.117 \end{pmatrix} \begin{pmatrix} R \\ G \\ B \end{pmatrix}. \tag{9}$$

The Y component is intentionally defined to match closely to luminance, while X and Z components give color information. As this system is a linear combination of the RGB components, it inherits all the dependencies on the imaging conditions from the RGB color system. Furthermore, it is commonly considered as a system of transition from RGB to another color system and rarely used directly.

3.1.2 Luminance-Chrominance Systems

Luminance-chrominance systems divide color into one luminance component and two chrominance components. The main advantage of these color models is that the luminance and the chrominance information are independent. Thus, the luminance component can be processed without affecting the color contents. Among the mostly used luminance-chrominance systems, we mainly distinguish between the perceptually uniform systems Luv and Lab [25], proposed by CIE, and the YC_rC_b system devoted for television and video transmission. Perceptual uniformity of Luv and Lab color systems means that the Euclidean distance between two colors in these spaces models the human perception of color differences. Chrominance components are (u,v) for the Luv space, and (a,b) in Lab. Both color spaces are derived from the XYZ color space, as follows:

$$L = \begin{cases} 116 \left(\frac{Y}{Y_n} \right)^{\frac{1}{3}} - 16, & \text{if } \frac{Y}{Y_n} > 0.01 \\ 903.3 \left(\frac{Y}{Y_n} \right)^{\frac{1}{3}}, & \text{otherwise,} \end{cases} \tag{10}$$

$$u = 13 L \left(u' - u'_n \right), \quad \text{where} \quad u' = \frac{4X}{X + 15Y + 3Z}, \tag{11}$$

$$v = 13 L \left(v' - v'_n \right), \quad \text{where} \quad v' = \frac{9Y}{X + 15Y + 3Z}, \tag{12}$$

$$a = 500 \left(f \left(\frac{X}{X_n} \right) - f \left(\frac{Y}{Y_n} \right) \right), \tag{13}$$

$$b = 200 \left(f \left(\frac{Y}{Y_n} \right) - f \left(\frac{Z}{Z_n} \right) \right), \tag{14}$$

$$\text{where } f(t) = \begin{cases} t^{\frac{1}{3}}, & \text{if } t > 0.008856 \\ 7.787\, t + \frac{16}{116}, & \text{otherwise.} \end{cases}$$

Here $Y_n = 1.0$ is the luminance, and $X_n = 0.312713$ and $Y_n = 0.329016$ are the chrominances of the white point of the system. The values of the L component are in the range $[0; 100]$, u component in the range $[-134; 220]$, and v component in the range $[-140; 122]$. The a and b component values are in the range $[-128; 127]$.

The YC_rC_b color model is the basic color model used in digital video applications [26]. The luminance component Y can be computed as a weighted average of Red, Green and Blue components. The color difference, or chrominance, components (C_r,C_b) are formed by subtracting luminance from Blue and Red. The equations for converting an RGB image to YC_rC_b color space are given by:

$$
\begin{aligned}
Y &= 0.299R + 0.587G + 0.114B, \\
C_r &= 0.564\,(B - Y), \\
C_b &= 0.713(R - Y)\,.
\end{aligned}
\tag{15}
$$

The details information in a digital image are mainly present in the image luminance component. Therefore, one can take advantage of the high sensibility of the human visual system to the luminance variation than to the chrominance variation, to represent the C_r and C_b components with a lower resolution than Y. This reduces the amount of data required to represent the chrominance information without having an obvious effect on visual quality.

3.1.3 The Perceptual HSV System

The HSV (hue, saturation, value) color system, introduced by Smith [27], models the human perceptual properties of hue, saturation, and value. It was developed to be more intuitive in manipulating color and was designed to approximate the way humans perceive and interpret color. Hue defines the basic color and is specified by an angle in degrees between 0 and 360. Saturation is the intensity of the color. Its values run from 0, which means no color saturation, to 1, which is the fullest saturation of a given hue at a given illumination. Value is the brightness of the color. It varies with color saturation and ranges from 0 to 1. The transformation from RGB to HSV is accomplished through the following equations [5]:

$$
H = \begin{cases}
\pi, & \text{if } R = G = B \\
\arccos\dfrac{\frac{1}{2}((R-G)+(R-B))}{\sqrt{(R-G)^2+(R-B)(G-B)}}, & \text{if } B \le G \\
2\pi - \arccos\dfrac{\frac{1}{2}((R-G)+(R-B))}{\sqrt{(R-G)^2+(R-B)(G-B)}}, & \text{otherwise,}
\end{cases}
\tag{16}
$$

$$
S = \begin{cases}
0, & \text{if } R = G = B \\
1 - \dfrac{3\min(R,G,B)}{R+G+B}, & \text{otherwise,}
\end{cases}
\tag{17}
$$

$$
V = \frac{R+G+B}{3}\,.
\tag{18}
$$

3.1.4 The Statistical Independent Component Systems

In primary systems, the three components are highly correlated because they have in common the luminance information. To overcome this limitation, statistical independent systems were proposed, using the principal component analysis that allows for uncorrelated components. The two commonly used systems are:

- The system $I_1 I_2 I_3$ of Ohta *et al.* [28] defined by

$$I_1 = \frac{R+G+B}{3},$$
$$I_2 = \frac{R-B}{2}, \tag{19}$$
$$I_3 = \frac{2G-R-B}{4}.$$

- The system $H_1 H_2 H_3$ whose components are given by

$$H_1 = R+G,$$
$$H_2 = R-G, \tag{20}$$
$$H_3 = B - \frac{R+G}{2}.$$

Both color spaces are a linear transformation of the RGB system.

3.2 A Survey of Color Based Stereo Methods

It is known that, in the human visual system, binocular vision is a key element of the three-dimensional perception: each eye is a sensor that provides to the brain its own image of the scene. Then, the spatial difference between the two retinal images is used to recover the three-dimensional (3-D) aspects of a scene. The experiment conducted in [29] on nine subjects demonstrate that the amount of perceived depth in 3-D stimuli was influenced by color, indicating therefore that color is one of the primitives used by the visual system to achieve binocular matching. This study confirms that color information may be used to solve the stereo matching problem. Different techniques proposed until now to deal with color stereo matching will be presented below. We will first describe color-based local methods and then we will focus our attention on global methods.

3.2.1 Local Approaches

We exposed in Section 2.3.1 the principle of local approaches to solve the correspondence problem based on gray level stereo images. The main idea is to perform a similarity check between two equal sized windows in the left and right images. A similarity measure between the pixel values inside the respective windows is computed for each disparity d within a search range Ω, and the disparity providing the minimum value is regarded as the optimal disparity value. The correlation measure is defined for gray value images as follows:

$$C_g(\mathbf{F}_l, \mathbf{F}_r, d) = \sum_{p \in \mathbf{B}} \varphi\left(\mathbf{F}_l(p), \mathbf{F}_r(p-d)\right), \tag{21}$$

where $\mathbf{F}_l$ and $\mathbf{F}_r$ are the gray levels of the left and right correlation windows, $\mathbf{B}$ is the support of the correlation window and φ is a similarity measure.

A number of different approaches have been proposed in the literature to extend window-based correlation methods taking into account color information available in the stereo images. In [30], Mühlmann et al. proposed a real time and efficient implementation for correlation-based disparity estimation from color stereo images. A combination of a hierarchical block matching technique with active color illumination is presented in [31]. To improve the quality of the matching results, especially in homogenous regions, the authors proposed to project a color code onto the scene. Al Ansari et al. [32] proposed a new region based method for matching color images based on the fact that regions contain much richer information than individual pixels. To guarantee a high similarity between corresponding regions, a color-based cost function that takes into account the local properties of region boundaries is used. Other color stereo techniques have been proposed in [33, 34]. It appears form this first review that there are two ways of incorporating color information into a stereo algorithm:

(1) Compute the correlation separately with each of the color components and then merge the results. In this case, the correlation measure becomes

$$C_c(\mathbf{F}_l, \mathbf{F}_r, d) = \chi\left(C_g(\mathbf{F}_l^{(1)}, \mathbf{F}_r^{(1)}, d), C_g(\mathbf{F}_l^{(2)}, \mathbf{F}_r^{(2)}, d), C_g(\mathbf{F}_l^{(3)}, \mathbf{F}_r^{(3)}, d)\right), \tag{22}$$

where $\mathbf{F}^{(k)}, k \in \{1,2,3\}$, represents the kth color channel in the selected color system, C_c is the color correlation, C_g is the gray level correlation defined in (21) and $\chi \in \{\min, \max, \mathrm{median}, \mathrm{mean}\}$ is the fusion operator. In [33], the fusion was performed by means of a weighted barycenter operator. It is clear that this first approach does not take advantage of all the color information because the color channels are considered separately. In addition, it is computationally inefficient since the matching is done several times.

(2) The second approach is to compute the correlation directly with colors. Here, in order to use color information, an appropriate metric has to be selected to measure color differences on the chosen color space. Some metrics have been proposed by Koschan [5], and generalized to L_p norm in [35], as follows

$$D_p(\mathbf{F}_l, \mathbf{F}_r) = \left(\sum_{k=1}^{3} (\mathbf{F}_l^{(k)} - \mathbf{F}_r^{(k)})^p\right)^{1/p}. \tag{23}$$

Notice that this norm is not suitable for the HSV color space. Assuming that the coordinates of two color images are $\mathbf{F}_l = (H_l, S_l, V_l)$ and $\mathbf{F}_r = (H_r, S_r, V_r)$, the color difference on this space can be defined by [5]:

$$D_{HSV}(\mathbf{F}_l, \mathbf{F}_r) = \sqrt{(D_V)^2 + (D_C)^2}, \tag{24}$$

$$\text{with} \quad D_V = |V_l - V_r|,$$

$$\text{and} \quad D_C = \sqrt{(S_l)^2 + (S_r)^2 - 2S_l S_r \cos(\theta)},$$

$$\text{where} \quad \theta = \begin{cases} |H_l - H_r| & \text{if} \quad |H_l - H_r| \leq \pi \\ 2\pi - |H_l - H_r| & \text{if} \quad |H_l - H_r| > \pi. \end{cases}$$

The correlation measure can now be easily extended to color images by using a color difference measure that is suitable for the selected color space:

$$C_c(\mathbf{F}_l, \mathbf{F}_r, d) = \sum_{p \in \mathbf{B}} D_m\Big(\mathbf{F}_l(p), \mathbf{F}_r(p - d)\Big), \tag{25}$$

where D_m is defined by (23) or (24) if the HSV color space is chosen.

3.2.2 Global Approaches

In recent years, global approaches have attracted much attention in the stereo vision community due to their excellent experimental results. These methods formulate stereo matching in terms of an energy function, which is typically the sum of a data term and a smoothness term, and solve the problem through various minimization techniques. Extension to color of gray based global approaches has involved in most cases the energy function. Alvarez and Sánchez [36] proposed a generalization of their work presented in [20], where they applied a PDE-based method for disparity estimation, by modifying the cost function so that to include all the three color components. The extended set theoretic variational approach proposed in [37] minimizes a global objective function, which is the sum of intensity differences over the three color channels, subject to three convex constraints. Disparity range and total variation regularization constraints proposed for gray value images remain available. However, the Nagel-Enkelmann constraint, which involves the left stereo image, has been extended to color images. The color evaluation study for global approaches, addressed in [7], investigates the role of color in stereo energy functions, believing also that real progress in global stereo matching can be achieved by improving the energy function rather than by investigating on the optimization component. Notice that for color based global approaches, the common idea is to estimate the disparity using jointly all the three color components, which is physically plausible since disparity must be consistent across channels. The data fidelity function that computes the color dissimilarity between a pixel p in one view and its matching point $p - d$ in the second view can be defined by

$$E_{\text{data}}(d) = \sum_{k=1}^{3} \sum_{p \in S} D_m\Big(I_l^{(k)}(p) - I_r^{(k)}(p - d)\Big), \tag{26}$$

where $S \subset \mathbb{N}$ is the image support and D_m is the color difference function that measures the distance between two points in the color space. It is defined by (23) or (24) as in the gray level case.

The second major approach to incorporate color information in global stereo methods is to use a segmentation algorithm that decomposes the image into homogeneous color regions [18, 38, 39]. Disparity smoothness constraint is then enforced inside each color segment, assuming that discontinuities only occur on the boundaries of homogeneous color segments. The use of color segmentation makes global stereo algorithms capable of handling large untextured regions, estimating precise depth boundaries and propagating disparity information to occluded regions. The comparative study conducted in [4] shows that these algorithms, based in general on graph cuts and belief propagation, are among the best performing. However, these methods require precise color segmentation that is very difficult when dealing with highly textured images.

4   Comparisons and Discussion

The previous survey on color based stereo methods has indicated that color information improves the estimation of binocular disparity to recover the three-dimensional scene structure from two-dimensional images. In addition, we have noticed that a large variety of color spaces exists, which raises the question which color system to use to solve the stereo matching problem. In this section, we will give some insights about the suitability of a color space for this application. From an intuitive point of view, a color system should exhibit perceptual uniformity, meaning that distances within the color space should model human perceptual color differences. Moreover, to achieve robust and discriminative image matching, color invariance is another important criterion. Indeed, stereo images are taken from different viewpoints and may be subject to photometric variations. The RGB color system is not perceptual uniform and depends on the imaging conditions and viewing direction. Therefore, RGB is not suitable for matching images taken under illumination variations. The luminance-chrominance systems are very close to human perception and are expected to achieve good performances. The color space $I_1 I_2 I_3$ can also offer suitable color features for stereo matching. The components I_2 and I_3 are invariant to the intensity variations and so systematic errors between the left and right images can be reduced. Moreover, since the color components of this system are statistically independent, color information can be fully used in the matching process.

To argue the above analysis, we made a comparison of two stereo algorithms which are among the most efficient in the literature: the convex variational approach based on the total variation (TV) regularization [37] and the global optimization algorithm (GC) of Kolmogorov and Zabih [40] based on graph-cuts. Four different color models RGB, Luv, Lab, $I_1 I_2 I_3$ are evaluated along with the gray level image representation and three stereo pairs taken from the Middlebury Database are considered (see Fig. 4). These stereo pairs have complex scene structures, wide disparity ranges and large occluded regions. As ground truth fields are available, results are

Fig. 4 Left images (Top) of the considered stereo pairs and corresponding ground truth images (Down). From left to right: Teddy, Baby, Dolls.

Table 1 Comparative results using different color spaces and the gray level representation

Color space	(TV) based method			(GC) based method		
	Teddy $\;$ MAE Err	Dolls $\;$ MAE Err	Baby $\;$ MAE Err	Teddy $\;$ MAE Err	Dolls $\;$ MAE Err	Baby $\;$ MAE Err
RGB	$0.49_3\;12_2$	$0.35_3\;7_1$	$0.51_3\;7_2$	$0.69_3\;20_4$	$0.82_3\;19_2$	$0.58_1\;23_2$
Luv	$0.43_1\;11_1$	$0.27_1\;8_2$	$0.44_1\;4_1$	$0.61_1\;11_1$	$0.75_2\;19_2$	$0.65_3\;21_1$
$I_1 I_2 I_3$	$0.47_2\;12_2$	$0.32_2\;10_3$	$0.49_2\;7_2$	$0.63_2\;14_2$	$0.66_1\;18_1$	$0.60_2\;24_3$
Lab	$0.56_4\;17_5$	$0.45_4\;15_5$	$0.73_4\;14_4$	$0.82_5\;25_5$	$0.89_4\;24_5$	$0.94_4\;26_4$
Gray	$0.57_5\;13_4$	$0.48_5\;11_4$	$0.91_5\;20_5$	$0.79_4\;15_3$	$0.92_5\;20_4$	$1.06_5\;26_4$

evaluated quantitatively using two error measures: the mean absolute error (MAE) between computed and ground truth fields and the percentage of bad matching pixels (Err) with absolute error larger than one pixel. The overall results are shown in Table 1, where the rank of the color spaces according to their MAE and Err errors is also indicated in red. As we can see, the precision of the matching has generally been improved when using the color information. The mean absolute error was significantly reduced when using the Luv, RGB and $I_1 I_2 I_3$ color spaces. However, no significant changes in the results have been noticed when using the Lab color space instead of the gray value information.

In Fig. 5, we show the disparity maps computed by the (TV) and (GC) based methods for the three stereo pairs, using gray values and the RGB and Luv color spaces. The obvious utility of color information in solving the stereo matching problem could be noticed when comparing the results of the gray value based matching and the color based matching. Indeed, many matching errors are reduced by using

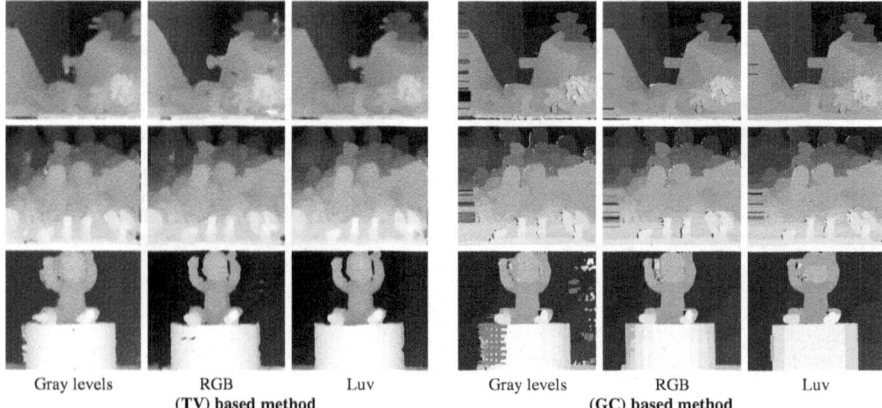

Gray levels   RGB   Luv   Gray levels   RGB   Luv
            (TV) based method   **(GC) based method**

Fig. 5 Disparity maps of the (TV) and (GC) based methods applied on Teddy, Dolls and Baby stereo pairs

the color information. Especially, we notice that the most precise results have been generally obtained by using the luminance-chrominance Luv space, which seems to be a suitable color space for stereo matching.

5 Conclusion

In sum, in this chapter has described a preliminary investigation into the utility of color information in solving the stereo correspondence problem. The results of this investigation strongly indicate that using color information can significantly improve the precision of color stereo matching, especially when a suitable color system is chosen. We found that the luminance-chrominance Luv color space offers the best performances. We also shown that the RGB space, which is the most popular color model in stereo matching, only achieves average results.

References

1. Fehn, C., Cooke, E., Schreer, O., Kauff, P.: 3D Analysis and Image-based Rendering for Immersive TV applications. Signal Processing: Image Communication 17(9), 705–715 (2002)
2. Carranza, J., Theobalt, C., Magnor, M.A., Seidel, H.P.: Free-Viewpoint Video of Human Actors. ACM Transactions on Graphics, 569–577 (2003)
3. Chen, Y., Wang, Y.K., Ugur, K., Hannuksela, M., Lainema, J., Gabbouj, M.: The Emerging MVC Standard for 3D Video Services. EURASIP Journal on Advances in Signal Processing, Article ID 786015 (2009)
4. Scharstein, D., Szeliski, R.: A Taxonomy and evaluation of dense two-frame stereo correspondence algorithms. Int. Journal of Computer vision 47, 7–42 (2002)

5. Koschan, A.: Dense Stereo Correspondence Using Polychromatic Block Matching. In: Chetverikov, D., Kropatsch, W.G. (eds.) CAIP 1993. LNCS, vol. 719, pp. 538–542. Springer, Heidelberg (1993)

6. Chambon, S., Crouzil, A.: Color stereo matching using correlation measures. In: Complex Systems Intelligence and Modern Technological Applications, Cherbourg, France, pp. 520–525 (2004)

7. Bleyer, M., Chambon, S., Poppe, U., Gelautz, M.: Evaluation of different methods for using colour information in global stereo matching approaches. In: The Congress of the International Society for Photogrammetry and Remote Sensing, Beijing, Chine (July 2008)

8. Hartley, R.I., Zisserman, A.: Multiple View Geometry in Computer Vision. Cambridge University Press, Cambridge (2004)

9. Fusiello, A., Roberto, V., Trucco, E.: A compact algorithm for rectification of stereo pairs. Machine Vision and Applications 12(1), 16–22 (2000)

10. Kanade, T., Okutomi, M.: A Stereo Matching Algorithm with an Adaptive Window: Theory and Experiment. IEEE Transactions on Pattern Analysis and Machine Intelligence 16(9), 920–932 (1994)

11. Veksler, O.: Stereo matching by compact windows via minimum ratio cycle. In: Proc. Int. Conf. on Computer Vision, Vancouver, BC, Canada, July 2001, vol. 1, pp. 540–547 (2001)

12. Kang, S.B., Szeliski, R., Chai, J.: Handling occlusions in dense multi-view stereo. In: Proc. Computer Vision and Pattern Recognition, Kauai Marriott, Hawaii, vol. 1, pp. 156–161 (2002)

13. Fusiello, A., Roberto, V., Trucco, E.: Symmetric stereo with multiple windowing. Int. Journal of Pattern Recognition and Artificial Intelligence 14(8), 1053–1066 (2000)

14. Wei, Y., Quan, L.: Region-based Progressive Stereo Matching. In: Proc. Computer Vision and Pattern Recognition, Washington, US, June 2004, vol. 1, pp. 106–113 (2004)

15. Kim, C., Lee, K.M., Choi, B.T., Lee, S.U.: A Dense Stereo Matching Using Two-Pass Dynamic Programming with Generalized Ground Control Points. In: Proc. Computer Vision and Pattern Recognition, San Diego, US, June 2005, vol. 2, pp. 1075–1082 (2005)

16. Bleyer, M., Gelautz, M.: A layered stereo algorithm using image segmentation and global visibility constraints. In: Proc. Int. Conf. on Image Processing, Singapour, October 2004, vol. 5, pp. 2997–3000 (2004)

17. Zhang, Y., Kambhamettu, C.: Stereo Matching with Segmentation-based Cooperation. In: Heyden, A., Sparr, G., Nielsen, M., Johansen, P. (eds.) ECCV 2002. LNCS, vol. 2351, pp. 556–571. Springer, Heidelberg (2002)

18. Bleyer, M., Gelautz, M.: Graph-based surface reconstruction from stereo pairs using image segmentation. In: Videometrics VIII, San José, US, January 2005, vol. SPIE–5665, pp. 288–299 (2005)

19. Tappen, M.F., Freeman, W.T.: Comparison of Graph Cuts with Belief Propagation for Stereo, using Identical MRF Parameters. In: Proc. Int. Conf. Computer Vision, Nice, France, October 2003, vol. 2, pp. 900–907 (2003)

20. Alvarez, L., Deriche, R., Sanchez, J., Weickert, J.: Dense disparity map estimation respecting image discontinuities: A PDE and scale-space based approach. Journal of Visual Communication and Image Representation 13, 3–21 (2002)

21. Slesareva, N., Bruhn, A., Weickert, J.: Optic flow goes stereo: A variational method for estimating discontinuity preserving dense disparity maps. In: Kropatsch, W.G., Sablatnig, R., Hanbury, A. (eds.) DAGM 2005. LNCS, vol. 3663, pp. 33–40. Springer, Heidelberg (2005)

22. Miled, W., Pesquet, J.-C., Parent, M.: A Convex Optimization Approach for Depth Estimation Under Illumination Variation. IEEE Transactions on Image Processing 18(4), 813–830 (2009)
23. Vandenbroucke, N., Macaire, L., Postaire, J.-G.: Color systems coding for color image processing. In: Int. Conf. Color in Graphics and Image Processing, Saint Etienne, France, October 2000, pp. 180–185 (2000)
24. Cie 15.2. Colorimetry, 2nd edn. Technical report, Commission Internationale de l'éclairage, Vienne, Autriche (1986)
25. Sharma, G., Trussel, H.J.: Digital color imaging. IEEE Transactions on Image Processing 6(7), 901–932 (1997)
26. Richardson, I.: H.264 and MPEG-4 Video Compression: Video Coding for Next-generation Multimedia. Wiley, Chichester (2003)
27. Smith, A.R.: Color gamut transform pairs. Computer Graphics 12(3), 12–19 (1978)
28. Ohta, Y.I., Kanade, T., Sakai, T.: Color Information for Region Segmentation. Computer Graphics and Image Processing 3(3), 222–241 (1980)
29. Den Ouden, H.E.M., van Ee, R., de Haan, E.H.F.: Colour helps to solve the binocular matching problem. The Journal of Physiology 567(2), 665–671 (2005)
30. Mühlmann, K., Maier, D., Hesser, J., Männer, R.: Calculating Dense Disparity Maps from Color Stereo Images, an Efficient Implementation. In: IEEE Workshop Stereo and Multi-Baseline Vision, Kauai, US, June 2001, pp. 30–36 (2001)
31. Koschan, A., Rodehorst, V., Spiller, K.: Color stereo vision using hierarchical block matching and active color illumination. In: Proc. Int. Conf. Pattern Recognition, Vienna, Austria, August 1996, vol. 1, pp. 835–839 (1996)
32. El Ansari, M., Masmoudi, L., Bensrhair, A.: A new regions matching for color stereo images. Pattern Recognit. Letters 28(13), 1679–1687 (2007)
33. Belli, T., Cord, M., Philipp-Foliguet, S.: Colour contribution for stereo image matching. In: Proc. Int. Conf. Color in Graphics and Image Processing, Saint-Etienne, France, October 2000, pp. 317–322 (2000)
34. Jordan, J.R., Bovik, A.C.: Computational stereo vision using color. IEEE Control Systems Magazine 8(3), 31–36 (1988)
35. Chambon, S., Crouzil, A.: Colour correlation-based matching. International Journal of Robotics and Automation 20(2), 78–85 (2005)
36. Alvarez, L., Sánchez, J.: 3-D geometry reconstruction using a color image stereo pair and partial differential equations. Cuadernos del Instituto Universitario de Ciencias y Tecnologas Cibernticase, Las Palmas. Technical report (2000)
37. Miled, W., Pesquet-Popescu, B., Pesquet, J.-C.: A Convex Programming Approach for Color Stereo Matching. In: Inter. Workshop on Multimedia Signal Processing, Queensland, Australie (2008)
38. Hong, L., Chen, G.: Segment-based stereo matching using graph cuts. In: Proc. IEEE Conf. Computer Vision and Pattern Recognition, Washington, DC, vol. 1, pp. 74–81 (2004)
39. Wei, Y., Quan, L.: Region-based progressive stereo matching. In: Proc. IEEE Conf. Computer Vision and Pattern Recognition, Washington, DC, vol. 1, pp. 106–113 (2004)
40. Kolmogorov, V., Zabih, R.: Computing visual correspondence with occlusions using graph cuts. In: Int. Conf. Computer Vision, Vancouver, Canada, vol. 2, pp. 508–515 (2001)

Chapter 14
3D Object Classification and Segmentation Methods

Martin Žagar, Mario Kovač, Josip Knezović, Hrvoje Mlinarić,
and Daniel Hofman

Abstract. Future multimedia high-quality systems will be, among all, based on improving 3D visual experience. To raise 3D visual content quality and interactivity it is necessary to enable segmentation and classification of content which involves dividing the scene into meaningful sub-regions with the same attributes. Partitioning the image into grouping objects has various different applications in a wide variety of areas, since distinctive features in raw images may appear unclear to the human eyes. Segmentation can be defined as the identification of meaningful image components. It is a fundamental task in image processing providing the basis for any kind of further high-level image analysis. There are many different ways of segmenting the 3D image, all of which can be considered as a good segmentations, depending on objects of interest on an image, and to a large extent, the user's own subjectivity. Key issues in this chapter include different techniques for segmentation of 3D object based on classification on different regions and shapes.

1 Introduction

In segmentation of an image, it is important to define the image features which will be the basis for segmentation. The term image feature refers to two possible entities: a global property of an image (e.g. the average grey level, an area in a voxel global feature); or a part of the image with some special properties (e.g. a circle, a line, or a textured region in an intensity image, a planar surface in a range image local feature). 3D image segmentation is, however, a difficult process, as it depends on a wide variety of factors such as the complexity of image content, the objects of interest or the number of classes required. While one mode of grouping may successfully segment the image into meaningful sub-regions, the same mode

Martin Žagar · Mario Kovač · Josip Knezović · Hrvoje Mlinarić · Daniel Hofman
Faculty of Electrical Engineering and Computing
Unska 3, 10 000 Zagreb, Croatia
e-mail: {martin.zagar, mario.kovac, josip.knezovic,
hrvoje.mlinaric, daniel.hofman}@fer.hr
http://www.fer.hr

of grouping may not work for another image. For this reason here are proposed different segmentation and classification techniques.

First part of this chapter involves 3D shape estimation algorithms based on edge and corner detection and surface extraction. Second part describes thresholding and a voxel context modelling as a base for classification. Third part describes neural network techniques and paradigms for 3D segmentation and classification. In a results section, segmentation and classification processes are described on practical segmentation of experimental volumetric neurodata. As a conclusion last part of this chapter provides some thoughts about future work and possible improvement of described methods.

2 Shape Estimation

The process of content segmentation and classification of 3D image begins with the detection and location of features in the input images that represent detectable parts of the image and that are meaningful. Detectable means that there must exist some algorithm that could detect a particular feature, otherwise this feature is not usable. Detectable parts of 3D image can be defined by using of 3D edge detection and 3D corners detection algorithms. Feature detection and extraction is an intermediate step and not the goal of the system [1]. Techniques that locate boundary voxels use the image gradient, which has high values at the edges of objects.

Meaningful features in the input images denote, in sense of human visual experience, the features that are associated to interesting scene elements and present some useful information for human perception. Typical examples of meaningful features are sharp intensity variations created by the contours of the objects in the scene, or image regions with uniform grey levels, for instance images of planar surfaces. Sometimes the image features that are looked for are not observably associated to any part or property of the scene, but reflect particular arrangements of image values with desirable properties. Meaningful image component can be identified in segmentation process described in Section 3.

2.1 3D Edge Detection

Edge points or simply edges, are voxels at or around which the intensity values of image (grey levels) undergo a sharp variation. An edge which represents an object boundary on an image is defined by the local voxel intensity gradient and can be computed by gradient components. According to this statement, the main problem in edge detection is locating the edges in a given image corrupted by acquisition noise. Main sharp variations correspond not only to significant contours, but also to image noise results in spurious edges [2]. These edges should be suppressed by the edge detection algorithm.

The main reason for taking interest in edges is generating the contours of object to segment it from the scene. The contours of potentially interesting scene elements such as solid objects, marks on surfaces, shadows, and other image contours which are often the basic elements for calibration, motion analysis and recognition, all generate intensity edges and can be detected from chains of edge points. Edge detection can be considered as a two-step process. Before edge localization it should be taken the noise smoothing process that suppresses as much of the image noise as possible, without destroying the true edges. In the absence of specific information about type of noise, it is usually assumed that the noise is white and Gaussian. After noise smoothing edges can be enhanced and localized. This process is done by designing a filter responding to edges whose output is large at edge voxels and low elsewhere, so that edges can be located as the local maxima in the filter's output and deciding which local maxima in the filer's output are edges and which are just caused by noise. This involves thinning wide edges to one voxel width and establishing the minimum value to declare a local maximum an edge (thresholding).

The Canny edge detector is at the moment the most widely used edge detection algorithm in multimedia systems. Constructing a Canny detector requires the formulation of a mathematical model of edges and noise and synthesizing the best filter once models and performance criteria are defined. Edges of intensity images can be modelled according to their intensity values and profiles. For most practical purposes, a few models are sufficient to cover all interesting edges. Description of regular Canny 2D edge detector can be found in [3].

The edge detection operator returns a value for the first derivative in horizontal and vertical direction. The depth dimension (the third dimension) can be achieved by modifying a regular Canny 2D detector. The edge gradient of a 3D image object G can be determined by computing the magnitude gradient components $G_x = \frac{\partial G}{\partial x}, G_y = \frac{\partial G}{\partial y}$ and $G_z = \frac{\partial G}{\partial z}$ for each voxel v(x,y,z) and can be displayed as an image which intensity levels are proportional to the magnitude of the local intensity changes [4]. The second step is to estimate the edge strength with $e_d(x, y, z) = \sqrt{G_x^2 + G_y^2 + G_z^2}$ as well as the orientation of the edge normal.

As shown in Fig. 1, the orientation of edge normal is specified by angles θ and φ that can be computed by equations

$$\tan \theta = \frac{G_y}{G_x} \qquad (1)$$

and

$$\tan \varphi = \frac{G_z}{\sqrt{(G_x)^2 + (G_y)^2}}. \qquad (2)$$

Weighted summations of the voxel intensities in local neighbourhoods can be listed as a numerical array in a form corresponding to the local image neighbourhood. The output of a gradient based edge detection is a binary image indicating

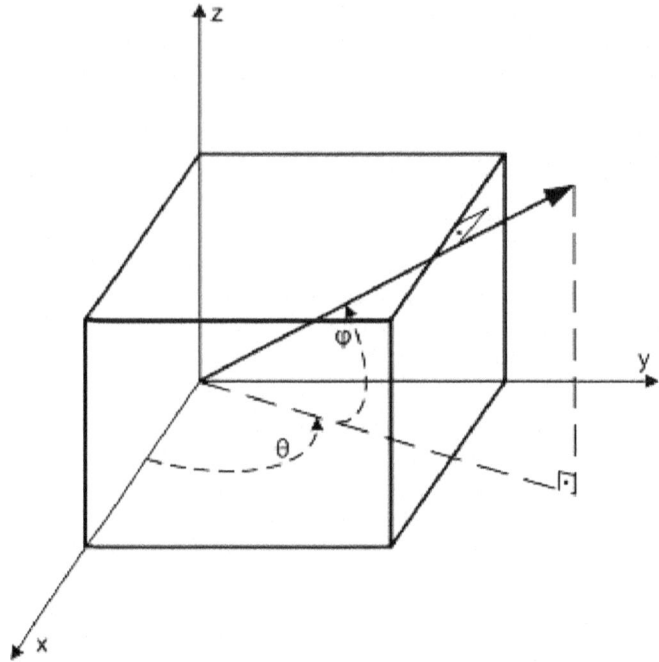

Fig. 1 3D orientation representation

where the edges are in order to decide whether an edge has been found, computing of gradient components is followed by a threshold operation on a gradient. More about thresholding can be found in Section 1.

Although that Canny 3D edge detection is computationally fast and easy to implement because no a-priori information about image features is needed, the object is often not well defined because detected edges do not surround and enclose the object completely. To form boundaries that enclose the object, a postprocessing step of grouping edges and linking them into one single boundary is required. Since the peaks in the first-order derivative correspond to zeros in the second - order derivative, the Laplacian operator (which approximates the second derivative) can be used to detect edges [4]. The Laplacian operator ∇^2 of a function $f(x,y,z)$ is defined as

$$\nabla^2 G(x, y, z) = \frac{\partial^2 G(x, y, z)}{\partial x^2} + \frac{\partial^2 G(x, y, z)}{\partial y^2} + \frac{\partial^2 G(x, y, z)}{\partial z^2} \tag{3}$$

The Laplacian operator can be approximated digital images by an $N \times N \times N$ convolution mask in each direction. The image edges are denoted with voxel positions where Lapalcain operator has zero value.

2.2 3D Corner Detection

Corners are easier to define in mathematical terms than edges, but they do not necessarily correspond to any geometric entity of the observed scene [5]. Corners detection of 3D object can be computed over matrix S that characterizes the structure values and is defined as

$$S = \begin{bmatrix} \sum G_x^2 & \sum G_x G_y & \sum G_x G_z \\ \sum G_x G_y & \sum G_y^2 & \sum G_y G_z \\ \sum G_x G_z & \sum G_y G_z & \sum G_z^2 \end{bmatrix} \tag{4}$$

where the sums are taken over the neighbourhood Q of a generic voxel v. The solution is building the eigenvalues of S and their geometric interpretation as described in [4] for two dimensions. Matrix S is symmetric, and can therefore be diagonalized as:

$$S = \begin{bmatrix} \lambda_1 & 0 & 0 \\ 0 & \lambda_2 & 0 \\ 0 & 0 & \lambda_3 \end{bmatrix} \tag{5}$$

where λ_1, λ_2 and λ_3 are the eigenvalues. If neighbourhood Q of v is perfectly uniform (e.g. image intensity is perfectly same in Q) image gradient components G_x, G_y and G_z vanish everywhere and eigenvalues are $\lambda_1 = \lambda_2 = \lambda_3 = 0$. Now assume that Q contains the corner of a black cube against a white background: as there are three principal directions in Q, it is expected that $\lambda_1 \geq \lambda_2 \geq \lambda_3 \geq 0$ and the larger the eigenvalues, the stronger (higher contrast) their corresponding image lines. Obviously, the eigenvectors describe the edge directions, and the eigenvalues the edge strength. A corner is identified by three strong edges, and as $\lambda_1 \geq \lambda_2 \geq \lambda_3$ it is a location where the smallest eigenvalue, λ_3 is large enough. In general terms, at corner voxels the intensity surface has three well-defined, distinctive directions, associated to eigenvalues of S, all of them significantly larger than zero [3]. If only one of them is not large enough ($\lambda_3 < \tau$, where τ is threshold) that voxel is not the corner voxel.

The procedure for locating the corners is as follows [6]. The input is formed by an image G and two parameters: the threshold τ on λ_3, and the linear size of a cube window (neighbourhood), for example $2N + 1$ voxels, where N is usually between 2 and 5. First the image gradient is computed over the entire image G, and then for each voxel v is formed the matrix S over a $(2N + 1) \times (2N + 1) \times (2N + 1)$ neighbourhood Q of v. In the next step λ_3, the smallest eigenvalue of S, is computed. If $\lambda_3 > \tau$ the coordinates of v are saved into a list of possible corner voxels L. The list L is then sorted in decreasing order of λ_3. The sorted list is scanned top to bottom: for each current point v, all points which belong to the neighbourhood of v and appear further on in the list are deleted. The output is a list of corner voxels for which $\lambda_3 > \tau$ and whose neighbourhoods do not overlap.

2.3 Surface Extraction

After edge and corners detection, it is necessary to determine surface of inbound object. Many 3D objects can be conveniently described in terms of the shape and position of the surfaces they are made of. Surface-based descriptions are used for object classification and motion estimation in the compression process [7]. This section presents a method of finding patches of various shapes which compose the visible surface of an object adapted to 3D.

For a given range image G, the goal is to compute a new image registered with G and with the same size in which each voxel is associated with a local shape class selected from a given kernel shapes. To solve this problem, two tools are needed: a dictionary of kernel shape classes and an algorithm determining which shape class gives the best approximation of the surface at each voxel.

2.3.1 Estimating the Local Shape

To estimate surface shape at each voxel, a local definition of shape is needed. The method called HK segmentation, described in [8], partitions a range image into regions of homogeneous shape, called homogeneous surface patches. The local surface shape can be classified using the sign of the mean curvature H and of the Gaussian curvature K.

In the Table 1, concave and convex are defined with respect to the viewing direction: a hole in the range surface is concave and its principal curvatures negative. At cylindrical points, one of the two principal curvatures vanishes, as, for instance, at any point of a simple cylinder or cone. At elliptic points, both principal curvatures have the same sign, and the surface looks locally like either the inside of a bowl (if concave) or the tip of a nose (if convex). At hyperbolic points, the principal curvatures are nonzero and have different signs; the surface is identified as a saddle.

This classification is qualitative in the sense that only the sign of the curvatures, not their magnitude, influences the result. This offers some robustness, as sign can often be estimated correctly even when magnitude estimates become noisy.

Table 1 Surface patches classification scheme

K	H	Local shape class
0	0	plane
0	+	concave cylindrical
0	-	convex cylindrical
+	+	concave elliptic
+	-	convex elliptic
-	any	hyperbolic

The basic HK segmentation algorithm works as follows: the input is a raw image G and a set of six kernel shape labels $\{s_1,\ldots,s_6\}$, associated to the classes of Table 1. The first and second order gradients of the input image, G_x, G_y, G_z, G_{xy}, G_{xz}, G_{yz}, G_{xx}, G_{yy} and G_{zz} should be computed first. The expressions of H and K are evaluated at each image point, with signs from Table 1. The Gaussian curvature operator K for 3D images can be computed as [6]:

$$K(x,y,z) = \frac{\left(\nabla G_1^\perp\right)^T S_{G1} \nabla G_1^\perp + \left(\nabla G_2^\perp\right)^T S_{G2} \nabla G_2^\perp + \left(\nabla G_3^\perp\right)^T S_{G3} \nabla G_3^\perp}{|\nabla G|^2} \tag{6}$$

using (subscripts x, y, z indicate partial differentiations)

$$G_1^\perp = \begin{bmatrix} -G_y \\ G_x \end{bmatrix},\ G_2^\perp = \begin{bmatrix} -G_z \\ G_x \end{bmatrix},\ G_3^\perp = \begin{bmatrix} -G_z \\ G_y \end{bmatrix} \tag{7}$$

and

$$S_{G1} = \begin{bmatrix} G_{xx} & G_{xy} \\ G_{yx} & G_{yy} \end{bmatrix},\ S_{G2} = \begin{bmatrix} G_{xx} & G_{xz} \\ G_{zx} & G_{zz} \end{bmatrix},\ S_{G3} = \begin{bmatrix} G_{yy} & G_{zy} \\ G_{yz} & G_{zz} \end{bmatrix}. \tag{8}$$

Mean curvature H of a 3D image can be extended from a 2D expression as [6]:

$$H(x,y,z) = \frac{G_x^2(G_{yy}+G_{zz}) + G_y^2(G_{zz}+G_{xx}) + G_z^2(G_{xx}+G_{yy})}{2\left(G_x^2 + G_y^2 + G_z^2\right)^{3/2}} - \\ - \frac{2\left(G_x G_y G_{xy} + G_x G_z G_{xz} + G_y G_z G_{yz}\right)}{2\left(G_x^2 + G_y^2 + G_z^2\right)^{3/2}}. \tag{9}$$

After computing the H and K, the shape image L can be computed by assigning a shape label l_i to each voxel, according to the rules in Table 1. The output is the shape image L. In order to be used by subsequent tasks, the output of segmentation algorithm can be converted into a list of symbolic patch descriptors. In each descriptor, a surface patch is associated with a number of attributes which may include a unique identifier position of the patch centre, patch area, information on normals and curvature contour representations, and pointers to neighbour patches. Closed form surface models (e.g. quadrics) are fitted to the surface patches extracted by the HK segmentation, and only the model's coefficients and type (e.g. cylinder, cone) are stored in the symbolic descriptors.

3   3 D Image Segmentation and Classification

The main goal of the segmentation process is to divide an image into subregions (also called subvolumes) that are homogenous with respect to one or more characteristics or features. There is a wide variety of segmentation techniques and processes depending on the input data. Computer segmentation is desirable to perform, but difficult to achieve, as the complex cognitive abilities can hardly be transferred to computer programs. That is why there is no one standard segmentation technique that can produce satisfactory results for all image applications. The definition of the goal of segmentation varies according to the goal of the study and type of data. Different assumptions about the nature of the analyzed images lead to the use of different techniques. These techniques can be classified into two main categories: region segmentation techniques that look for the regions satisfying a given homogeneity criteria, and edge-based techniques that look for edges between regions with different characteristics [9]. Thresholding is a region-based method in which a threshold is selected and an image is divided into groups of voxels with values less than the threshold and groups of voxels with values greater than or equal to the threshold.

3.1 Thresholding

Since segmentation requires classification of voxels, it is often treated as a pattern-recognition problem and addressed with related techniques. The most intuitive approach for segmentation is global thresholding, when only one threshold based on the image histogram is selected for the entire image. If the threshold depends on local properties of some image regions, it is called local. If local thresholds are selected independently for each voxel (or groups of voxels), thresholding is called dynamic and adaptive.

For images that have biomodal histogram (i.e. grey levels grouped into two dominant sets, object and background), the object can be extracted from the background by a simple operation that compares image values with the threshold value τ. Suppose an image $G(x,y,z)$ with a histogram shown on the Fig. 2. The threshold image $L(x,y,z)$ is defined as

$$L(x, y, z) = \begin{cases} 1, & G(x, y, z) > \tau \\ 0, & G(x, y, z) \leq \tau \end{cases}. \tag{10}$$

The result of thresholding is a binary image, where voxels with threshold value 1 correspond to objects, while voxels with value 0 correspond to the background. There are a number of selection methods for threshold τ based on classification model that minimizes the probability of an error. With the semiautomated version, an expert (operator) selects two voxels – one inside an object and one from the background. By comparing the distribution of voxel intensities in the circular regions around the selected voxels, the threshold is calculated automatically. It corresponds to the least number of misclassified voxels between two distributions. The result of the thresholding operation is displayed as a contour map and superimposed on the original image. If needed, the operator can manually modify any part of the border.

If an image contains just two types of regions, objects with uniform intensity values and a contrasting background, in most cases good segmentation is obtained when the background area and the objects are minimally sensitive to small variations of the selected threshold level. In this case global thresholding can be

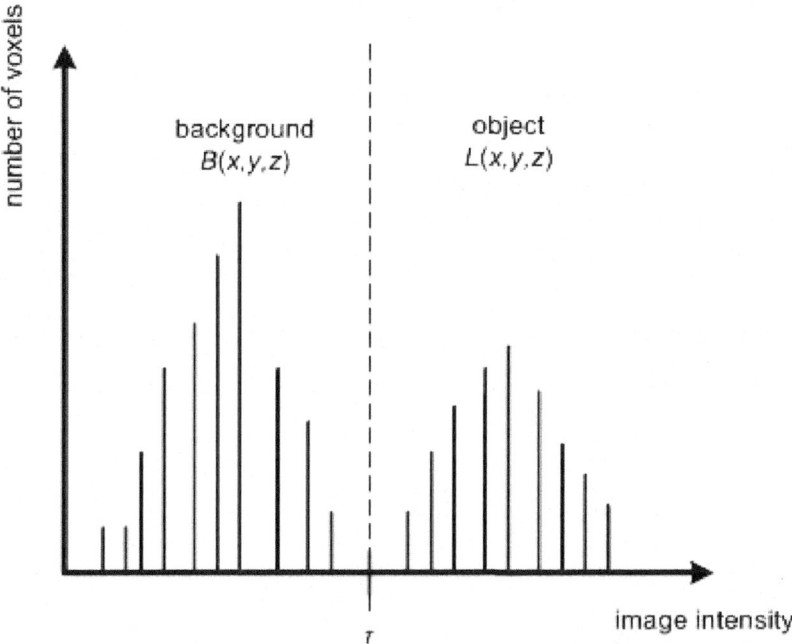

Fig. 2 An example of biomodal histogram with the selected threshold τ

applied. The object area $L(x,y,z)$ and the background $B(x,y,z)$ are functions of the threshold τ. Therefore, the threshold level that minimizes either $dL/d\tau$ or $dB/d\tau$ is often a good choice, especially in the absence of operator guidance and when prior information on object locations is not available.

Global thresholding fails if there is a low contrast between the object and the background, if the image is noisy, or if the background intensity varies significantly across the image. In this case adaptive thresholding can be applied. Adaptive thresholding is locally-oriented and is determined by splitting an image into subvolumes and by calculating thresholds for each subvolume, or by examining the image intensities in the neighbourhood of each voxel. The splitting method divides an image into rectangular overlapping subvolumes the histograms are calculated for each subimage. The subvolumes used should be large enough to include both object and background voxels. If a subvolume has a biomodal histogram, then the minimum between the histogram peeks should determine a local threshold. If a histogram is unimodal, the threshold can be assigned by interpolation from the local thresholds found for nearby images. In the final step, a second interpolation is necessary to find the correct thresholds for each voxel. Although local thresholding is computationally more expensive, it is generally very useful for segmenting objects from a varying background, as well as for extraction of regions that are very small and sparse.

3.2 Context Modelling and Classification of 3D Objects

Once the borders and surface shape of object are determined, next step to analyze is object context (i.e. voxel labels). Suppose a 3D image object $G \equiv N_x \times N_y \times N_z$ voxels. Assume that this image contains K subvolumes and that each voxel v is decomposed into a voxel object o and a context (label) l. By ignoring information regarding the spatial ordering of voxels, we can treat context as random variables and describe them using a multinomial distribution with unknown parameters π_k. Since this parameter reflects the distribution of the total number of voxels in each region, π_k can be interpreted as a prior probability of voxel labels determined by the global context information.

The finite mixtures distribution for any voxel object can be obtained by writing the joint probability density of o and l and then summing the joint density over all possible outcomes of l, i.e., by computing $p(o_v) = \sum_l p(o_v, l)$, resulting in a sum of the following general form [4]:

$$p_\Gamma(o_v) = \sum_{k=1}^{K} \pi_k p_k(o_v), v \in G \tag{11}$$

where o_v is the grey level of voxel v, $p_k(o_v)$s are conditional region probability density functions with the weighting factor π_k, if $\pi_k > 0$, and $\sum_{k=1}^{K} \pi_k = 1$.

Index k denotes one of subvolumes K. The whole object can be closely approximated by an independent and identically distributed random field O. The corresponding joint probability density function is

$$P(o) = \prod_{i=1}^{N} \sum_{k=1}^{K} \pi_k p_k(o_v) \tag{12}$$

where $o = [o_1, o_2, ..., o_G]$, and $o \in O$.

Disadvantage of this method is that it does not use local neighbourhood information in the decision. To improve this, the following should be done. Let Q be the neighbourhood of voxel v with an $N \times N \times N$ template centred at voxel v. An indicator function $I(l_v, l_w)$ is used to represent the local neighbourhood constraints, where l_v and l_w are labels of voxels v and w, respectively with $v, w \in Q$. The pairs of labels are now either compatible or incompatible, and the frequency of neighbours of voxel v, which has the same label values k as at voxel v, can be computed as:

$$\pi_k^{(v)} = p(l_v = k \mid l_Q) = \frac{1}{N^3 - 1} \sum_{w \in Q, w \neq v} I(k, l_w) \tag{13}$$

where l_Q denotes the labels of the neighbours of voxel v. Since $\pi_k^{(v)}$ is a conditional probability of a region, the localized probability density function of grey-level o_v at voxel v is given by:

$$p_\Gamma\left(o_v \mid l_Q\right) = \sum_{k=1}^{K} \pi_k^{(v)} p_k\left(o_v\right), v \in G. \tag{14}$$

Assuming grey values of the image are condition-independent, the joint probability density function of o, given the context labels l, is

$$P\left(o \mid l\right) = \prod_{v=1}^{G} \sum_{k=1}^{K} \pi_k^{(v)} p_k\left(o_v\right) \tag{15}$$

where $l = \left(l_v; v = 1,...,G\right)$.

Instead of mapping the whole data set using a single complex network, it is more practical to design a set of simple class subnets with local mixture clusters, each of which represents a specific region of the knowledge space. It can be assumed that there is more data classes with more class clusters. Since the true cluster membership for each voxel is unknown, cluster labels of the data can be treated as random variables. There is a difference between the mixture model for modelling the voxel image distribution over the entire image where the voxel objects are scalar valued quantities, whereas here a mixture distribution within each class is assumed and the class index in the formulation and modelling of the feature vector distribution is specified. Also, all data points in a class are identically distributed from a mixture distribution.

3.3 Neural Network Based Segmentation

In previous sections there are described different algorithms based on thresholding, region growing, edge detection or voxel classification based on kernels. Neural networks with applications to various stages of image processing can also be addressed to solve the image segmentation problem [10, 11]. This method involves mapping the problem into a neural network by means of an energy function, and allowing the network to converge so as to minimize the energy function. The iterative updating of the neuron states will eventually force the network to converge to a stable and preferably valid state, with the lowest energy. The final state should ideally correspond to the optimum solution.

The neural network model for the 3D object, with size $N_x \times N_y \times N_z$, segmentation problem can be described with $N_x \times N_y \times N_z \times C$ neurons, where C is number of classes, i.e. regions on which the scene is segmented. As previous described methods, this method of segmentation also uses spatial information and is thus image-size dependent. To optimize the algorithm in terms of the computational time and resources when the image size or the number of classes required is large, neural network based segmentation method can employ the grey level intensities distribution instead of spatial information, and thus has the advantage of being image-size independent, using fewer neurons and therefore requiring lesser computational time and resources. Such method uses O neurons representing O grey levels and C classes of subvolumes, so the neural network consisting of $O \times C$ neurons, with each row representing one grey level of the image and each column a particular class. Values of voxel object $o_{v,i}$ are represented by one neuron and defined with

$$o_{v,i} = \begin{cases} 1, & if \ v \in C_i \\ 0, & otherwise \end{cases} \qquad (16)$$

Each row should consist of only one '1', and the column which this '1' falls under will indicate that particular grey level class C_i. The connection strength between two neurons $o_{v,i}$ and $o_{w,j}$ (where neuron $o_{w,j}$ represents some other voxel object that is element of grey level class C_j) is denoted as $W_{vi,wj}$. A neuron in this network would receive input from all other neurons weighted by $W_{vi,wj}$. Mathematically, the total network input to the neuron (v,i) is given as [12]

$$Net_{vi} = \sum_{w}^{O} \sum_{j}^{C} W_{vi,wj} v_{wj} \ . \qquad (17)$$

The neuron with the largest net-input in the same row is declared as the winner, according to winner-takes-all rule [13] , and the updating of all neurons in the same row is as follows:

$$o_{v,i} = \begin{cases} 1, & if \ Net_{v,i} = \max\{Net_v\} \\ 0, & otherwise. \end{cases} \qquad (18)$$

All network states upon convergence will be valid, i.e., no grey levels will be classified into two or more different classes, and each grey level will be assigned a class.

The implementation of the neural network in 3D object segmentation on grey level intensity classes is as follows. At input O grey levels of an image and the number of classes C must be defined. Initializing of the neural network is done by randomly assigning one '1' per row, and setting the rest of neurons in the same row to '0'. Same procedure is done for all other rows, while ensuring that there is at least one '1' per column. The next step is loop done for O times: one row is randomly chosen and $Net_{v,i}$ is calculated to all neurons in the same row and a winner-takes-all rule is applied to update all neuron states within the same row. After performing the whole loop, one epoch is constituted and algorithm is repeating from initialization step until convergence, i.e. until the network state, $V = (o_{v,i})$, for previous epoch is the same as for the current epoch.

4 Results

Experimental input volumetric neurodata are shown in Fig. 3 and aim is to segment it into subvolumes. The segmentation and classification procedure is shown in Fig. 4. Main task is to detect and isolate brain mass and separate it from the background. First, region properties are detected. Region properties are denoted

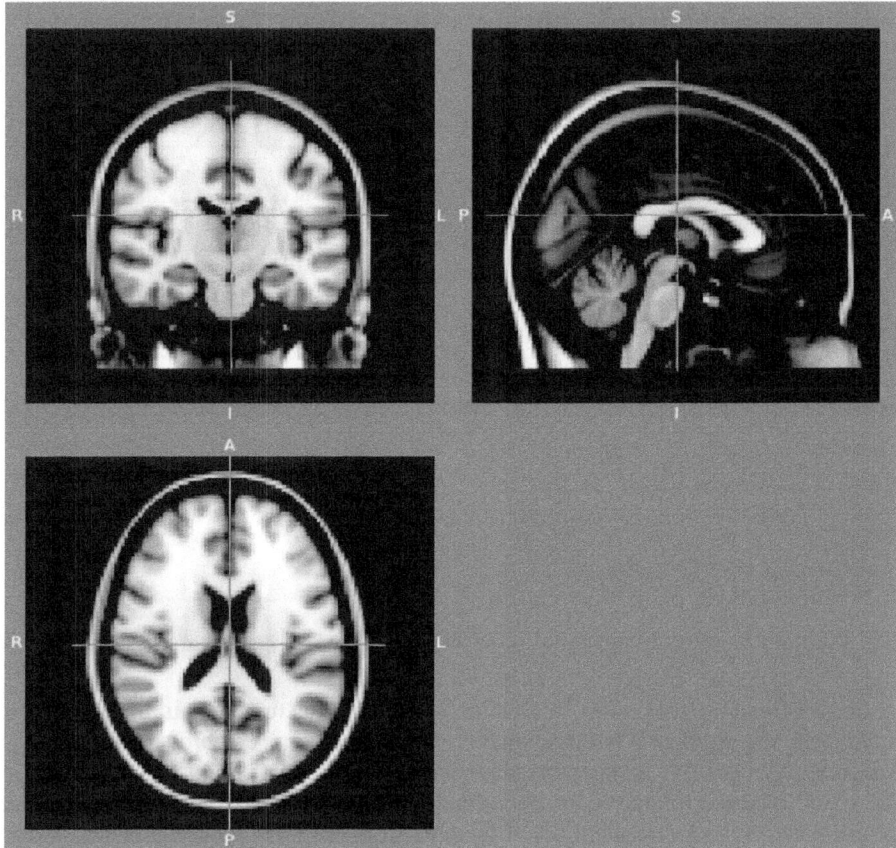

Fig. 3 Experimental input volumetric neurodata

with the anatomical location (voxel location) and accurate intensity values. Neurodata can be segmented into three main tissue types, i.e. initial regions of interest: grey matter, white matter and cerebro-spinal fluid. Regions of interest are set according to intensity values.

To detect the subvolumes it is necessary to use one of the segmentation techniques. This work introduces the usage of automated image segmentation techniques, based on the modified Canny edge and corner detection along with global thresholding techniques, since segmentation requires the classification of voxels and it is done based on kernel shapes. The strategy of edge-based segmentation is to find object boundaries and segment regions enclosed by the boundaries. Separating head from background is accurate enough because there is a great intensity difference between voxels associated to the skull, which represents the boundary, and those from the background. Results of edge and corner detection are shown in Fig. 5. As a final step of the segmentation, head voxels inside the boundary are labelled and prepared for classification step.

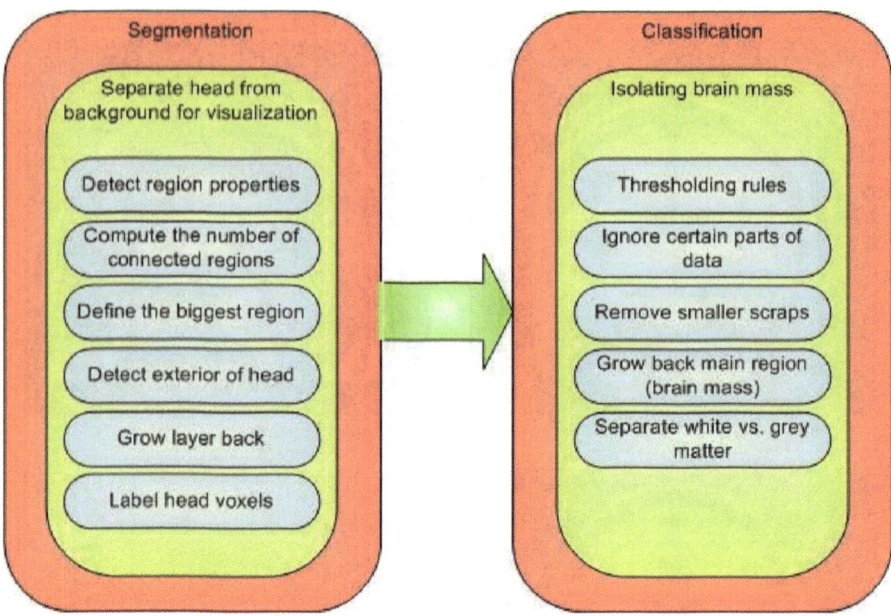

Fig. 4 Segmentation and classification procedure

In classification step, after removing parts of no interest, aim is to isolate the brain mass and segment it into grey matter, white matter and cerebro-spinal fluid. By assigning each voxel to an intensity class and each intensity class to a tissue class, all the voxels of the 3D data set can be attributed to a tissue class and local shapes can be estimated. It first extracts a set of features from the input database and then designs a neural network classifier for the specific detection task using those features. Results of brain segmentation and classification are shown in Fig. 6.

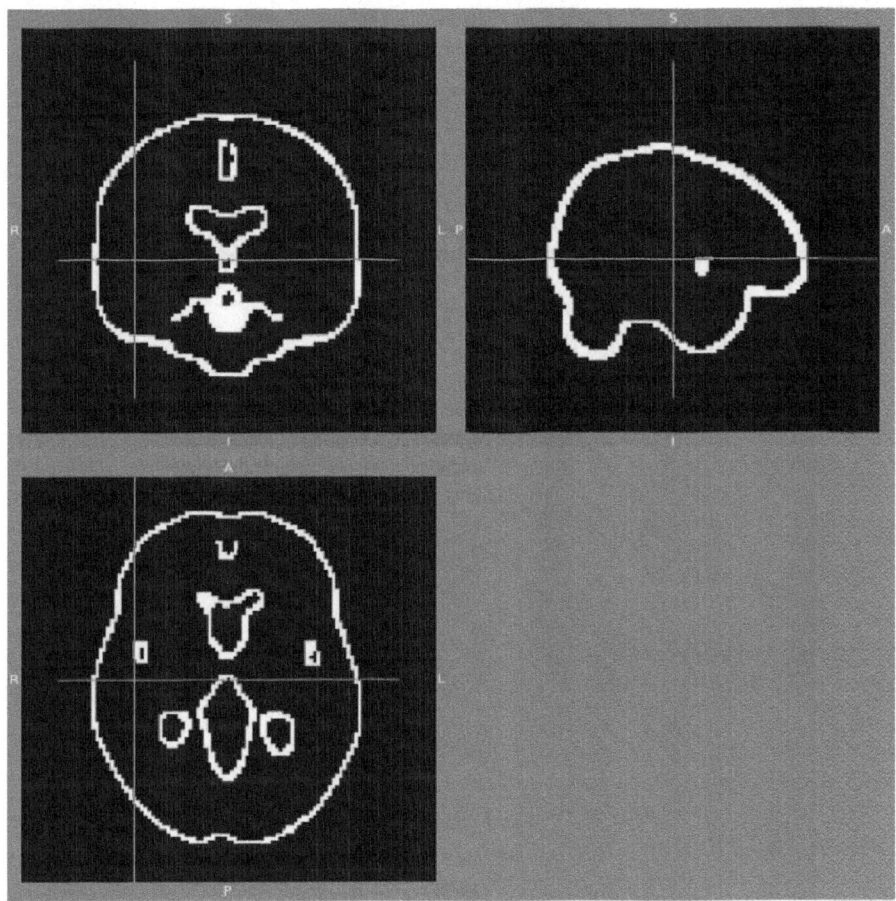

Fig. 5 Edge and corner detection of experimental volumetric neurodata

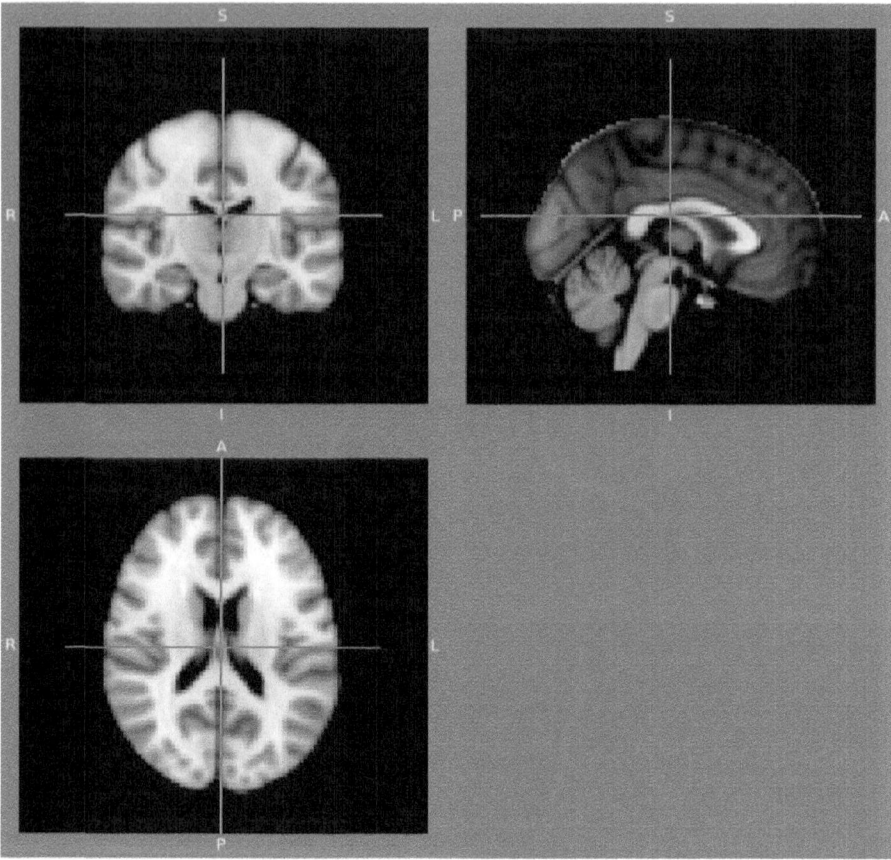

Fig. 6 Results of brain segmentation and classification

5 Conclusion

Segmentation is an essential analysis function for which numerous algorithms have been developed in the field of image processing. Segmentation can also be used as an initial step for visualization and can improve presentation content quality and interactivity. Typically, segmentation of a 3D object is achieved either by identifying all voxels that belong to an object or by locating those that form its boundary. The former is based primarily on the intensity of voxels, but other attributes, such as texture, that can be associated with each voxel, can also be used in segmentation.

Building more sophisticated models of regions, allows more direct and interesting questions to be asked at the voxel level, as well as improving quality of visual experience. Such segmentation and classification processes can obtain detecting and visualizing regions of high interest (in this example: brain) and other regions of small interest (in this example: skull and noisy background). This is very important step in achieving high-quality visual experience in future multimedia content quality and interactivity improvement.

Object modelling and segmentation is also very important in motion estimation and prediction when several matched shapes can appear in one object, so the motion can be tracked for each shape separately. Depending on the number of kernel shapes and the match precision, motion prediction can be improved as well as content quality and interactivity in visual experience of 3D motion objects.

References

1. Trucco, E., Verri, A.: Introductory Techniques for 3-D Computer Vision. Prentice Hall, Upper Saddle River (2003)
2. Tankus, A., Yeshurun, Y.: Detection of regions of interest and camouflage breaking by directconvexity estimation. In: IEEE Workshop on Visual Surveillance, pp. 42–48 (1998)
3. Hwang, T., Clark, J.J.: A Spatio-Temporal Generalization of Canny's Edge Detector. In: Proc. of 10th International Conference on Pattern Recognition, pp. 314–318 (1990)
4. Bankman, I.M.: Handbook of medical imaging: Processing and analysis. Academic Press, San Diego (2000)
5. Žagar, M., Knezović, J., Mlinarić, H.: 4D Data Compression Methods for Modelling Virtual Medical Reality. In: Proc. of the 18th International Conference on Information and Intelligent Systems, Varaždin, Croatia, pp. 155–161 (2007)
6. Žagar, M.: 4D Medical Data Compression Architecture. PhD Thesis, Faculty of Electrical Engineering and Computing, Zagreb, Croatia (2009)
7. Sarris, N., Strintzis, M.G.: 3D Modelling and Animation: Synthesis and Analysis Techniques for the Human Body. IRM Press, Hershey (2005)
8. Corrochano, E.B.: Handbook of Geometric Computing Applications in Pattern Recognition, Computer Vision, Neuralcomputing, and Robotics. Springer, Berlin (2005)
9. Žagar, M., Kovač, M., Bosnić, I.: Lossless and Lossy Compression in 4D Biomodelling. In: Proc. of the First International Conference in Information and Communication Technology & Accessibility: New Trends in ICT & Accessibility, Hammamet, Tunis, pp. 271–277 (2007)
10. Galushkin, A.I.: Neural Networks Theory. Springer, Berlin (2007)
11. Veelenturf, L.P.J.: Analysis and Applications of Artificial Neural Networks. Prentice Hall, Hempstead (1999)
12. Tang, H., Tan, K.C., Yi, Z.: Neural Networks: Computational Models and Applications. Springer, Berlin (2007)
13. Medseker, L.R., Jain, L.C.: Recurrent Neural Networks, Design and Applications. CRC Press, Boca Raton (2001)

Chapter 15
Three-Dimensional Video Contents Exploitation in Depth Camera-Based Hybrid Camera System

Sung-Yeol Kim, Andreas Koschan, Mongi A. Abidi, and Yo-Sung Ho

Abstract. *Video-plus-depth* is an image sequence of synchronized color and depth images. As importance of video-plus-depth increases as an essential part of the next-generation multimedia applications, it is crucial to estimate accurate depth information from a real scene and to find a practical framework to use the immersive video in industry. In this chapter, we introduce a hybrid camera system composed of a stereoscopic camera and a time-of-flight depth camera to generate high-quality and high-resolution video-plus-depth. We also handle a hierarchical decomposition method of depth images to render a dynamic 3D scene represented by video-plus-depth rapidly. Finally, we present a method to generate streamable 3D video contents based on video-plus-depth and computer graphic models in the MPEG-4 multimedia framework. The MPEG-4-based 3D video contents can support a variety of user-friendly interactions, such as free viewpoint changing and free composition with computer graphic images.

1 Introduction

As immersive multimedia services are expected to be available in the near future through a high-speed optical network, a three-dimensional (3D) video is recognized as an essential part of the next-generation multimedia applications. As one of 3D video representations, it is widely accepted that an image sequence of synchronized color and depth images, which is often called as *video-plus-depth* [1], provides the groundwork for the envisaging 3D applications. For a practical use of the immersive video in future interactive 3D applications, such a 3D TV [2], it is very important to estimate accurate depth information from a real scene.

In order to obtain reliable depth data, a variety of depth estimation algorithms have been presented in the fields of computer vision and image processing [3].

Sung-Yeol Kim Andreas Koschan Mongi A. Abidi
The University of Tennessee, Knoxville, Tennessee, USA, 37996
e-mail: {sykim,akoschan,abidi}@utk.edu

Yo-Sung Ho
Gwangju Institute of Science and Technology, Gwangju, 500-712, Republic of Korea
e-mail: hoyo@gist.ac.kr

However, accurate measurement of depth information from a natural scene still remains problematic due to the difficulty of depth estimation on textureless, depth discontinuous and occluded regions.

In general, we can classify depth estimation methods into two categories: active depth sensing and passive depth sensing. Passive depth sensing methods calculate depth information indirectly from 2D images [4]. Contrarily, active depth sensing methods employ physical sensors for depth acquisition, such as laser sensors, infrared (IR) sensors [5], or light pattern sensors [6]. Although current direct depth estimation tools are expensive and produce only low-resolution depth images, they can obtain more accurate depth information in a shorter time than passive depth sensing methods.

We can obtain depth images from a natural scene in real time using an IR-based time-of-flight (TOF) depth camera, such as Z-Cam, developed by 3DV Systems, Ltd. [7] or NHK Axi-vision HDTV camera [8]. The depth camera simultaneously captures color images and the associated depth images by integrating a high-speed pulsed IR light source with a conventional video camera. The ATTEST project has shown us a possibility of realizing a 3D TV system using a depth camera [9]. In addition, 3D contents generated by a depth camera were demonstrated for future broadcasting [10].

In spite of these successful activities using a depth camera, we still suffer from handling depth information captured by current depth cameras due to their inherent problems. The first problem is that *a depth image captured by the depth camera usually includes severe noise.* This noise usually occurs as a result of differences in reflectivity of IR sensors according to color variation in objects. The second problem is that *the measuring distance of the depth camera to get depth information from a real scene is limited.* In practice, the measuring distance is approximately from 1m to 4m. Thus, we cannot obtain depth information from far objects. The last problem is that *the current depth camera can only produce low-resolution depth images.* Most depth cameras usually have a resolution of less than 320×240 depth pixels due to many challenges in real-time distance measuring systems. The maximum image resolution of depth images acquired by Z-Cam is 720×486.

In order to solve these built-in problems, we introduce a system to generate high-quality and high-resolution video-plus-depth by combining a high-resolution stereoscopic camera and a low-resolution depth camera, called a *depth camera-based hybrid camera system* or a *hybrid camera system* shortly [11]. There are three questions to be addressed in this chapter related to the hybrid camera system:

1) How can we obtain high-quality and high-resolution video-plus-depth using a hybrid camera system?
2) How can we render consecutive 3D scenes generated by high-resolution video-plus-depth rapidly using a mesh representation?
3) How can we stream 3D video contents including video-plus-depth data and computer graphic images?

In this chapter, we first introduce a method to obtain high-quality and high-resolution video-plus-depth using a hybrid camera system [12]. The hybrid camera system provides region-of-interest (ROI) enhanced depth images by regarding depth information captured by a depth camera as ROI depth information on the left image captured by a stereoscopic camera. Then, we handle a scheme to render a dynamic 3D scene represented by video-plus-depth using meshes, called hierarchical decomposition of depth images [13]. In the hierarchical decomposition, we create three disjoint layers from a depth image according to the existence of edge information: regular mesh, boundary, and feature point layers. Finally, we present a method to generate streamable 3D video contents based on video-plus-depth in the MPEG-4 multimedia framework. Since traditional multimedia frameworks merely deal with efficient coding issues and synchronization problems between video and audio, we pay attention to the MPEG-4 multimedia framework that supports streaming functionality for various media objects and provides flexible interactivity.

The rest of this chapter is organized as follows. Section 2 introduces a hybrid camera system and a depth estimation method using the hybrid camera system. Section 3 describes a hierarchical decomposition method to render a dynamic 3D scene with video-plus-depth. Section 4 presents MPEG-4-based 3D video contents generation. The performance of generated video-plus-depth and its 3D video contents is shown in Section 5. The chapter is concluded in Section 6.

2 Video-Plus-Depth Generation

2.1 Hybrid Camera System

We introduce a hybrid camera system combining a number of video cameras and a depth camera. The video camera set can be single, stereoscopic, or multiview cameras. In this work, we set up a hybrid camera system composed of a HD stereoscopic camera and a SD depth camera, Z-Cam. Figure 1 shows the hybrid

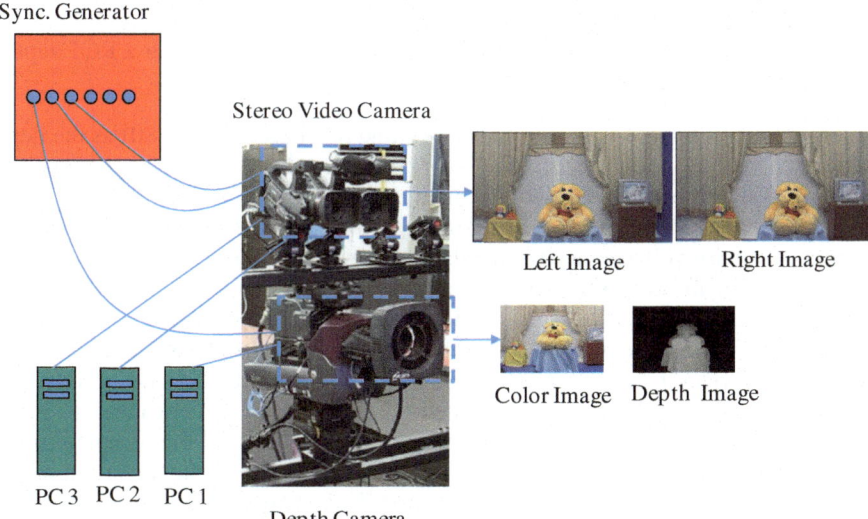

Fig. 1 Depth camera-based hybrid camera system

camera system and its output images. Notably, each camera is connected to a personal computer equipped with a video capturing board, and a clock generator is linked to the cameras to provide synchronization signals constantly. Table 1 shows the specification of the hybrid system.

Table 1 Specification of the hybrid camera system

Device	Specification	Detail
Stereo Camera	Output Format	NTSC or PAL(16:9 ratio, High Definition)
Depth Camera	Depth Range	0.5 to 7.0m (In practice, 1.0 to 4.0m)
	Field of View	40 degrees
	Output Format	NTSC or PAL(4:3 ratio, Standard Definition)
Sync. Generator	Output Signal	SD/HD Video Generation

In the hybrid camera system, we capture four synchronized 2D images in each frame: left and right images from the stereoscopic camera, and color and depth images from the depth camera. In order to clearly explain the methodology, we define image terminologies used in the rest of this chapter as follows.

- Left image: a color image captured by the left camera.
- Right image: a color image captured by the right camera.
- Color image: a color image captured by the depth camera.
- Depth image: a depth image captured by the depth camera.
- ROI depth image: a spatially-extended depth image from a depth image captured by the depth camera.
- ROI enhanced depth image: the final depth image combining a ROI depth image and its background depth image.

Color and depth images naturally have the same resolution (720×480) as the depth camera. On the other hand, the other images have the same resolution (1920×1080) as the HD stereoscopic cameras.

Since we are employing two different types of cameras to construct the hybrid camera system, it is necessary to calculate relative camera information using camera calibration. In order to carry out relative camera calibration, we measure the projection matrices P_s, P_l, and P_r of the depth, left and right cameras induced by their camera intrinsic matrices K_s, K_l, and K_r, rotation matrices R_s, R_l, and R_r, and transition matrices t_s, t_l, and t_r, respectively. Then, the left and right images are rectified by rectification matrices induced by the changed camera intrinsic matrices K_l' and K_r', the changed rotation matrices R_l' and R_r', and the changed transition matrices t_r'. Thereafter, we convert rotation matrix R_s and the transition matrix t_s of the depth camera into the identity matrix I the zero matrix O by multiplying inverse rotation matrix R_s^{-1} and subtracting the transition matrix itself. Hence, we can redefine the new relative projection matrices for the left and right cameras on the basis of the depth camera as Eq. 1.

$$P_s' = K_s[I \mid O]$$
$$P_l' = K_l'[R_l'R_s^{-1} \mid t_l - t_s]$$
$$P_r' = K_r'[R_r'R_s^{-1} \mid t_r' - t_s]$$

(1)

where P_s', P_l', and P_r' indicate the modified projection matrices of the depth, left, and right cameras, respectively.

2.2 Depth Calibration

In practice, depth information obtained from the depth camera has three critical problems. The first problem is that *the captured depth images are very noisy*. The acquired depth images usually contain quantization errors and optical noise, mainly due to the reflectivity or color variation of objects. The second problem is that *depth data on shiny and dark surface regions can be lost or the boundary of color images does not match well with its depth images*. The depth camera does not capture shiny and dark surfaces well, such as black leather and black hair. Especially, for a 3D human actor it often causes the loss of hair region. In addition, when we calibrate the depth camera using an auto calibration tool, it does not guarantee an exact match between the boundaries of both images. The last problem is that *the measured depth information by a depth camera is not equal to the real one*. Even though the distance from the depth camera to the object is constant, depth information obtained by a depth camera depends on the capturing environment. In general, the depth camera system has its own depth calibration tool, but it is very poorly calibrated. Figure 2 shows the problem of raw depth image.

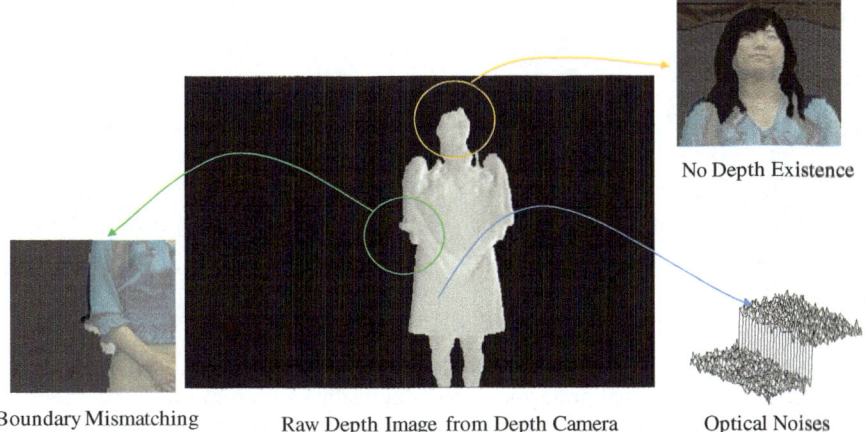

No Depth Existence

Boundary Mismatching   Raw Depth Image from Depth Camera   Optical Noises

Fig. 2 Noisy depth image

In order to reduce optical noise and boundary matching in raw depth images, a
novel method was introduced for depth image enhancement [14]. For noise reduction, a newly-designed joint bilateral filtering method was presented. The bilateral
filter reduces the noise while preserving important sharp edges. By applying the
joint bilateral filter onto the raw depth image, we can reduce the artifact in it.
Formally, for some position p of the synchronized color image I' and depth image
I'', the filtered value J_p is represented by Eq. 2.

$$J_p = \frac{1}{k_p} \sum_{q \in \Omega} G_s\left(\|p-q\|\right) \cdot G_{r1}\left(\|I'_p - I'_q\|\right) \cdot G_{r2}\left(\|I''_p - I''_q\|\right) I_q \tag{2}$$

where G_s, G_{r1}, and G_{r2} are the space weight, color difference in the color image and
depth difference in the depth image at the position p and q, respectively. The term of Ω
is the spatial support of the weight G_s, and the term of K_p is a normalizing factor.

For lost depth region recovery, especially, to recover the lost hair region in human
modeling, a novel modeling algorithm was developed using a series of methods including detection of the hair region, recovery of the boundary, and estimation of the
hair shape [15]. In addition, in order to fix the boundary mismatches between color
and depth information, we compensate the boundary of a human actor using a digital
image matting technique considering color and depth information at the same time.
Figure 3 shows one of the results of the depth image enhancement.

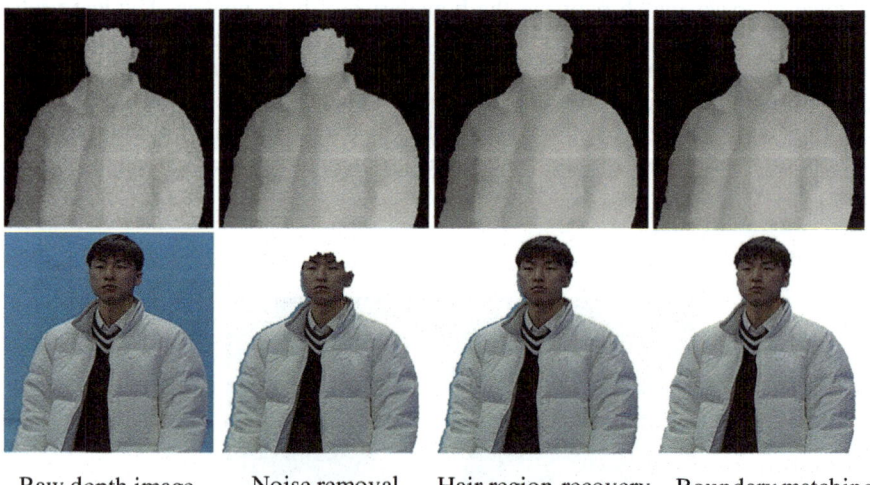

    Raw depth image   Noise removal   Hair region recovery   Boundary matching

Fig. 3 Depth image enhancement

Finally, in order to calibrate measured depth data into real ones, we check the
depth of the planar image pattern within the limited space by increasing the distance (10cm) from the image pattern to the depth camera [16]. Since we already
know the camera parameters of each camera, the real depth values are calculated by

$$D_n(p_x,p_y)=\frac{K\cdot B}{d_n(p_x,p_y)} \tag{3}$$

where K is the focal length of the video camera, B is the baseline distance between two video cameras. $D_n(p_x, p_y)$ is the real depth value corresponding to the measured disparity value $d_n(p_x, p_y)$ at pixel position (p_x, p_y) in the image pattern depth image.

Thereafter, we generate a mapping curve between real depths and measured depths from the depth camera and find the fitting curve using the cubic equation $y=a+bx+cx^2+dx^3$. The cross small rectangular points on the x–y plane are formed by the measured depths x and real depths y that minimizes the sum of squared distances to these points.

2.3 Depth Image Generation

Since the measuring distance of the depth camera is approximately from 1m to 4m, we cannot obtain depth information from far objects. In the video-plus-depth generation, we regard the near objects captured by the depth camera as an ROI or a foreground. First, we move the depth data captured by the depth camera to the world coordinate, and then reproject the warped depth data onto the image plane of the left camera.

When $D_s(p_{sx}, p_{sy})$ is the depth information at the pixel position (p_{sx}, p_{sy}) in the depth image, we can regard the pixel p_s $(p_{sx}, p_{sy}, D_s(p_{sx}, p_{sy}))$ as a 3D point. The corresponding point p_l of the left image is calculated by

$$p_l = P_l'\cdot P_s^{-1}\cdot p_s \tag{4}$$

where P_l' and P_s^{-1} are the relative projection matrix of the left camera and the inverse relative projection matrix of the depth camera, respectively.

Figure 4 shows an initial ROI depth image that is overlapped onto the left image. When we compare the original depth image with it, we can notice that the body region is extended to fit with the spatially high-resolution left image. We can also notice that holes occur in the initial ROI depth image due to the warping operation.

3D image warping

Original depth image

Initial ROI depth image

Fig. 4 Generation of initial ROI depth image

ROI of the left image and the initial ROI depth image do not match correctly on the region of ROI boundaries. The main reason of the mismatch is the slight incorrectness of the camera calibration. We solve the mismatch problem using image segmentation for the left image. In order to correctly detect ROI of the left image, we overlap the color-segmented left image onto the initial ROI depth image. Then, we measure the color segment set for ROI from color segments of the left image by

$$R(s_i) = \begin{cases} 1, & if\ \dfrac{n(A(s_i))}{n(s_i)} \geq 0.5 \\ 0, & otherwise \end{cases} \tag{5}$$

where $R(s_i)$ indicates whether the ith color segment s_i of the color segmented left image is included in ROI of the left image or not. When $R(s_i)$ is 1, the corresponding color segment is included in the color segment set for ROI. The term of $n(s_i)$ is the total count of pixels in s_i, and $n(A(s_i))$ is the total count of pixels on the region of initial ROI depth image $A(s_i)$ that is matched with the region of s_i. Figure 5(a) and Figure 5(b) show the left image and its color segment set for ROI, respectively.

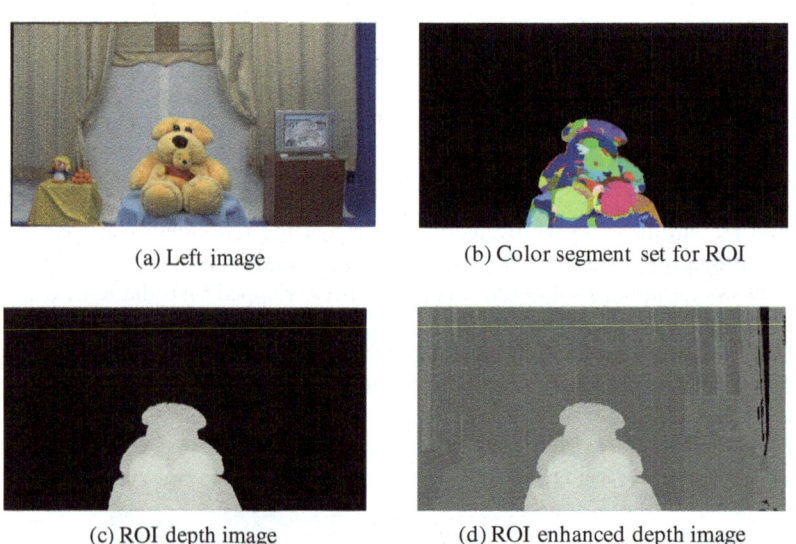

(a) Left image	(b) Color segment set for ROI
(c) ROI depth image	(d) ROI enhanced depth image

Fig. 5 ROI enhanced depth image generation

After ROI detection, we refine the initial ROI depth image from the color segment set by eliminating outside pixels on the former with comparison to the letter. Then, we fill holes in the ROI depth image with the pixels generated by linearly interpolating with their neighboring pixels [17]. The hole-filling algorithm is performed by the unit of a color segment applying

$$R(x, y)_k = \frac{1}{n} \cdot \sum_{i=-\lfloor W/2 \rfloor}^{\lfloor W/2 \rfloor} \sum_{j=-\lfloor W/2 \rfloor}^{\lfloor W/2 \rfloor} R(x+i, y+j)_k \qquad (6)$$

where $R(x, y)_k$ is the interpolated pixel value at the hole position (x, y) of the kth color segment in the initial ROI depth image R using the valid neighboring pixel value $R(x+i, y+j)_k$ in the kth color segment. The term n is the valid number of pixels within a $W \times W$ window. Since the hole-filling algorithm is performed in each color segment, the valid depth pixels in its neighboring segments will not affect the holes in the color segment. Figure 5(c) shows an ROI depth image.

Finally, we obtain the background depth image applying a stereo matching algorithm, such a belief propagation method [18]. Then, we combine the background depth image with the ROI depth image to generate an ROI enhanced depth image. Figure 5(d) shows the ROI enhanced depth image. The pair consisting of the left image and its ROI enhanced depth image becomes a frame of video-plus-depth.

3 Hierarchical Decomposition of Depth Images

For rendering a 3D scene with video-plus-depth data, we employ depth image-based rendering using meshes [19]. In this chapter, we introduce the hierarchical decomposition of depth images to represent a dynamic 3D scene represented by video-plus-depth. In the hierarchical decomposition, we decompose a depth image into three layers: regular mesh, boundary, and feature point layers. The main benefit of hierarchical decomposition is to maintain geometric regularity by using 3D shape patterns induced by these three layers so that we can reconstruct a 3D surface rapidly.

First, we extract edge information by applying the Sobel filter to a depth image vertically and horizontally. The reason using a depth image instead of its color image for edge extraction is that it is not disturbed by lights or surroundings. Thereafter, we divide the region of the depth image uniformly into pixel blocks or grid cells. According to the existence of edge information in a grid cell, we divide the depth image [4] into regions of edges and regions without edges, as shown in Fig. 6. The region of edges is the set of grid cells that includes edge information, referred to as edge-grid cells; similarly, the region without edges is the set of grid cells excluding edge information, referred to as no-edge-grid cells.

We define the size of a grid cell as $2^m \times 2^n$ resolution, such as 16×16, 8×8, or 16×8. Once we choose the size of a grid cell, we should maintain it for each depth image during the hierarchical decomposition. In addition, we should be careful to select the size of a grid cell, because it is inversely proportional to the amount of distortion of generated 3D scenes. We usually set the size of a grid cell as 4×4 or 8×8.

A regular mesh layer is obtained by downsampling the depth image. When the size of a grid cell is $p \times q$, the regular mesh layer is generated by downsampling its depth image with the horizontal sampling rate p and the vertical sampling rate q. In other words, we gather the four depth pixels at the corner of each grid cell to

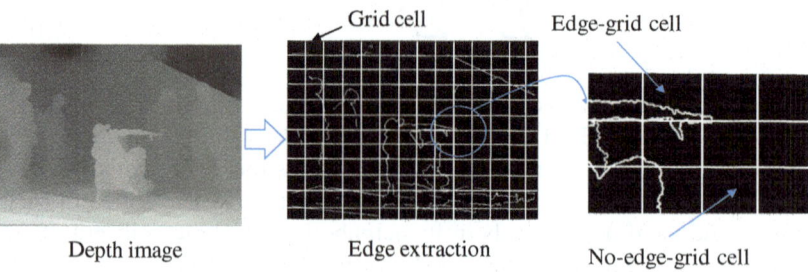

Fig. 6 Edge-grid cell and no-edge grid cell

make regular mesh layers in each frame. Figure 7(a) shows the rendering result of the wire frame mode and the rendering result for a 3D surface. However, it is not enough for the surface of a 3D scene, because there are serious distortions in the region of edges as shown in the rendering result.

A boundary layer includes depth pixels in the edge-grid cell. In order to regularly extract depth pixels, we employed four quad-tree modes and a full modeling mode. After uniformly dividing each edge-grid cell into four sub-grid cells, we use the full modeling mode when more than two sub-grid cells include edges. Otherwise, one of the quad-tree modes is selected according to the location of the sub-grid cell that includes edges. Table 2 shows the full modeling mode and four quad-tree modes in the boundary layer: up-left, up-right, down-left, and down-right quad-tree modes. Here, we extracted 10 depth pixels in the quad-tree mode and 21 depth pixels in the full modeling mode. It should be noted that we can handle serious distortions, holes, close to the region of edges due to the difference of depth values in the boundary layer. For preventing holes, additional processing is required to fill out them, as shown in Table 2.

A boundary layer is used to refine a 3D surface generated by a regular mesh layer for the region of edges. Since most of the serious distortions are mainly occurred in their areas, we should deal with the region of edges carefully. Figure 7(b) shows the rendering result of the wire frame mode for a 3D surface with both layers and its rendering result for a 3D surface.

A feature point layer includes depth pixels in the no-edge-grid cell. While we deal with the region of edges to generate a boundary layer in each frame, feature point layers is to handle the region of no edges. Feature point layers are used to enhance the visual quality of the region of no edges in the 3D scene. In order to determine the influential depth pixels in the no-edge-grid cells, scores of all pixels in the no-edge-grid cell are estimated using a maximum distance algorithm. The most influential depth pixels are then gathered into the 1st feature point layer. Likewise, the second influential depth pixels are also gathered into the 2nd feature point layer; this process is repeated for all subsequent points. Figure 7(c) shows the wire frame for the final surface generated by regular mesh, boundary, and feature point layers. We can notice the visual quality of the 3D surfaces is enhanced by adding the depth information in feature point layers.

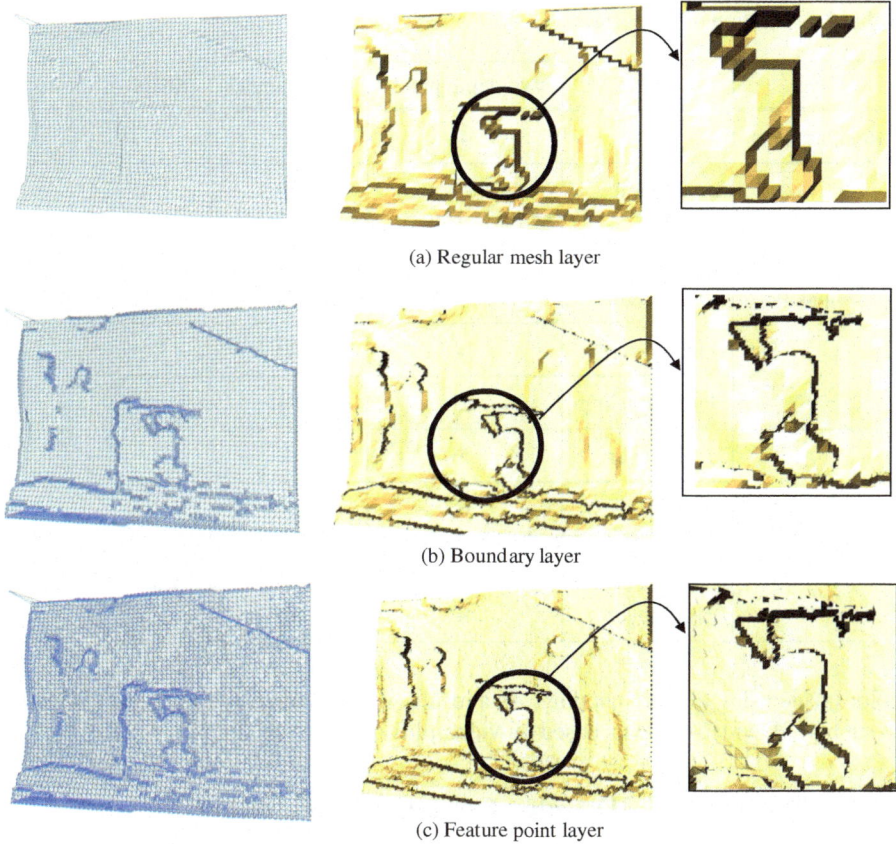

(a) Regular mesh layer

(b) Boundary layer

(c) Feature point layer

Fig. 7 3D scene rendering using hierarchical decomposition

Table 2 shows the number of feature points, the number of triangles, and the shape of the reconstructed 3D surface for grid cells according to layers. In the table, we consider only four feature point layers and the shape of surface generated from them is dependent on the location of their depth pixels. When a grid cell has a regular mesh layer only, we create the 3D surfaces with only 2 triangles. For the other no-edge-grid cells, of which is represented by the regular mesh and feature points layers, we generate the 3D surface using from 4 to 10 triangles with from 5 to 8 depth pixels extracted from the regular mesh and feature point layers. For edge-grid cells represented by quad-tree modes in the boundary layer, we generate the 3D surface using 20 triangles with 14 depth pixels extracted from the regular mesh and boundary layers. For edge-grid cells represented by the full modeling mode, we generate the 3D surface using 44 triangles with 25 depth pixels from a regular mesh and boundary layers.

Likewise, we can generate a dynamic 3D scene rapidly by assigning regularly-predefined 3D shape patterns according to layers into the grid cells in each frame. The generated 3D surface by hierarchical decomposition is covered by the corresponding

Table 2 Generation of 3D surface

Layer	Mode	# of depth pixels	# of triangles	Shape of surface
Regular mesh	-	4	2	
Boundary	Up-left	14	20	
	Down-left	14	20	
	Up-right	14	20	
	Down-right	14	20	
	Full modeling	25	44	
Feature Points	1st layer	5	4	
	2nd layer	6	6	
	3rd layer	7	8	
	4th layer	8	10	

color image using texture mapping. As a result, we can realize a fast rendering system to support 3D video contents based on video-plus-depth in real time.

4 MPEG-4-Based 3D Video Contents Exploitation

In order to deliver 3D video contents, a multimedia framework is needed. Traditional multimedia frameworks, such as MPEG-1 and MPEG-2, merely deal with efficient coding issues and synchronization problems between video and audio. In addition, they do not provide any framework to support interactive functionalities to users. Hence, we direct attention to the MPEG-4 multimedia framework [20] that supports streaming functionality for a variety of media objects and provides flexible interactivity.

In this chapter, we design a new node for a depth image sequence in the MPEG-4 system to provide a practical solution to stream 3D video contents while supporting a variety of user-friendly interactions. Figure 8 illustrates the overall system architecture to generate the 3D video contents based on video-plus-depth in the MPEG-4 multimedia framework.

At the sender side, we generate high-resolution video-plus-depth using the hybrid camera system as introduced in Section 2. Video-plus-depth is then spatio-temporally combined with other multimedia, such as audio and computer graphics models, using the MPEG-4 Binary Format for Scene (BIFS). The MPEG-4 BIFS is a scene descriptor that contains the spatio-temporal relationship between each multimedia object and some interactivity information. The video-plus-depth data are compressed by two video coders; one for color image sequence and the other

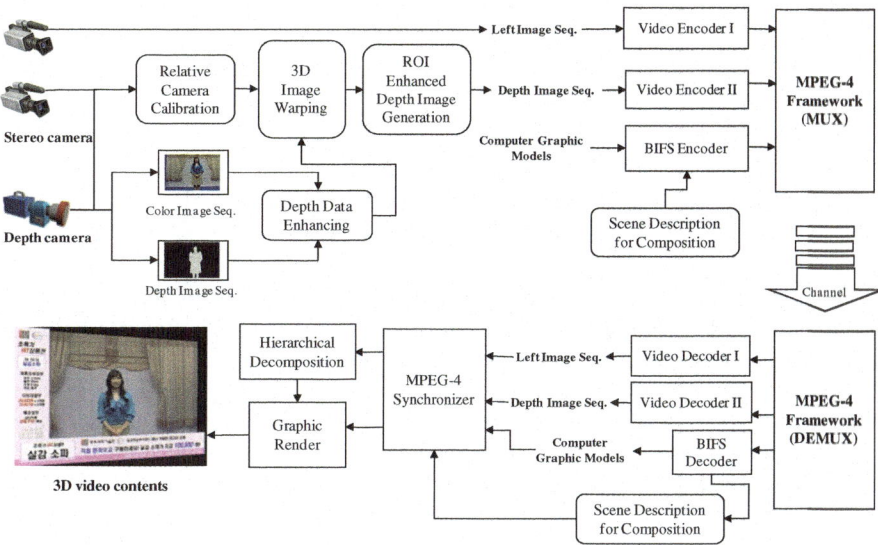

Fig. 8 3D Video Contents Generation

for depth image sequence. Other multimedia data and the scene description information are compressed by their coders, respectively. These compressed bitstreams are multiplexed into one bitstream in the MPEG-4 system.

At the client side, we extract video-plus-depth, scene description information, and other multimedia data from the transmitted bitstream by their decoders. Thereafter, we construct 3D scenes from video-plus-depth using the hierarchical decomposition method as introduced in Section 3. Other multimedia data are combined with the dynamic 3D scene by referring the scene description information. Finally, we can experience various interactions with the immersive content.

A major difference in MPEG-4, with respect to previous audio-visual standards, is the object-based audio-visual representation model. In the MPEG-4 multimedia framework, an object-based scene is built using individual objects that have relationships in space and time. Based on this relationship, the MPEG-4 system allows us to combine a variety of media objects into a scene. The MPEG-4 BIFS defines how the objects are spatio-temporally combined for presentation. All visible objects in the 3D scene are described within the *Shape* node in MPEG-4 BIFS. The *Shape* node should have both appearance and geometry information; the appearance is expressed by the color image sequence through a *MovieTexture* node.

However, although the geometry should be expressed by the depth image sequence, MPEG-4 BIFS does not support a node related to this geometry. Therefore, we design a new node representing the depth image sequence, referred to as a *DepthMovie* node. A new *DepthMovie* node that can be stored in the geometry field is designed as follows.

DepthMovie
```
{
    field  SFVec2f        fieldOfView        0.785398 0.785398
    field  SFFloat        nearPlane 10
    field  SFFloat        farPlane   100
    field  SFBool         orthographic       TRUE
    field  SFTextureNode texture             NULL
}
```

The upper four fields of the *DepthMovie* node are the same as the fields of the *DepthImage* node [21] that indicates the camera parameters. The texture field can store a depth image sequence as geometry through a *MovieTexture* node that usually indicates the 2D video. Then, the corresponding color image sequence is stored on the texture field of the *Appearance* node. In this way, these nodes can describe a 3D surface. Following shows an example describing video-plus-depth using the *DepthMovie* node. In this example, "colorVideo.h264" and "depthVideo.h264" are the compressed versions of color and depth image sequences, respectively.

Shape
```
{
    appearance Appearance{
        texture MovieTexture { url "colorVideo.h264" }
    }
    geometry DepthMovie {
        texture MovieTexture { url "depthVideo.h264"}
    }
}
```

In general, computer graphic models are represented by the mesh structure and described using predefined nodes in MPEG-4 BIFS. The MPEG-4 BIFS data including scene description information and computer graphic model data are coded by the BIFS encoder provided by the MPEG-4 system. Thereafter, the compressed video-plus-depth and MPEG-4 BIFS bitstreams are multiplexed into a MP4 file that is designed to contain the media data by the MPEG-4 representation. The MP4 file can be played from a local hard disk and over existing IP networks. Hence, users can enjoy the 3D video contents in the context of a video-on-demand concept.

5 Experimental Analysis

5.1 Evaluation of Depth Accuracy

For this experiment, as shown in Fig. 1, we set up a hybrid camera system with two HD cameras (Canon XL-H1) as a stereoscopic camera and one Z-Cam as a depth camera. The measuring distance of the depth camera was from 1.75m to 2.15m and the baseline distance between HD left and right cameras was 20cm. For test images, we captured BEAR and ACTOR images using the hybrid camera system. Especially, since the BEAR images included a scene that a big yellow bear doll embraced a small yellow bear doll, they were good to evaluate depth accuracy of ROI enhanced depth image generated by hybrid camera system. Figure 9 shows the test images.

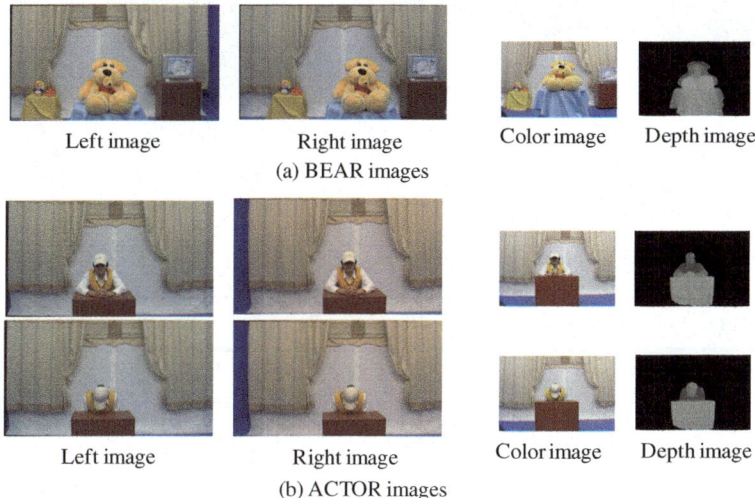

| Left image | Right image | Color image | Depth image |

(a) BEAR images

| Left image | Right image | Color image | Depth image |

(b) ACTOR images

Fig. 9 Test images

For comparison with the previous depth estimation methods, we estimated ROI depth images of left images by applying the state-of-the-arts stereo matching methods, which are belief propagation [18], graph cuts [22], dynamic programming [23], and scan-line optimization [3]. For background depths, we only used SAD method based on color segmentation. We have also made a ROI ground truth depth image of BEAR images by projecting the depth data acquired by a 3D laser range finder [24] onto the camera plane of the left camera. Figure 10 shows the ground truth depth image and the results of depth estimation for BEAR images.

As objective evaluation methodology, we used two quality measures based on known ground truth data [3]: the root-mean squared error R_E and the percentage of bad matching pixels B_A. Here, bad matching means that the depth value is different from the corresponding ground truth depth value by more than one pixel value. Table 3 shows the result of R_E, B_A, and the B_A difference between the stereo matching algorithms and the hybird camera method, B_{Diff}.

When we compared the accuracy of ROI depths generated by the hybrid camera system with belief propagation, which was the best among previous methods, our method was more accurate by approximately 2.1 for R_E and 11.2 % for B_A than belief propagation for BEAR images.

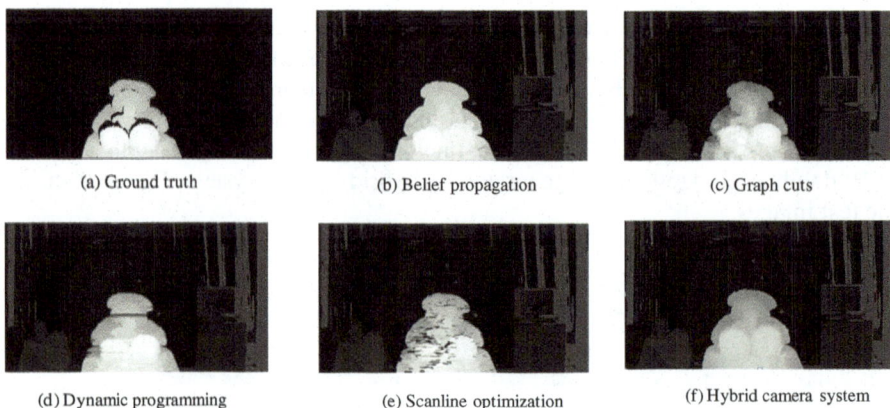

(a) Ground truth	(b) Belief propagation	(c) Graph cuts

(d) Dynamic programming	(e) Scanline optimization	(f) Hybrid camera system

Fig. 10 Results of depth estimation for BEAR image

Table 3 ROI depth quality evaluation

Methods	R_E	B_A	B_{Diff}
Belief gropagation	26.5	50.1%	+11.2%
Graph cuts	62.1	83.3%	+44.4%
Dyanaimc programimg	46.1	76.7%	+37.8%
Scanline optimization	67.7	79.5%	+40.6%
Hybrid camera system	24.4	38.9%	-

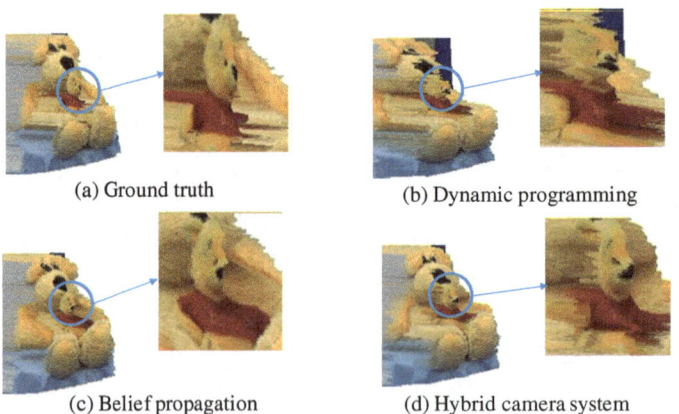

(a) Ground truth	(b) Dynamic programming

(c) Belief propagation	(d) Hybrid camera system

Fig. 11 3D scene reconstruction on ROI of BEAR images

Figure 11 shows the results of 3D scene reconstruction on ROI of BEAR images using hierarchical decomposition. When we subjectively compare the scenes with the one generated with the ground truth for BEAR images, the 3D scene generated by our method more closely resembled the original scene than other methods. Especially, the regions of the big bear doll's leg and the small bear doll in the original scene were much similar with ours. Hence, we subjectively notice that the depth image obtained by the hybrid camera has more reliable depth data than the other methods. Figure 12 shows the result of depth estimation with the ACTOR images and the side view of its 3D scene reconstruction. We could notice that the regions of a table and a hat marked by circles were reconstructed into 3D scenes better from depth information generated by the hybrid camera system than the other methods.

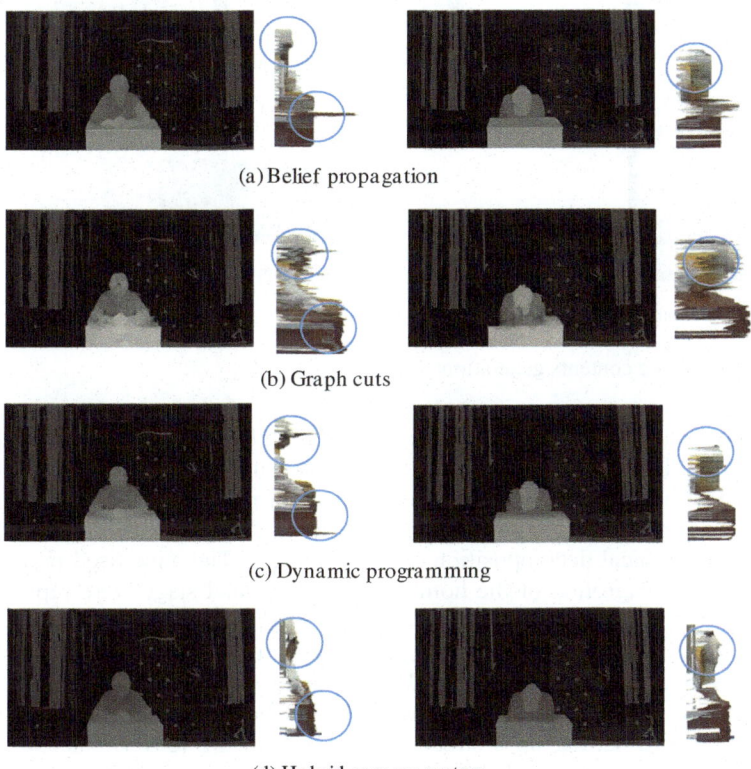

(a) Belief propagation

(b) Graph cuts

(c) Dynamic programming

(d) Hybrid camera system

Fig. 12 Results of depth estimation for ACTOR image

5.2 3D Video Contents Generation

In this example, we have created 3D video contents with the main theme of a home-shopping channel scenario using the hybrid camera system in the MPEG-4 multimedia framework. Figure 13 depicts each stage to create the 3D home-shopping contents for future broadcasting.

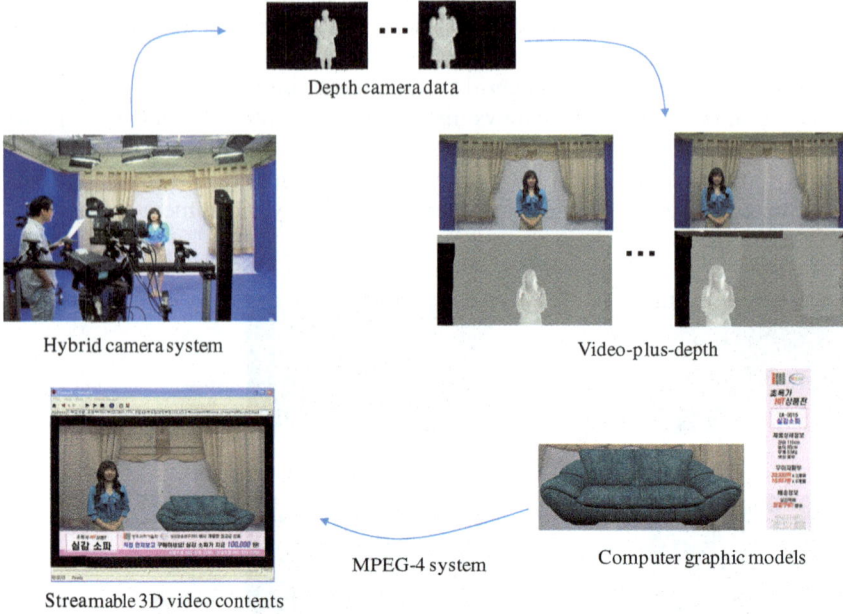

Depth camera data

Hybrid camera system   Video-plus-depth

MPEG-4 system   Computer graphic models

Streamable 3D video contents

Fig. 13 3D home-shopping contents generation

In the 3D home-shopping contents, the geometric and photometric information of a home-shopping host had been derived from video-plus-depth data generated by the hybrid camera system. The video-plus-depth information was encoded by two H.264/AVC coders [25]. In order to render the home-shopping host into a 3D scene, we used hierarchical decomposition of depth images. The advertised product, a sofa, and the background of the home-shopping channel stage were represented by computer graphic models. The sofa was represented by a 3D mesh model composed of 4,774 vertices and 2,596 triangular faces. All computer graphic models were encoded by a MPEG-4 BIFS coder.

In this experiment, it was possible to stream the 3D home-shopping contents through a network after setting up a streaming server [26]. Moreover, since we successfully represented a real and natural object, a home-shopping host, with video-plus-depth and rapidly rendered it into a dynamic 3D scene, we could serve the immersive contents to users and support a variety of interactive functionality in the MPEG-4 multimedia framework.

As shown in Fig. 14(a), the home-shopping contents could provide a 3D view to users by freely reallocating the position of a virtual camera in 3D space. In addition, as shown in Fig. 14(b), since the home-shopping host based on video-plus-depth data was described by a newly-designed MPEG-4 BIFS node *DepthMovie*, the natural 3D actor could be easily combined with computer graphic images, a

(a) Free viewpoint changing

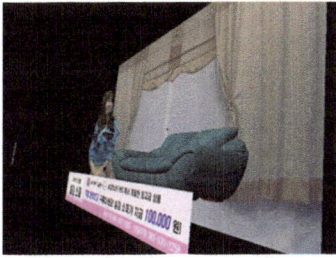

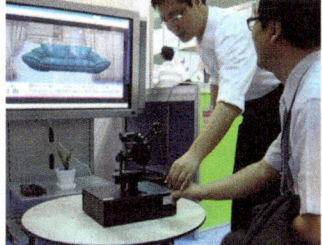

(b) Free composition with computer graphics

(c) Haptic interaction

Fig. 14 User interactions

sofa and a subtitle, described by existing MPEG-4 BIFS nodes. Furthermore, as shown in Fig. 14(c), when the sofa included haptic information represented by a bump map, we could feel its surface property wearing a haptic device using a haptic rendering algorithm [27]. The home-shopping video clip is available in the web site. (*http://www.imaging.utk.edu/people/sykim/*).

6 Conclusions

We addressed three problems in this chapter. First, we introduced a depth camera-based hybrid camera system to generate high-quality and high-resolution video-plus-depth. With the hybrid camera system, we intended to minimize inherent technical problems in current depth cameras and generate reliable depth information. Second, we talked about a hierarchical decomposition technique to render a 3D dynamic scene with video-plus-depth data. Finally, as one of possible applications of the hybrid camera system, we introduced a method to generate streamable MPEG-4-based 3D video contents for the future home-shopping channel. The 3D video contents including video-plus-depth and computer graphics images could support various user-friendly interactions. We believe that the 3D video contents exploitation system can present new directions for further researches related to interactive 3D multimedia applications.

Acknowledgments. This work was supported by the National Research Foundation of KoreaGrant funded by the Korean Government (NRF-2009-352-D00277) and in part by ITRC through RBRC at GIST (IITA-2008-C1090-0801-0017).

References

1. Fehn, C.: A 3D-TV System Based on Video Plus Depth Information. In: Proc. Asilomar Conference on Signals, Systems and Computers, Pacific Grove, CA, USA (2003)
2. Kauff, P., Atzpadin, N., Fehn, C., Müller, M., Schreer, O., Smolic, A., Tanger, R.: Depth Map Creation and Image-based Rendering for Advanced 3DTV Services Providing Interoperability and Scalability. Signal Processing: Image Communication 22(2), 217–234 (2007)
3. Scharstein, D., Szeliski, R.: A Taxonomy and Evaluation of Dense Two-frame Stereo Correspondence Algorithms. International Journal of Computer Vision 47(1-3), 7–42 (2002)
4. Zitnick, C., Kang, S., Uyttendaele, M., Winder, S., Szeliski, R.: High-quality Video View Interpolation Using a Layered Representation. ACM Trans. on Graphics 23(3), 600–608 (2004)
5. Kim, S.M., Cha, J., Ryu, J., Lee, K.H.: Depth Video Enhancement of Haptic Interaction Using a Smooth Surface Reconstruction. IEICE Trans. on Information and System E89-D(1), 37–44 (2006)
6. Waschbüsch, M., Würmlin, S., Cotting, D., Gross, M.: Point-sampled 3D Video of Real-world Scenes. Signal Processing: Image Communication 22(2), 203–216 (2007)
7. Iddan, G.J., Yahav, G.: 3D Imaging in the Studio and Elsewhere. In: Proc. SPUE Videometrics and Optical Methods for 3D Shape Measurements, San Jose, CA, USA (2001)
8. Kawakita, M., Kurita, T., Kikuchi, H., Inoue, S.: HDTV Axi-vision Camera. In: Proc. International Broadcasting Conference, Amsterdam, Netherlands (2002)
9. Redert, A., Beeck, M., Fehn, C., IJsselsteijn, W., Pollefeys, M., Gool, L., Ofek, E., Sexton, I., Surman, P.: ATTEST – Advanced Three-Dimensional Television System Technologies. In: Proc. International Symposium on 3D Data Processing Visualization and Transmission, Padova, Italy (2002)
10. Cha, J., Kim, S.M., Kim, S.Y., Kim, S., Oakley, I., Ryu, J., Lee, K.H., Woo, W., Ho, Y.S.: Client System for Realistic Broadcasting: a First Prototype. In: Ho, Y.-S., Kim, H.-J. (eds.) PCM 2005. LNCS, vol. 3768, pp. 176–186. Springer, Heidelberg (2005)
11. Ho, Y.S., Kim, S.Y., Lee, E.K.: Three-dimensional Video Generation for Realistic Broadcasting Services. In: Proc. International Technical Conference on Circuits, Systems, Computers and Communications, Shimonoseki, Japan (2008)
12. Kim, S.Y., Lee, E.K., Ho, Y.S.: Generation of ROI Enhanced Depth Maps Using Stereoscopic Cameras and a Depth Camera. IEEE Trans. on Broadcasting 54(4), 732–740 (2008)
13. Kim, S.Y., Ho, Y.S.: Hierarchical Decomposition of Depth Map Sequences for Representation of Three-dimensional Dynamic Scenes. IEICE Trans. on Information and Systems E90-D(11), 1813–1820 (2007)
14. Cho, J.H., Kim, S.Y., Ho, Y.S., Lee, K.H.: Dynamic 3D Human Actor Generation Method Using a Time-of-flight Depth Camera. IEEE Trans. on Consumer Electronics 54(4), 1514–1521 (2008)

15. Cho, J.H., Chang, I.Y., Kim, S.M., Lee, K.H.: Depth Image Processing Technique for Representing Human Actors in 3DTV Using Single Depth Camera. In: Proc. IEEE 3DTV Conference, pp. 1–4 (2007)
16. Zhu, J., Wang, L., Yang, R., Davis, J.: Fusion of Time-of-flight Depth and Stereo for High Accuracy Depth Maps. In: Proc. IEEE Conference on Computer Vision and Pattern Recognition, Anchorage, Alaska, USA (2008)
17. Yoon, S.U., Ho, Y.S.: Multiple Color and Depth Video Coding Using a Hierarchical Representation. IEEE Trans. on Circuits and Systems for Video Technology 17(11), 1450–1460 (2007)
18. Felzenszwalb, P.F., Huttenlocher, D.P.: Efficient Belief Propagation for Early Vision. International Journal of Computer Vision 70(1), 41–54 (2006)
19. Farin, D., Peerlings, R., With, P.: Depth-image Representation Employing Meshes for Intermediate-view Rendering and Coding. In: Proc. IEEE 3DTV conference, Kos Island, Greece (2007)
20. Pereira, F.: MPEG-4: Why, What, How and When. Signal Processing: Image Communication 15(4), 271–279 (2000)
21. Levkovich-Maslyuk, L., Ignatenko, A., Zhirkov, A., Konushin, A., Park, I., Han, M., Bayakovski, Y.: Depth Image-Based Representation and Compression for Static and Animated 3-D Objects. IEEE Trans. on Circuits and Systems for Video Technology 14(7), 1032–1045 (2004)
22. Kolmogorov, V., Zabih, R.: Computing Visual Correspondence with Occlusions Using Graph Cuts. In: Proc. International Conference on Computer Vision, Vancouver, Canada (2001)
23. Bobick, A.F., Intille, S.S.: Large Occlusion Stereo. International Journal of Computer Vision 33(3), 181–200 (1999)
24. LMS-Z390i, http://www.riegl.com/
25. Wiegand, T., Lightstone, M., Mukherjee, D., Campbell, T.G., Mitra, S.K.: Rate-distortion Optimized Mode Selection for Very Low Bit Rate Video Coding and the Emerging H.263 Standard. IEEE Trans. on Circuit and System for Video Technology 6(9), 182–190 (1996)
26. Darwin Streaming Server, http://developer.apple.com/
27. Cha, J., Kim, S.Y., Ho, Y.S., Ryu, J.: 3D Video Player System with Haptic Interaction Based on Depth Image-Based Representation. IEEE Trans. on Consumer Electronics 52(2), 477–484 (2006)

Chapter 16
Improving 3D Visual Experience by Controlling the Perceived Depth Distortion

Jessica Prévoteau, Sylvia Chalençon-Piotin, Didier Debons, Laurent Lucas, and Yannick Remion

Abstract. A fundamental element of stereoscopic and/or autostereoscopic image production is the geometrical analysis of shooting and viewing conditions in order to obtain a qualitative 3D perception experience. Starting from the usual multiscopic rendering geometry and the classical off-axis coplanar multipoint 3D shooting geometry, we firstly compare the perceived depth with the shot scene depth, for a couple of shooting and rendering devices. This yields a depth distortion model whose parameters are expressed from the geometrical characteristics of shooting and rendering devices. Then, we explain how to invert these expressions in order to design the appropriate shooting layout from a chosen rendering device and a desired effect of depth. Thus, thanks to our scientific know-how, we based our work on the link between the shooting and rendering geometries, which enables to control the distortion of the perceived depth. Finally, thanks to our technological expertise, this design scheme provides three patented shooting technologies producing qualitative 3D content for various kinds of scenes (real or virtual, still or animated), complying with any pre-chosen distortion when rendered on any specific multiscopic technology and device as specified previously.

Jessica Prévoteau · Didier Debons · Laurent Lucas · Yannick Remion
TéléRelief, 2 Allée Albert Caquot, 51100 Reims France
e-mail: j.prevoteau@telerelief.com, d.debons@telerelief.com,
{laurent.lucas,yannick.remion}@univ-reims.fr

Jessica Prévoteau · Sylvia Chalençon-Piotin · Laurent Lucas · Yannick Remion
CReSTIC-SIC EA3804, rue des Crayères BP 1035 - 51687 Reims Cedex 2 France
e-mail: sylvia.chalencon@univ-reims.fr

1 Introduction

Extending visual content with a third dimension, or capturing a dynamic scene in 3D and generating an optical duplicate of it in real-time, has been a dream over decades. All components (hardware and software) related to this viewing experience are collectively referred to as three-dimensional television (3DTV). Often presented as the next evolution of television, this new area of research holds tremendous potential for many applications in entertainment, telepresence, medicine, visualization and remote manipulation to name just few. From a technological point of view, creating the illusion of a real environment, is necessary condition over the whole 3DTV chain, including 3D image acquisition, 3D representation, compression, transmission, signal processing and interactive rendering. We now have numerous multiscopic rendering systems with or without glasses. Different technologies support all these systems; stereoscopy with colorimetric or temporal mixing such as anaglyph [1, 2], occultation and polarization [3], for example for projections with glasses as in some movie theaters; autostereoscopy [4, 5] such as parallax barrier and lenticular lens; or again for 3D advertising billboards, autostereoscopic displays or lenticular printing.

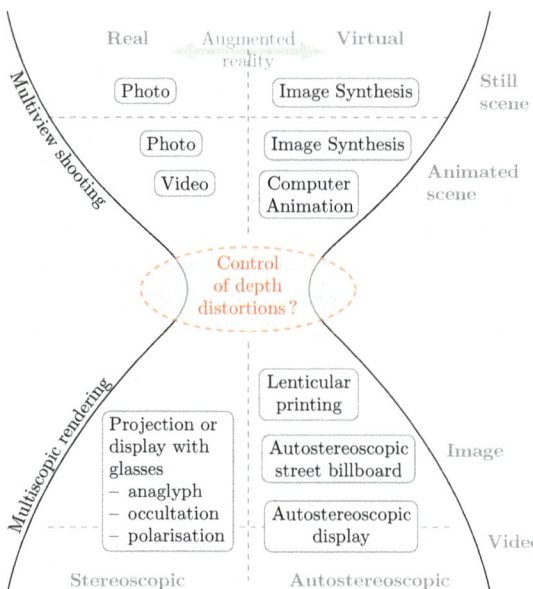

Fig. 1 From shooting to rendering process: the different rendering modes and kinds of scenes (real or virtual, still or animated) to be shot.

As shown in Figure 1, the different rendering modes and kinds of scenes to be shot are well-known, but all these systems need content; up to now, there has been no 3D shooting system specifically designed to acquire a qualitative 3D content. Some works [6, 7, 8, 9] present efficient algorithms for stereoscopic display to obtain a 3D content from a 2D-plus-depth content. Their global and common idea lies in the slight difference between the right view and the left view. So, they generate the right eye image by a transformation of the left eye image conforming to the stereo disparity and then they reduce the processing cost for stereoscopic displays. The main disadvantage of these methods lies in the lack of information in occluded areas which is impossible to overcome in a generic way. Yet, to comply with our demand of qualitative 3D content we focus on multi-shooting technologies.

Other works [10, 11] have been published in the general case of multi-cameras. They define the projective relations between the images shot by multi-cameras in order to calibrate the different cameras and then to reconstruct the 3D shot scene from these multiple views. There is no link with any viewing device since the target is a reconstruction module. In our case, flat multiscopic viewing requires a simplified shooting layout also called "rectified geometry". Moreover the control of the viewer's 3D experience implies to connect shooting and viewing geometries in order to model and set the geometrical distortions between shot and perceived scenes.

Furthermore, some works have been done to improve the control of the viewer's 3D experience in stereoscopy and computer graphics fields [12, 13]. They usually compare shooting and viewing geometries in order to choose a shooting layout fitting a given depth range in virtual space to the "comfortable" depth range of the display. We believe that choices that can be made in the shooting design may be richer than a simple mapping of depth and could differ for each observation position in the multi-view case. This requires a detailed model and a precise analysis of possible distortions for the multiscopic shooting/viewing couple. Indeed, such a model will provide the characteristics of shooting which will generate the chosen distortions on the chosen viewing device. If some authors have described the transformation between the shot and the real scene [12] in the stereoscopic case, none of them has been interested in producing an analytic multi-observer and reversible model allowing to pilot the shooting for all kinds of possible distortions. Thus, we suggest a solution to produce 3D content according to the chosen rendering mode and the desired depth effect.

Additionally, it is important to consider limits of the human visual system upon the perceived quality of stereoscopic images. Some publications on human factors [14, 15, 16] have studied in detail the issue of viewer comfort for stereoscopic displays. All these studies lead to a similar conclusion: the amount of disparity in stereoscopic images should be limited so as to be within a defined comfortable range. The main reason given for this is that the human visual system normally operates so that the convergence of the eyes and the focus are linked. For all stereoscopic displays this relationship is thought to be stressed by requiring the viewers eyes to converge to a perceived point much deeper than the display plane while still being required to focus on the display plane itself. Limiting disparity ensures that the viewers perceived depth is controlled and the convergence/accommodation link is not stressed.

So we'll explain how to model and quantify the depth distortion from given rendering and shooting geometries and also from a chosen rendering device and a desired depth effect and how to design the appropriate shooting layout.

This chapter introduces a complete analysis of the geometrical quality of 3D content based on distortion analysis by linking shooting and viewing geometries. Starting from previous related knowledge (*i.e.*, viewing and shooting geometries), we'll show remaining the problems and model the possibilities of depth distortions between the scene perceived by a viewer and the scene shot initially. Next, we will present a shooting layout design scheme ensuring a desired depth effect (controlled depth distortion or perfect depth effect) upon a pre-determined rendering device. Finally, we will introduce derived shooting technologies (which are patent pending) complying with this scheme and thus achieve qualitative 3D content on the previously given rendering device: 3D computer graphics software and 3D devices. We will present these prototypes and some of their results.

2 Previous Related Knowledge

2.1 Viewing

3D image rendering, with or without glasses, is known to require "stereoscopic" or "autostereoscopic" devices. All these devices make a spatial, colorimetric and/or temporal mixing over a single region of interest (ROI area physically filled by the displayed image on the rendering device) of $n \times m$ so-called "initial images" of one scene shot from several distinct viewpoints. These systems allow to optically and/or temporally separate the images reaching each eye of one or more viewers. In case of stereoscopic systems, both images are emitted in a single optical beam independently of the viewer's position in this beam [1, 2, 17]. However, autostereoscopic systems separate the images in several distinct optical beams, organized for example, in horizontal "range" of n images ($n \geq 2$ and $m = 1$) [4, 5]. We can also imagine optical beams organized in both horizontal and vertical ranges. Then we dispose of a matrix disposition of $n \times m$ optical beams ($n \geq 2$ and $m \geq 2$), each one transporting a different image. Thus, all known devices broadcast alternately and/or simultaneously $n \times m$ images ($n \geq 2$ and $m \geq 1$) within one or several optical beams in such a way that both eyes of a correctly-placed viewer get different consistent images (*i.e.*, initial images and not combinations of them). Thereby the viewer's brain rebuilds his depth perception by stereoscopy [18]. Even if the human visual system has a tolerance as for epipolar alignment, ideal positions within this tolerance correspond in particular to the eyes line which has to be parallel to the display's rows. Despite this human tolerance, we calculate our images in such a way that they have a perfect epipolar alignment for a well-placed eyes line.

So let's analyse the geometry of these devices the "viewing geometry" (Fig. 2) which will constrain the compatible shooting layout.

A 3D rendering device mixes $n \times m$ images sharing out a ROI of dimension W (width) and H (height). Each image (image's index $i = (i_1, i_2) \in \{1, n\} \times \{1, m\}$) is supposed to be "correctly" visible (without much mixing with others) at least from the chosen preferential position E_i. These positions are aligned upon m lines parallel to the rows of the ROI located at distance d_{i_2}, $i_2 \in \{1, ...m\}$, from the device

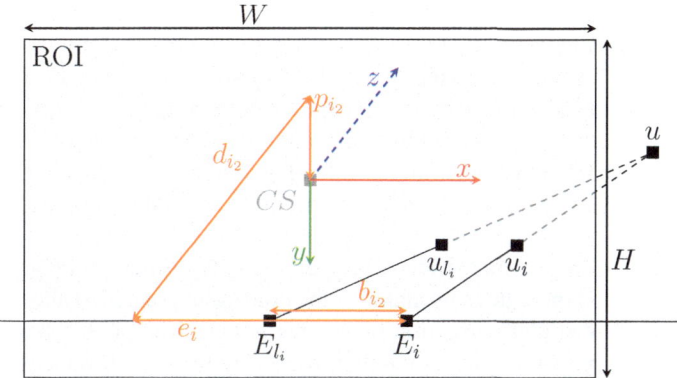

Fig. 2 The viewing geometry

ROI. The preferential positions E_i are placed on those lines according to their second index i_2 in order to guarantee that a viewer whose binocular gap is b_{i_2} (often identical to human medium binocular gap 65 mm, but possibly different according to the expected public: children, etc.), with the eyes line parallel to the device rows, would have his right eye in E_i and his left eye in E_{l_i}. The right eye in E_i would catch image number i, while the left eye in E_{l_i} would catch image number l_i knowing that $l_i = i - (q_{i2}, 0)$ with q_{i2} being the gap between the indexes of images which compose the visible consistent stereoscopic couples with binocular gap b_{i_2} at distance d_{i_2}. Hence, associated left and right eye preferential positions E_{l_i} and E_i verify $E_i = E_{l_i} + b_{i_2}x$ and $e_i = e_{l_i} + b_{i_2}$.

We also define the lines positions vertically (because viewers of various sizes use the device) by p_{i_2} which represents the "overhang", *i.e.*, the vertical gap of eyes positioning compared with the ROI center CS. If we don't know p_{i_2}, we use a medium overhang corresponding to a viewer of medium size, which has to be chosen at design stage. Assuming u_i and u_{l_i} are stereoscopic homologous for images i and l_i, their perception by the right and left eye of a viewer from E_i and E_{l_i} leads this spectator's brain to perceive a 3D point u. The viewing geometry analysis is expressed thanks to a global reference frame $r = (CS, x, y, z \equiv x \times y)$, chosen at the ROI center CS, with x parallel to its rows and turned towards the right of the spectators, and y parallel to its columns and turned towards the bottom.

2.2 Shooting

In order to "feed" such devices with 3D content, we need sets of $n \times m$ images from a single scene acquired from several distinct and judicious viewpoints and with specific projective geometry as the rendering upon flat multiscopic devices involves coplanar mixing of these images. This major issue is well known in multiscopy.

The image viewing is achieved according to distorted pyramids whose common base corresponds to the device ROI and the tops are the viewer's eyes or

E_i positions. Given that vision axes are not necessarily orthogonal to the observed images area (ROI), the viewing of these images induces trapezoidal distortion if we don't take into account this slanted viewing during the shooting. This has an immediate consequence in order to achieve depth perception. If the trapezoidal distortions are not similar for the two images seen by a spectator, the stereoscopic matching by the brain is more delicate, or even impossible. This reduces or cancels the depth perception. This constraint, well-known in stereoscopy, is called the "epipolar constraint".

Solutions (also called toe-in camera model) of convergent systems have been proposed [19, 20], but such convergent devices manifest the constraint presented above. So, unless a systematic trapezoidal correction of images is performed beforehand (which might not be desirable as it loads down the processing line and produces a qualitative deterioration of the images) such devices do not afford to produce a qualitative 3D content. As demonstrated by [14, 21], we must use devices with shooting pyramids sharing a common rectangular base (off-axis camera model) and with tops arranged on a line parallel to the rows of this common base in the scene. For example Dodgson *et al.* use this shooting layout for their time-multiplexed autostereoscopic camera system [22].

Thus, aiming axes are necessarily convergent at the center of the common base and the tops of the shooting pyramids must lie on m lines parallel to the rows of the common base. Figure 3(a) shows a perspective representation of such a shooting geometry. This figure defines the layout of the capture areas (CA_i), and the centers (C_i) and specifies a set of parameters describing the whole shooting geometry completely. Figures 3(b) and 3(c) show top and full-face representations of this geometry, respectively.

The shooting geometry analysis is expressed using a shooting global reference frame $R = (CP, X, Y, Z \equiv X \times Y)$ centered at the desired convergence point CP (which is also the center of the common base CB of the scene) and oriented in such a way that the first two vectors of the reference frame are parallel to the main directions of the common base CB of the scene and so, parallel to the main directions of the capture areas. The physical size of CB is Wb and Hb. Furthermore, the first axis is supposed to be parallel to the rows of the capture areas and the second axis is supposed to be parallel to the columns of these areas.

The $n \times m$ pyramids, representative of a shooting layout, according to the principles explained before to resolve the known issue, are specified by:

- an optical axis of Z direction,
- optical centers C_i (*i.e.*: principal points) aligned on one or more (m) line(s) parallel to the rows of the common base (so on X direction) and
- rectangular capture areas CA_i.

These capture areas must be orthogonal to Z, so parallel between them and parallel to CB and to centers lines (which are defined by their distances from CB, D_{i_2} along Z, P_{i_2} along Y and c_i along X). These capture areas are also placed at distances f_i along Z, β_i along Y and α_i along X from their respective optical center C_i. Their physical size is given by w_i and h_i. They are centered on points I_i in such a way that lines I_iC_i defining the axes of sight are convergent at CP. The centers C_i and C_{l_i} must be on the same "centers line" and with a spacing of B_i ($C_i = C_{l_i} + B_iX$ and $c_i = c_{l_i} + B_i$).

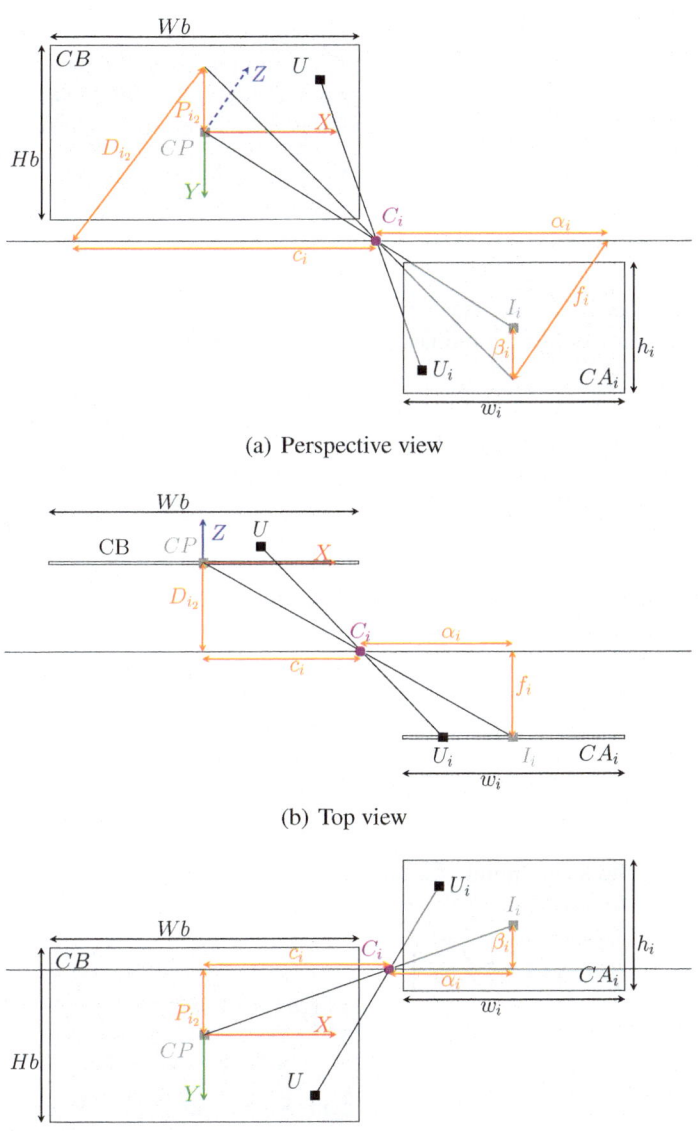

(a) Perspective view

(b) Top view

(c) Full-face view

Fig. 3 Implied shooting geometry

Such a shooting layout is necessary to obtain a depth effect on a multiscopic device. Nevertheless, its does not ensure that the perceived scene will not be distorted compared with the shot scene. Non distortion implies that viewing pyramids are perfect counterparts of shooting pyramids (*i.e.*, have exactly the same opening and main axis deviation angles in both horizontal and vertical directions). In case of pyramids dissimilarity, the 3D image corresponds to a complex distortion of the scene acquired initially. This can be desirable in some applications to carry out some special effects, as it can be undesirable in others. This requires, that shooting and viewing must be designed in a consistent way whether we desire a depth distortion or not. Let's now model those distortion effects generated by a couple of shooting and viewing geometries.

3 Distortion Analysis and Model

In this section, we consider that we use perfect sensors and lenses, without any distortion. This assumption implies some issues which will be presented for each derived technology.

Thanks to the previous analysis of the shooting and viewing geometries, and assuming that pixels U_i and U_{l_i} from shot images are displayed at u_i and u_{l_i} positions of the ROI, we can link the coordinates (X,Y,Z), in reference frame R, of point U of the scene, shot by the sensors defined previously, with the coordinates (x_i, y_i, z_i) in reference frame r, of its counterparts u seen by an observer of the viewing device placed in a preferential position (right eye in E_i and left eye in E_{l_i}).

Assuming that the scene point U is visible on image number i, its projection U_i verifies:

$$C_i U_i = \frac{-f_i}{Z + D_{i_2}} C_i U \quad i \in \{1,n\} \times \{1,m\} \tag{1}$$

Knowing that I_i, centers of CA_i, verifies:

$$C_i I_i = \frac{-f_i}{D_{i_2}} C_i CP \quad i \in \{1,n\} \times \{1,m\}, \tag{2}$$

The relative position of the scene point U's projections in the various images are expressed as:

$$I_i U_i = \frac{f_i}{Z + D_{i_2}} \begin{bmatrix} -X - Z\frac{c_i}{D_{i_2}} \\ -Y + Z\frac{P_{i_2}}{D_{i_2}} \\ 0 \end{bmatrix}_R \quad i \in \{1,n\} \times \{1,m\} \tag{3}$$

As the images are captured behind the optical centers, the projection reverses up/down and left/right axes, and the implicit axes of the images are opposite of those of the global shooting reference frame R. Moreover, the images are then scaled on the whole ROI of the rendering device. This relates U_i projections of U to their "rendered positions" u_i on the ROI:

$$CSu_{i|_r} = - \begin{bmatrix} \frac{W}{w_i} & & \\ & \frac{H}{h_i} & \\ & & 1 \end{bmatrix} I_i u_{i|_R} \quad \forall i \tag{4}$$

Remarking that $f_i Wb = D_{i_2} w_i$ and $f_i Hb = D_{i_2} h_i$, u_i is expressed in reference frame r as:

$$u_{i|_r} = \frac{D_{i_2}}{Z + D_{i_2}} \begin{bmatrix} \left(X + Z\frac{c_i}{D_{i_2}}\right)\frac{W}{Wb} \\ \left(Y - Z\frac{P_{i_2}}{D_{i_2}}\right)\frac{H}{Hb} \\ 0 \end{bmatrix} \quad \forall i \tag{5}$$

By this time, and assuming U was visible on both images l_i and i, we notice that u_{l_i} and u_i lie on the same row of the ROI. This fulfills the epipolar constraint and thus permits stereoscopic reconstruction of $u = [x_i, y_i, z_i]_r^t$ from E_{l_i} and E_i according to:

$$u_{l_i} u_i = \frac{z_i}{z_i + d_{i_2}} b_{i_2} x, \quad \text{which yields } z_i \text{ and} \tag{6}$$

$$E_i u = \frac{z_i + d_{i_2}}{d_{i_2}} E_i u_i, \quad \text{which then gives } x_i, y_i \tag{7}$$

Thus, after some calculus, the relation between the 3D coordinates of the scene points and those of their images perceived by a viewer may be characterized under homogeneous coordinates by:

$$a_i \begin{bmatrix} x_i \\ y_i \\ z_i \\ 1 \end{bmatrix} = \begin{bmatrix} \mu_i & \gamma_i & 0 \\ k_{i_2} & \rho\mu_i & \delta_i & 0 \\ & 1 & 0 \\ 0 & 0 & \frac{k_{i_2}(\varepsilon_i - 1)}{d_{i_2}} & \varepsilon_i \end{bmatrix} * \begin{bmatrix} X \\ Y \\ Z \\ 1 \end{bmatrix} \tag{8}$$

The above equation can be seen as the analytic distortion model for observer position i which matches the stereoscopic transformation matrix given in [12]. As such this model clearly exhibits the whole set of distortions to be expected in any multiscopic 3D experience, whatever the number of views implied or the very nature of these images (real or virtual). It shows too that these distortions are somehow independent from one another and may vary for each observer position i. The following detailed analysis of this model and its further inversion will offer a novel multiscopic shooting layout design scheme acting from freely chosen distortion effects and for any specified multiscopic rendering device.

The above model exhibits some new parameters quantifying independent distortion effects. Those parameters may be analytically expressed from geometrical parameters of both shooting and rendering multiscopic devices. Their relations to geometrical parameters and impact on distortion effects are now presented:

- k_{i_2} control(s) the global enlarging factor(s),

$$k_{i_2} = \frac{d_{i_2}}{D_{i_2}} \tag{9}$$

- ε_i control(s) the potential nonlinear distortion which transforms a cube into a pyramid trunk according to the global reducing rate $a_i = \varepsilon_i + k_{i_2}(\varepsilon_i - 1)\frac{Z}{d_{i_2}}$ possibly varying along Z,

$$\varepsilon_i = \frac{b_{i_2}}{B_i}\frac{W_b}{W} \tag{10}$$

- μ_i control(s) width over depth relative enlarging rate(s), or the horizontal/depth anamorphose factor,

$$\mu_i = \frac{b_{i_2}}{k_{i_2}B_i} \tag{11}$$

- ρ control(s) height over width relative enlarging rate(s), or the vertical/horizontal anamorphose factor,

$$\rho = \frac{W_b}{H_b}\frac{H}{W} \tag{12}$$

- γ_i control(s) the horizontal"shear" rate(s) of the perceived depth effect,

$$\gamma_i = \frac{c_i b_{i_2} - e_i B_i}{d_{i_2}B_i} \tag{13}$$

- δ_i control(s) the vertical "shear" rate(s) of the perceived depth effect by an observer whose overhanging position complies with what is expected,

$$\delta_i = \frac{p_{i_2}B_i - P_{i_2}b_{i_2}\rho}{d_{i_2}B_i} \tag{14}$$

Thus we have defined the depth distortion possibilities using the previously established shooting and viewing geometries. Moreover, this model makes the quantifying of those distortions possible for any couple of shooting and viewing settings by simple calculus based upon their geometric parameters.

4 Shooting Design Scheme for Chosen Distortion

One can use any multiscopic shooting device with any multiscopic viewing device while giving an effect of depth to any well-placed viewer (3D movie theater for example) but section 3 shows that distortions will not be similar for each couple of technologies. In this section, we will design the shooting geometry needed to obtain a desired distortion on a given viewing device: whether perfect depth or chosen distortion effect of a shot scene.

Knowing how distortions, shooting and viewing parameters are related, it becomes possible to derive the shooting layout from former distortion and viewing choices.

We will describe two shooting layout design schemes complying to this use of the distortion model:

- a generic scheme allowing for a precise control of each distortion parameter and
- a more dedicated one of huge interest as it is focused on "non distortion" or "perfect depth", allowing the user to control of global enlarging factor(s) k_{i_2} as any other distortion parameter is set to its "non distortion value".

4.1 Controlled Depth Distortion

To define the shooting layout using this scheme, we control global enlargement (by k_{i_2}) and 4 potential depth distortions:

1. when $\varepsilon_i \neq 1$, a global nonlinearity which results in a deformation of the returned volume onto a "pyramid trunk" (as a_i varies along Z axis) (cf. Fig. 4(b)),

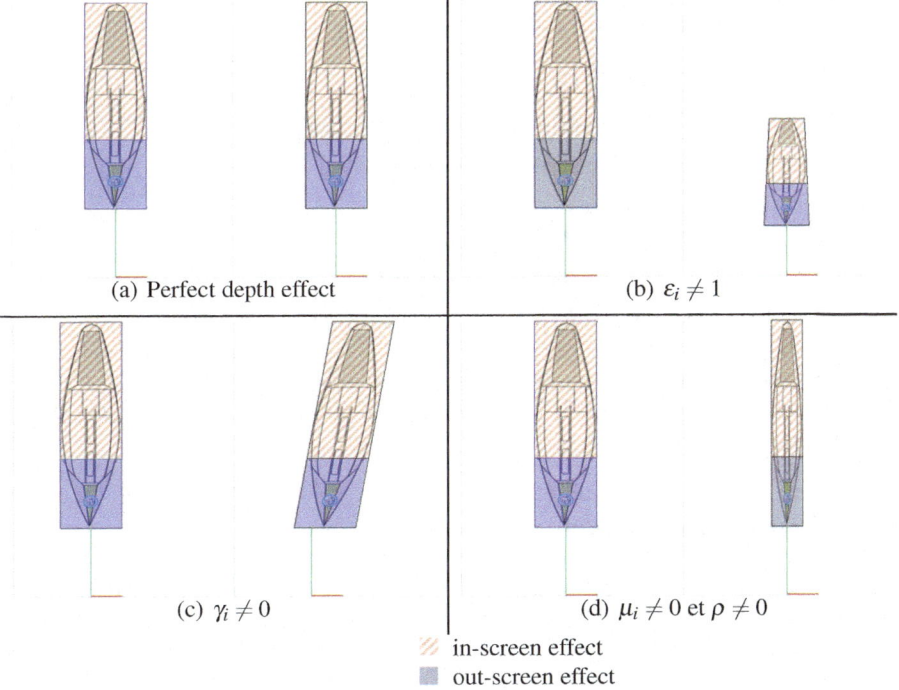

(a) Perfect depth effect   (b) $\varepsilon_i \neq 1$

(c) $\gamma_i \neq 0$ (d) $\mu_i \neq 0$ et $\rho \neq 0$

in-screen effect
out-screen effect

Fig. 4 Illustration of the different distortions according to each parameter. For each couple, the image on the left corresponds to the top view of the real scene whereas the right one corresponds to the top view of the viewed scene.

2. when $\gamma_i \neq 0$, a sliding or "horizontal shear" of the returned volume according to the depth (cf. Fig. 4(c)),
3. when $\delta_i \neq 0$ and/or when the actual overhanging position of the observer differs from the optimal one, a sliding or "vertical shear" of the returned volume according to the depth and
4. when $\mu_i \neq 1$ and/or $\rho \neq 1$, an anamorphose producing uneven distortions of the 3 axis (X versus Z for μ_i and Y versus X for ρ) (cf. Fig. 4(d)).

The depth controlled-distortion is obtained by adjusting the enlarging factor(s) k_{i_2} and the distortion parameters ε_i (and so $a_i = \varepsilon_i * k_{i_2} (\varepsilon_i - 1)/d_{i_2}$), μ_i, ρ, γ_i and δ_i. This latter condition on δ_i is more delicate because it depends on the height of the viewer which inevitably affects the effective position towards the device. So the chosen vertical sliding δ_i can be reached only for an observer whose overhanging position is defined in the viewing settings for this observation position.

Thus, given the viewing settings and the desired distortion parameters, the shooting parameters can be calculated as follows:

$$P_{i_2} = (p_{i_2} - \delta_i d_{i_2})/(k_{i_2} \rho \mu_i) \quad D_{i_2} = d_{i_2}/k_{i_2}$$
$$W_b = W \varepsilon_i /(k_{i_2} \mu_i) \qquad\qquad H_b = H \varepsilon_i /(k_{i_2} \rho \mu_i)$$

$$c_i = (e_i + \gamma_i d_{i_2})/(k_{i_2} \mu_i)$$

f_i imposed or chosen, individually $\forall i \in \{1...n\} \times \{1...m\}$,
by lot $\forall i_2 \in \{1...n\}$ or on the whole. $\hfill (15)$

$$w_i = W_b f_i / D_{i_2} = W f_i \varepsilon_i /(\mu_i d_{i_2})$$
$$h_i = H_b f_i / D_{i_2} = H f_i \varepsilon_i /(\mu_i \rho d_{i_2})$$

$$\alpha_i = c_i f_i / D_{i_2} = f_i (e_i + \gamma_i d_{i_2})/(\mu_i d_{i_2})$$
$$\beta_i = P_{i_2} f_i / D_{i_2} = f_i (p_{i_2} - \delta_i d_{i_2})/(\mu_i \rho d_{i_2})$$

This depth controlled-distortion scheme allows to obtain the parameters of a shooting layout producing desired 3D content for any rendering device and any depth distortions combination.

4.2 Perfect Depth Effect

A particular case of the depth controlled-distortion is a perfect depth effect (depth perception without any distortion compared with the depth of the shot scene). To produce a perfect depth effect (whatever the enlarging factor(s) k_{i_2}), we should configure the shooting in order to avoid the 4 potential distortions. This is obtained by making sure that the distortion parameters verify $\varepsilon_i = 1$, $\mu_i = 1$, $\rho = 1$, $\gamma_i = 0$ and $\delta_i = 0$ (cf. Fig. 4(a)). The latter condition $\delta_i = 0$ is more delicate, as it can be assured only for an observer complying to the defined overhanging position.

In case of shooting for perfect depth effect, the shooting parameters can be calculated as below:

$$P_{i_2} = p_{i_2}/k_{i_2} \quad D_{i_2} = d_{i_2}/k_{i_2}$$
$$W_b = W/k_{i_2} \quad H_b = H/k_{i_2}$$

$$c_i = e_i/k_{i_2}$$

f_i imposed or chosen, individually $\forall i \in \{1...n\} \times \{1...m\}$,
by lot $\forall i_2 \in \{1...n\}$ or on the whole.   (16)

$$w_i = W_b f_i/D_{i_2} = W f_i/d_{i_2}$$
$$h_i = H_b f_i/D_{i_2} = H f_i/d_{i_2}$$

$$\alpha_i = c_i f_i/D_{i_2} = e_i f_i/d_{i_2}$$
$$\beta_i = P_{i_2} f_i/D_{i_2} = p_{i_2} f_i/d_{i_2}$$

This particular case is very interesting for its realism *i.e.* in order to convince financiers or deciders, it may be important to give the real volumetric perception of a building, or a mechanical piece, in a computer aided design (CAD) application, or medical visualization software, in a surgical simulation application.

5   Derived Shooting Technologies

This work is the result of a collaboration between a research laboratory and a company. In this context and thanks to these design schemes, we have created 3 different technologies to shoot 3D scenes: multi-viewpoint computer graphics software, photo rail and camera system. These products have been developed under the "3DTV Solutions" brand and patents are pending for each of them. We have developed 3 solutions to obtain qualitative photo or video content for any relevant kind of scene, still or animated, real or virtual. We use anaglyph to illustrate our results even if their viewing on paper or 2D screen media is not optimum because the images have been calculated to be rendered on specific devices.

5.1   3D Computer Graphics Software

Thanks to the previous shooting design scheme, we are able to place the virtual sensors around a standard monocular camera according to the chosen viewing device in order to obtain the desired depth effect. In this case, virtual cameras are perfect and there is no issue with distortions due to sensors or lenses.

Thus we have developed plugins and software (3DVizCAD, 3DVizMED and 3DTricks) to visualize and handle in real-time files from CAD software such as AutoCAD, Archicad, Pro/Engineer, etc. as well as medical data, such as MRI. We are going to apply this technology to other virtual scenes. In those software pieces, we choose the rendering device parameters and the desired depth effect, and the software computes and uses its virtual shooting layout. It is possible to record different rendering devices and depth effect distortions so as to switch easily between these

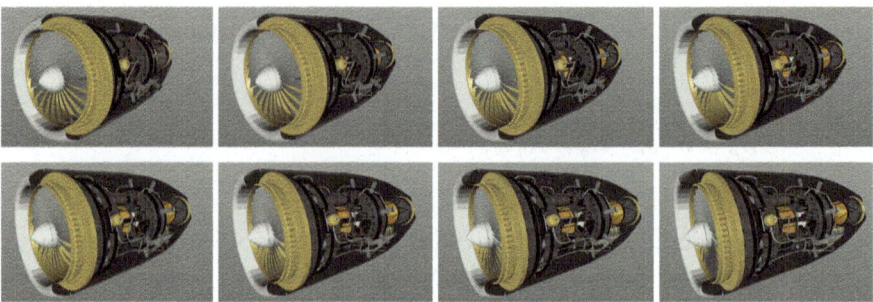

(a) One temporal frame: a set of 8 images

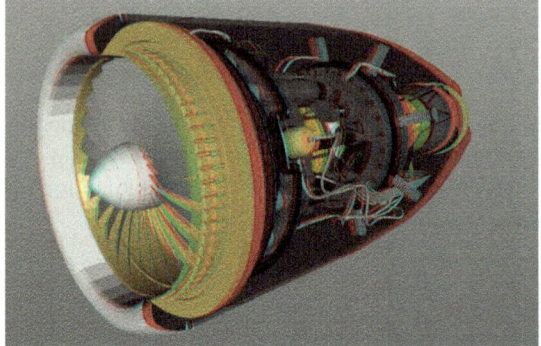

(b) Anaglyph of images 3 and 4[1]

Fig. 5 Image synthesis of a jet engine

devices and these distortions. Those pieces of software currently handle scenes up to 7 million polygons at interactive rate. Figure 5 shows an example of images of a jet engine part shot as the software was tuned to achieve a perfect depth effect on an autostereoscopic parallax display 57"[1] (optimal viewing distance 4.2 m).

We have also provided existing 3D software with a new functionality of stereo-vision: wrapping graphics stream for multiview rendering. For example, let us mention Google Earth (cf. Figure 6) which could be transformed into a virtual tourism application by means of stereo-vision. This way all the tools offered by this application, like land use impact analysis or location-based data representation, will be improved in their realism and relevance.

---

[1] The 3D content has been produced for autostereoscopic display. Obviously, it can only be experimented with the chosen device and in no way upon 2D media such as paper or conventional display. Nevertheless, anaglyph helps the reader to notice the depth effect on such 2D media.

(a) Statue of Liberty, New-York.

(b) New-York.

(c) The Dome, New-Orleans.

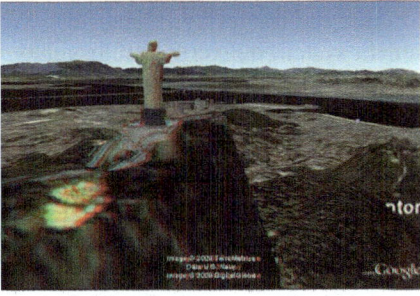

(d) Corcovado, Rio de Janeiro.

Fig. 6 Different anaglyph views of 3D models included in Google Earth[1]

5.2 3D Devices

These devices are complying with the scheme to produce qualitative 3D content on a given rendering device and ensuring a desired depth effect. The goal of these devices is to shoot different real scenes, such as photos of still or animated scenes as well as video of animated scenes. For this, we have created two types of devices: photo rail and camera system.

By using the photo rail (Fig. 7(a)) with its controlling software it is possible to control both the usual operations of a professional digital camera and its movement along a linear axis parallel to its sensor rows. This allows us to carry out any shooting

(a) Photo rail

(b) First camera system

(c) Second camera system

Fig. 7 3D Devices

configuration whatever the chosen depth distortion settings and viewing device if we crop the needed capture area CA_i in each digital photo in order to comply with the needed shooting geometry. With this photo rail, there is a possibility of distortion due to the digital camera, but distortions will be consistent for all images and will be of negligible magnitude, as it is professional equipment. We have not tried to correct those possible distortions but such a task could be done easily.

Thanks to the previous shooting design method, we know how to create a camera system containing several couples of lenses and image sensors in order to produce simultaneously the multiple images required by an autostereoscopic display with a desired depth effect. As these couples are multiple, their induced distortions can be different. We have introduced a couple-by-couple process of calibration/correction based upon the model by Zhang [23]. We have already produced two prototypes of camera system delivering multi-video stream in real-time (25 Hz). Their layout parameters have been defined for no distortion of specific scenes (see below) and set at manufacturing. The first camera system (Fig. 7(b)) allows to shoot a life size scene (ratio $k_i = 1$) of the bust of a person to be viewed on an autostereoscopic parallax display 57" (optimal viewing distance 4.2 m). The second camera system (Fig. 7(c)) enables to shoot small size objects (in the order of 10-20 cm) and to display them on an autostereoscopic lenticular display 24" (optimal viewing distance 2.8 m) with an enlargement factor set to $k_i = 1,85$. According to numerous viewers both novice and expert, the 3D perception is really good.

For example, Figure 8 illustrates the shooting of a room in "Musée Automobile Reims Champagne" [24] in Reims with a perfect depth effect for autostereoscopic parallax display 57" (optimal viewing distance 4.2 m). We made a 3D shooting of a big hall with a significant depth[1].

5.3   Combination of Real and Virtual 3D Scenes

The work reported in this chapter is included in an overall project, which carries the combination of real and virtual 3D scenes. Then, one speaks about 3D augmented reality. This could be applied to autostereoscopic displays in a straightforward way by adding virtual objects on each image. However it is much more interesting to use the depth information of the real scene so that virtual objects could be hidden by real ones. To that end, it is necessary to obtain one depth map for each view. The particular context of images destined to autostereoscopic displays allows working on a simplified geometry: no rectification is needed, epipolar couples are horizontal lines of same rank and disparity vectors are thus aligned along the abscissa. The aim is to obtain a good estimation of depth in any kind of scene, without making any assumption about its content. In our project, Niquin et al. [25] have been working on this subject and have presented their first results on accurate multi-view depth reconstruction with occlusions handling. They have worked on a new approach to handle occlusions in stereovision algorithms in the multiview context using images destined to autostereoscopic displays. It takes advantage of information from all views and ensures the consistency of their disparity maps. For example, Figure 9

(a) One temporal frame: a set of 8 images

(b) Anaglyph of images 3 and 4[1]

Fig. 8 The shooting in "Musée Automobile Reims Champagne" in Reims

Fig. 9 3D augmented reality image resulting from Niquin's works [25], the real scene is a room in "Palais du Tau" with courtesy of Monum[1]

illustrates the shooting of a room in "Palais du TAU" [26] in which we had a virtual rabbit[1].

6 Conclusion

This work models geometrical distortions between the shot scene and its multiscopically viewed avatar. These distortions are related to geometrical parameters of both the shooting and rendering devices or systems. This model enables quantitative objective assessments on the geometrical reliability of any multiscopic shooting and rendering couple.

The formulas expressing distortion parameters from geometrical characteristics of the shooting and rendering devices have been inverted subsequently in order to express the desirable shooting layout yielding a chosen distortion scheme upon a chosen rendering device. This design scheme *a priori* insures that the 3D experience will meet the chosen requirements for each expected observer position. Such a scheme may prove highly valuable for applications needing reliable accurate 3D perception or specific distortion effects.

From this design scheme we have produced several shooting technologies ensuring the desired depth effect upon a pre-determined rendering device. The proposed shooting technologies cover any needs of multi-viewpoint scene shooting (real/virtual, still/animated, photo/video).

This work proposes several perspectives. We are developing a configurable camera with flexible geometric parameters in order to adapt to a chosen rendering device and a desired depth effect. Thus, we could test different depth distortions for the same scene. Moreover, we could produce qualitative 3D content for several rendering devices from a single camera box.

We will need to do some experiments on a demanding subject as we have to validate that the perception is geometrically conform to our expectations. This will require a significant panel of viewers but also to define and set up the perception test which will permit to precisely quantify the distances between some characteristic points of the perceived scene.

Acknowledgements. We would you like to thank the ANRT, the French National Agency of Research and Technology for its financial support. The work reported in this chapter was supported as part of the CamRelief project by the French National Agency of Research. This project is a collaboration between the University of Reims Champagne-Ardenne and TéléRelief. We would like to thank Michel Frichet, Florence Debons and the staff for their contribution to the project.

References

1. Sanders, W.R., McAllister, D.F.: Producing anaglyphs from synthetic images. In: Proc. SPIE Stereoscopic Displays and Virtual Reality Systems X, Santa Clara, CA, USA (2003)

2. Dubois, E.: A projection method to generate anaglyph stereo images. In: Proc. IEEE Int. Conf. Acoustics Speech Signal Processing, Salt Lake City, UT, USA (2001)

3. Blach, R., Bues, M., Hochstrate, J., Springer, J., Fröhlich, B.: Experiences with Multi-Viewer Stereo Displays Based on LC-Shutters and Polarization. In: IEEE VR Workshop Emerging Display Technologies, Bonn, Germany (2005)

4. Perlin, K., Paxia, S., Kollin, J.S.: An autostereoscopic display. In: SIGGRAPH 2000 Proceedings of the 27th annual conference on Computer graphics and interactive techniques, New York, NY, USA (2000)

5. Dodgson, N.A.: Analysis of the viewing zone of multi-view autostereoscopic displays. In: Proc. SPIE Stereoscopic Displays and Applications XIII, San Jose, California, USA (2002)

6. Müller, K., Smolic, A., Dix, K., Merkle, P., Kauff, P., Wiegand, T.: View Synthesis for Advanced 3D Video Systems. EURASIP Journal on Image and Video Processing (2008)

7. Güdükbay, U., Yilmaz, T.: Stereoscopic View-Dependent Visualization of Terrain Height Fields. IEEE Transactions on Visualization and Computer Graphics 8(4), 330–345 (2002)

8. Yilmaz, T., Gudukbay, U.: Stereoscopic urban visualization based on graphics processor unit. SPIE: Optical Engineering 47(9), 097005 (2008)

9. Sheng, F., Hujun, B., Qunsheng, P.: An accelerated rendering algorithm for stereoscopic display. Computers & graphics 20(2), 223–229 (1996)

10. Faugeras, O., Luong, Q.T., Papadopoulou, T.: The Geometry of Multiple Images: The Laws That Govern The Formation of Images of A Scene and Some of Their Applications. MIT Press, Cambridge (2001)

11. Hartley, R., Zisserman, A.: Multiple view geometry in computer vision. Cambridge University Press, Cambridge (2000)

12. Jones, G.R., Lee, D., Holliman, N.S., Ezra, D.: Controlling Perceived Depth in Stereoscopic Images. In: Proc. SPIE Stereoscopic Displays and Virtual Reality Systems VIII, San Jose, CA, USA (2001)

13. Held, R.T., Banks, M.S.: Misperceptions in stereoscopic displays: a vision science perspective. In: APGV 2008 Proceedings of the 5th symposium on Applied perception in graphics and visualization, Los Angeles, CA, USA (2008)

14. Woods, A.J., Docherty, T., Koch, R.: Image distortions in stereoscopic video systems. In: Proc. SPIE Stereoscopic Displays and Applications IV, San Jose, CA, USA (1993)

15. Wöpking, M.: Viewing comfort with stereoscopic pictures: An experimental study on the subjective effects of disparity magnitude and depth of focus. Journal of the Society for Information Display 3(3), 101–103 (1995)

16. Yeh, Y.Y., Silverstein, L.D.: Using electronic stereoscopic color displays: limits of fusion and depth discrimination. In: Proc. SPIE Three-Dimensional Visualization and Display Technologies (1989)

17. Peinsipp-Byma, E., Rehfeld, N., Eck, R.: Evaluation of stereoscopic 3D displays for image analysis tasks. In: Proc. SPIE Stereoscopic Displays and Applications XX, San Jose, CA, USA (2009)

18. Hill, A.J.: A Mathematical and Experimental Foundation for Stereoscopic Photography. SMPTE journal (1953)

19. Son, J.Y., Gruts, Y.N., Kwack, K.D., Cha, K.H., Kim, S.K.: Stereoscopic image distortion in radial camera and projector configurations. Journal of the Optical Society of America A 24(3), 643–650 (2007)

20. Yamanoue, H.: The relation between size distortion and shooting conditions for stereoscopic images. SMPTE journal 106(4), 225–232 (1997)

21. Yamanoue, H.: The Differences Between Toed-in Camera Configurations and Parallel Camera Configurations in Shooting Stereoscopic Images. In: IEEE International Conference on Multimedia and Expo., pp. 1701–1704 (2006)
22. Dodgson, N.A., Moore, J.R., Lan, S.R.: Time-multiplexed autostereoscopic camera system. In: Proc. SPIE Stereoscopic Displays and Virtual Reality Systems IV, San Jose, CA, USA (1997)
23. Zhang, Z.: A flexible new technique for camera calibration. IEEE Transactions on Pattern Analysis and Machine Intelligence 22(11), 1330–1334 (2000)
24. Musée Automobile Reims Champagne, http://www.musee-automobile-reims-champagne.com/ (accessed October 29, 2009)
25. Niquin, C., Prévost, S., Remion, Y.: Accurate multi-view depth reconstruction with occlusions handling. In: 3DTV-Conference 2009 The True Vision - Capture, Transmission and Display of 3D Video, Postdam, Germany (2009)
26. Palais du Tau, http://palais-tau.monuments-nationaux.fr/ (accessed October 29, 2009)

Chapter 17
3D Visual Experience

Péter Tamás Kovács and Tibor Balogh

Abstract. The large variety of different 3D displaying techniques available today can be confusing, especially since the term "3D" is highly overloaded. This chapter introduces 3D display technologies and proposes a categorization that can help to easily grasp the essence of specific 3D displays that one may face, regardless of the often confusing and ambiguous descriptions provided by manufacturers. Different methods for creating the illusion of spatial vision, along with the advantages and disadvantages will be analyzed. Specific examples of stereoscopic, autostereoscopic, volumetric and light-field displays emerging or already available in the market are referenced. Common uncompressed 3D image formats preferred by each display technology are also discussed.

1 Introduction

The chapter will go through the main technologies used for implementing 3D displays using the four top level categories of the "3D display family tree" created by the 3D@Home Consortium, Steering Team 4 [1]. It will take a different approach from that of the family tree detailing the main categories based on selected driving technologies that the authors think the most important. Other categorizations of 3D displays might exist, hopefully this one helps to understand the main trends and easily grasp the technology underlying different 3D displays.

The chapter strictly focuses on technologies that generate spatial vision, so it does not cover for example displays that project a floating 2D image using a fresnel lens, or displays that project 2D images on some surface(s).

2 Stereoscopic Displays

Stereoscopic displays [2,3] simulate 3D vision by showing different images to the eyes. The two images are either shown on a traditional 2D display, projected onto

Péter Tamás Kovács · Tibor Balogh
Holografika Kft
Baross u. 3. H-1192 Budapest
Hungary
e-mail: p.kovacs@holografika.com, t.balogh@holografika.com

a special surface, or projected separately to the eyes. Stereoscopic displays by definition all require some kind of eyewear to perceive 3D (otherwise they are called autostereoscopic, as seen later). Separation of the two images, corresponding to the left and right eye happens either time-sequentially, or by means of differentiating wavelength or polarization.

2.1 Time Sequential Separation

In the time sequential case, left and right images are displayed on LCD or PDP or projected one after the other, and then separated by shutter glasses that block incoming light to one eye at a time, alternating the blocked eye with the same frequency as the display changes the images. Such shutter glasses are usually implemented with LCDs, which become transparent and opaque synchronized with the display. Several companies provide shutter glasses based 3D solutions including LG [4], Panasonic [5], Toshiba [4], eDimensional [6] and NVIDIA[9], projectors with high refresh rate for stereoscopic operation [7,8], and NVIDIA also provides a stereo driver to use the glasses with PC games [9]. A stylish NVIDIA shutter glass can be seen in Fig. 1, with the IR sensor used for synchronization in the frame of the glasses.

Fig. 1 NVIDIA 3D Vision Glasses. Image courtesy of NVIDIA Corporation.

2.2 Wavelength Based Separation

Wavelength based separation is achieved by tinting the left and right images using different colours, overlaying the two and displaying the resulting 2D image. Separation is done by glasses with corresponding colour filters in front of the eyes, as done in the well known red-blue or red-green glasses. This method of creating stereoscopic vision is often referred to as the anaglyph method. The main advantage of anaglyph is that all signals and displaying requirements match 2D displaying requirements, thus existing storage, transmission and display systems can readily be used to show 3D imagery, only coloured glasses are needed (which is inexpensive, and often packaged together with an anaglyph "3D" DVD). This is possible because the left and right images are overlapped and separated by means of colour differences. A sample anaglyph image is shown in Fig. 2 where the two differently tinted overlapped images are clearly visible. This causes the main disadvantage of this technology, that is, colours are not preserved correctly, and ghosting artefacts are also present. Because of its simplicity, anaglyph stereoscopic videos are appearing on YouTube, and also hundreds of games support anaglyph mode using NVIDIA 3D Vision™ Discover.

Fig.2 Anaglyph image. Image courtesy of Kim Scarborough.

A similar method better preserving colours apply narrow-band colour filters, separating the left and right images with wavelength triplets biased in a few 10 nm range, less visible to human perception [10].

2.3 Polarization Based Separation

Polarization based separation exploits the possibility of polarizing light and filtering them with polar filters. The two images are projected through different polarization filters onto a surface that reflects light toward viewers, keeping the polarization of the incoming light (mostly) unmodified. Viewers wearing glasses with the respective filters in front of the eyes can then perceive a stereoscopic view. A popular example of this technology can be experienced in most 3D cinemas [11,12].

Light can be polarized either linearly or circularly. In the first case, the left and right images pass through two perpendicular linear polarizers and then projected onto a surface. The reflected images then pass through the respective polarizing filters that are embedded into glasses, separating the left and right images. The downside of linear polarization is that the image degrades when a user tilts her head, as separation does not work as intended with this orientation. Circular polarization overcomes this problem being invariant to head tilt. In this case one image is polarized with clockwise, the other with counter-clockwise direction.

The advantage of the polarization based stereoscopic technique is that it keeps image colours intact (unlike anaglyph), with glasses that are relatively cheap, however the overall brightness is challenged and some cross-talk is always present.

One way of generating a pair of polarized images is by using two projectors, one projecting the left eye image with a polarizing filter in front of it, the other projecting the right eye image with orthogonal polarization [13,14]. There is also a single-projector technique, in which a rotating polarizator wheel or an LCD polarization modulator is used in the projector to change the direction of polarization of every second frame [15]. One needs a special projection screen to reflect polarized images, as surfaces used for 2D projection do not maintain the polarization of the reflected light. Previously silver screens have been used, now specialized materials are available for this purpose [16]. Polarized stereo images can also be created using two LCD monitors with perpendicular polarization arranged with a passive beamsplitter (half-mirror) at a bisecting angle between the displays. The resulting stereo image pair can be seen directly with polarizing glasses [17,18], as shown in. Fig. 3.

Another approach to create polarized images is using a patterned micro-polarizer sheet (also called x-pol or micro-pol), which is placed on the surface of a 2D LCD panel. The sheet is aligned with the rows on the LCD panel so that pixels in the even row will be polarized clockwise, pixels in the odd row will polarized in reverse, as shown in Fig. 4. Providing corresponding line interleaved stereoscopic images for the display will result in a 3D effect when using circularly polarized glasses (although with resolution reduced by half). Some manufacturers providing such displays are LG [4] and Zalman [19], but 3D laptops using this technology also appeared from Acer [20].

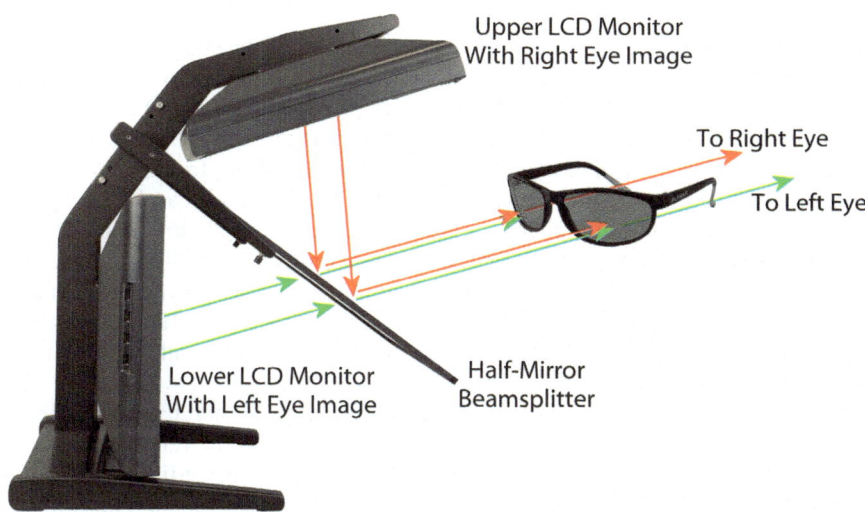

Fig. 3 Polarization based stereoscopy using two flat screens. Image courtesy of Planar Systems, Inc.

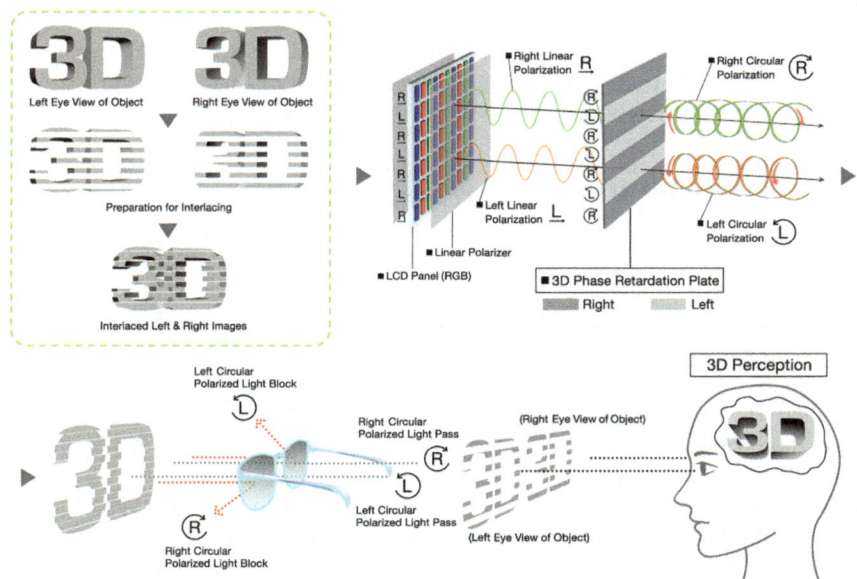

Fig. 4 Principle of micro-polarization. Image courtesy of Zalman Tech Co., Ltd.

2.4 Discussion of Stereoscopic Systems

Stereoscopic techniques are definitely the simplest and cheapest, thus the most widespread methods to generate 3D vision. On the other hand, they come with several drawbacks. A stereo image with glasses provides correct 3D images only from a single point of view. Observing the same image from other locations results in distorted views, which is most visible while moving in front of the screen, when the image "follows" the viewer. Although this limitation can be overcome by tracking the position / orientation / gaze of the user and updating images in response to movements [21], some latency will inherently be introduced [22], significantly compromising immersiveness and limiting the correct view to a single (tracked) user. This and other missing 3D cues result in effects like discomfort, sea sickness, nausea and headache which make them inconvenient for long-term use according to some users [23].

One possible explanation comes from neuroscientists' research in the field of human perception of 3D. They found that showing each eye its relevant image is not enough for the brain to understand the 3D space [24]. For getting the 3D picture of the environment, humans rely on two main visual cues: the slightly different image seen by each eye and the way the shape of an object changes as it moves. A brain area, the anterior intraparietal cortex (AIP), integrates this information [25]. With a stereoscopic display the image becomes 3D, but as soon the brain thinks that it does see a 3D image, it starts working like in a normal 3D world, employing micro head movements to repeatedly and unconsciously check the 3D model built in our brain. When an image on a stereo display is checked and the real 3D world mismatches the 3D image, the trick is revealed. Presumably the AIP cortex never got used to experience such 3D cue mismatch during its evolution and this produces glitches which result in unwanted effects.

2.5 Stereoscopic 3D Uncompressed Image Formats

Stereoscopic displays need two images as input (left eye and right eye image), which seems to be simple, yet various formats exist. The most straightforward solution is having two different images making up a 3D frame (see Fig. 5.), but this requires double bandwidth compared to the 2D case.

Another common approach uses an image and a corresponding depth image often called 2D + Depth (see Fig. 6.), which may consume less bandwidth depending on the bit depth of the depth map, but needs metadata to map depth information to the 3D context, and still consumes more than a 2D image.

The 2D + Delta format stores the left (or right) video stream intact, and adds the stereo disparity or delta image that is used to reconstruct the other view. The advantage is that compressed Delta information can be embedded into an MPEG stream in a way that does not affect 2D players, but provides stereoscopic information to compatible 3D decoders [26].

To make the transition from 2D to 3D easier, broadcasters and manufacturers preferred stereoscopic image formats that can be fit into a 2D frame compatible

Fig. 5 Left-right image pair. Image courtesy of NXP Semiconductors.

Fig. 6 Image plus depth map. Image courtesy of NXP Semiconductors.

format, in order to defer the upgrade of the transmission infrastructure. Some examples of such formats include frame doubling, side-by-side, interleaved and checkerboard, which can be seen in Fig. 7.

The frame doubling approach uses a single 2D stream to transmit alternating left and right images, halving the effective frame rate. This is the most suitable format for shutter-glass based systems and 3D projectors using rotating polarizers.

Side-by-side places the left and right images next to each other. This either requires doubled horizontal resolution, or halves the horizontal resolution of left and right images, fitting them in the original 2D image size. A very similar image configuration is over/under.

Interleaving places rows of the left view into even lines, and rows of the right view into odd lines (or the same reversed). As with side-by-side, two possibilities are doubling image size and keeping the resolution of the images or halving the resolution of the component images to fit them into a 2D frame with the same size. Interleaving can also work in a vertical configuration. This representation is the best choice for a 3D display based on micro-polarizers.

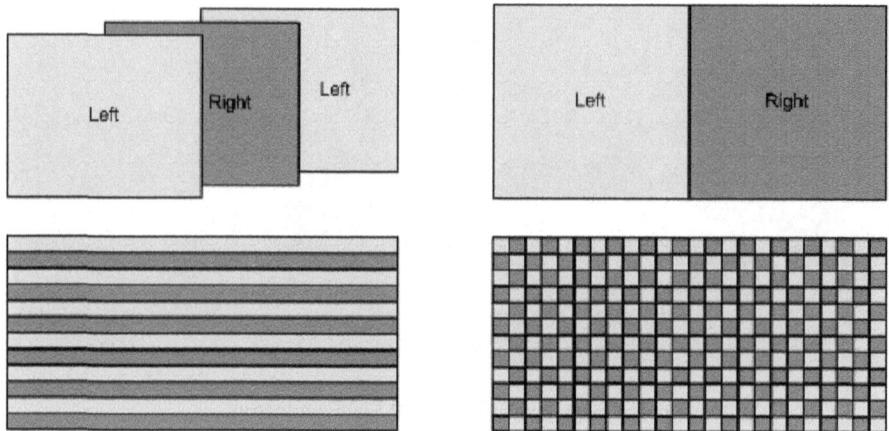

Fig. 7 Stereoscopic image formats (from left to right, top to bottom): Frame doubling, Side-by-side, Interleaved and Checkerboard. Image courtesy of NXP Semiconductors.

The checkerboard format mixes pixels of the left and right images so that they alternate in one row, and alternate the reverse way in the next row. This makes better interpolation of the missing pixels possible when reconstructing the left and right images. This representation is used by Texas Instruments DLPs.

The High-Definition Multimedia Interface (HDMI) supports stereoscopic 3D transmission starting from version 1.4 of the HDMI specification. It defines common 3D formats and resolutions for supporting 3D up to 1080p resolution and supports many 3D formats including frame doubling, side-by-side, interleaving and 2D+depth. There are two mandatory 3D formats defined, which must be supported by all 3D display devices: 1080p@24Hz and 720p@50/60Hz [27].

2.6 Multi-user Stereo and CAVE Systems

A common extension of stereoscopic projection systems is using them in CAVEs [28] that use three to six walls (possibly including the floor and ceiling) as stereoscopic 3D projection screens. The users entering the CAVE wear glasses for stereoscopic viewing, one of them (commonly referred to as "leader" or "driver") wearing extra equipment for tracking. Since the stereo pairs are generated for a single point of view that of the driver, using stereoscopic 3D for multiple users is problematic, as only the driver will perceive a correct 3D image, all others will see a distorted scene. Whenever the driver moves, the images are updated, thus all other users will see the scene moving (according to the movement of the driver), even is they stay at the same place not doing any movements, resulting in disturbing effects. Stereoscopic CAVEs are widely used for providing immersive 3D experience, but unfortunately carry all the drawbacks of stereoscopic systems.

2.7 Head Mounted Displays

A head mounted display [29] is a display device worn on the head or as part of a helmet that has a small display optic in front of both eyes in case of a binocular HMD (monocular HMDs also exist but unable to produce 3D images). A typical HMD has two small displays with lenses embedded in a helmet or eye-glasses. The display units are miniaturized and may include CRT, LCDs, LCOS, or OLED. Some HMDs also allow partial see-through thus super-imposing the virtual scene on the real world. Most HMDs also have head tracking functionality integrated. From the 3D vision point of view, they are equivalent to glasses based systems. HMD manufacturers include Cybermind [30], I-O [31], Rockwell Collins [32], Trivisio [33], Lumus [34].

3 Autostereoscopic Displays

Autostereoscopic displays provide 3D perception without the need for wearing special glasses or other head-gear, as separation of left / right image is implemented using optical or lens raster techniques directly above the screen surface. In case of two views, one of the two visible images consists of even columns of pixels; the second image is made up of odd columns (other layouts also exist). The two displayed images are visible in multiple zones in space. If the viewer stands at the ideal distance and in the correct position he or she will perceive a stereoscopic image (sweet spot). Such passive autostereoscopic displays require the viewer to be carefully positioned at a specific viewing angle, and with her head in a position within a certain range, otherwise there is a chance of the viewer being in the wrong position (invalid zone) and seeing an incorrect image. This means that the viewer is forced to a fixed position, reducing the ability to navigate freely and be immersed.

To overcome the problem of invalid zones head and/or eye tracking systems can be used to refresh the images whenever the viewer is about to enter such a zone and experience an incorrect 3D image [35]. Even though there could be latency effects, such a system provides the viewer with parallax information and it is, therefore, a good solution for single user applications. Multi-user extensions of this technique are also developed [36].

Some autostereoscopic displays show stereoscopic 3D (consisting of two images), others go beyond that and display multiview 3D (consisting of more than two views). Multiview displays [37] project different images to multiple zones in space. In each zone only one image (view) of the scene is visible. The viewer's two eyes are located in different zones, seeing different images thus 3D perception is enabled. When the user moves, entering different zones will result in different views, thus a somewhat limited horizontal motion parallax effect is achieved. As the number of views ranges from 4 to 9 in current multiview displays, the transition to adjacent zones is discrete, causing „jumps" as the viewer moves. Multiview displays allow multiple simultaneous viewers, restricting them, however, to be within a limited viewing angle. The image sequences are periodically repeated in most multi-view displays, thus enabling more diamond shaped viewing positions at the expense of invalid zones in between.

Autostereoscopic displays typically use parallax barrier, lenticular sheet or wavelength selective filter which divide the pixels of the underlying, typically LCD display into two or more sets corresponding to the multiple directions.

3.1 Parallax Barrier

Parallax barrier [38] is an array of slits spaced at a defined distance from a high resolution display panel. The parallax effect is created by this lattice of very thin vertical lines, causing each eye to view only light passing through alternate image columns, allowing the well-positioned viewer to perceive stereoscopic 3D, as shown In Fig. 8. Parallax barrier-based displays typically show stereoscopic 3D made up of two images, but with the proper choice of distance and width of the slit multi-view effect can be provided. Parallax barrier systems are less efficient in terms of light output, thus the image gets darker than in 2D, especially in case of multiple views.

Parallax barrier displays are making their way to mobile devices, as they can be easily implemented in small size. One example is a 3.07" size WVGA 3D LCD

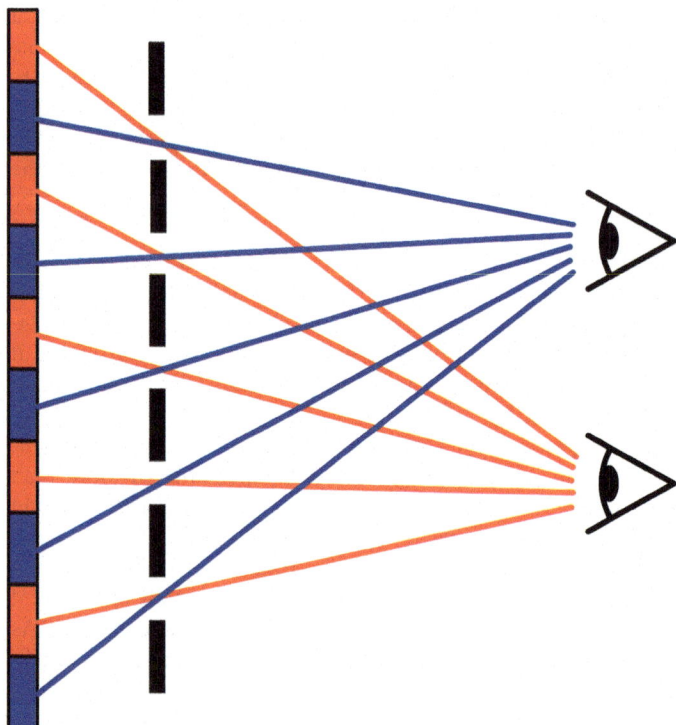

Fig. 8 Principle of parallax barrier based stereoscopic vision

from Masterimage with an integrated, configurable parallax barrier layer on top of the LCD (2D, portrait 3D, landscape 3D). Such displays make the manufacturing of 3D-enabled handheld devices like the Hitachi Wooo H001 possible [39].

Parallax barrier display manufacturers include Spatial View [40], Tridelity [41] and NewSight [42].

3.2 Lenticular Lens

Lenticular lens [37] based displays, which are the most common for implementing multiview 3D, use a sheet of cylindrical lens array placed on top of a high resolution LCD in such a way that the LCD image plane is located at the focal plane of the lenses. The effect of this arrangement is that different LCD pixels located at different positions underneath the lenticular fill the lenses when viewed from different directions. Provided these pixels are loaded with suitable 3D image information, 3D effect is obtained in which left and right eyes see different but matching information, as shown in Fig. 9. Both parallax barrier and lenticular lens based 3D displays require the user to be located at a specific position and distance to

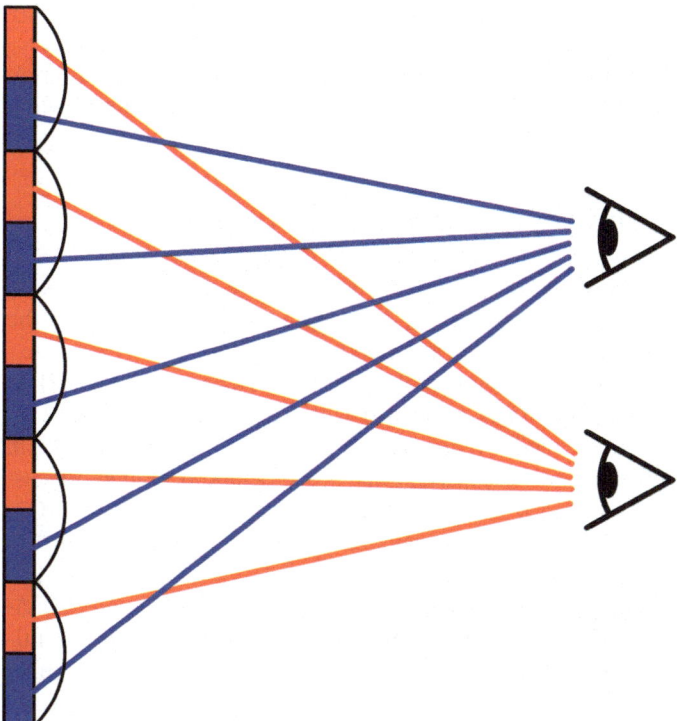

Fig. 9 Principle of lenticular lens based stereoscopic vision

correctly perceive the stereoscopic image, as incorrect positioning results in incorrect images reaching the eye. A major disadvantage of lenticular lens based systems is their inability to use the displays in 2D with full resolution.

Lenticular 3D display manufacturers include Alioscopy [43], Philips (now retired from 3D display business) [44], NEC [4] and Tridelity [41].

Since both parallax barrier and lenticular lens based displays require a flat panel display underneath, the size of these 3D displays is always limited by the maximum size of such panels manufactured. As of November 2009, the maximum size is slightly more than 100 inches diagonal. Since tiling such displays is not seamless, these technologies are not scalable to arbitrary large sizes.

3.3 Wavelength Selective Filters

Another possible implementation is using wavelength selective filters for the multi-view separation. The wavelength-selective filter array is placed on a flat LCD panel oriented diagonally so that each of the three colour channels correspond to a different direction, creating the divided viewing space necessary for 3D vision. A combination of several perspective views (also combining colour channels) is displayed. The filter array itself is positioned in front of the display and transmits the light of the pixels from the combined image into different directions, depending on their wavelengths. As seen from the viewer position different spectral components are blocked, filtered or transmitted, separating the viewing space into several zones where different images can be seen [45].

3.4 Multiview 3D Uncompressed Image Formats

Common image formats used by multi-view displays include multiple images on multiple links, 2D+Depth (described earlier), 2D+Depth with two layers, and the extension of frame-doubling, side-by-side and interleaving to the multi-view case.

Using multiple links, the same number of display interfaces are provided as many views the display have (possibly combined with side-by-side or similar, reducing the number of links needed). When used for multi-view, the 2D + Depth approach is often criticized for missing parts of the scene behind occluded objects. This effect is somewhat reduced by using two layers, that is 2D + Depth + Occluded 2D + Occluded depth, what Philips calls Declipse format. An example 3D image in Declipse format can be seen in Fig. 10.

Frame doubling, side-by-side and interleaving (either horizontal or vertical), as described at stereoscopic displays can be naturally extended for using with multiple views. However, if the resolution of the image is to be kept, even more significant reduction in the resolution of the component images is required. We have to note that in case of multi-view displays, the resolution of the individual views is divided anyway as it cannot have more pixels than the underlying LCD panel.

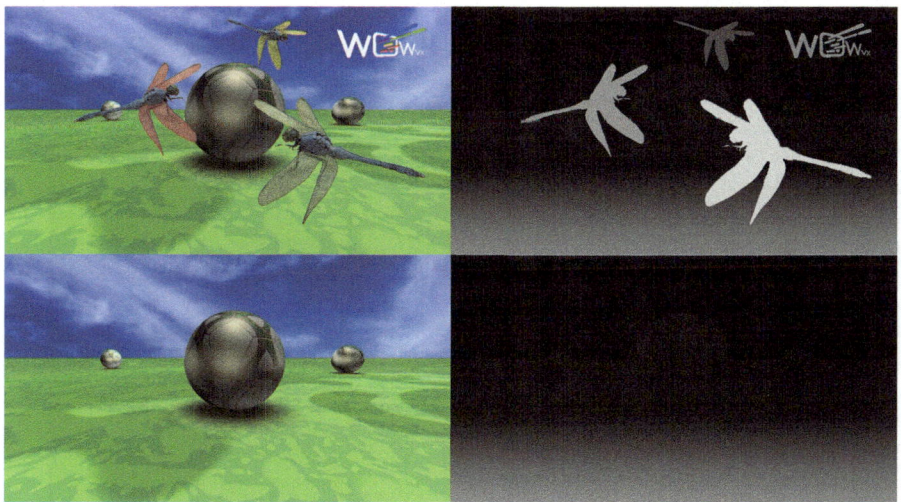

Fig. 10 2D image + depth + occluded image + occluded depth. Image courtesy of Philips Electronics N.V.

As a general rule for multi-view systems, the resolution seen in a direction is equal to the native resolution of the underlying display panel divided by the number of views.

4 Volumetric Displays

Volumetric displays use a media positioned or moved in space on which they project/reflect light beams so they are scattered/reflected from that point of space. The media used is generally a semi-transparent or diffuse surface. Among volumetric displays there are exotic solutions like the laser induced plasma explosions [46]. In general they are less conform to displaying conventions and in most cases follow the "looking into" instead of "looking out" philosophy.

One possible solution is a moving screen on which different perspectives of the 3D object are projected. A well known solution [47] is a lightweight screen sheet that is rotated at very high speed in a protecting globe and the light beams from a DLP microdisplay are projected onto it. Such a display is shown in Fig. 11. Employing proper synchronization it is possible to see 3D objects in the globe [48]. Such systems can be considered time-multiplexing solutions, where number of the displayable layers or voxels is determined by the speed of the projection component. A similar solution is the usage of rotated LED arrays as the emissive counterpart of the reflective moving media.

Another technique in volumetric display technology is using two or more LCD layers as a projection screen, creating the vision of depth. Deep Video Imaging and PureDepth [49] produced a display consisting two LCDs. The depth resolution

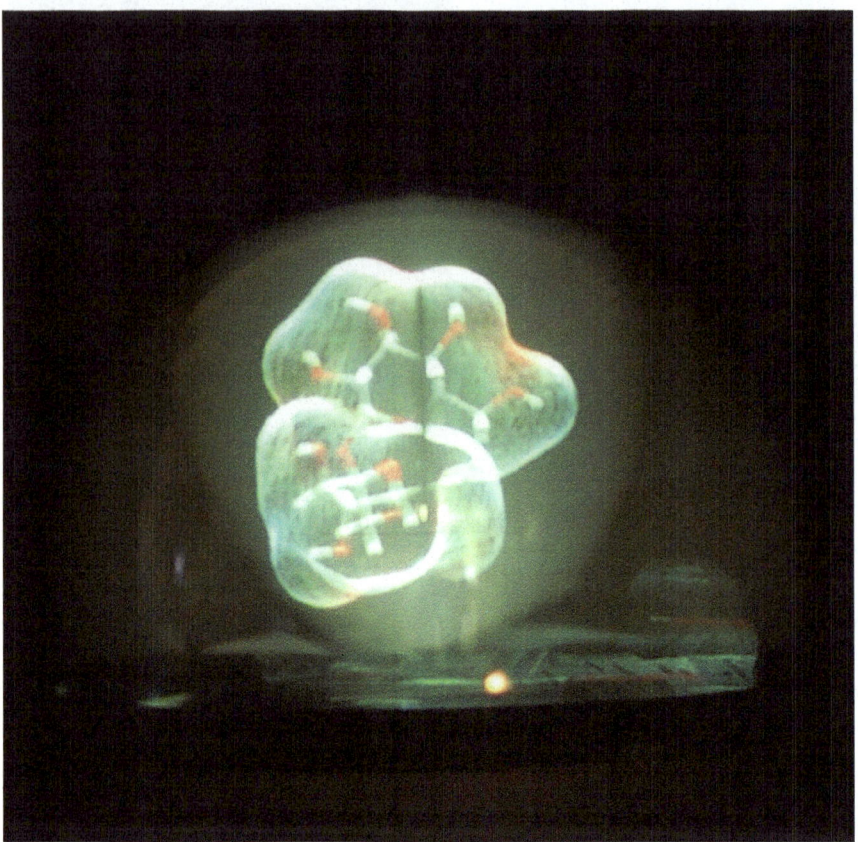

Fig. 11 Perspecta volumetric display from Actuality. Image courtesy of Actuality Systems, Inc.

equals 2, enabling special foreground-background style content only, which is hard to qualify as 3D. The DepthCube display [50] from LightSpace Technologies shown in Fig. 12 has 20 layers inside. The layers are LCD sheets that are transparent / opaque (diffuse) when switched on/off, and are acting as a projection screen positioned in 20 positions. Switching the 20 layers is synchronized to the projection engine, inside which an adapting optics is keeping the focus.

Disadvantages of volumetric displays are scalability and the ability to display occlusion, since the light energy addressed to points in space cannot be absorbed by foreground pixels. The problem of occlusion has been recently solved by using an anisotropic diffuser covering a rapidly spinning mirror [51]. As of advantages, both vertical and horizontal parallax is provided by principle.

The natural data format for volumetric displays is layered images (in the layered case) or image sequence showing the scene from all around (in the rotating case).

Fig. 12 DepthCube volumetric display. Image courtesy of LightSpace Technologies Inc

5 Light Field Displays

5.1 Integral Imaging

Integral imaging [52] 3D displays use a lens array and a planar display panel. Each elemental lens constituting the lens array forms each corresponding elemental image based on its position, and these elemental images displayed on the panel are integrated forming a 3D image. Integral imaging can be though of as a 2D extension of lenticular lens based multiview techniques, providing both horizontal and vertical parallax. Real-time generation of integral images from live images has been demonstrated [53].

Its disadvantages are narrow viewing angle and reduced resolution. The viewing angle within which observers can see the complete image is limited due to the restriction of the area where each elemental image can be displayed. Each elemental lens has its corresponding area on the display panel. To prevent image flipping the elemental image that exceeds the corresponding area is discarded optically in direct pick up method or electrically in computer-generated integral imaging method. Therefore the number of the elemental images is limited and observers outside the viewing zone cannot see the integrated image.

5.2 Holographic Displays

Pure holographic systems [54] have the ability to store and reproduce the properties of light waves. Techniques for creating such holographic displays include the use of acusto-optic material and optically addressed spatial light modulators [55]. Pure hologram technology utilises 3D information to calculate a holographic pattern [56], generating true 3D images by computer control of laser beams and a system of mirrors. Compared to stereoscopic and multi-view technologies the main advantage of a hologram is the good quality of the generated 3D image. Practical application of this technology today is hampered by the huge amount of information contained in the hologram which limits its use to mostly static 3D models, in limited size and narrow viewing angle.

5.3 HoloVizio Type Light-Field Displays

Such displays follow hologram geometry rules, however direction selective light emission is obtained by directly generating the light beams instead of interference. In this way the huge amount of redundant information present in a hologram (phase, speckle) is removed and only those light beams are kept which are needed to build up the 3D view. Each point of the holographic screen emits light beams of different colour and intensity to the various directions in a controlled manner. The light beams are generated through a specially arranged light modulation system and the holographic screen makes the necessary optical transformation to compose these beams into a 3D view. The light beams cross each other in front of the screen or they propagate as if they were emitted from a common point behind the screen, as shown in Fig. 13. With proper control of the light beams viewers see objects behind the screen or floating in the air in front of the screen just like with a hologram.

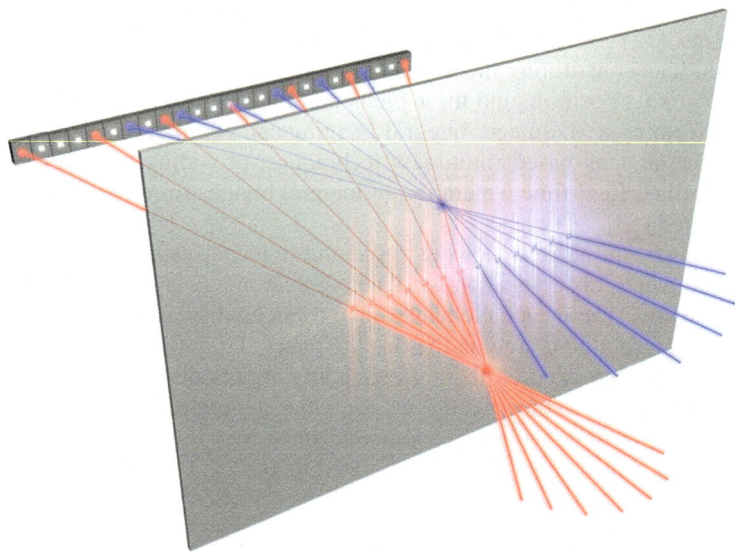

Fig. 13 Principle of HoloVizio light-field displays

The main advantage of this approach is that, similarly to the pure holographic displays, it is able to provide all the depth cues for multiple freely moving users within a reasonably large field of view. Being projection based and using arbitrary number of projection modules, this technique is well scalable to very high pixel count and display size, not being limited to the resolution of a specific display technology (like the ones using a single LCD panel). 2D compatibility is implicitly solved here, as light rays making up a 2D image are also easy to emit without any reconfiguration. These systems are fairly complex because of the large number of optical modules and the required driving/image generation electronics.

The natural data format for this kind of display is the light field [57] that is present in a natural 3D view. HoloVizio 3D displays are an implementation of this technology [58, 59].

6 Conclusion

Very different ideas have been used so far to achieve the goal of displaying realistic 3D scenes, the ultimate goal being a virtual 3D window that is indistinguishable from a real window. Most implementations of the approaches mentioned have found their specific application areas where they perform best, and they are gaining an ever growing share in visualization. 3D data is already there in a surprisingly large number of industrial applications, still visualized in 2D in most cases.

As for home use, the natural progression of technology will bring the simplest technologies to the mainstream first, and with advances in technology, cost effectiveness and increased expectations regarding 3D will eventually bring more advanced 3D displays currently only used by professionals to the homes.

References

1. 3D@Home Consortium Display Technology Steering Team 4: 3D Display Technology Family Tree. Motion Imaging Journal 118(7) (2009)
2. Holmes, O.W.: The Stereoscope and Stereoscopic Photographs. Underwood & Underwood, New York (1906)
3. Ezra, D., Woodgate, G.J., Omar, B.A., Holliman, N.S., Harrold, J., Shapiro, L.S.: New autostereoscopic display system. In: Stereoscopic Displays and Virtual Reality Systems II, Proc. SPIE 2409, pp. 31–40 (1995)
4. Insight Media: 3D Displays & Applications at SID 2009,
   http://www.insightmedia.info (accessed November 10, 2009)
5. Panasonic: Full HD 3D Technology,
   http://www.panasonic.com/3d/default.aspx
   (accessed November 10, 2009)
6. eDimensional 3D Glasses,
   http://www.edimensional.com/images/demo1.swf
   (accessed November 10, 2009)
7. Barco: Stereoscopic projectors,
   http://www.barco.com/en/productcategory/15
   (accessed November 10, 2009)

8. Viewsonic: PJD6241 projector,
   `http://ap.viewsonic.com/in/products/productspecs.php?id=382`
   (accessed November 10, 2009)
9. NVIDIA 3D Vision Experience,
   `http://www.nvidia.com/object/3D_Vision_Overview.html`
   (accessed November 10, 2009)
10. Jorke, H., Fritz, M.: INFITEC a new stereoscopic visualisation tool by wavelength multiplex imaging. In: Proc. Electronic Displays, Wiesbaden, Germany (2003)
11. RealD, `http://www.reald.com/Content/about-reald.aspx` (accessed November 10, 2009)
12. IMAX, `http://www.imax.com/corporate/theatreSystems/` (accessed November 10, 2009)
13. Valley View Tech.: VisDuo,
    `http://www.valleyviewtech.com/immersive.htm#visduo`
    (accessed November 10, 2009)
14. Barco: Passive 3D display systems with two projectors,
    `http://www1.barco.com/entertainment/en/stereoscopic/passiv` `e.asp` (accessed November 10, 2009)
15. DepthQ Polarization Modulator, `http://www.depthq.com/modulator.html` (accessed November 10, 2009)
16. Brubaker, B.: 3D and 3D Screen Technology (Da-Lite 3D Screen Whitepaper), `http://www.3dathome.org/files/products/product.aspx?produc` `t=1840` (accessed November 10, 2009)
17. Fergason, J.L., Robinson, S.D., McLaughlin, C.W., Brown, B., Abileah, A., Baker, T.E., Green, P.J.: An innovative beamsplitter-based stereoscopic/3D display design. In: Stereoscopic Displays and Virtual Reality Systems XII, Proc. SPIE 5664, pp. 488–494 (2005)
18. Robinson, S.D., Abileah, A., Green, P.J.: The StereoMirror^{TM}: A High Performance Stereoscopic 3D Display Design. In: Proc. SID Americas Display Engineering and Applications Conference (ADEAC 2005), Portland, Oregon, USA (2005)
19. Zalman Trimon 2D/3D Convertible LCD Monitor,
    `http://www.zalman.co.kr/ENG/product/Product_Read.asp?idx=219`
    (accessed November 10, 2009)
20. Insight Media: 3D Displays & Applications (August 2009),
    `http://www.insightmedia.info/` (accessed November 10, 2009)
21. Woodgate, G.J., Ezra, D., Harrold, J., Holliman, N.S., Jones, G.R., Moseley, R.R.: Observer-tracking autostereoscopic 3D display systems. In: Stereoscopic Displays and Virtual Reality Systems IV, Proc. SPIE 3012, pp. 187–198 (2004)
22. Wul, J.R., Ouhyoung, M.: On latency compensation and its effects on headmotion trajectories in virtual environments. The Visual Computer 16(2), 79–90 (2000)
23. Takada, H., Fujikake, K., Miyao, M.: On a Qualitative Method to Evaluate Motion Sickness Induced by Stereoscopic Images on Liquid Crystal Displays. In: Proc. 3rd International Conference on Virtual and Mixed Reality, San Diego, CA, USA (2009)
24. Baecke, S., Lützkendorf, R., Hollmann, M., Macholl, S., Mönch, T., Mulla-Osman, S., Bernarding, J.: Neuronal Activation of 3D Perception Monitored with Functional Magnetic Resonance Imaging. In: Proc. Annual Meeting of Deutsche Gesellschaft für Medizinische Informatik, Biometrie und Epidemi-ologie e.V. (gmds), Leipzig, Germany (2006)

25. Vanduffel, W., Fize, D., Peuskens, H., Denys, K., Sunaert, S., Todd, J.T., Orban, G.A.: Extracting 3D from motion: differences in human and monkey intraparietal cortex. Science 298(5592), 413–415 (2002)
26. TDVision Knowledgebase contribution: 3D Ecosystem, http://www.tdvision.com/WhitePapers/ TDVision_Knowledgbase_Public_Release_Rev_2.pdf (accessed November 19, 2009)
27. Park, J.: 3D over HDMI – New feature of the HDMI 1.4 Specification. In: Proc. DisplaySearch TV Ecosystem Conference, San Jose, CA, USA (2009)
28. Cruz-Neira, C., Sandin, D.J., DeFanti, T.A., Kenyon, R.V., Hart, J.C.: The CAVE: Audio Visual Experience Automatic Virtual Environment. Communications of the ACM 35(6), 65–72 (1992)
29. Heilig, M.L.: Stereoscopic-television apparatus for individual use. United States Patent 2955156 (1960)
30. Cybermind Interactive Nederland, http://www.cybermindnl.com/index.php?option=com_content&ta sk=view&id=19&Itemid=49 (accessed November 11, 2009)
31. i-O Display Systems, http://www.i-glassesstore.com/hmds.html (accessed November 11, 2009)
32. Rockwell Collins: Soldier Displays, http://www.rockwellcollins.com/products/gov/surface/soldie r/soldier-systems/index.html (accessed November 11, 2009)
33. Trivisio Prototyping, http://trivisio.com/index.php/products/hmdnte/options/hmd-options (accessed November 11, 2009)
34. Lumus http://www.lumus-optical.com/index.php?option=com_content&task=view&id=5&Itemid=8 (accessed November 11, 2009)
35. Boev, A., Raunio, K., Georgiev, M., Gotchev, A., Egiazarian, K.: Opengl-Based Control of Semi-Active 3D Display. In: Proc. 3DTV Conference: The True Vision - Capture, Transmission and Display of 3D Video, Istanbul, Turkey (2008)
36. HELIUM 3D: High Efficiency Laser-Based Multi-User Multi-Modal 3D Display, http://www.helium3d.eu (accessed November 10, 2009)
37. van Berkel, C., Parker, D.W., Franklin, A.R.: Multiview 3D-LCD. In: Stereoscopic Displays and Virtual Reality Systems III, Proc. SPIE 2653, pp. 32–39 (1996)
38. Sandin, D., Margolis, T., Ge, J., Girado, J., Peterka, T., DeFanti, T.: The Varrier autostereoscopic virtual reality display. ACM Transactions on Graphics, Proc. ACM SIGGRAPH 24(3), 894–903 (2005)
39. Insight Media: 3D Displays & Applications (April 2009), http://www.insightmedia.info/ (accessed November 10, 2009)
40. SpatialView, http://www.spatialview.com (accessed November 10, 2009)
41. Tridelity Display Solutions, http://www.tridelity.com (accessed November 11, 2009)
42. Newsight Advanced Display Solutions, http://www.newsight.com/ support/faqs/autostereoscopic-displays.html (accessed November 11, 2009)
43. Alioscopy – glasses-free 3D displays, http://www.alioscopyusa.com/content/technology-overview (accessed November 19, 2009)

44. Philips 3D Display Products, `http://www.business-sites.philips.com/3dsolutions/3ddisplayproducts/index.page` (accessed November 19, 2009)

45. Schmidt, A., Grasnick, A.: Multiviewpoint autostereoscopic dispays from 4D-Vision GmbH. In: Stereoscopic Displays and Virtual Reality Systems IX, Proc. SPIE 4660, pp. 212–221 (2002)

46. Advanced Industrial Science and Technology: Three Dimensional Images in the Air, Visualization of real 3D images using laser plasma, `http://www.aist.go.jp/aist_e/latest_research/2006/20060210/20060210.html` (accessed November 19, 2009)

47. Favalora, G.E., Napoli, J., Hall, D.M., Dorval, R.K., Giovinco, M., Richmond, M.J., Chun, W.S.: 100 Million-voxel volumetric display. In: Cockpit Displays IX: Displays for Defense Applications, Proc. SPIE 4712, pp. 300–312 (2002)

48. Jones, A., Lang, M., Fyffe, G., Yu, X., Busch, J., McDowall, I., Bolas, M., Debevec, P.: Achieving Eye Contact in a One-to-Many 3D Video Teleconferencing System. ACM Transactions on Graphics, Proc. SIGGRAPH 28(3), 64 (2009)

49. PureDepth multi layer display, `http://www.puredepth.com/technologyPlatform_ip.php?1=en` (accessed December 03, 2009)

50. Sullivan, A.: 3 Deep, `http://www.spectrum.ieee.org/computing/hardware/3-deep` (accessed October 05, 2009)

51. Jones, A., McDowall, I., Yamada, H., Bolas, M., Debevec, P.: Rendering for an interactive 360° light field display. ACM Transactions on Graphics 26(3), 40 (2007)

52. Davies, N., McCormick, M.: Holoscopic Imaging with True 3D-Content in Full Natural Colour. Journal of Photonic Science 40, 46–49 (1992)

53. Taguchi, Y., Koike, T., Takahashi, K., Naemura, T.: TransCAIP: Live Transmission of Light Field from a Camera Array to an Integral Photography Display. IEEE Transactions on Visualization and Computer Graphics 15(5), 841–852 (2009)

54. Lucente, M.: Interactive three-dimensional holographic displays: seeing the future in depth. ACM SIGGRAPH Computer Graphics 31(2), 63–67 (1997)

55. Benton, S.A.: The Second Generation of the MIT Holographic Video System. In: Proc. TAO First International Symposium on Three Dimensional Image Communication Technologies, Tokyo, Japan (1993)

56. Yaras, F., Kang, H., Onural, L.: Real-time color holographic video display system. In: Proc. 3DTV-Conference: The True Vision Capture, Transmission and Display of 3D Video, Potsdam, Germany (2009)

57. Levoy, M., Hanrahan, P.: Light Field Rendering. In: Proc. ACM SIGGRAPH, New Orleans, LA, USA (1996)

58. Balogh, T.: Method & apparatus for displaying 3D images. U.S. Patent 6,201,565, EP0900501 (1997)

59. Balogh, T.: The HoloVizio System. In: Stereoscopic displays and virtual reality systems XIII, Proc. SPIE 6055, 60550U (2006)

Chapter 18
3D Holoscopic Imaging Technology for Real-Time Volume Processing and Display

Amar Aggoun

Abstract. 3D holoscopic imaging is employed as part of a three-dimensional imaging system, allowing the display of full colour images with continuous parallax within a wide viewing zone. A review of the 3D holoscopic imaging technology from the point of view optical systems and 3D image processing including 3D image coding, depth map computation and computer generated graphics is discussed.

1 Background

Content creators always look for new forms and ways for improving their content and adding new sensations to the viewing experience. High Definition video has been the latest innovation in the area of content enrichment. 3D is the next single greatest innovation in film-making. There has been a trend in cinema in producing films with 3D enriched content such the latest animated adventure film "Beowulf".

Many different approaches have been adopted in attempts to realise free viewing 3D displays [1, 2]. Several groups [3, 4] have demonstrated stereoscopic 3D displays, which work on the principle of presenting multiple images to the viewer by use of temporal or spatial multiplexing of several discrete viewpoints to the eyes. This is achieved using either colour, polarisation or time separation techniques requiring special glasses or by creating separate optical paths to provide directional selectivity in respect of the viewed images. Some sophisticated systems additionally provide eye-tracking capability to allow the viewer to move position. Considerable effort has been invested in providing electronic stereoscopic displays suitable for entertainment and NHK Japan and Sony demonstrated a stereoscopic TV system using both 2 and 6 views in the early 1990's. Since this time autostereoscopic 3D displays are now being launched on the market by several companies worldwide such as Philips and Sharp, for use in niche applications. Multiview autostereoscopic content is captured using several cameras which renders 3D autostereoscopic video production very difficult. More recently, a combination of

Amar Aggoun
School of Engineering and Design, Brunel University, Uxbridge, UB8 3PH, (UK)
email: amar.aggoun@brunel.ac.uk

conventional 2D video capture with depth map generation have been used for the capture of multiview auto-stereoscopic 3D content. However, the display of multiview autostereoscopic 3D content relies upon the brain to fuse two disparate images to create the 3D sensation. A particularly contentious aspect for entertainment applications is the human factors issue. For example, in stereoscopy the viewer needs to focus at the screen plane while simultaneously converging their eyes to locations in space producing unnatural viewing [5, 6]. This can cause eyestrain and headaches in some people. Consequently content producers limit the depth of scene to be viewed to minimise this problem. The transmission of stereoscopic content in Korea and Japan during the 2002 World Cup showed that fast moving action caused nausea in some viewers. With recent advances in digital technology, some human factors which result in eye fatigue have been eliminated. However, some intrinsic eye fatigue factors will always exist in stereoscopic 3D technology [4, 7]. Furthermore, due to the lack of perspective continuity in 2D view systems, objects in the scene often lack solidity (cardboarding) and give rise to an 'unreal' experience.

Creating a truly realistic 3D real-time viewing experience in an ergonomic and cost effective manner is a fundamental engineering challenge. Holographic techniques demonstrate true 3D and are being researched by different groups in an effort to produce full colour images with spatial content [7, 8, 9]. Holography is a technology that overcomes the shortcomings of stereoscopic imaging and offers the ultimate 3D viewing experience, but their adoptions for 3D TV and 3D cinema are still in its infancy. Holographic recording requires coherent light which makes holography, at least in the near future, unsuitable for live capture.

3D Holoscopic imaging (also referred to as Integral Imaging) is a technique that is capable of creating and encoding a true volume spatial optical model of the object scene in the form of a planar intensity distribution by using unique optical components [10, 12, 18]. It is akin to holography in that 3D information recorded on a 2-D medium can be replayed as a full 3D optical model, however, in contrast to holography, coherent light sources are not required. This conveniently allows more conventional live capture and display procedures to be adopted. Furthermore, 3D holoscopic imaging offers fatigue free viewing to more than one person independently of the viewer's position. With recent progress in the theory and microlens manufacturing, holoscopic imaging is becoming a practical and prospective 3D display technology and is attracting much interest in the 3D area [7, 12, 20]. It is now accepted as a strong candidate for next generation 3D TV [7].

2   3D Holoscopic Content Generation

The first 3D holoscopic imaging method was "Integral Photography". It was first proposed by G. Lippmann [10] in 1908. To record an integral photograph Lippmann used a regularly spaced array of small lenslets closely packed together in contact with a photographic emulsion as shown in figure 1a. Each lenslet views the scene at a slightly different angle to its neighbour and therefore a scene is captured from many view points and parallax information is recorded. After processing, if the photographic transparency is re-registered with the original recording array and illuminated by diffuse white light from the rear, the object will be constructed in space by the intersection of ray bundles emanating from each of the lenslets as shown in figure 1b. It is the integration of the pencil beams, which renders 3D holoscopic imaging (integral imaging) unique and separates it from Gaussian imaging or holography. The biggest drawback, however, to the Lippmann method was that replay of the reconstructed images were pseudoscopic, or depth reversed, where the foreground becomes the background and vice versa, as shown in figure 1b. H. E. Ives was the first to recognize the problem in 1931 [11], and proposed a secondary exposure solution to invert the depth. This is known as a "two step" method, where a secondary exposure of the original photographic plate through another lens sheet was made. He demonstrated this solution by using a secondary array of pin-hole apertures. This proposal does not constitute an effective solution for the pseudoscopic to orthoscopic conversion problem. This is because the two-step recording introduces significant amounts of noise, due to degradation in the sampling caused by aberrations in the lenses. Since, optical and digital techniques to convert the pseudoscopic images to orthoscopic images have been proposed by several researchers [12, 18, 20].

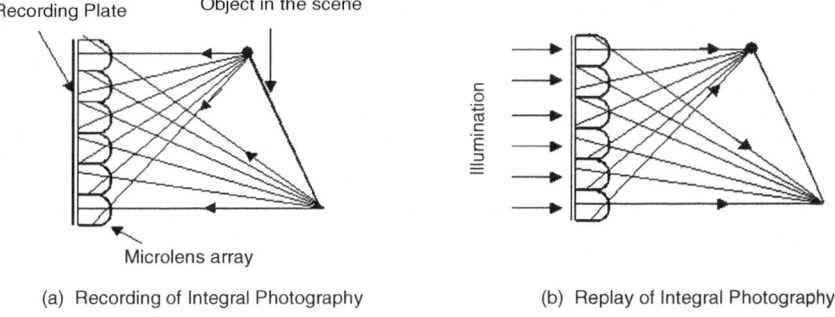

(a) Recording of Integral Photography   (b) Replay of Integral Photography

Fig. 1 Recording and replay of the Integral Photography

An optical configuration necessary to record one stage orthoscopic 3D holoscopic images has been proposed by Davies *et. al.* [12-16] and is shown in figure 2. This employs a pair of microlens arrays placed back to back and separated by their joint focal length, which produces spatial inversion. The arrangement allows a pseudoscopic image to be transferred such that it can straddle a separate microlens recording array (close imaging). The recording micro-lens array can be put anywhere in the transferred image space to allow the desired effect to be achieved freely: The object can be entirely inside of the display, outside of the display, or even straddling the display. The space transfer imaging scheme offers the flexibility of recording the object at a desired depth.

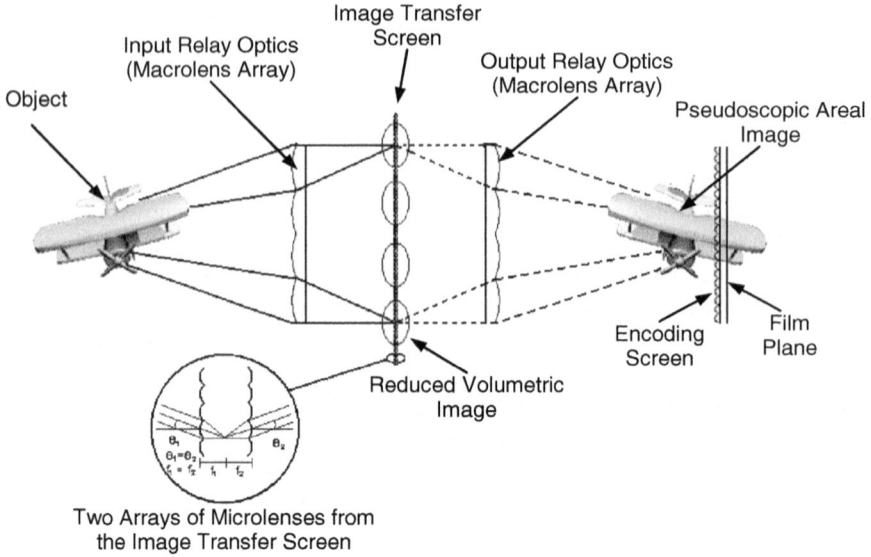

Fig. 2 3D Holoscopic Imaging camera optical system

The system uses an additional lens array, which images the object space around the plane of the microlens combination. This arrangement has been termed a two-tier optical combination. Effectively the first macro array produces a number of pseudoscopic, laterally inverted, images around the double microlens screen. This image is transmitted effectively negating the sign of the input angle such that each point in object space is returned to the same position in image space. The arrangement performs pseudo phase conjugation, i.e. transfer of volumetric data in space. The image is transmitted with equal lateral longitudinal magnification, and the relative spatial co-ordinates, are preserved i.e. there is no inversion in the recorded image and no scale reduction in depth.

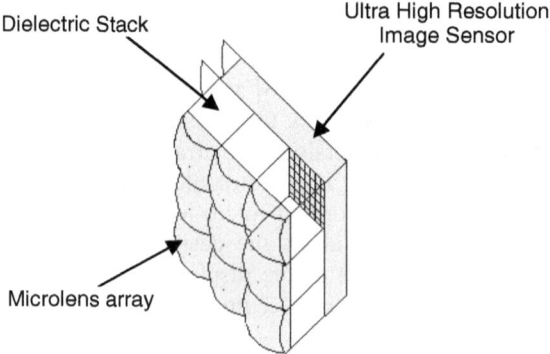

Fig. 3 3D Holoscopic image sensor

It is possible to capture 3D Holoscopic images electronically using an image sensor. This form of capture requires a high resolution image sensor together with specialised optical components to record the micro-images fields produced by precision micro-optics (see figure 3). The two-tier system shown in figure 2 has been used for the capture of the 3D holoscopic images used in this work. The object/scene is recorded on a film placed behind the recording microlens array through a rectangular aperture. The recorded data is then scanned using a high resolution scanner. The aperture greatly affects the characteristics of the micro-images recorded. Since each micro-image is an image of the object seen through the aperture independently, its shape and size is determined by the aperture. If the field of a sub-image is fully covered by the image, it is said to be *fully-filled*, otherwise it is said to be *under-filled* or *over-filled*.

The system will record live images in a regular block pixel pattern. The planar intensity distribution representing a 3D holoscopic image is comprised of 2D array of M×M micro-images due to the structure of the microlens array used in the capture and replay. The resulting 3D images are termed omnidirection 3D holoscopic images and have parallax in all directions. The rectangular aperture at the front of the camera and the regular structure of the hexagonal microlenses array used in the hexagonal grid (recording microlens array) gives rise to a regular 'brick structure' in the intensity distribution as illustrated in Figure 4.

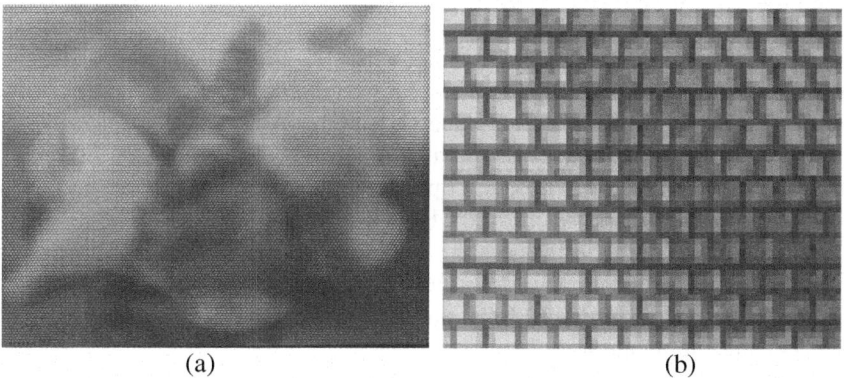

(a)   (b)

Fig. 4 (a) Example of the nature of sub-image field. (b) Magnified section.

Unidirectional 3D holoscopic images are obtained by using a special case of the 3D holoscopic imaging system where 1D cylindrical microlens array is used for capture and replay instead of a 2D array of microlenses. The resulting images contain parallax in the horizontal direction only. Figure 5(a) shows an electronically captured unidirectional 3D holoscopic image and figure 5(b) shows a magnified section of the image. The M vertically running bands present in the planar intensity distribution captured by the 3D Holoscopic imaging camera are due to the regular structure of the 1D cylindrical microlens array used in the capture process.

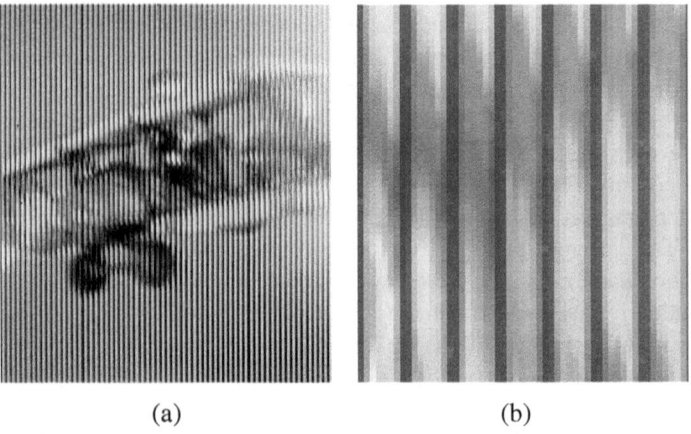

(a)   (b)

Fig. 5 An electronically captured unidirectional 3D holoscopic image a) Full. b) Magnification.

Other optical techniques proposed for the pseudoscopic to orthoscopic conversion by using a convergence lens as a substitute to the image transfer screen in figure 2 [17]. However, due to the non-constant lateral magnification of the converging lens, the reconstructed image appears clearly distorted.

Among digital methods proposed for pseudoscopic to orthoscopic conversion, Okano *et. al.* [18-19] demonstrated a system which captures the micro-images using a microlens array in front of a high resolution video camera and electronically inverts the each micro-image at the plane of capture. Although the image is acceptable the presentation reduces the parallax angle for close points in the scene. Another digital technique for pseudoscopic to orthoscopic conversion has been proposed by Martinez-Corral *et. al.* [20]. However, to avoid typical aliasing problems in the pixels mapping, it is necessary to assume that the number of pixels per lenslet is a multiple of the number of lenslets. This makes the number of pixels per micro-image very large (order of 100s) and renders the procedure impractical for many 3D display applications.

3   Computer Generation of 3D Holoscopic Images

In recent years several research groups have proposed techniques for generating 3D holoscopic graphics [27-35]. However, most of the work concentrated on reproducing the various physical setups using computer generation software packages. To produce computer generated 3D holoscopic image content a software model capable of generating rendered orthoscopic 3D holoscopic images is needed. The general properties of projective transformations were used in a variety of methods which evolved from micro-images containing spherical aberrations, defocus and field curvature to micro-images generated using an approach which alleviated many problems associated with the previous attempts. The greatest hurdle to overcome is not the production of the computer generated 3D holoscopic images themselves but the computational overhead required to attain real-time speeds on high-resolution displays. 3D holoscopic images are by definition the re-integration of multiple disseminated intensity values to produce the attribute of all-round viewing. To view a complete replayed volumetric scene from any arbitrary viewing position and hence mimic the original scene exactly with depth and parallax requires a high sampling rate that is dependent upon the scene depth. Small pixel sizes are required to hold a satisfactory depth without compromising viewer comfort and to generate a large enough display inevitably equates to a very high number of calculated intensity values for each frame. Adding to the complexities is the increased computations required when using spherical or hexagonally packed microlens arrays that generate 3D holoscopic images with omni-directional parallax. A less computationally severe option is to use semicylindrical lens arrays that generate unidirectional 3D holoscopic images with parallax in horizontal direction.

There has been a small amount of work that focused on the efficiency of the execution time required for the generation of photo realistic 3D Holoscopic images. One of the techniques reported in literature is based on parallel group rendering [33] where rather than rendering each perspective micro-image; each group of parallel rays is rendered using orthographic projection. A slightly modified version termed viewpoint vector rendering was later proposed to make the rendering performance independent of the number of the micro-images [34]. In this method each micro-image is assembled from a segmented area of the directional scenes. Both techniques are based on rasterization rendering technique and hence do not produce photo-realistic images.

A technique used to generate fast photo-realistic 3D Holoscopic images was reported by Youssef *et. al.* [35]. The technique of accelerating ray tracing is to reduce the number of intersection tests for shadow rays using a shadow cache algorithm. The image-space coherence is analysed describing the relation between rays and projected shadows in the scene rendered. Shadow cache algorithm has been adapted in order to minimise shadow intersection tests in ray tracing of 3D Holoscopic images. Shadow intersection tests make the majority of the intersection tests in ray tracing. The structure of the lenses and the camera model in the 3D Holoscopic image ray-tracing affects the way primary rays are spawned and traced as well as the spatial coherence among successive rays. As a result various pixel-tracing styles can be developed uniquely for 3D Holoscopic image ray tracing to improve the image-space coherence and the performance of the shadow cache algorithm. Examples of grouping of pixels tracing styles are shown in figure 6. Acceleration of the photo-realistic 3D Holoscopic images generation using the image-space coherence information between shadows and rays in 3D holoscopic ray tracing has been achieved with up to 41% of time saving [35]. Also, it has been proven that applying the new styles of pixel-tracing does not affect the scalability of 3D Holoscopic image ray tracing running over parallel computers.

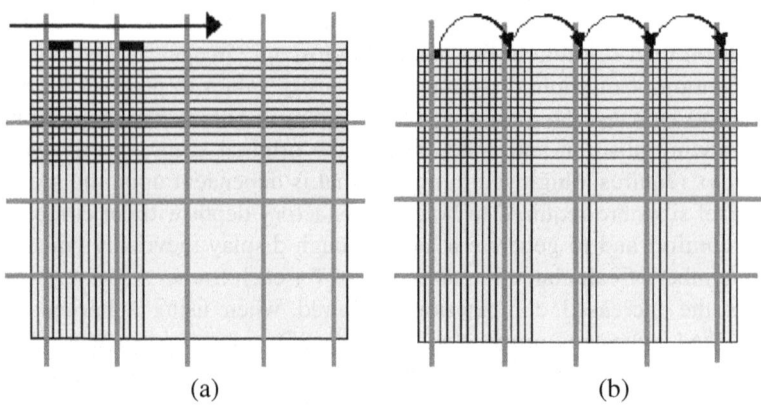

(a)   (b)

Fig. 6 (a) Grouping in the horizontal direction. (b) Grouping using adjacent lenses.

4 3D Object Segmentation

Research on depth extraction from multi-view imaging systems has been extensive. However, the depth extraction from 3D holoscopic imaging systems is still in
its infancy. The first reported work is that of Manolache *et. al.* [36], where the
point-spread function of the optical recording is used to describe the associated integral image system and the depth estimation task is tackled as an inverse problem. In the practical case, the image inverse problem proves to be ill posed and the
discrete correspondents are ill conditioned due to the inherent loss of information
associated with the model in the direct process. Therefore, the method can only be
applied on simulation using numerical data [37, 41].

A practical approach for obtaining depth by viewpoint image extraction and disparity analysis was explored and presented by Wu, *et. al.* [37]. The viewpoint image was formed by sampling pixels from different micro-images rather than a
macro block of pixels corresponding to a microlens unit. Each "viewpoint image"
presented a two-dimensional (2D) parallel recording of the 3D scene from a
particular direction. Figure 7 graphically illustrates how the viewpoint images are
extracted.

Figure 8 shows one example of unidirectional 3D holoscopic image and the extracted viewpoint images. The number of pixels of the formed viewpoint image
(along horizontal direction) will depend on the size of the 3D holoscopic image. In
the typical case described in figure 8, where the pitch size of the microlens sheet is
600 μm and the image has a size of 12cm, there will be 200 pixels in one formed
viewpoint image along the horizontal direction. Assuming that the intensity distribution has been sampled so that each micro-image comprises 8 pixels in the horizontal direction. This results in eight viewpoint images being extracted as shown
in figure 8b.

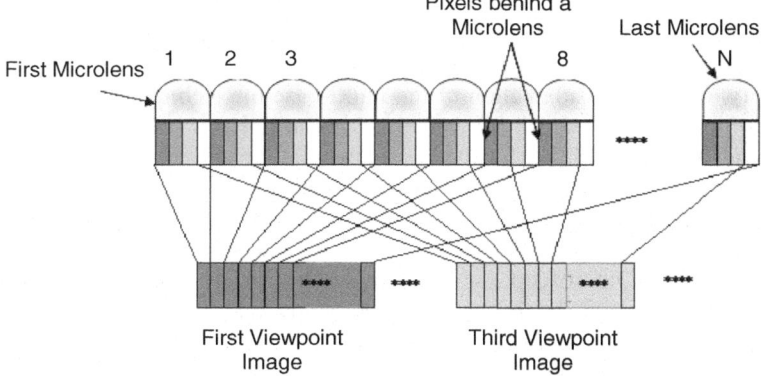

Fig. 7. Illustration of viewpoint image extraction (For simplicity, assume there are only four pixels under each microlens. Pixels in the same position under different microlenses, represented by the same pattern, are employed to form one viewpoint image.)

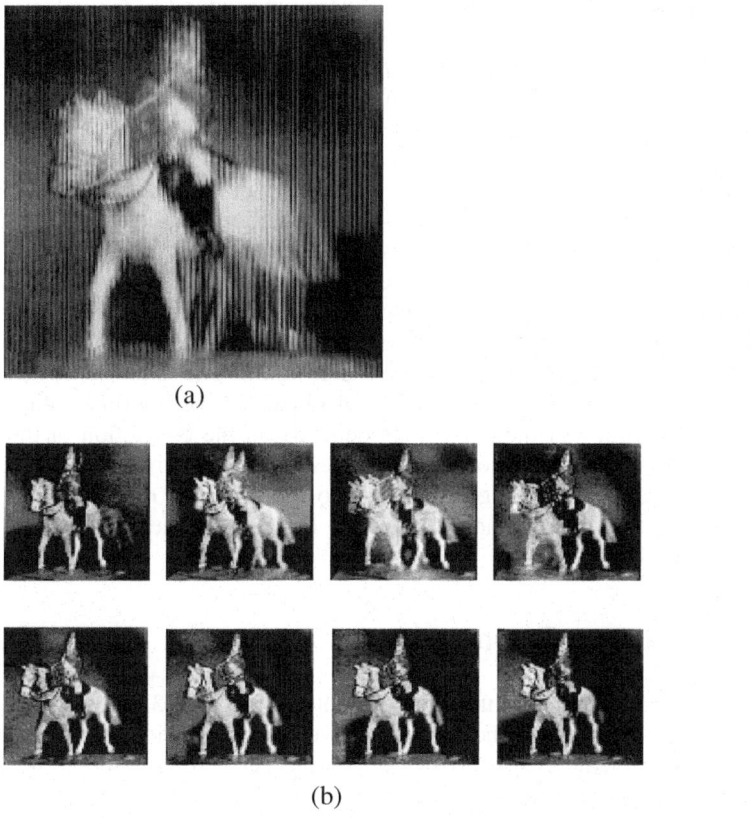

Fig. 8 (a) One captured Unidirectional 3D Holoscopic image and (b) the extracted viewpoint images (b) (The images have been scaled for illustration purpose)

Object depth was calculated from the viewpoint image displacement using a depth equation, which gave the mathematical relationship between object depth and correspondence viewpoint image displacement. To improve the performance of the disparity analysis, an adaptation of the multi-baseline technique taking advantage of the information redundancy contained in multiple viewpoint images of the same scene was used [37]. The idea of viewpoint image extraction on 3D object reconstruction was also reported by Arimoto and Javidi [38].

The 3D holoscopic imaging requires only one recording in obtaining 3D information and therefore no calibration is necessary to acquire depth values. The compactness of using 3D holoscopic imaging in depth measurement was soon attracting attention as a novel depth extraction technique [39, 40]. In the conventional stereo matching system, the quantization error is increased with the object depth and a considerable quantization error will be caused when the object depth is large. While different to the conventional stereo vision method, the quantization error obtained from the extracted viewpoint images is maintained at a constant value and irrelevant with the depth [37, 40]. To take the advantage of both, Park, et. al. proposed a method for extracting depth information using a specially designed lens arrangement [40]. A drawback of the work reported in [37, 38, 39, 40] is that the window size for matching has to be chosen experimentally. In general, a smaller matching window gives a poor result within the object/background region while a larger window size gives a poorer contour of the object.

More recently, a method was reported which addresses the problem of choosing an appropriate window size, where a neighbourhood constraint and relaxation technique is adapted by considering the spatial constraints in the image [41]. The hybrid algorithm combining both multi-baseline and neighborhood constraint and relaxation techniques with feature block pre-selection in disparity analysis has been shown to improve the performance of the depth estimation [41].

Another method which uses a blur metric-based depth extraction technique was proposed [42]. It requires the estimation of plane objects images using the computational 3D holoscopic imaging reconstruction algorithm. The algorithm was shown to extract the position of a small number of objects is well defined situations. However, the accuracy of the depth map depends on the estimation of the blur metric which is prone to large errors, as these metrics are sensitive not only to the threshold used to classify the edges, but also to the presence of noise. The scope of the 3D holoscopic imaging application also has been further extended to 3D object recognition [43, 44].

5 3D Holoscopic Image Compression

Due to the large amount of data required to represent the captured 3D holoscopic image with adequate resolution, it is necessary to develop compression algorithms tailor to take advantage of the characteristics of the recorded 3D holoscopic image. The planar intensity distribution representing 3D holoscopic image is comprised of 2D array of micro-images due to the structure of the microlens array used in the capture and replay. The structure of the recorded 3D holoscopic image intensity distribution is such that a high cross correlation in a third domain, i.e.

between the micro-images produced by the recording microlens array, is present. This is due to the small angular disparity between adjacent microlenses. In order to maximise the efficiency of a compression scheme for use with the 3D holoscopic image intensity distribution, both inter and intra micro-image correlation should be evaluated.

In the last decade, a lossy compression scheme for use with 3D holoscopic images, making use of a three dimensional discrete cosine transform (3D-DCT) has been developed [45, 46]. It was shown that the performance with respect to compression ratio and image quality is vastly improved compared with that achieved using baseline JPEG for compression of 3D holoscopic image data. More recently a wavelet-based lossy compression technique for 3D holoscopic images was reported [47]. The method requires the extraction of different viewpoint images from the 3D holoscopic image. A single viewpoint image is constructed by extracting one pixel from each micro-image, then each viewpoint image is decomposed using a two dimensional discrete wavelet transform (2D-DWT). The lower frequency bands of the viewpoint images are assembled and compressed using a 3D-DCT followed by Huffman coding. It was found that the algorithm achieves better rate-distortion performance, with respect to compression ratio and image quality at very low bit rates when compared to the 3D DCT based algorithms [47].

The 3D wavelet decomposition is computed by applying three separate 1D transforms viewpoint images. The spatial wavelet decomposition on a single viewpoint is performed using the biorthogonal Daubechies 9/7 filter bank while the inter-viewpoint image decomposition on the sequence is performed using the lifting scheme by means of the 5/3 filter bank [48]. All the resulting wavelet coefficients from the application of the 3D wavelet decomposition are arithmetic encoded.

5.1 Preprocessing of 3D Holoscopic Images

Prior to computation of the forward DWT, different viewpoint images are extracted from the original 3D Holoscopic image. The viewpoint image comprises pixels of the recorded object scene corresponding to a unique recording direction as discussed in section 4. The post-processing stage at the decoder essentially undoes the effects of pre-processing in the encoder. The original nominal dynamic range is restored and each pixel from each reconstructed viewpoint image is put back into its original position within the microlens to reconstruct the whole 3D holoscopic image. The intensity distribution of an omnidirectional 3D holoscopic image consists of an array of micro-images as shown in figure 4. The intensity distribution is sampled so that each micro-image comprises (8×7) pixels. Since a viewpoint image is obtained by extracting one pixel from each micro-image provided by the 2D array arrangement, a total of 56 different viewpoint images are constructed. It is important to point out that the viewpoint image is different from the traditional 2D image. It is a parallel projection recording of the 3D space rather than a perspective projection as in the common 2D recording.

5.2 2D WDT Based Compression Algorithm

The general structure of the 2D wavelet based compression algorithm is shown in Figure 9. The input to the encoding process is a 3D holoscopic image. Prior to computation of the 2D DWT, different viewpoint images are extracted from the original 3D holoscopic image. The viewpoint image components are then decomposed into different decomposition levels using a 2D WDT. The 2-D transform is performed by two separate 1-D transforms along the rows and the columns of the viewpoint image data, resulting in four frequency subbands.

The lowest frequency subband is a coarse scale approximation of the original viewpoint image and the rest of the frequency bands are detail signals. The 2D transform can be applied recursively to the lowest frequency subband to obtain decomposition at coarser scales. In [47] a two-level of decomposition was applied by means of the Daubechies 9/7 filter.

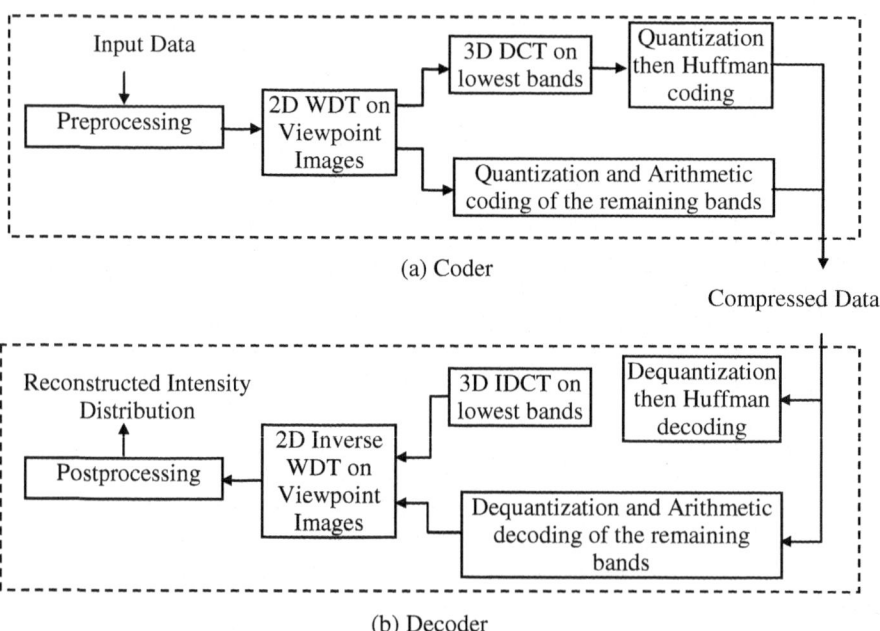

(a) Coder

(b) Decoder

Fig. 9 The general structure of the proposed scheme: (a) Encoder, (b) Decoder

After decomposition of the viewpoint images using the 2D-DWT (Figure 10(a)), the resulting lowest frequency subbands are assembled as shown in Figure 10(b) and compressed using a 3D-DCT. This will achieve de-correlation within and between the lowest frequency subbands from the different viewpoint images. The 3D DCT is performed on an 8×8×8 volume. Hence, the 56 lowest frequency subbands are assembled together, giving seven groups of eight viewpoint images each. The size of the lowest frequency subbands for all 3D holoscopic test images used here are a multiple of 8×8, which simplifies the computation of the 3D-DCT transform. Hence blocks of 8×8 pixels from eight successive view point images are grouped together to form an 8×8×8 volume as input to the 3D-DCT unit. The coefficients resulting from the application of the 3D-DCT are passed to a Huffman encoder while all the other coefficients are passed to an arithmetic encoder.

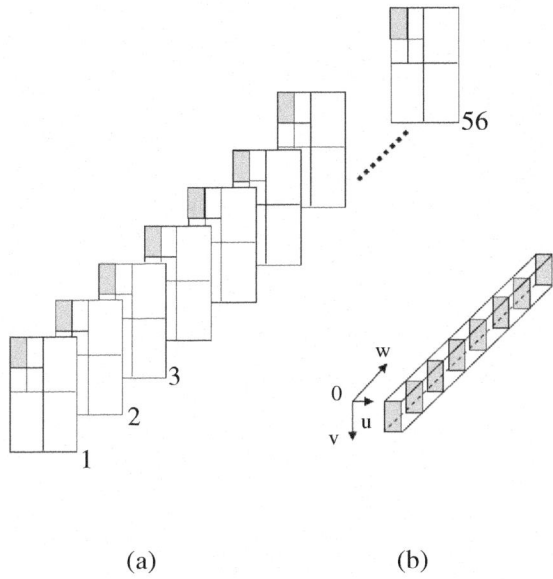

(a)   (b)

Fig. 10 (a) 2-levels 2D-DWT on viewpoint images, (b) Grouping of the lower frequency bands into 8×8×8 blocks

5.3 Proposed Compression Algorithm

The 56 extracted viewpoint images are DC level shifted. Then, a forward 1D DWT is applied on the whole sequence of viewpoint images. This results in 28 low frequency bands and 28 high frequency bands. The same procedure is then repeated on the resulting 28 low frequency bands only. This leads to 14 low frequency bands and 14 high frequency bands. The procedure is repeated to the low frequency bands at each decomposition level until only two low frequency bands are reached. The procedure is depicted in Figure 11 for five levels of inter-viewpoint image decomposition. Next, a forward 2-levels 2D DWT on the last two low frequency bands is carried out. After quantization, all the resulting quantized samples are Arithmetic encoded. Finally, the decoder undoes all these operations allowing the reconstruction of the intensity distribution.

Prior to Arithmetic coding, de-correlated sub-image groups resulting from application of the spatial decomposition on the last two low frequency bands, are

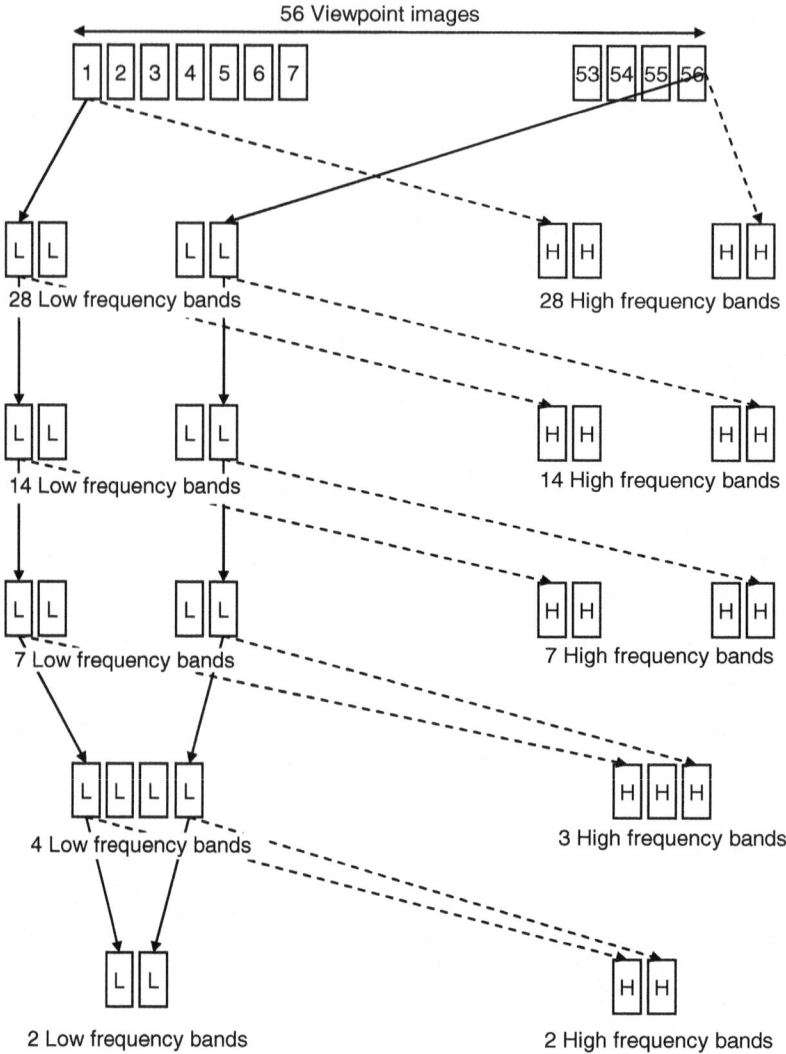

Fig. 11 Inter-viewpoint image decomposition of an input sequence of fifty-six viewpoint images

presented to a dead-zone scalar quantizer [47, 48]. The remaining fifty-four high frequency bands are quantized using a uniform scalar quantizer with a step-size varying from $\Delta=1$ to $\Delta=70$. After coding, the resulting coded bit stream is sent to the decoder. The optimal reconstruction bias used in the dequantization process is $r=0.4$. In the next section, the results of the 3D DWT based algorithm are presented and compared to those of previously experimented 2D DWT scheme on omnidirectional 3D holoscopic image data.

5.4   Simulation Results and Discussions

The 3D DWT based compression algorithm has been implemented for simulation using the several 3D holoscopic test images. The performance of the encoder and decoder was measured in terms of the Peak Signal to Noise Ratio (PSNR) and the compression achieved expressed in bits per pixel (bpp). Figure 12 shows plots of PSNR versus bit rate for the proposed scheme and the previously experimented 2D DWT based model. From Figure 12, it can be seen that the 3D DWT based algorithm shows a higher improvement in PSNR for all bit rate values compared to the previous reported 2D DWT based scheme [47]. Table 1 shows bit rate values resulting from both methods simulated for a typical quality requirement. As we can see from Table 1, 1.42 dB are gained at 0.1 bpp and an improved performance by an average of 0.94 dB (average of PSNR values at 0.1, 0.2, and 0.3 bpp) is achieved by the proposed compression algorithm when compared to the previously experimented 2D DWT based scheme.

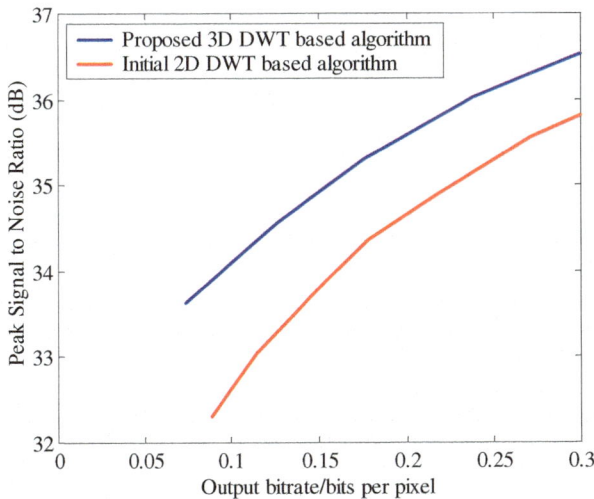

Fig. 12 Performance of the proposed 3D DWT based compression algorithm and the 2D DWT based scheme for compression of 3D Holoscopic images

Table 1 Summary of performance of both compression schemes tested on 3D holosopic image

Encoder	Peak Signal to Noise Ratio (dB)		
	0.1 bpp	0.2 bpp	0.3 bpp
Proposed 3D DWT based algorithm	34.09	35.67	36.52
2D DWT based algorithm [47]	32.67	34.91	35.87

6 Conclusions

A review of 3D Holoscopic imaging technology is provided from the point of view of 3D content capture, computer generation, depth computation and 3D content coding. 3D Holoscopic imaging technology is receiving a lot of interest in recent years and is a candidate for consideration for several applications including 3D TV in terms of human factors, cost of conversion of studios, production of content, decoders and current cost of display technology. A 3D discrete wavelet based compression scheme is discussed in more details to show that dedicated algorithms are required to take full advantage of the data structure inherent in 3D holoscopic images and hence achieve the best possible performance.

References

1. Okoshi, T.: Three-Dimensional Imaging Techniques. Academic Press, Inc., London (1976)
2. Javidi, B., Okano, F.: Special Issue on 3-D Technologies for Imaging and Display. Proceedings of IEEE 94(3) (2006)
3. Dodgson, N.A.: Autostereoscopic 3D Displays. IEEE Computer 38(8), 31–36 (2005)
4. Onural, L., et al.: An Assessment of 3dtv Technologies. In: Proc. NAB Broadcast Engineering Conference, pp. 456–467 (2006)
5. Yamazaki, T., Kamijo, K., Fukuzumi, S.: Quantative evaluation of visual fatigue. In: Proc. Japan Display, pp. 606–609 (1989)
6. Lambooij, T.M., IJsselsteijn, W. A., Heynderickx, I.: Visual Discomfort in Stereoscopic Displays: A Review. In: Proc. IS&T/SPIE Electronic Imaging Symposium, San Jose, California, USA (2007)
7. Onural, L.: Television in 3-D: What are the Prospects. Proc. IEEE 95(6) (2007)
8. Benton, S.A.: Elements of Holographic video imaging. In: Proc. SPIE (1991)
9. Honda, T.: Holographic Display for movie and video. In: Proc. ITE 3D images Symposium (1991)
10. Lippmann, G.: Eppreuves Reversibles Donnat Durelief. J. Phys. Paris 821 (1908)
11. Ives, H.E.: Optical properties of a Lippmann lenticulated sheet. J. Opt. Soc. Am. 21, 171–176 (1931)
12. Davies, N., et al.: Three-dimensional imaging systems: A new development. Applied Optics 27, 4520 (1988)

13. McCormick, M., Davies, N., Aggoun, A., Brewin, M.: Resolution requirements for autostereoscopic full parallax 3D-TV. In: Proc. IEE Broadcasting Conference, Amsterdam (1994)
14. Stevens, R., Davies, N., Milnethorpe, G.: Lens arrays and optical systems for orthoscopic three-dimensional imaging. The Imaging Science Journal 49, 151–164 (2001)
15. Monaleche, S., Aggoun, A., McCormick, M., Davies, N., Kung, S.Y.: Analytical model of a 3d recording camera system using circular and hexagonal based spherical microlenses. Journal of Optical Society America A 18(8), 1814–1821 (2001)
16. Aggoun, A.: Pre-processing of Integral Images for 3D Displays. IEEE Journal of Display Technology 2(4), 393–400 (2006)
17. Jang, J.S., Javidi, B.: Formation of orthoscopic three dimensional real images in direct pickup one step integral imaging. Optical Engineering 42(7), 1869–1870 (2003)
18. Okano, F., et al.: Real time pickup method for a three-dimensional image based on integral photography. Applied Optics 36(7), 1598–1603 (1997)
19. Arai, J., et al.: Gradient-index lens-array method based on real time integral photography for three-dimensional images. Applied Optics 11, 2034–2045 (1998)
20. Martínez-Corral, M., et al.: Formation of real, orthoscopic integral images by smart pixel mapping. Optics Express 13, 9175–9180 (2005)
21. Javidi, B., et al.: Orthoscopic, long-focal-depth integral imaging by hybrid method. In: Proc. of SPIE, vol. 6392, p. 639203 (2006)
22. Javidi, B., Okano, F.: Three-dimensional television, video, and display technologies. Springer, New York (2002)
23. Jang, J.S., Javidi, B.: Time-Multiplexed Integral Imaging. Optics & Photonics News, 36–43 (2004)
24. Kim, Y., et al.: Integral imaging with variable image planes using polymer-dispersed liquid crystal layers. In: Proc. of SPIE, vol. 6392, p. 639204 (2006)
25. Kim, Y., et al.: Depth-enhanced three-dimensional integral imaging by use of multi-layered display devices. Applied Optics 45(18), 4334–4343 (2006)
26. Hong, J., et al.: A depth-enhanced integral imaging by use of optical path control. Opt. Lett. 29(15), 1790–1792 (2004)
27. Milnthorpe, G., McCormick, M., Aggoun, A., Davies, N., Forman, M.C.: Computer generated Content for 3D TV Displays. In: International Broadcasting Convention, Amsterdam, Netherlands (2002)
28. Ren, J., Aggoun, A., McCormick, M.: Computer generation of integral 3D images with maximum effective viewing angle. Journal of Electronic Imaging 14(2), 023019.1–023019.9 (2005)
29. Eljadid, M., Aggoun, A., Yousef, O.: Computer Generated Content for 3D TV. In: Proc. of 3DTV conference, Greece (2007)
30. Min, S., et al.: Three-Dimensional Display System Based On Computer-Generated Integral Imaging. In: Proc. SPIE, Stereoscopic Displays and Virtual Reality Systems VIII, Conference, San Jose, CA, USA (2001)
31. Halle, M.W., et al.: Fast Computer Graphics Rendering for Full Parallax Spatial Displays. In: Proc. SPIE, Practical Holography XI and Holographic Materials III Conference (1997)
32. Naemura, T., et al.: 3D Computer Graphics Based on Integral Photography. Optics express 8(2), 255–262 (2001)
33. Yang, R., Huang, X., Chen, S.: Efficient Rendering of Integral images. In: Proc. SIGGRAPH Computer Graphics and Interactive Techniques Conference, LA, USA (2005)

34. Park, K.S., Min, S.W., Cho, Y.: Viewpoint vector rendering for efficient elemental image generation. IEICE Trans. Inf. & Syst. E90-D(1), 233–241 (2007)
35. Youssef, O., Aggoun, A., Wolf, W., McCormick, M.: Pixels Grouping and Shadow Cache for Faster Integral 3D Ray-Tracing. In: Proc. SPIE Electronic Imaging Symposium, San Jose, California, USA (2002)
36. Manolache, S., Kung, S.Y., McCormick, M., Aggoun, A.: 3D-object space reconstruction from planar recorded data of 3D-integral images. Journal of VLSI Signal Processing Systems 35, 5–18 (2003)
37. Wu, C., et al.: Depth extraction from unidirectional integral image using a modified multi-baseline technique. In: Proc. SPIE Electronic Imaging Symposium, San Jose, California, USA (2002)
38. Arimoto, H., Javidi, B.: Integral three-dimensional imaging with digital reconstruction. Optics Letters 26(3), 157–159 (2001)
39. Park, J., Jung, S., Choi, H., Lee, B.: A novel depth extraction algorithm incorporating a lens array and a camera by reassembling pixel columns of elemental images. In: Proc. SPIE Photonics Asia, Shanghai, China (2002)
40. Park, J., Jung, S., Choi, H., Kim, Y., Lee, B.: Depth extraction by use of a rectangular lens array and one-dimensional elemental image modification. Applied Optics 43(25), 4882–4895 (2004)
41. Wu, C., et al.: Depth Map from Unidirectional Integral Images using a Hybrid Disparity Analysis Algorithm. IEEE Journal of Display Technology 4(1), 101–108 (2008)
42. Hwang, D.C., et al.: Extraction of location coordinates of 3D objects from computationally reconstructed integral images basing on a blur metric. Optics Express 16(6), 3623–3635 (2008)
43. Kishk, S., Javidi, B.: Improved resolution 3D object sensing and recognition using time multiplexed computational integral imaging. Optics Express 11, 3528–3541 (2003)
44. Yeom, S., Javidi, B.: Three-dimensional distortion-tolerant object recognition using integral imaging. Optics Express 12, 5795–5808 (2004)
45. Forman, M.C., Aggoun, A., McCormick, M.: Compression of 3D integral TV pictures. In: Proc. IEE 5th Image processing and its application Conference (1995)
46. Zaharia, R., Aggoun, A., McCormick, M.: Adaptive 3D-DCT compression algorithm for continuous parallax 3D integral imaging. Signal Processing: Image Communications 17(3), 231–242 (2002)
47. Aggoun, A., Mazri, M.: Wavelet-based Compression Algorithm for Still Omnidirectional 3D Integral Images. Signal, Image and Video Processing, SIViP 2, 141–153 (June 2008)
48. Rabbani, M., Joshi, R.: An overview of the JPEG 2000 still image compression standard. Signal Processing: Image Communications 17, 3–48 (2002)

Part V
Accessing Technologies for Visual Content

Chapter 19
Video Streaming with Interactive Pan/Tilt/Zoom

Aditya Mavlankar and Bernd Girod

Abstract. High-spatial-resolution videos offer the possibility of viewing an arbitrary region-of-interest (RoI) interactively. The user can pan/tilt/zoom while watching the video. This chapter presents spatial-random-access-enabled video compression that encodes the content such that arbitrary RoIs corresponding to different zoom factors can be extracted from the compressed bit-stream. The chapter also covers RoI trajectory prediction, which allows pre-fetching relevant content in a streaming scenario. The more accurate the prediction the lower is the percentage of missing pixels. RoI prediction techniques can perform better by adapting according to the video content in addition to simply extrapolating previous moves of the input device. Finally, the chapter presents a streaming system that employs application-layer peer-to-peer (P2P) multicast while still allowing the users to freely choose individual RoIs. The P2P overlay adapts on-the-fly for exploiting the commonalities in the peers' RoIs. This enables peers to relay data to each other in real-time, thus drastically reducing the bandwidth required from dedicated servers.

1 Introduction

High-spatial-resolution digital video will be widely available at low cost in the near future. This development is driven by increasing spatial resolution offered by digital imaging sensors and increasing capacities of storage devices. Furthermore, there exist algorithms for stitching a comprehensive high-resolution view from multiple cameras [1, 2]. Some currently available video-conferencing systems stitch a large panoramic view in real-time [3]. Also, image acquisition on spherical, cylindrical or hyperbolic image planes via multiple cameras can record scenes with a wide

Aditya Mavlankar · Bernd Girod
Stanford University, 350 Serra Mall, Stanford CA 94305, USA
e-mail: maditya@stanford.edu, e-mail: bgirod@stanford.edu

field-of-view while the recorded data can be warped later to the desired viewing format [4]. An example of such an acquisition device is [5].

Imagine that a user wants to watch a high-spatial-resolution video that exceeds the resolution of his/her display screen. If the user were to watch a downsampled version of the video that fits the display screen then he/she might not be able to view local regions with the recorded high resolution. A possible solution to this problem is a video player that supports interactive pan/tilt/zoom. The user can thus choose to watch an arbitrary region-of-interest (RoI). We refer to this functionality as interactive region-of-interest (IRoI). Figure 1 shows screen-shots of a video player supporting IRoI. Such a video player could also offer to track certain objects, whereby the user is not required to control pan and tilt, but could still control the zoom factor.

Some practical scenarios where this kind of interactivity is well-suited are: interactive playback of high-resolution video from locally stored media, interactive TV for watching content captured with very high detail (e.g., interactive viewing of sports events), providing virtual pan/tilt/zoom within a wide-angle and high-resolution scene from a surveillance camera, and streaming instructional videos captured with high spatial resolution (e.g., lectures, panel discussions). A video clip that showcases interactive viewing of soccer in a TV-like setting can be seen here [6].

Consider the first example mentioned above, i.e., playback from locally stored media. In this case, the video content is encoded offline before storing it on the relevant media, for example, a high-capacity portable disk. Note that the RoI trajectory is not known while encoding the content. An RoI trajectory is determined each time a user watches the video with interactive pan/tilt/zoom. This leads us to two design choices; 1) the video player can be designed to decode the entire high spatial resolution while displaying only the RoI or 2) the adopted compression format could allow decoding only relevant regions, possibly with some overhead. Depending on the resolution of the video and the hardware capability of the player, the first design choice might be prohibitive. Other application scenarios mentioned above entail streaming from a remote source. In most cases, streaming the full spatial extent of the video to a user can be ruled out due to prohibitive bandwidth requirement. If RoI-specific portions can be streamed to the remote user, the RoI dimensions could be adapted to suit the available data rate for communication apart from the user's display screen as noted above.

Now let us consider the difficulty of employing a standard video encoder in the streaming scenario. A standard video encoder generally does not provide efficient spatial random access, i.e., the ability to extract regions from the compressed bitstream. The video streaming can be for live content or for pre-stored content. For live content, the server can crop out an RoI sequence on-the-fly considering the user's pan/tilt/zoom commands and compress it as a video sequence using standard video encoding. The load of encoding might get prohibitively large with increasing numbers of users. Pre-stored content might not be stored in raw format implying that the server has to decode the high-spatial-resolution video prior to cropping the RoI sequence. Not only does the load of encoding increase, but if multiple users watch the content asynchronously then even the decoding load at the server increases. On the other hand, if a spatial-random-access-enabled video coding scheme is employed, the server needs to encode the recorded field-of-view only once, possibly with multiple resolution layers to support different zoom factors. The encoding load can thus be upper-bounded both for live content as well as pre-stored content irrespective of the number of users.

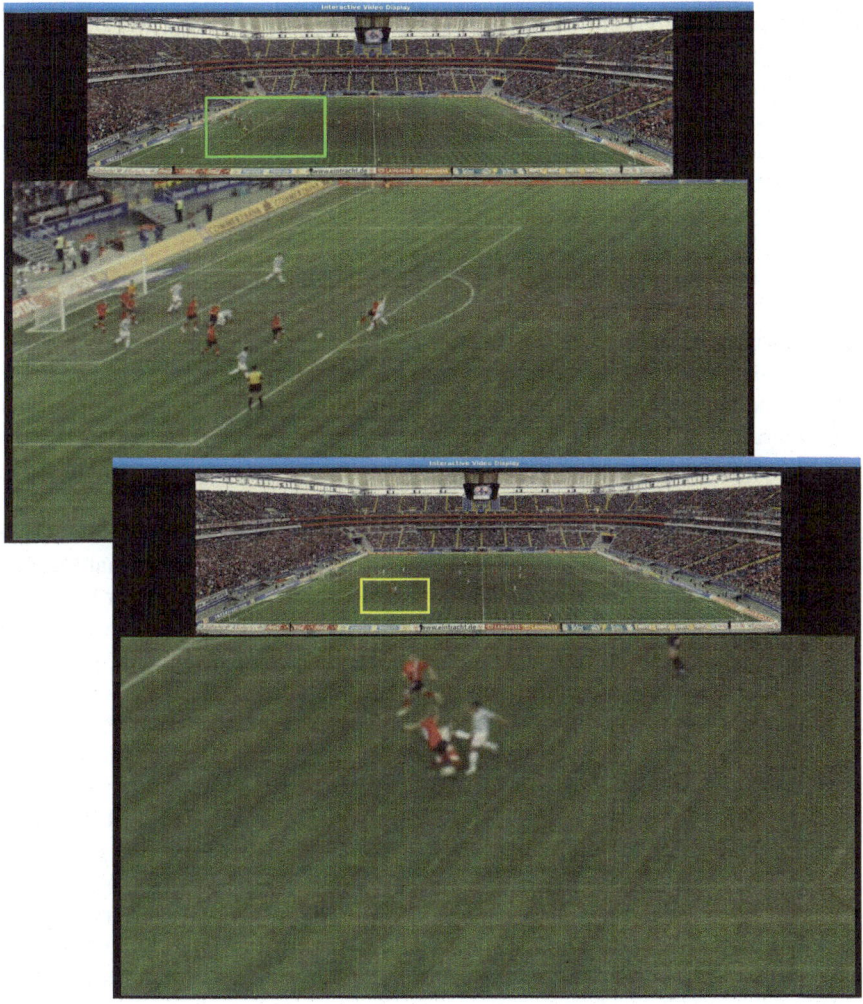

Fig. 1 Screen-shots of a video player supporting interactive pan/tilt/zoom. Apart from displaying the RoI, the video player can display a thumbnail overview to aid navigation in the scene. The player could also offer to track certain objects, for example, the soccer ball and/or the soccer players. In the tracking mode, the user is not required to control pan and tilt, but could still control the zoom factor.

In addition to limiting the encoding load, if the streaming bandwidth required from the server can also be limited then the streaming system can scale to large numbers of users. This chapter presents a solution that can be employed when several users are synchronously watching arbitrary regions of a high-spatial-resolution video. It hinges on employing application-layer peer-to-peer (P2P) multicast for delivering the streams to the users. The solution exploits the commonalities in the peers' regions such that they relay data to each other in real-time. This allows limiting the bandwidth required at the server by making use of the forwarding capacities of the peers. The main challenge is that user-interaction determines real-time which regions are commonly wanted by which peers. The P2P overlay needs to adapt quickly and in a distributed manner, i.e., peers take most of the action necessary for acquiring the data they need, without much central intervention. Larger dependence on central intervention represents another hurdle in scaling. The second challenge is that peers can switch off randomly, taking away the resources they bring with them.

Ideally, the changing RoI should be rendered immediately upon user input; i.e., without waiting for new data to arrive. If the client would delay the rendering until new data arrive, the induced latency might hamper the experience of interactivity. In both client-server unicast streaming as well as P2P multicast streaming, predicting the user's navigation path ahead of time helps pre-fetch relevant sub-streams. The more accurate the RoI prediction the lower is the percentage of pixels that have to be error-concealed.

This chapter is organized as follows. Section 2 provides a sampling of interactive streaming systems found in the literature. The goal is to highlight the challenges as well as earlier proposed approaches for providing random access, enabling pre-fetching and P2P design for other interactive applications that are similar in spirit to IRoI video. Section 3 discusses several approaches for providing spatial random access within videos. It elaborates one video coding scheme in particular. This scheme builds a multi-resolution pyramid comprising slices. It is shown how background extraction can be used to improve the coding efficiency of such a scheme. The trade-off in choosing the slice size is also analyzed. The slice size can be chosen to strike the right balance between storage requirement and transmission bit-rate. Section 4 describes variants of pre-fetching schemes. In one of the variants, the RoI prediction is based on analyzing the motion of objects in the video in addition to extrapolating moves of the input device. RoI prediction can be carried out at the client, at the server or collectively. Section 5 presents the P2P multicasting system in which peers can control their individual RoIs. Key aspects of the design are presented that enable peers to receive and relay respective regions despite the challenges outlined above.

2   Related Work

This section draws on interactive streaming systems found in the literature. A brief survey of the challenges in designing such systems and the solutions found in the literature sets the stage for the discussion on video streaming with interactive pan/tilt/zoom appearing in later sections. The later sections particularly aim at building a system that scales to large numbers of users.

2.1 Coding for Random Access

Images. Remote image browsing with interactive pan/tilt/zoom is very similar in spirit. It is generally used for high-resolution archaeological images, aerial or satellite images, images of museum exhibits, online maps, etc. Online maps provide about 20 zoom levels. The image corresponding to each zoom level is coded into tiles. Generally, the images corresponding to different zoom levels are coded independently. This so-called Gaussian pyramid fails to exploit redundancy across zoom levels but provides easy random access. The server accesses the tiles intersecting the selected view and sends these tiles to the user. Generally, after a zoom operation, the relevant part from the current zoom level is interpolated to quickly render the newly desired view. As the tiles from the new zoom level arrive, the graphics become crisper. Note that this cursory rendering based on earlier received data might not be possible for some portions due to lack of received data.

Interactive browsing of images using JPEG2000 is explored in [7, 8]. This leverages the multi-resolution representation of an image using wavelets. This representation is not overcomplete unlike the Gaussian and Laplacian pyramids that generate more coefficients than the high-resolution image. JPEG2000 encodes blocks of wavelet transform coefficients independently. This means that every coded block has influence on the reconstruction of a limited number of pixels of the image. Moreover, the coding of each block results in an independent, embedded sub-bitstream. This makes it possible to stream any given block with a desired degree of fidelity. A transmission protocol, called JPEG2000 over Internet Protocol (JPIP), has also been developed. The protocol governs communication between a client and a server to support remote interactive browsing of JPEG2000 coded images [9]. The server can keep track of the RoI trajectory of the client as well as the parts of the bit-stream that have already been streamed to the client. Given a rate of transmission for the current time interval, the server solves an optimization problem to determine which parts of the bit-stream need to be sent in order to maximize the quality of the current RoI.

Video. The video compression standard H.264/AVC [10, 11] includes tools like Flexible Macroblock Ordering (FMO) and Arbitrary Slice Ordering (ASO). These tools were primarily created for error resilience, but can also be used to define an RoI prior to encoding [12]. The RoI can either be defined through manual input or through automatic content analysis. Slices corresponding to the RoI (or multiple RoIs) can be encoded with higher quality compared to other regions. Optionally, the scalable extension of H.264/AVC, called SVC [13, 14], can be used for adding fine or coarse granular fidelity refinements for RoI slices. The user experiences higher quality for the RoI if the refinement packets are received. The RoI encoding parameters can be adapted to the network and/or the user [15]. Note that these systems transmit the entire picture while delivering the RoI with higher quality. Among the class of such systems, some employ JPEG2000 with RoI support and conditional replenishment for exploiting correlation among successive frames [16]. Parts of the image that are not replenished can be copied from the previous frame or a background store.

In our own work, we have proposed a video transmission system for interactive pan/tilt/zoom [17]. This system crops the RoI sequence from the high-resolution video and encodes it using H.264/AVC. The RoI cropping is adapted to yield efficient motion compensation in the video encoder. The RoI adjustment is confined to ensure that the user does not notice the manipulation and experiences accurate RoI control. The normal mode of operation for this system is streaming live content but we also allow the user to rewind and play back older video. Note that in the second mode of operation, the high-resolution video is decoded prior to cropping the RoI sequence. Although efficient in terms of transmitted bit-rate, the drawback is that RoI video encoding has to be invoked for each user, thus limiting the system to few users. This system targets remote surveillance in which the number of simultaneous users is likely to be less than other applications like interactive TV.

Video coding for spatial random access presents a special challenge. To achieve good compression efficiency, video compression schemes typically exploit correlation among successive frames. This is accomplished through motion-compensated interframe prediction [18, 19, 20]. However, this makes it difficult to provide random access for spatial browsing within the scene. This is because the decoding of a block of pixels requires that other reference frame blocks used by the predictor have previously been decoded. These reference frame blocks might lie outside the RoI and might not have been transmitted and/or decoded earlier.

Coding, transmission and rendering of high-resolution panoramic videos using MPEG-4 is proposed in [21, 22]. A limited part of the entire scene is transmitted to the client depending on the chosen viewpoint. Only intraframe coding is used to allow random access. The scene is coded into independent slices. The authors mention the possibility of employing interframe coding to gain more compression efficiency. However, they note that this involves transmitting slices from the past if the current slice requires those for its decoding. A longer intraframe period entails significant complexity for slices from the latter frames in the group of pictures (GOP), as this "dependency chain" grows.

Multi-View Images/Videos. Interactive streaming systems that provide virtual fly-around in the scene employ novel-view generation to render views of the scene from arbitrary viewpoints. With these systems, the user can experience more free interactive navigation in the scene [23, 24, 25]. These systems typically employ image-based rendering (IBR) which is a technique to generate the novel view from multiple views of the scene recorded using multiple cameras [26, 27]. Note that in these applications, the scene itself might or might not be evolving in time. Transmitting arbitrary views from the multi-view data-set on-the-fly also entails random access issues similar to those arising for transmitting arbitrary regions in interactive pan/tilt/zoom. Interframe coding for compressing successive images in time as well as from neighboring views can achieve higher compression efficiency but can lead to undesirable dependencies for accessing random views. There exists a large body of works that employs hybrid video coding for compressing multi-view data-sets [28, 29, 30, 31, 32]. These studies highlight the trade-off in storage requirement, mean transmission bit-rate and decoding complexity. Recently, an analytical frame-work was proposed for optimizing the coding structure for coding multi-view data-sets [33]. The framework allows multiple representations of a picture, for example, compressed using different reference pictures. The optimization not only finds the best coding structure but also determines the best set of coded pictures to transmit corresponding to a viewing path. The framework can accommodate constraints like limited step-size for view switching, permitting view switching only during certain frame-intervals and capping the length of the burst of reference frames that are used for decoding a viewed frame but are not themselves displayed. The framework can minimize a weighted sum of expected transmission bit-rate and storage cost for storing the compressed pictures.

The video compression standard H.264/AVC defines two new slice types, called SP and SI slices. Using these slice types, it is possible to create multiple repre-sentations of a video frame using different reference frames. Similar to the solu-tions described above, the representation to be streamed is chosen according to the reference frames available at the decoder. However, the novelty is that the recon-struction is guaranteed to be identical. This drastically reduces the total number of multiple representations required to be stored. SP frames have been used for in-teractive streaming of static light fields [34, 35]. Another solution to the random access problem associated with multi-view data-sets is based on distributed source coding (DSC) [36, 37]. In this solution, an interframe coded picture is represented using enough parity bits which leads to an identical reconstruction irrespective of the reference frame used by the decoder. This implies that multiple representations are not required to be stored, however, the number of parity bits is determined by the reference frame having the least correlation to the frame to be coded. Similar to some prior work based on hybrid video coding for multi-view data-sets mentioned above, recent work based on DSC also explores the trade-off between transmission bit-rate and storage requirement [38].

2.2 Navigation Path Prediction

A simple user-input device, for example a computer mouse, typically senses po-sition. More sophisticated devices like game-controllers can also measure velocity and/or acceleration. Studies on view trajectory prediction have been conducted in the context of Virtual Reality [39] and networked multi-player video games [40]. A common navigation path prediction technique, dead reckoning, predicts the fu-ture path by assuming that the user maintains the current velocity. The velocity can be either read from the input device or computed from successive position measurements.

In their work on interactive streaming of light fields, the authors predict the (x, y) mouse co-ordinates based on dead reckoning and translate these into the viewpoint [41]. The use of a Kalman filter for head movement prediction in scenarios where head movements can control the application have been proposed in [42]. In prior work on dynamic light fields, six Kalman filters have been used for predicting the 3-D co-ordinates and the 3 Euler angles that define the viewpoint [43, 44]. The viewpoint and the rendering algorithm together determine the number of views that need to be streamed to the client. The authors mention two possible system design choices. Viewpoints exceeding the bit-rate threshold can be disallowed or those viewpoints can be rendered with lower quality by not streaming all the views demanded by that viewpoint. The authors also note that if the streaming system allows tuning into a view-stream only during certain frame-intervals, one can choose an appropriately long prediction lookahead and tune into new view-streams beforehand to avoid missing the join opportunities.

2.3 Multicasting

Multicasting can drastically reduce the bandwidth required from dedicated media servers. IP multicast, specified decades ago [45], allows sending an IP datagram to a group of hosts identified by a single IP destination address [46]. Hosts may join and leave a multicast group at any time. This requires multicast-capable routers that replicate packets as required. Even though IP multicast is extremely efficient at distributing data to multiple interested receivers, most routers on the Internet keep this functionality turned off due to reasons related to security, billing and the size of the data-structures to be maintained by the router. Nevertheless, the bandwidth conservation benefits of IP multicast have resulted in rising deployment for corporate communications and, more recently, IPTV service.

The seminal work on receiver-driven layered multicast (RLM) [47] focuses on video streaming without interactive pan/tilt/zoom. The authors propose compressing the multimedia signal in hierarchical layers and letting individual receivers choose the layers to join. Receiving more layers leads to better quality. Each layer is delivered using a different multicast group. Note that if a receiver joins too many layers and creates congestion on a link then packets can be dropped indiscriminately from all layers affecting received quality, possibly for multiple receivers that share the congested link. A receiver performs regular tests to decide if it should unsubscribe already joined layers or subscribe new layers. "Shared learning" among receivers can reduce the number of tests and hence the convergence time.

Recently, the RLM framework was adapted for interactive dynamic light field streaming [43]. Depending on the chosen viewpoint, the client decides which views and consequently which multicast groups to subscribe. The latency for joining a new multicast group is generally low with IP multicast [48]. As in the case of RLM, it is the client's responsibility to avoid congestion on intermediate links. The source does not adapt transmission to curtail congestion; it keeps transmitting IP datagrams to the multicast groups' addresses.

Contrary to network-layer IP multicast, P2P streaming implements the multicasting logic in software at the end-hosts rather than routers inside the network [49]. Unlike IP multicast, the application-layer software can be widely deployed with little investment. Although the P2P approach generally results in more duplication of packets and inefficient routing compared to IP multicast, the benefits outweigh the inefficiencies. The source as well as each peer can respond to local retransmission requests as well as perform sophisticated packet scheduling to maximize the experience of downstream peers [50].

P2P streaming systems can be broadly classified into mesh-pull vs. tree-push systems [51]. The design of mesh-pull systems evolved from P2P file-sharing systems. In these systems, a peer advertises the chunks of data that it has and complies with requests to relay chunks to other peers. Tree-push systems, on the other hand, distribute data using one or more complementary trees. After finding its place inside a distribution tree, a peer generally persists to keep its association with the parent and its children and relays data without waiting for requests from children. Generally, tree-push systems result in fewer duplicate packets, lower end-to-end delay and less delay-jitter [52, 53]. These traits are beneficial for interactive streaming systems where select sub-streams of the coded content are required on-the-fly. A tree-based P2P protocol has been recently proposed for interactive streaming of dynamic light fields [54, 55]. Early results demonstrate the capability of the system to support many more users with the same server resources as compared to traditional unicast client-server streaming [55].

3   Spatial-Random-Access-Enabled Video Coding

We have proposed a spatial-random-access-enabled video coding scheme, shown in Fig. 2, in our earlier work [56]. The coded representation consists of multiple resolution layers. The thumbnail video constitutes a base layer and is coded with H.264/AVC using I, P and B pictures. The reconstructed base layer video frames are upsampled by a suitable factor and used as prediction signal for encoding video corresponding to the higher resolution layers. Each frame belonging to a higher resolution layer is coded using a grid of rectangular P slices. Employing upward prediction from only the thumbnail enables efficient random access to local regions within any spatial resolution. For a given frame-interval, the display of the client is rendered by transmitting the corresponding frame from the base layer and few P slices from exactly one higher resolution layer. Slices are transmitted from the resolution layer that corresponds closest to the user's current zoom factor. At the client's side, the corresponding RoI from this resolution layer is resampled to correspond to the user's zoom factor. Thus, smooth zoom control can be rendered despite storing only few dyadically spaced resolution layers at the server. Note that the encoding takes place once and generates a repository of slices. Relevant slices can be served to several clients depending on their individual RoIs. The encoding can either take place live or offline beforehand.

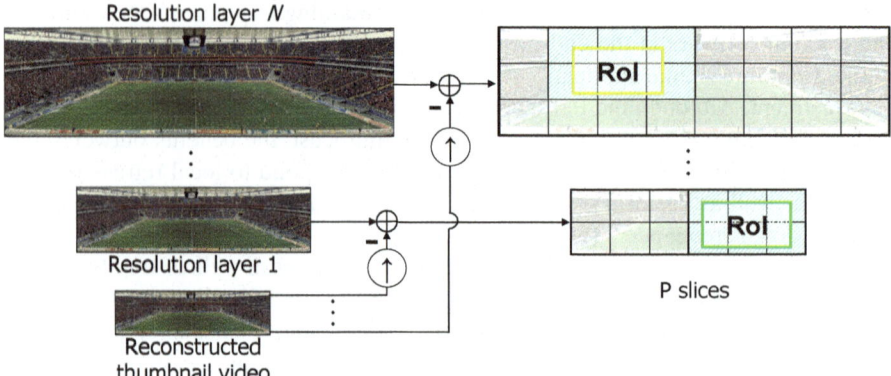

Fig. 2 The thumbnail video constitutes a base layer and is coded with H.264/AVC using I, P and B pictures. The reconstructed base layer video frames are upsampled by a suitable factor and used as prediction signal for encoding video corresponding to the higher resolution layers. Higher resolution layers are coded using P slices.

With the above-mentioned coding scheme, the thumbnail is transmitted continuously. As shown in Fig. 1, the video player can display it to aid navigation. Moreover, the thumbnail can be used for error concealment, in case parts of the RoI do not arrive in time. Ideally, the video delivery system should react to the client's changing RoI with as little latency as possible. The described coding scheme enables access to a new region, with an arbitrary zoom factor, during any frame-interval instead of having to wait for the end of a GOP or having to transmit extra slices from previous frames. The coding scheme described above uses H.264/AVC building blocks, but it is neither AVC-compliant nor SVC-compliant.

Compliance with State-of-the-Art Video Compression Standards. Current video compression standards provide tools like slices but no straightforward method for spatial random access since their main focus has been compression efficiency of full-frame video and resilience to losses. SVC supports both slices as well as spatial resolution layers. Alas, SVC allows only single-loop decoding whereas upward prediction from intercoded base-layer frames implies multiple-loop decoding, and hence is not supported by the standard. If the base layer frame is intercoded, then SVC allows predicting the motion-compensation residual at the higher-resolution layer from the residual at the base layer. However, interframe prediction dependencies across slices belonging to a high-resolution layer hamper spatial random access. Note that the motion vectors (MVs) can be chosen such that they do not point outside slice boundaries. Also note that instead of SVC, AVC can be employed separately for the high-resolution layers and the MVs can be similarly restricted to eliminate inter-slice dependencies. However, this is very similar to treating the slices as separate video sequences. An obvious drawback is the redundancy between the high-resolution slices and the base layer. A second drawback is that after RoI change, a newly joined slice can only be decoded starting from an intracoded frame. However, if the video player stops displaying the thumbnail video, the transmission of the base layer can be discontinued.

Coding Slices with Multiple Representations. Prior work on view random access, mentioned in Sect. 2.1 employs multiple representations for coding an image. Similarly, we can use multiple representations for coding a high-resolution slice. This will allow us to use interframe coding among successive high-resolution layer frames and transmit the appropriate representation for a slice depending on the slices that have been transmitted earlier. For some representations, the MVs can be allowed to point outside slice boundaries. Note that this might lower the transmission bit-rate but more storage will be required for multiple representations. The benefit of the scheme in Fig. 2 is that knowing the current RoI is enough to decide which data need to be transmitted unlike the case of multiple representations where the decision is conditional on prior transmitted data.

Improvement Based on Background Extraction. Now let us see how the coding scheme from Fig. 2 can be improved for higher coding efficiency without employing multiple representations. Although the coding scheme of Fig. 2 enables efficient random access, upward prediction using the reconstructed thumbnail frames might result in substantial residual energy for high spatial frequencies. We propose creating a background frame [57, 58] for each high-resolution layer and employing long-term memory motion-compensated prediction (LTM MCP) [59] to exploit the correlation between this frame and each high-resolution frame to be encoded [60]. The background frame is intracoded. As shown in Fig. 3, high-resolution P slices have two references to choose from, upward prediction and the background frame. If a transmitted high-resolution P slice refers to the background frame, relevant I slices from the background frame are transmitted only if they have not been transmitted

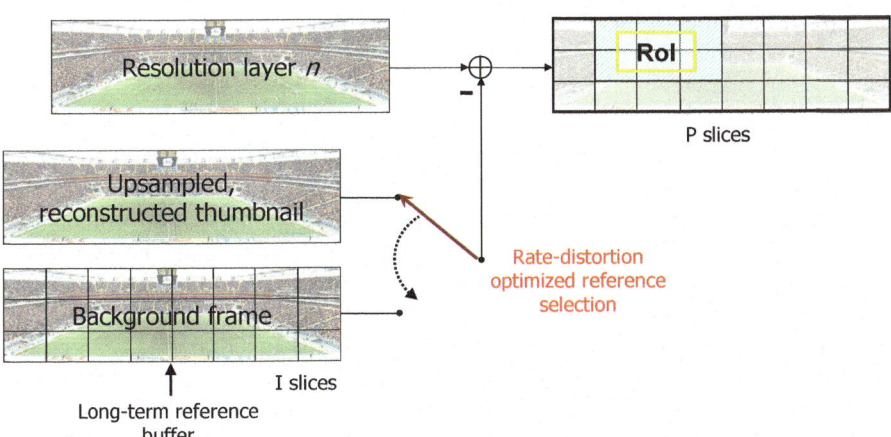

Fig. 3 Each high-resolution layer frame has two references to choose from, the frame obtained by upsampling the reconstructed thumbnail frame and the background frame from the same layer in the background pyramid.

earlier. This is different from prior work [61] employing a background pyramid, in which the encoder uses only those parts of the background for prediction that exist in the decoder's multi-resolution background pyramid. In [61], the encoder mimics the decoder which builds a background pyramid out of all previously received frames. Note that the camera is likely to be static in such applications since a moving camera might hamper the interactive browsing experience. Background extraction is generally easier with a static camera. Background extraction algorithms as well as detection and update of changed background portions have been previously studied, for example in [62]. Note that the improved coding scheme entails transmitting some I slices from the background frame that might be required for decoding the current high-resolution P slice. Nevertheless, the cost of doing this is amortized over the streaming session. Bit-rate reduction of 70–80% can be obtained with this improvement while retaining efficient random access.

Optimal Slice Size. Generally, whenever tiles or slices are employed, choosing the tile size or slice size poses the following trade-off. On one hand, a smaller slice size reduces the overhead of transmitted pixels. The overhead is constituted by pixels that have to be transmitted due to the coarse slice grid but are not used for rendering the display. On the other hand, reducing the slice size worsens the coding efficiency. This is due to increased number of headers and inability to exploit correlation across the slices. The optimal slice size depends on the RoI display dimensions, the dimensions of the high-spatial-resolution video, the content itself and the distribution of the user-selected zoom-factor. Nevertheless, we have demonstrated in prior work that stochastic analysis can estimate the expected number of transmitted pixels per frame [56]. This quantity, denoted by $\psi(s_w, s_h)$, is a function of the slice width, s_w and the slice height, s_h. The average number of bits per pixel required to encode the high-resolution video frame, denoted by $\eta(s_w, s_h)$, can also be observed or estimated as a function of the slice size. The optimal slice size is the one that minimizes the expected number of bits transmitted per frame,

$$(s_w^{\mathrm{opt}}, s_h^{\mathrm{opt}}) = \arg \min_{(s_w, s_h)} \eta(s_w, s_h) \times \psi(s_w, s_h). \tag{1}$$

The results in our earlier work show that the optimal slice size can be determined accurately without capturing user-interaction trajectories [56]. Although the model predicts the optimal slice size accurately, it can underestimate or overestimate the transmitted bit-rate. This is because the popular slices that constitute the salient objects in the video might entail high or low bit-rate compared to the average. Also, the location of the objects can bias the pixel overhead to the high or low side, whereas the model uses the average overhead. Note that the cost function in (1) can be replaced with a Lagrangian cost function that minimizes the weighted sum of the average transmission bit-rate and the incurred storage cost. The storage cost can be represented by an appropriate constant multiplying $\eta(s_w, s_h)$.

4 Pre-fetching Based on RoI Prediction

The rationale behind pre-fetching is lowering the latency of interaction. Imagine that frame number n is being rendered on the screen. At this point, the user's RoI selection up to frame n has been observed. The goal is to predict the user's RoI at frame $n + d$ ahead of time and pre-fetch relevant slices.

Extrapolating the Navigation Trajectory. In our own work [63, 64], we have used an autoregressive moving average (ARMA) model to estimate the velocity of the RoI center:

$$v_t = \alpha v_{t-1} + (1 - \alpha)(p_t - p_{t-1}), \tag{2}$$

where, the co-ordinates of the RoI center, observed up to frame n, are given by $p_t = (x_t, y_t)$ for $t = 0, 1 \ldots, n$. The predicted RoI center co-ordinates $\hat{p}_{n+d} = (\hat{x}_{n+d}, \hat{y}_{n+d})$ for frame $n + d$ are given by

$$\hat{p}_{n+d} = p_n + dv_n, \tag{3}$$

suitably adjusted if the RoI happens to veer off the extent of the video frame. The prediction lookahead, d frames, should be chosen by taking into account network delays and the desired interaction latency. The parameter α above trades off responsiveness to the user's RoI trajectory and smoothness of the predicted trajectory.

Video-Content-Aware RoI Prediction. Note that the approach described above is agnostic of the video content. We have explored video-content-aware RoI prediction that analyzes the motion of objects in the video to improve the RoI prediction [63, 64]. The transmission system in this work employs the multi-resolution video coding scheme presented in Sect. 3. The transmission system ensures that some future thumbnail video frames are buffered at the client's side. Figure 4 illustrates client-side video-content-aware RoI prediction. Following are some approaches explored in [63]:

1. Optical flow estimation techniques, for example the Kanade-Lucas-Tomasi (KLT) feature tracker [65], can find feature points in buffered thumbnail frames and track the features in successive frames. The feature closest to the RoI center in frame n can be followed up to frame $n + d$. The location of the tracked feature point can be made the center of the predicted RoI in frame $n + d$ or the predicted RoI can be chosen such that the tracked feature point appears in the same relative location. Alternatively, a smoother trajectory can be obtained by making a change to the RoI center only if the feature point moves more than a certain distance away from the RoI center.

2. Depending on the chosen optical flow estimation technique, the above approach can be computationally intensive. An alternative approach exploits MVs contained in the buffered thumbnail bit-stream. The MVs are used to find a plausible propagation of the RoI center pixel in every subsequent frame up to frame $n + d$. The location of the propagated pixel in frame $n + d$ is deemed to be the center of the predicted RoI. Although the MVs are rate-distortion optimized and might not reflect true motion, the results are competitive to those obtained with the KLT

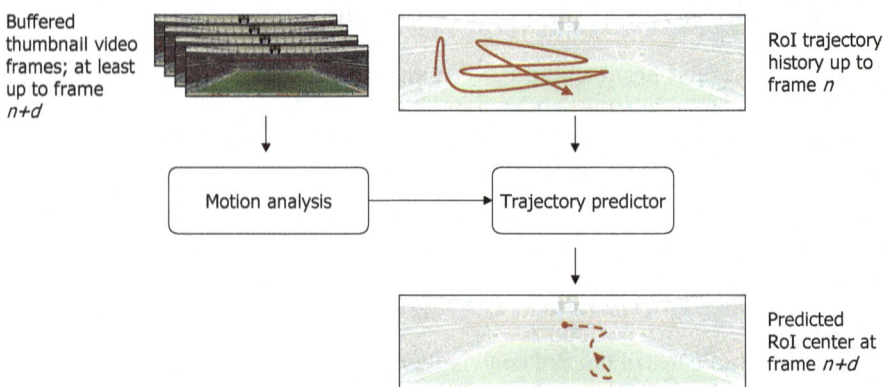

Fig. 4 Video-content-aware RoI prediction analyzes motion in the buffered thumbnail video frames. The video transmission system ensures that some thumbnail video frames are sent ahead of time. Although not shown in the figure, RoI prediction can alternatively be performed at the server. In this case, the server can analyze motion in the high-resolution frames, however, the available trajectory history might be older than current due to network delays. Also, the load on the server increases with the number of clients.

feature tracker [63]. The work in [66] is related in spirit, although the context, mobile augmented reality, is different. In this work, MVs are used to track multiple feature points from one frame to the next while employing homography testing to eliminate outliers among tracked feature points. The algorithm also considers the case of B frames.

3. One can employ multiple RoI predictors and combine their results, for example, through a median operation. This choice guarantees that for any frame-interval, if one of the predictors performs poorly compared to the rest, then the median operation does not choose that predictor. In general, the more diversity among the predictors the better.

Compared to the video-content-agnostic schemes, the gain obtained through video-content-aware RoI prediction is higher for longer prediction lookahead d [63]. Moreover, unlike the above approaches that are generic, the motion analysis can be domain-specific [64]. For example, for interactive viewing of soccer, certain objects-of-interest like the ball, the players, the referees, etc. can be tracked and their positions can drive the RoI prediction.

In the approaches above, the user actively controls the input device and the goal of the system is to predict the future path as accurately as possible. In another mode of operation, the system offers to track a certain object-of-interest for the user such that it relieves navigation burden. In this case, a user-selected trajectory might not be available for comparison or as trajectory history input. In this mode, the goal of the algorithm is to provide a smooth trajectory without deviating from the object.

Figure 5 reproduces a result from [64] that shows the tracking of a soccer player over successive frames of the thumbnail video. The algorithm is based on

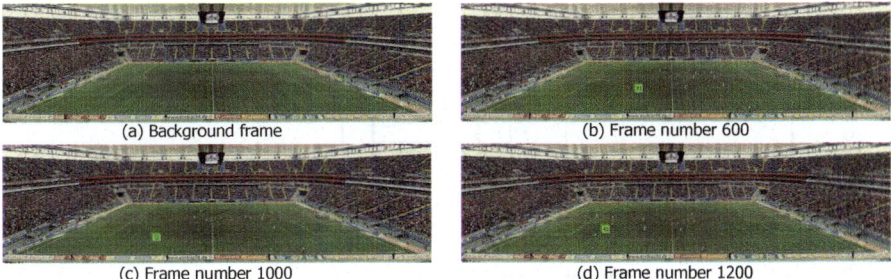

Fig. 5 (a) Background frame. (b)–(d) Example of player tracking. Tracked player is highlighted for better visibility. The frames belong to the decoded thumbnail video having resolution 640×176 pixels. Note that a player is typically less than 5 pixels wide in the thumbnail video.

background subtraction and blob tracking using MVs. Note that a player is typically less than 5 pixels wide in the thumbnail video. Alternatively, the server can process the high-resolution video or the tracking information can be generated through human assistance and trajectories of certain objects-of-interest can be conveyed to the clients to aid their pre-fetching modules.

5 P2P Multicasting for Interactive Region-of-Interest

From the perspective of allowing the system to scale to large numbers of users, it is important to limit both the encoding load as well as the bandwidth required at the server. The video compression approach presented in Sect. 3 limits the encoding load on the server irrespective of the number of users. The goal of this section is to limit the bandwidth required from dedicated servers. We assume that several users are concurrently watching the video, however, each user enjoys independent control of the region to watch. In this section, we review our IRoI P2P streaming system, first introduced in [67, 68], that can achieve P2P live multicast of IRoI video.

5.1 System Architecture

We employ the compression scheme from Sect. 3, illustrated in Figs. 2 and 3, for compressing the thumbnail video and the high-resolution layers. IRoI P2P aims to exploit overlaps among the users' RoIs. Figure 6 shows overlaps among RoIs of three users. The P2P protocol builds on top of the Stanford Peer-to-Peer Multicast (SPPM) protocol [69, 50] which operates in tree-push manner. SPPM was originally developed for P2P video streaming without any pan/tilt/zoom functionality. Nevertheless, we can leverage SPPM for building and maintaining distribution trees in a distributed manner. Each high-resolution slice, also called enhancement layer slice, is delivered using a separate set of multicast trees. Similarly, multiple complementary multicast trees deliver the thumbnail video, called the base layer. Each

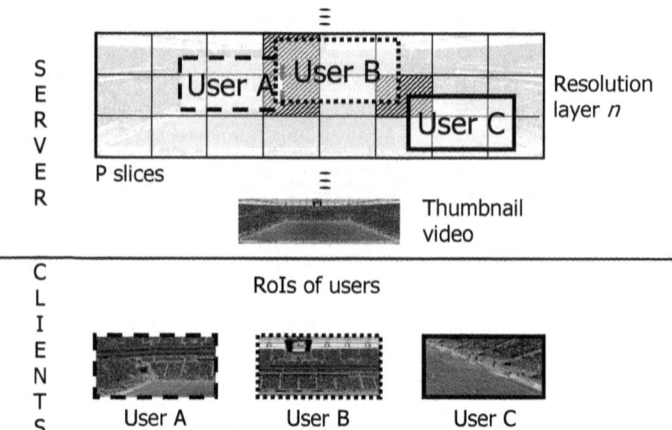

Fig. 6 Example illustrating RoIs of three users within the multi-resolution video representation. The slices shown shaded are commonly wanted by more than one user and represent the "overlaps" exploited by the IRoI P2P system.

peer subscribes the base layer at all times and additionally some enhancement layer slices that are required to render the RoI. Peers also dynamically unsubscribe slices that are no longer required to render the RoI. The RoI prediction lookahead accounts for the latency in joining new trees as well as the playout delay that is employed to mitigate delay jitter among the high-resolution slice packets.

5.2 P2P Protocol

The server maintains a database of slices that each peer is currently subscribed to. Whenever the RoI prediction indicates a change of RoI, the peer sends an RoI-switch request to the server. This consists of the top-left and bottom-right slice IDs of the old RoI as well as the new RoI. In response to the RoI-switch request, the server sends a list of potential parents for every new multicast tree that the peer needs to subscribe. Corresponding to every multicast tree, there is a limit on the number of peers the server can directly serve, and the server includes itself in the list if this quota is not yet full. The server also updates its database assuming that the peer will be successful in updating its subscriptions. After receiving the list from the server, the peer probes potential parents for every new multicast tree it needs to join. If it receives a positive reply, it sends an attach request for that tree. If it still fails to connect, the peer checks for positive replies from other probed peers and tries attaching to one of them. Once connected to any multicast tree corresponding to a slice, the peer checks if it has previously received the corresponding background I slice. If it has not then the peer obtains the background I slice from one of the peers in the list or the server.

When the RoI prediction indicates a change of RoI, the peer waits a while before sending leave messages to its parents on trees that its RoI no longer requires. This ensures that slices are not unsubscribed prematurely. On the other hand, the peer sends leave messages to its children immediately but keeps forwarding data as long as it receives data from its parent. Upon receiving leave messages, the respective children request potential parents' lists from the server for the respective multicast trees and try finding new parents. The delay in unsubscribing is chosen such that the children experience a smooth handoff from old parent to new parent. In rare cases, a child peer takes longer than the handoff deadline to find a new parent and experiences disruption on that tree. The cumulative distribution function (cdf) of slice subscription durations shown in Fig. 7 indicates how long peers attach to a multicast tree. For the shown example, there are two high-resolution layers apart from the thumbnail video. The RoI is 480×240 pixels whereas the highest resolution layer is 2560×704 pixels. The total number of slices, counting the thumbnail video as one slice and counting slices of the two high-resolution layers, is 382. Each peer subscribes about 24 slices on average corresponding to about 1.1 Mbps bit-rate, whereas the collective bit-rate of all the slices is about 14.1 Mbps.

In addition to leaving multicast trees gracefully, peers can also switch off altogether leading to ungraceful departures. If a child peer does not receive data for a particular tree for a timeout interval, it assumes that the parent is unavailable and tries to rejoin the tree by enquiring about other potential parents. To monitor the online status of parents, peers send Hello messages regularly to their parents and the parents reply back. Since most tree disconnections are graceful and occur due to RoI change, the interval for sending Hello messages can be large to limit the protocol overhead. Similar to SPPM, a loop-avoidance mechanism on individual distribution trees ensures that a descendant is not chosen as a parent [69, 70, 71, 72, 50]. For additional details on peer state transitions and timeouts associated with sending and receiving control messages, the reader may refer to [73].

The server advances the base layer transmission slightly compared to the transmission of the enhancement layer slices. This way peers can buffer some base layer

Fig. 7 Cumulative distribution function (cdf) of slice subscription durations for the *Soccer* sequence. The cdf is computed from 1000-second-long user-interaction trajectories of 100 peers. Peer lifetimes themselves are exponentially distributed with an average of 90 seconds. The average slice subscription duration for this sequence is about 16.5 seconds.

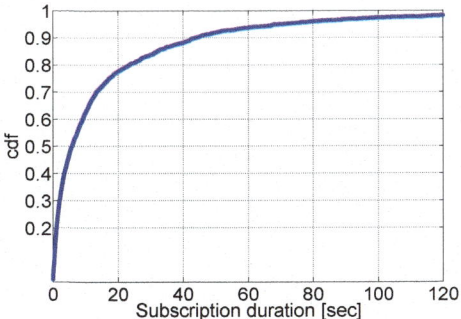

frames as well as request retransmissions of lost base layer packets. The stringent latency constraint associated with interactive RoI makes retransmissions of enhancement layer packets difficult. Recall that the base layer can be used to fill in missing parts while rendering the RoI. The error-concealed parts might appear blurry but the user experiences low-latency RoI control.

5.3 Protocol Performance

A simulation with 100 peers was carried out by implementing the IRoI P2P protocol within the NS-2 network simulator. The shape of the cdf of peer uplink capacities was modeled after the one presented in [74], however, the average of the peer uplink capacities was set to 2 Mbps, slightly higher than the 1.7 Mbps average reported in [74]. A single tree was built per slice. The average upper bound of PSNR among the peers was 41.9 dB. This corresponds to the hypothetical case when each peer receives all the slices that it needs. The average lower bound of PSNR among the peers was 30.6 dB assuming that the base layer is successfully received. The lower bound corresponds to the case when no high-resolution slices are received by the peers and the RoI is rendered only using the base layer. The average PSNR was found to be 38.6 dB, indicating that peers receive most of the enhancement layer slices required to render respective RoIs.

Figure 8 shows the trace of received, required, and missing slices collectively for the 100 peers. The percentage of missing slices is about 8.3%. The server was limited to directly serve up to 3 peers per multicast tree. Note that without such a limit, the server's capacity might be exhausted and the system might not be able to supply a new slice that no peer currently subscribes. Interestingly, the average number of slices with non-zero fan-out is only about 172 indicating that all slices are not streamed all the time. The load on the server was about 13.7 Mbps which is less than the 14.1 Mbps bit-rate of the multi-resolution representation. Another simulation was carried out in which two multicast trees were built per slice delivering odd and even frames respectively. The percentage of missing slices remained roughly the

Fig. 8 Trace of received, required and missing slices shown collectively for 100 peers watching the *Soccer* sequence. The percentage of missing slices is about 8.3%. The server was limited to directly serve up to 3 peers per multicast tree. One multicast tree was built per slice. Note that due to the unsubscription delay, peers can receive more slices than required.

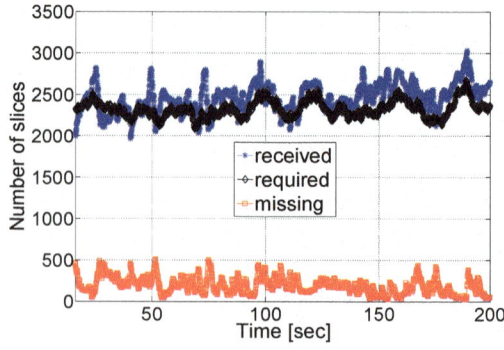

same, however, it was observed that for about 65% missing slices, the corresponding slice from the previous frame was available. This allows error-concealment using pixels from the previous frame in most cases, thus maintaining high spatial resolution, which is important for virtual pan/tilt/zoom. The picture quality was better even though the average PSNR improved by only about 0.05 dB. In experiments with other high-spatial-resolution video sequences, the average PSNR improved by about 1–2 dB compared to single tree per slice. The protocol overhead due to control messages was observed to be between 5–10% of the total traffic.

5.4 Server Bandwidth Allocation

The slices hosted by the server constitute a set of P2P multicast streams which generally vary in popularity. A framework for server bandwidth allocation among multiple P2P multicast streams has been proposed in a related thread of our research [68, 75]. The framework accommodates multiple multicast trees per stream and can take into account the popularity, the rate-distortion operating point as well as the peer churn rate associated with each stream. The framework allows minimizing different metrics like mean distortion among the peers, number of frame-freezes overall, etc. When the available server bandwidth is scarce, it is very important to judiciously allocate rate to the most important slices. For the above example with 100 peers and 2 trees per slice, the server capacity was set to 10 Mbps and the limits on the numbers of direct children associated with the multicast trees were computed by minimizing expected mean distortion. Note that the 10 Mbps server capacity is less than the 14.1 Mbps bit-rate of the multi-resolution representation. The optimized rate allocation among the slices was compared against a heuristic scheme that sequentially allocates rate to slices with ascending slice IDs, stopping when the capacity exhausts. The optimized rate allocation resulted in about 21% missing slices whereas the heuristic scheme resulted in about 82% missing slices.

6 Conclusions

Interactive pan/tilt/zoom allows watching user-selected portions of high-resolution video even on displays of lower spatial resolution. In this chapter, we have reviewed the technical challenges that must be overcome for watching IRoI video and possible solutions. From the gamut of solutions, we have elaborated those that facilitate scaling to large numbers of users.

In the remote streaming scenario, the transmission of the entire high-resolution video is generally not possible due to bandwidth limitations. Broadly speaking, there are two approaches to provide a video sequence as controlled by the user's pan/tilt/zoom commands. The RoI video sequence can either be cropped from the raw high-resolution video and encoded prior to transmission or the adopted compression format can allow easy extraction of the relevant portions from the compressed representation. The first approach possesses the drawback that RoI video encoding has to be performed for each user separately. Additionally, if the high-spatial-resolution video is not available in the raw format and the users watch the sequence asynchronously, the high-spatial-resolution video has to be decoded before cropping the RoI sequence for each user.

Spatial-random-access-enabled video compression limits the load of encoding to a one-time encoding of the video, possibly with multiple resolution layers to support continuous zoom. This is beneficial for streaming both live content as well as pre-stored content to multiple users. Even when the video is played back from locally stored media, a different RoI trajectory has to be accommodated each time a user watches the content. Spatial random access allows the video player to selectively decode relevant regions only. This chapter presents a spatial-random-access-enabled video coding scheme in detail that allows the receiver to start decoding a new region, with an arbitrary zoom factor, during any frame-interval instead of having to wait for the end of a GOP or having to transmit extra slices from previous frames. Background extraction can be used with such a coding scheme to reduce transmission bit-rate as well as the size of the stored video. We also show how to choose the slice size to attain the right balance between storage and mean transmission bit-rate.

Pre-fetching helps to reduce the latency of interaction. Irrespective of whether pre-fetching is employed or not, having a base layer helps render missing parts of the RoI. This way, the system can always render the RoI chosen by the user, thus offering accurate and low-latency RoI control. The chapter presents several RoI prediction techniques for pre-fetching. Some techniques are employed at the client, some at the server and some are distributed between the server and the client. For example, the server can send the trajectories of key objects in the video to the clients to aid their RoI prediction modules.

This chapter also shows how to use P2P streaming to drastically reduce the bandwidth required from the server for supporting increasing numbers of users. It is crucial for this approach that the P2P overlay reacts quickly to the changing RoIs of the peers and limits the disruption due to the changing relationships among the peers. The IRoI P2P protocol presented in this chapter makes sure that a child-peer experiences smooth transition from old parent to new parent when the old parent willfully unsubscribes a multicast tree that is no longer required for its RoI. Typically, when users choose regions from a high-spatial-resolution video, some regions are more popular than others. It is very important, especially when the server has limited bandwidth, to judiciously allocate the available rate among the regions streamed by the server.

Spatial-random-access-enabled video coding plays an important role in the P2P distribution system. It simplifies the peer's task of choosing which multicast trees to join on-the-fly. A scheme based on multiple representations coding of the slices might further reduce the download rate required by each peer. However, such a scheme might reduce the degree of overlaps and affect the gains possible from the P2P approach apart from requiring more storage space at the server. The IRoI P2P system presented in this chapter assumes that peers watch the video synchronously. If peers can watch any time-segment, i.e., rewind and fast-forward then the relevant data could still be retrieved from each others' cache. Since storage is becoming cheaper, the cache size employed by the peers for storing previously received content can be assumed to be reasonably large. Such kind of "time-shifted P2P streaming" can be looked upon as the temporal counterpart of the spatial freedom provided by pan/tilt/zoom. A system providing both functionalities would be a natural extension of the system presented here.

Acknowledgements. Fraunhofer Heinrich-Hertz Institute (HHI), Berlin, Germany generously provided the *Soccer* sequence.

References

1. Fehn, C., Weissig, C., Feldmann, I., Mueller, M., Eisert, P., Kauff, P., Bloss, H.: Creation of High-Resolution Video Panoramas of Sport Events. In: Proc. IEEE 8th International Symposium on Multimedia, San Diego, CA, USA (2006)
2. Kopf, J., Uyttendaele, M., Deussen, O., Cohen, M.F.: Capturing and Viewing Gigapixel Images. In: Proc. ACM SIGGRAPH, San Diego, CA, USA (2007)
3. Halo: Video conferencing product by Hewlett-Packard, http://www.hp.com/halo/index.html (accessed November 5, 2009)
4. Smolic, A., McCutchen, D.: 3DAV Exploration of Video-based Rendering Technology in MPEG. IEEE Transactions on Circuits and Systems for Video Technology 14(3), 348–356 (2004)
5. Dodeca 2360: An omni-directional video camera providing over 100 million pixels per second by Immersive Media, http://www.immersivemedia.com (accessed November 5, 2009)
6. Video clip showcasing interactive TV with pan/tilt/zoom, http://www.youtube.com/watch?v=Ko9jcIjBXnk (accessed November 5, 2009)
7. ISO/IEC 15444-1:2004, JPEG 2000 Specification. Standard (2004)
8. Taubman, D., Rosenbaum, R.: Rate-Distortion Optimized Interactive Browsing of JPEG 2000 Images. In: Proc. IEEE International Conference on Image Processing, Barcelona, Spain (2000)
9. Taubman, D., Prandolini, R.: Architecture, Philosophy and Performance of JPIP: Internet Protocol Standard for JPEG 2000. In: Proc. SPIE International Symposium on Visual Communications and Image Processing, Lugano, Switzerland (2003)
10. H.264/AVC/MPEG-4 Part 10 (ISO/IEC 14496-10: Advanced Video Coding). Standard (2003)
11. Wiegand, T., Sullivan, G., Bjontegaard, G., Luthra, A.: Overview of the H.264/AVC Video Coding Standard. IEEE Transactions on Circuits and Systems for Video Technology 13(7), 560–576 (2003)
12. Dhondt, Y., Lambert, P., Notebaert, S., Van de Walle, R.: Flexible macroblock ordering as a content adaptation tool in H.264/AVC. In: Proc. SPIE Conference on Multimedia Systems and Applications VIII, Boston, MA, USA (2005)
13. Annex G of H.264/AVC/MPEG-4 Part 10: Scalable Video Coding (SVC). Standard (2007)
14. Schwarz, H., Marpe, D., Wiegand, T.: Overview of the Scalable Video Coding Extension of the H.264/AVC Standard. IEEE Transactions on Circuits and Systems for Video Technology 17(9), 1103–1120 (2007)

15. Baccichet, P., Zhu, X., Girod, B.: Network-Aware H.264/AVC Region-of-Interest Coding for a Multi-Camera Wireless Surveillance Network. In: Proc. Picture Coding Symposium, Beijing, China (2006)
16. Devaux, F., Meessen, J., Parisot, C., Delaigle, J., Macq, B., Vleeschouwer, C.D.: A Flexible Video Transmission System based on JPEG 2000 Conditional Replenishment with Multiple References. In: Proc. IEEE International Conference on Acoustics, Speech, and Signal Processing, Honolulu, HI, USA (2007)
17. Makar, M., Mavlankar, A., Girod, B.: Compression-Aware Digital Pan/Tilt/Zoom. In: Proc. 43rd Annual Asilomar Conference on Signals, Systems, and Computers, Pacific Grove, CA, USA (2009)
18. Girod, B.: The Efficiency of Motion-Compensating Prediction for Hybrid Coding of Video Sequences. IEEE Journal on Selected Areas in Communications 5(7), 1140–1154 (1987)
19. Girod, B.: Motion-Compensating Prediction with Fractional-Pel Accuracy. IEEE Transactions on Communications 41(4), 604–612 (1993)
20. Girod, B.: Efficiency analysis of multihypothesis motion-compensated prediction for video coding. IEEE Transactions on Image Processing 9(2), 173–183 (2000)
21. Gruenheit, C., Smolic, A., Wiegand, T.: Efficient Representation and Interactive Streaming of High-Resolution Panoramic Views. In: Proc. IEEE International Conference on Image Processing, Rochester, NY, USA (2002)
22. Heymann, S., Smolic, A., Mueller, K., Guo, Y., Rurainsky, J., Eisert, P., Wiegand, T.: Representation, Coding and Interactive Rendering of High-Resolution Panoramic Images and Video using MPEG-4. In: Proc. Panoramic Photogrammetry Workshop, Berlin, Germany (2005)
23. Kauff, P., Schreer, O.: Virtual Team User Environments - A Step from Tele-cubicles towards Distributed Tele-collaboration in Mediated Workspaces. In: Proc. IEEE International Conference on Multimedia and Expo., Lausanne, Switzerland (2002)
24. Tanimoto, M.: Free Viewpoint Television — FTV. In: Proc. Picture Coding Symposium, San Francisco, CA, USA (2004)
25. Smolic, A., Mueller, K., Merkle, P., Fehn, C., Kauff, P., Eisert, P., Wiegand, T.: 3D Video and Free Viewpoint Video - Technologies, Applications and MPEG Standards. In: Proc. IEEE International Conference on Multimedia and Expo., Toronto, ON, Canada (2006)
26. Shum, H.Y., Kang, S.B., Chan, S.C.: Survey of Image-based Representations and Compression Techniques. IEEE Transactions on Circuits and Systems for Video Technology 13(11), 1020–1037 (2003)
27. Levoy, M., Hanrahan, P.: Light Field Rendering. In: Proc. ACM SIGGRAPH, New Orleans, LA, USA (1996)
28. Bauermann, I., Steinbach, E.: RDTC Optimized Compression of Image-Based Scene Representations (Part I): Modeling and Theoretical Analysis. IEEE Transactions on Image Processing 17(5), 709–723 (2008)
29. Bauermann, I., Steinbach, E.: RDTC Optimized Compression of Image-Based Scene Representations (Part II): Practical Coding. IEEE Transactions on Image Processing 17(5), 724–736 (2008)
30. Kimata, H., Kitahara, M., Kamikura, K., Yashima, Y., Fujii, T., Tanimoto, M.: Low-Delay Multiview Video Coding for Free-viewpoint Video Communication. Systems and Computers in Japan 38(5), 14–29 (2007)
31. Liu, Y., Huang, Q., Zhao, D., Gao, W.: Low-delay View Random Access for Multi-view Video Coding. In: Proc. IEEE International Symposium on Circuits and Systems, New Orleans, LA, USA (2007)

32. Flierl, M., Mavlankar, A., Girod, B.: Motion and Disparity Compensated Coding for Multi-View Video. IEEE Transactions on Circuits and Systems for Video Technology 17(11), 1474–1484 (2007) (invited Paper)

33. Cheung, G., Ortega, A., Cheung, N.M.: Generation of Redundant Frame Structure for Interactive Multiview Streaming. In: Proc. IEEE 17th Packet Video Workshop, Seattle, WA, USA (2009)

34. Ramanathan, P., Girod, B.: Rate-Distortion Optimized Streaming of Compressed Light Fields with Multiple Representations. In: Proc. IEEE 14th Packet Video Workshop, Irvine, CA, USA (2004)

35. Ramanathan, P., Girod, B.: Random Access for Compressed Light Fields using Multiple Representations. In: Proc. IEEE 6th International Workshop on Multimedia Signal Processing, Siena, Italy (2004)

36. Jagmohan, A., Sehgal, A., Ahuja, N.: Compression of Lightfield Rendered Images using Coset Codes. In: Proc. 37th Annual Asilomar Conference on Signals, Systems, and Computers, Pacific Grove, CA, USA (2003)

37. Aaron, A., Ramanathan, P., Girod, B.: Wyner-Ziv Coding of Light Fields for Random Access. In: Proc. IEEE 6th Workshop on Multimedia Signal Processing, Siena, Italy (2004)

38. Cheung, N.M., Ortega, A., Cheung, G.: Distributed Source Coding Techniques for Interactive Multiview Video Streaming. In: Proc. Picture Coding Symposium, Chicago, IL, USA (2009)

39. Azuma, R., Bishop, G.: A Frequency-domain Analysis of Head-motion Prediction. In: Proc. ACM SIGGRAPH, Los Angeles, CA, USA (1995)

40. Singhal, S.K., Cheriton, D.R.: Exploiting Position History for Efficient Remote Rendering in Networked Virtual Reality. Presence: Teleoperators and Virtual Environments 4, 169–193 (1995)

41. Ramanathan, P., Kalman, M., Girod, B.: Rate-Distortion Optimized Interactive Light Field Streaming. IEEE Transactions on Multimedia 9(4), 813–825 (2007)

42. Kiruluta, A., Eizenman, M., Pasupathy, S.: Predictive Head Movement Tracking using a Kalman Filter. IEEE Transactions on Systems, Man and Cybernetics, Part B: Cybernetics 27(2), 326–331 (1997)

43. Kurutepe, E., Civanlar, M.R., Tekalp, A.M.: A Receiver-driven Multicasting Framework for 3DTV Transmission. In: Proc. 13th European Signal Processing Conference, Antalya, Turkey (2005)

44. Kurutepe, E., Civanlar, M.R., Tekalp, A.M.: Interactive Transport of Multi-view Videos for 3DTV Applications. Journal of Zhejiang University - Science A (2006)

45. Deering, S.: Host Extensions for IP Multicasting. RFC 1112 (1989)

46. Albanna, Z., Almeroth, K., Meyer, D., Schipper, M.: IANA guidelines for IPv4 multicast address assignments. RFC 3171 (2001)

47. McCanne, S., Jacobson, V., Vetterli, M.: Receiver-Driven Layered Multicast. In: Proc. ACM SIGCOMM, Stanford, CA, USA (1996)

48. Estrin, D., Handley, M., Helmy, A., Huang, P., Thaler, D.: A Dynamic Bootstrap Mechanism for Rendezvous-based Multicast Routing. In: Proc. IEEE INFOCOM, New York, USA (1999)

49. Chu, Y.H., Rao, S., Seshan, S., Zhang, H.: A Case for End System Multicast. IEEE Journal on Selected Areas in Communications 20(8), 1456–1471 (2002)

50. Setton, E., Baccichet, P., Girod, B.: Peer-to-Peer Live Multicast: A Video Perspective. Proceedings of the IEEE 96(1), 25–38 (2008)

51. Magharei, N., Rejaie, R., Guo, Y.: Mesh or Multiple-Tree: A Comparative Study of Live Peer-to-Peer Streaming Approaches. In: Proc. IEEE INFOCOM (2007)

52. Agarwal, S., Singh, J., Mavlankar, A., Baccichet, P., Girod, B.: Performance of P2P Live Video Streaming Systems on a Controlled Test-bed. In: Proc. 4th International Conference on Testbeds and Research Infrastructures for the Development of Networks and Communities, Innsbruck, Austria (2008)
53. Mavlankar, A., Baccichet, P., Girod, B., Agarwal, S., Singh, J.: Video Quality Assessment and Comparative Evaluation of Peer-to-Peer Video Streaming Systems. In: Proc. IEEE International Conference on Multimedia and Expo., Hanover, Germany (2008)
54. Kurutepe, E., Sikora, T.: Feasibility of Multi-View Video Streaming Over P2P Networks. In: 3DTV Conference: The True Vision - Capture, Transmission and Display of 3D Video (2008)
55. Kurutepe, E., Sikora, T.: Multi-view video streaming over p2p networks with low start-up delay. In: Proc. IEEE International Conference on Image Processing (2008)
56. Mavlankar, A., Baccichet, P., Varodayan, D., Girod, B.: Optimal Slice Size for Streaming Regions of High Resolution Video with Virtual Pan/Tilt/Zoom Functionality. In: Proc. 15th European Signal Processing Conference, Poznan, Poland (2007)
57. Massey, M., Bender, W.: Salient Stills: Process and Practice. IBM Systems Journal 35(3&4), 557–573 (1996)
58. Farin, D., de With, P., Effelsberg, W.: Robust Background Estimation for Complex Video Sequences. In: Proc. IEEE International Conference on Image Processing, Barcelona, Spain (2003)
59. Wiegand, T., Zhang, X., Girod, B.: Long-Term Memory Motion-Compensated Prediction. IEEE Transactions on Circuits and Systems for Video Technology 9(1), 70–84 (1999)
60. Mavlankar, A., Girod, B.: Background Extraction and Long-Term Memory Motion-Compensated Prediction for Spatial-Random-Access-Enabled Video Coding. In: Proc. Picture Coding Symposium, Chicago, IL, USA (2009)
61. Bernstein, J., Girod, B., Yuan, X.: Hierarchical Encoding Method and Apparatus Employing Background References for Efficiently Communicating Image Sequences. US Patent (1992)
62. Hepper, D.: Efficiency Analysis and Application of Uncovered Background Prediction in a Low Bit Rate Image Coder. IEEE Transactions on Communications 38(9), 1578–1584 (1990)
63. Mavlankar, A., Varodayan, D., Girod, B.: Region-of-Interest Prediction for Interactively Streaming regions of High Resolution Video. In: Proc. IEEE 16th Packet Video Workshop, Lausanne, Switzerland (2007)
64. Mavlankar, A., Girod, B.: Pre-fetching based on Video Analysis for Interactive Region-of-Interest Streaming of Soccer Sequences. In: Proc. IEEE International Conference on Image Processing, Cairo, Egypt (2009)
65. Tomasi, C., Kanade, T.: Detection and Tracking of Point Features. Tech. Rep. CMU-CS-91-132, Carnegie Mellon University, Pittsburgh, PA (1991)
66. Takacs, G., Chandrasekhar, V., Girod, B., Grzeszczuk, R.: Feature Tracking for Mobile Augmented Reality Using Video Coder Motion Vectors. In: Proc. IEEE and ACM 6th International Symposium on Mixed and Augmented Reality, Nara, Japan (2007)
67. Mavlankar, A., Noh, J., Baccichet, P., Girod, B.: Peer-to-Peer Multicast Live Video Streaming with Interactive Virtual Pan/Tilt/Zoom Functionality. In: Proc. IEEE International Conference on Image Processing, San Diego, CA, USA (2008)
68. Mavlankar, A., Noh, J., Baccichet, P., Girod, B.: Optimal Server Bandwidth Allocation for Streaming Multiple Streams via P2P Multicast. In: Proc. IEEE 10th Workshop on Multimedia Signal Processing, Cairns, Australia (2008)

69. Setton, E., Noh, J., Girod, B.: Rate-Distortion Optimized Video Peer-to-Peer Multicast Streaming. In: Proc. Workshop on Advances in Peer-to-Peer Multimedia Streaming at ACM Multimedia, Singapore (2005) (invited Paper)
70. Setton, E., Noh, J., Girod, B.: Low Latency Video Streaming over Peer-To-Peer Networks. In: Proc. IEEE International Conference on Multimedia and Expo., Toronto, Canada (2006)
71. Setton, E., Noh, J., Girod, B.: Congestion-Distortion Optimized Peer-to-Peer Video Streaming. In: Proc. IEEE International Conference on Image Processing, Atlanta, GA, USA (2006)
72. Baccichet, P., Noh, J., Setton, E., Girod, B.: Content-Aware P2P Video Streaming with Low Latency. In: Proc. IEEE International Conference on Multimedia and Expo., Beijing, China (2007)
73. Setton, E.: Congestion-Aware Video Streaming over Peer-to-Peer Networks. Ph.D. thesis, Stanford University, Stanford, CA, USA (2006)
74. Noh, J., Baccichet, P., Girod, B.: Experiences with a Large-Scale Deployment of Stanford Peer-to-Peer Multicast. In: Proc. IEEE 17th Packet Video Workshop, Seattle, WA, USA (2009)
75. Mavlankar, A., Noh, J., Baccichet, P., Girod, B.: Optimal Server Bandwidth Allocation among Multiple P2P Multicast Live Video Streaming Sessions. In: Proc. IEEE 17th Packet Video Workshop, Seattle, WA, USA (2009)

Chapter 20
End-to-End Management of Heterogeneous Environments Enabling Quality of Experience

Christian Timmerer, Maria Teresa Andrade, and Alberto Leon Martin

Abstract. End-to-end support for Quality of Service (QoS) or Quality of Experience (QoE) has been broadly discussed in the literature. Many technologies have been proposed, each focusing on specific aspects for providing QoS/QoE guarantees to the end user. However, the integrated management of the end-to-end chain preserving QoS/QoE in heterogeneous environments is still an aspect insufficiently addressed to date, regardless the fact that it significantly impacts the overall quality of the service paid by the end-user. In this chapter we propose an integrated management supervisor that takes into account the requirements from all stakeholders along the multimedia content delivery chain. It provides an end-to-end management solution enabling QoS/QoE to the end user. Furthermore, we describe a QoS/QoE model which allows one to measure the perceptual quality of video transmissions by exploiting metrics from different layers (service, application and network) in an interoperable way. As such we are able to keep the quality as experienced by the end user at a satisfactory level, even when facing adverse delivery conditions, without cost-intensive subjective tests. Therefore, we propose a detailed QoS/QoE model for video transmission following the philosophy of the ITU-T's E-model for audio, and show how this can be translated into interoperable description formats offered by the MPEG-21 Multimedia Framework, as a contribution to balance the current network neutrality debate among its key players.

Christian Timmerer
Klagenfurt University, Universitätsstrasse 65-67, A-9020 Klagenfurt, Austria
e-mail: `christian.timmerer@itec.uni-klu.ac.at`

Maria Teresa Andrade
INESC Porto / FEUP, Rua Dr. Roberto Frias 378, P-4200-465 Porto, Portugal
e-mail: `maria.andrade@inescporto.pt`

Alberto Leon Martin
Telefónica I+D, 6, Emilio Vargas Street, 28043 Madrid, Spain
e-mail: `alm@tid.es`

1 Introduction

Many different technologies are currently being used to enable end-to-end Quality of Service for advanced multimedia services [1]. However, for the most part, these technologies are neither integrated within an interoperable framework nor provide a means to effectively manage the end-to-end multimedia delivery chain. As such, their scope of use is limited to specific applications or situations. Thus, there exists a need for solutions based on an interoperable multimedia framework supporting the end-to-end management of heterogeneous contents, networks, and terminals while enabling Quality of Service (QoS), or even Quality of Experience (QoE), for the end user.

This topic is also linked with network neutrality [2], which shall not provide restrictions on content, sites, or platforms, on the kinds of equipment that may be attached, and on the modes of communication allowed, as well as one where communication is not unreasonably degraded by other communication streams. In general, network neutrality is already provided in a sense that telecommunications companies rarely offer different (QoS/QoE) rates to broadband and dial-up Internet consumers. However, there are no clear legal restrictions against allowing certain service providers to intentionally slow-down peer-to-peer (P2P) communications or to perform deep packet inspection in order to discriminate against P2P, FTP and online games, instituting a cell-phone style billing system of overages, free-to-telecom value-added services, and anti-competitive tying (i.e., "bundling").

Quality of Experience (QoE) [3, 4], some times also known as Quality of User Experience, is a subjective measure from the user's perspective of the overall value of the service provided. Although QoE is perceived as subjective, it is the only measure that counts for customers of a service. Being able to measure it in a controlled manner helps operators to understand what may be wrong with their services. A framework that can be used for this purpose is the MPEG-21 multimedia framework, which enables the transparent and augmented use of multimedia resources across a wide range of networks, devices, user preferences, and communities [5]. In particular, MPEG-21 provides means for the transaction of Digital Items (i.e., multimedia resources and metadata within a standardized structure) among Users and whose functions can be categorized into six categories: declaration (and identification), digital rights management, adaptation, processing, systems, and miscellaneous aspects (i.e., reference software, conformance, etc.).

In this chapter we describe an architecture for the integrated management of the end-to-end multimedia delivery chain that utilizes the MPEG-21 multimedia framework [6], and that enables QoS/QoE for the end-user by adopting cross-layer techniques [7]. As this architecture has been developed in the course of the ENTHRONE II project [8], it is referred to as the ENTHRONE Integrated Management Supervisor (EIMS). Furthermore, we describe a QoS/QoE model enabling one to measure the perceptual quality of video transmissions by exploiting metrics from different layers (service, application and network) in an interoperable way. As such we are able to keep the quality as experienced by the end user at a satisfactory level, even when facing adverse delivery conditions, without cost-intensive subjective tests.

The remainder of this chapter is organized as follows. The high-level architecture of ENTHRONE for end-to-end management enabling QoS/QoE is described in Section 2. This section also highlights the cross-layer QoS adaptation concept. Section 3 describes the EIMS with its functional building blocks and interfaces. Finally, Section 4 describes how to measure the QoS/QoE for an MPEG-21-based cross-layer multimedia content adaptation, and Section 5 concludes the chapter.

2 Architecture and Overall Concept

2.1 End-to-End Management Enabling QoS/QoE

The ENTHRONE high-level architecture for end-to-end management enabling QoS is given in Fig. 1 which comprises three layers:

The top layer is the supervision layer which role is to manage and to monitor the services that participate in the content delivery chain. It is implemented by the ENTHRONE Integrated Management Supervision (EIMS), which uses MPEG-21

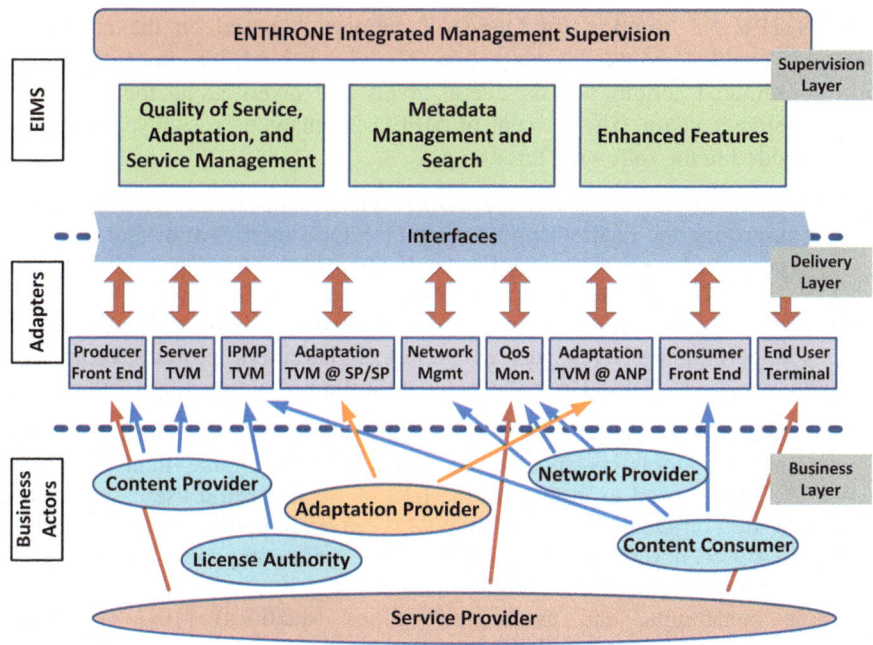

Fig. 1 ENTHRONE high-level architecture for end-to-end management enabling QoS

interfaces to the components at the underlying service delivery layer. The EIMS is composed of several subsystems that can be grouped in three main classes:

- QoS/QoE-based service management and multimedia content adaptation;
- Metadata management and content search; and
- Multicast, service monitoring, content caching, and content distribution networks, specified as enhanced features.

The end-to-end management performed by the EIMS relies on functionality available in the delivery layer to actually provide QoS/QoE to the end user throughout the service lifecycle. This middle layer thus comprises all functionalities required for a proper content delivery, with end-to-end management enabling QoS. An adapter is the implementation, in hardware or software, of some of this functionality. Fig. 1 shows the main adapters but this is not intended to be exhaustive.

The business layer (bottom) specifies the concerned actors in the end-to-end QoS/QoE service delivery chain. Each actor relies on the services of one or more components of the delivery layer. This chapter focuses on the top layer and does not detail further the other two layers.

2.2   Cross-Layer QoS/QoE Adaptation Concept

The ENTHRONE solution for QoS/QoE support is based on the concept of MPEG-21-enabled cross-layer adaptation. The idea behind this concept is to perform coordinated actions across several levels and layers along the end-to-end content delivery chain. The concept of MPEG-21-enabled cross-layer adaptation can be divided in the following three steps:

1. The *ENTHRONE Cross-Layer Model (EXLM)*: The EXLM provides means for describing the relationship between QoS/QoE metrics at different levels – i.e., perceived QoS (PQoS/QoE), application QoS (AppQoS), and network QoS (NQoS) – and layers – i.e., according to the well-known ISO/OSI reference model – which may cooperate to improve the ability of applications to ensure certain objectives, such as QoS/QoE guarantees, power savings, users preferences, etc. This also includes the definition of optimization criteria.

2. The *instantiation of the EXLM by utilizing MPEG-21 metadata*: Description formats (i.e., tools) as specified within MPEG-21 Digital Item Adaptation (DIA) [9] are used to instantiate the EXLM for a specific use case scenario, e.g., Video-on-Demand or conversional services. In particular, the Adaptation QoS (AQoS) description tool is used as the main component to describe the relationship between constraints, feasible adaptation operations satisfying these constraints, and associated utilities (qualities) [10]. The Usage Environment Description (UED) tools are used to describe the context where Digital Items are consumed in terms of network conditions, terminal capabilities, user preferences, and conversion capabilities. Finally, the Universal

Constraints Description (UCD) tools are used to express limitation and optimization constraints that apply to these context conditions.

3. The *Cross-Layer Adaptation Decision-Taking Engine (XL-ADTE)*: The **XL-ADTE** is part of an EIMS subsystem which provides the optimal parameter settings for media resource engines (e.g., encoder, transcoder, streaming server, etc. which are collectively referred to as television and multimedia (TVM) processors), according to the EXLM by processing the metadata compliant to MPEG-21 DIA. In other words, the XL-ADTE is a generic (software) module that solves optimization problems [11] expressed by using MPEG-21 DIA-based metadata according to the EXLM.

3 ENTHRONE Integrated Management Supervisor (EIMS)

3.1 Architecture Overview

The ENTHRONE Integrated Management Supervisor (EIMS) enables the deployment of multimedia services allowing for end-to-end management with QoS/QoE support across heterogeneous environments. Therefore, the EIMS provides a set of management subsystems, i.e., EIMS Managers, with predefined functionalities and interfaces – based on Web Services and interoperable payload formats – which enable the construction of ENTHRONE-based services according to the requirements of various scenarios.

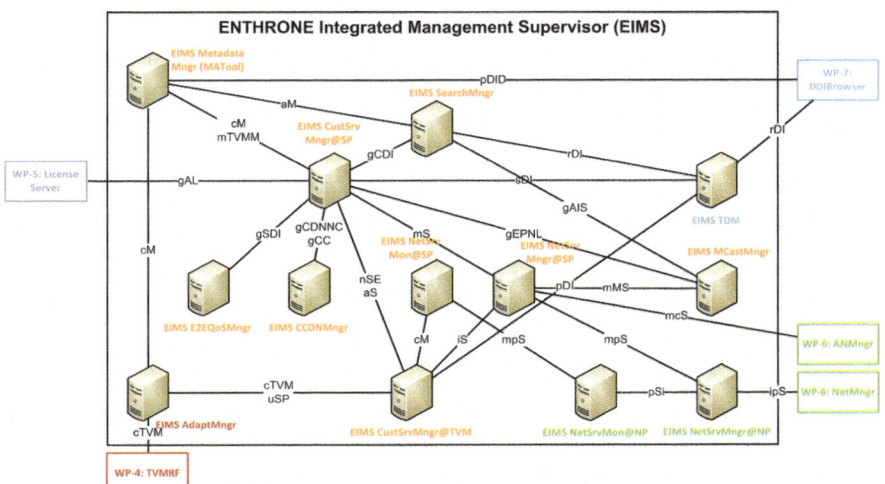

Fig. 2 Overview of the ENTHRONE Integrated Management Supervision (EIMS) architecture

An overview of the architecture of the EIMS is depicted in Fig. 2 highlighting the key EIMS Managers and its main interfaces. Note however, that due to space constraints it is not possible to describe in detail every EIMS Manager and every interface shown in the figure and, thus, for further details the interested reader is referred to [12].

The different EIMS subsystems cooperate among them, implementing functionality to:

- receive and process user requests, contacting various sources of content to find useful results for the user queries;
- determining the restrictions imposed by the context of usage by collecting relevant contextual metadata;
- selecting the best source(s) of content which are able to provide the content in a format that suits the sensed restrictions;
- determine the actual service parameters to pass to the source of the content; and
- subscribing to the required resources to support the selected service.

3.2 End-to-End QoS Manager

The aim of the End-to-End QoS Manager (E2E-QoS-Mngr) is to provide the best Digital Item configuration towards the content consumer taking into account various information coming from different business actors along the delivery chain. This information is ideally represented as MPEG-21 compliant metadata and encapsulated within the Digital Item Declaration which will be configured according to the requirements of the content consumer. The various metadata assets are briefly described in the following:

- The *Digital Item Declaration (DID)* comprising the content variations from which the E2E-QoS-Mngr may choose. The DID provides a representation of the Digital Item compliant to MPEG-21 Part 2 [13] and is usually provided by the content provider (CP).
- The characteristics of the available television and multimedia (TVM) processors which may be selected by the E2E-QoS-Mngr based on the network characteristics or conditions (i.e., the information from the provider-to-provider Service Level Specification (pSLS)) between the content consumer and the various TVMs. The TVMs are described by a unified description model featuring MPEG-7 and MPEG-21 tools and is generally referred to as TVM Resource Function (TVMRF) [14].
- The capabilities and network information from the terminal of the content consumer, including the class of service for the requested multimedia service. This kind of information is described in a format compliant to MPEG-21 DIA UED tools.

The E2E-QoS-Mngr implements a process of automated construction of a possibly suitable sequence of TVMs for a given set of multimedia content variations and a set of environmental requirements as described in [15].

The output of the E2E-QoS-Mngr is a DID configured with the chosen content variation including the location of the selected TVMs and the corresponding QoS characteristics of the actual multimedia content.

3.3 Service Manager

The Service Manager (SrvMngr) is responsible for service management and can be further divided into four subsystems with distinguished functionalities:

- The *Customer Service Manager (CustSrvMngr)* acts as a central component which provides the actual service towards the customer by implementing the service logic according to the service requirements. It instantiates other EIMS managers and coordinates the information flow within the EIMS.
- The *Network Service Manager (NetSrvMngr)* is responsible for managing network connectivity services used to transport the multimedia content with the requested QoS/QoE guarantees from its source to its consumers through several established pSLSs. Specifically, it encompasses the functionalities for service planning, provisioning, offering and fulfillment of the connectivity service, in a multi-domain context. The NetSrvMngr also strongly interacts with the Service Monitoring as described in the next bullet.
- *Service Monitoring (ServMon)* provides a means for monitoring the service with the aim to keep track of the end-to-end QoS level of a particular service [16]. Service monitoring is mainly provided within the network on aggregated streams and within the CustSrvMngr for a particular service stream. Therefore, service monitoring provides means for mapping network QoS (NQoS) – monitored on aggregated streams within the core network – to perceived QoS (PQoS/QoE) that is relevant on a per stream level [17].
- The *Terminal Device Manager (TDM)* enables the management of heterogeneous end-user devices in terms of capturing the capabilities of the terminal, PQoS probe configuration including handling its alarms, and license handling [18].

3.4 Adaptation Manager

The Adaptation Manager (AdaptMngr) aims to provide adaptation decisions according to dynamically varying context conditions coming from various sources across service and network layers. Thus, the adaptation manager hosts the Cross-Layer Adaptation Decision-Taking Engine (XL-ADTE) – see also Section 2.2 – and steers exactly one TVM (in a possible chain of TVMs). That is, the E2E-QoS-Mngr selects one (or more) TVM(s) during the service request and basically establishes the QoS-enabled end-to-end multimedia service. The role of the AdaptMngr is actually to configure each TVM involved within the chain. Furthermore, it adjusts this configuration according to possible changing usage environment properties, dynamically received from ServMons and the TDM through the CustSrvMngr. Finally, it may provide updates for the various QoS probes along the multimedia service delivery chain.

3.5 Metadata Manager

The EIMS Metadata Manager is responsible for performing metadata related tasks within the scope of the EIMS. These tasks include:

- *Aggregation and enrichment of metadata* from different metadata sources;
- *Contextual metadata collection* from and retrieval by different EIMS components; and
- *Metadata conversion between different formats.*

For further details of this EIMS subsystem the interested reader is referred to [19].

3.6 Search Manager

The Search Manager supports searching of Digital Items and browsing of DIDs in the DID database via a well-defined interface. The search and browsing is based on the data model specifically developed within ENTHRONE [20] and adopts the query format as defined by MPEG [21].

The data model supports high-level and low-level features associated to audio-visual content. That is, the Search Manager provides the following main functionalities:

- Search by making use of *high-level features* (e.g., keywords) as well as *low-level features* (e.g., color, shape, etc.);
- *Relevance feedback* by the user, through the submission of user's annotations upon browsing or consuming the content;
- *Query by relevance feedback* where feedback provided by users, who previously annotated Digital Items, is considered in the search process.

3.7 Multicast Manager

The Multicast Manager is responsible for all the tasks related to the management of multicast communication services in ENTHRONE. That is, the multicast overlay network and possibly the cross-layer multicast agent, which takes advantage of IP multicast in the last core network domain towards the customer [22]. The tasks that the Multicast Manager carries out can be grouped in two main categories:

- The *overlay network configuration and administration*, which includes the definition of so-called E-Nodes (i.e., a network component that is part of the multicast tree which performs packet forwarding and replication) as well as the subscription and invocation of contracts within network provides (pSLSs).

- The *handling of multicast service requests*, which includes the subscription and invocation of a multicast customer subscription (cSLA) in order to enable an end user to consume a particular QoS-enabled multicast stream.

3.8 Caching and CDN Manager

The Caching and CDN Manager (CCDNMngr) is the EIMS subsystem responsible for all the tasks related to supporting caching and CDNs in ENTHRONE. The aim of the CCDNMngr is to transparently supplement existing EIMS functionality by providing alternative content sources that will either improve the performance of the content streaming or improve the robustness and scalability of the architecture.

Within ENTHRONE, CDNs are considered to be a discrete entity holding complete, static copies of content and are managed either within a Service Provider (via the EIMS CCDNMngr) or externally by a separate entity. In contrast, Caching Nodes are managed as semi-autonomous standalone entities that dynamically cache portions of content. The main functionalities of the CCDNMngr include to:

- *manage the provisioning of the content*, i.e., inject content into the CDN or upload content to the local caches;
- *manage the placement of content in different CDN Nodes* including collecting statistics used as input to the content placement algorithms;
- *manage the cache policy* which tunes the performance of Caching Nodes (caching method, replacement policy, etc) based on content usage statistics; and
- *select a Caching Node and/or CDN Node* to stream/deliver content in response to a consumer request, taking into account different factors such as the consumer location and the state of the nodes.

3.9 Summary

The EIMS defines a set of management subsystems, i.e., EIMS Managers, with predefined functionalities and interfaces. The functionalities pertain to:

- the *end-to-end QoS including QoE*;
- the actual *service* for managing the customer needs, networks, terminal devices, and the monitoring thereof;
- the *adaptation* of the service according to the usage environment context;
- the aggregation and collection of metadata (including the conversion between different formats);
- the *search* for appropriate Digital Items;
- *multicasting* including IP multicast within the last core network domain towards the customer; and, finally,
- *caching* and *content distribution networks*.

The interfaces between the EIMS managers have been defined in a generic way and implemented as traditional Web Services. However, these management services are just the foundation enabling end-to-end QoS/QoE and the management thereof. The next section will describe means how to effectively measure QoE and its mapping to standardized description formats enabling interoperability among the involved entities.

4   Measuring Quality of Experience for MPEG-21-Based Cross-Layer Multimedia Content Adaptation

4.1   Introduction

The requirement to access multimedia content such as video and audio streams during everyday's life is omnipresent. Research and standardization efforts around to what is commonly known as Universal Multimedia Access (UMA) has gained momentum and offer a rich set of tools enabling such an access from a technological point of view. However, most of these techniques exclude the human end user who is actually the source of the above mentioned requirements and ultimately wants to consume multimedia content independent of his/her context. The issues resulting from a more user-centric perspective are collectively referred to as Universal Multimedia Experience (UME) [23] where the user takes a center stage.

An important aspect with regard to UME is to measure the quality experienced by the user in an objective way and to signal the required quality metrics by standardized, i.e., interoperable, description formats. As the objective measures may require quality metrics coming from various layers (i.e., service, application, and network) we propose to adopt cross-layer interactions, especially when transmitting multimedia content over wireless channels [24].

Therefore, we propose a QoS/QoE model enabling to measure the perceptual quality of video transmissions taking into account quality metrics coming from different layers and following the philosophy of the ITU-T's E-model for audio. In order to enable interoperability among the involved parties – mainly the service provider and the content consumer – we propose to adopt description formats (i.e., tools) from the MPEG-21 DIA standard (introduce earlier already). In particular, we demonstrate how our QoS/QoE model can be instantiated using MPEG-21 DIA tools enabling a generic metadata-driven decision-taking component to determine which parameters of the content needs to be adjusted and how in order to provide a satisfactory quality experienced by the end user.

4.2   Probing Quality of Service

For audio streams, in 1996 the European Telecommunications Standards Institute (ETSI) and the ITU-T published a model that estimates the perceived quality experimented by users in phone calls. This model (E-model) is based on the premise that: "Psychological factors on the psychological scale are additive". The E-model takes into account factors such as packet loss, delay, and others like the equipment impairment and the packet loss robustness factors that depends on the codec used in the connection, as described by the recommendation ITU-T G.113.

When we are dealing with video streams, there are multiple parameters that can change between two videos even if they are coded with the same codec. Most of up-to-date codecs define different profiles with several resolutions, frames per second, bit rates, etc. Some approaches have used automatic measures of the video quality based on the Peak Signal to Noise Ratio (PSNR) [25]. However, it is impossible to have the original picture at the destination and, therefore, the PSNR can only be used to extract some parameters of the Mean Opinion Square (MOS) in function of the packet loss.

If we analyze the PSNR picture by picture, it does not take into account the transition between pictures, and the human perception is very sensitive to this transitions. Thus, this method does not give a good perceptual approach of a video sequence. Other solutions such as the Video Quality Metric (VQM) offer an objective quality approach to help in the design and control of digital transmission systems [26]. In [27] appears a study which takes into account the perceptual impression of packet loss, variation in the frame rate and synchronization with the audio signal. Finally, a more sophisticated study considering the opinion of the users is also explained in [25].

4.3   An Interoperable QoS Model for Video Transmission Exploiting Cross-Layer Interactions

4.3.1   QoS Probes and Mapping

In our study we made an analysis of the perceptual quality for video transmission with different ranges of parameters. The video sequences have encoded using AVC/H.264 with a frame rate from 6.25 to 25 frames per second. The video bandwidth used is between 150 and 1500 kbps, and finally, we introduced simulated random packet loss up to 10%. With all this ranges we build a huge repository of videos with some of their parameters modified. In this way, we can observe not only the effect on the subjective quality of the video (QoE) when varying only one parameter, but also can simultaneously study the cumulative effect on the quality of several of them (QoS).

A public survey [28] has been distributed in order to include as wide and heterogeneous audience as possible in both, internal and external approach, in a national and European environment. Each person watched a minimum of 10 videos randomly selected from the repository and rated the quality of the video between a value of 1 for a bad quality and 5 for a perfect quality. From this evaluation and the corresponding content parameters, we were able to derive the formulas as presented in the subsequent sections (bottom-up) to be used in a video transmission system.

4.3.2 Impact of Packet Loss

For loss distribution we have used a Bernoulli model. All the packet loss introduced in the videos were made assuming random losses distributed over an uniform probability density function which means that all the packets have the same probability to be dropped.

In the real world, packet dropping used to appear in the form of bursts of random length. A burst is a period with a high density of losses with independent probability which produces a larger distortion than isolated losses. Other models like Gilbert and Markov describe state models that transition between gap (good) states and bad (burst) states which have high or low density of independent losses. For this reason when we calculate the quality in short intervals, the packet loss density distribution can be considered uniform even if we are inside a burst.

The first step to process all the gathered data was to remove the atypical qualities, i.e., those values that are too different from the majority. These atypical data can be explained due to mistakes made during the input of the data, the videos were watched in non-optimum conditions, the user had a bad day, etc. We consider atypical (outliers values) all the data that were deviated from the mean more than 3/2 of the standard deviation. Then we found the equations that fit the curves described by the data clouds as shown in Fig. 3.

The equation that minimizes the mean error among all the opinions (once we discard all the atypical data) for the different values of the analyzed bandwidth is shown in Equation (1).

$$f_{150}(l) = \frac{-0.06914 \cdot l^2 + 1.545 \cdot l + 1.719}{l + 0.447}$$

$$f_{300}(l) = \frac{-0.1343 \cdot l^2 + 2.061 \cdot l + 2.041}{l + 0.4721}$$

$$f_{600}(l) = \frac{-0.1591 \cdot l^2 + 2.333 \cdot l + 2.342}{l + 0.5274} \tag{1}$$

$$f_{900}(l) = \frac{-0.1757 \cdot l^2 + 2.496 \cdot l + 2.5}{l + 0.5391}$$

$$f_{1500}(l) = \frac{-0.21 \cdot l^2 + 2.73 \cdot l + 3.012}{l + 0.6336}$$

4.3.3 The Impact of Bandwidth

Another factor that can influence in the video stream quality is the bandwidth, for this reason, many bandwidth rates has been studied. In order to obtain a model that fits all the curves generated (one for each bandwidth) we calculated the equations

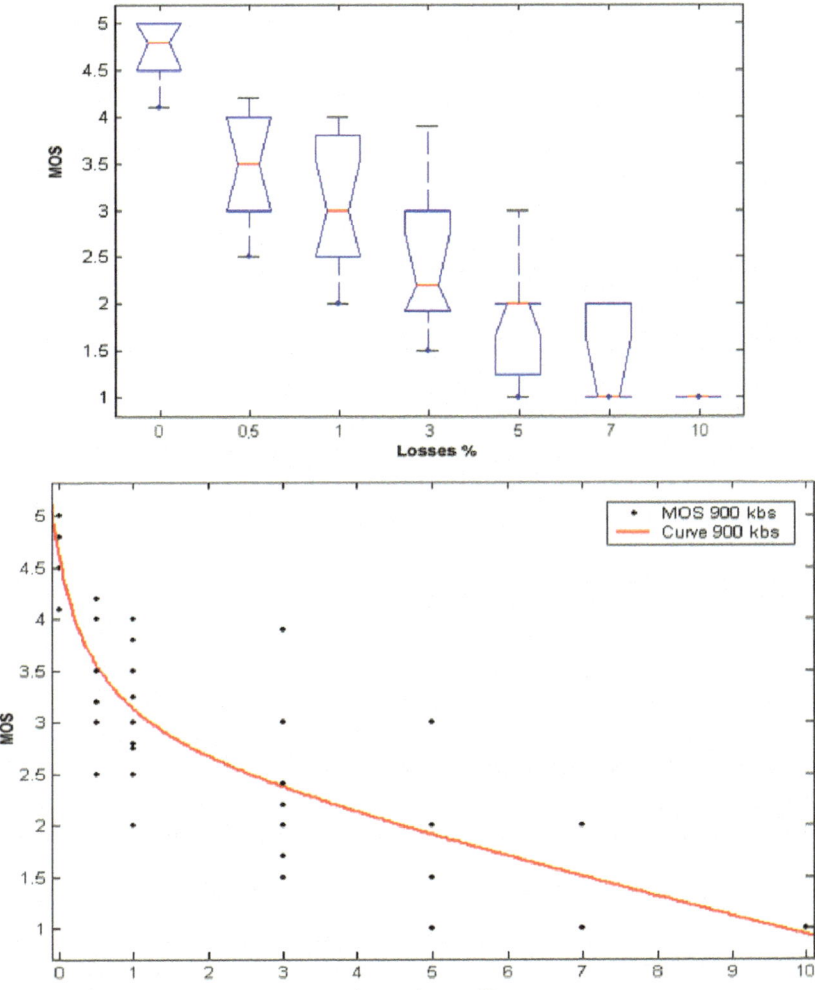

Fig. 3 MOS vs. Packet Loss for 900 kbps

that describe the different constant parameters from Equation (1) as shown in Equation (2).

$$f_L(l) = \frac{P_1 \cdot l^2 + P_2 \cdot l + P_3}{l + Q_1} \quad \forall\, l\,[\,0,10\,]\%$$

(2)

The values of the constants P1, P2, P3, and Q1 for the different bit rates analyzed when represented describe increasing/decreasing curves that can be easily approximated (cf. Equation (3)).

$$P_1(br) = -0.1387 \cdot e^{\frac{2.721 \cdot br}{10,000}} + 0.2823 \cdot e^{\frac{-8.885 \cdot br}{1,000}}$$

$$P_2(br) = 2.154 \cdot e^{\frac{1.584 \cdot br}{10,000}} - 2.125 \cdot e^{\frac{-7.8 \cdot br}{1,000}}$$

$$P_3(br) = 1.95 \cdot e^{\frac{2.887 \cdot br}{10,000}} - 1.307 \cdot e^{\frac{-9.414 \cdot br}{1,000}}$$

$$Q_1(br) = \frac{1.75 \cdot br^3}{10^{10}} - \frac{4.327 \cdot br^2}{10^7} + \frac{4.19 \cdot br}{10^4} + 0.3876$$

(3)

This way we can describe the relationship between the bit rate and the packet loss and how they influence in the obtaining of the quality perceived by the users as shown in Equation (4).

$$f_L(l,br) = \frac{P_1(br) \cdot l^2 + P_2(br) \cdot l + P_3(br)}{l + Q_1(br)}$$

(4)

4.3.4 The Impact of Frame Rate

An interesting study in how the quality perception changes as a function of the frame rate is shown in [29]. This study categorizes the media streams using three parameters: temporal nature of the data (i.e., soccer match vs. interview), audio (auditory) and visual content. Based on this categorization the watchability of the media streams is analyzed for all the possible combinations and making a classification based on the perception of the users between 1 and 7 (where 7 is the best quality).

We extrapolated this study to the case of a video on demand scenario where the video sequences have a high temporal nature. Considering that 30 fps has the best quality (i.e., 7) and normalizing the values, we can see the degradation factor that suffers the media rating in function of its frame rate in Fig. 4.

The equation that describes the curve of degradation of quality in function of the frame rate and minimizes the mean quadratic error is shown in Equation (5) where fps is the number of frames per second.

$$f_R(fps) = \frac{-0.00102 \cdot fps^2 + 1.164 \cdot fps + 1.704}{fps + 5.714} \quad \forall fps \in [5,30]$$

(5)

The curves that describe the different frame rates studied were calculated using an initial curve obtained – as explained in Section 4.3.2 – and the degradation factor of Equation (5) were applied. In order to check the accuracy of the proposed equation, the different curves have been compared with the data obtained from the real users as is shown in Fig. 5.

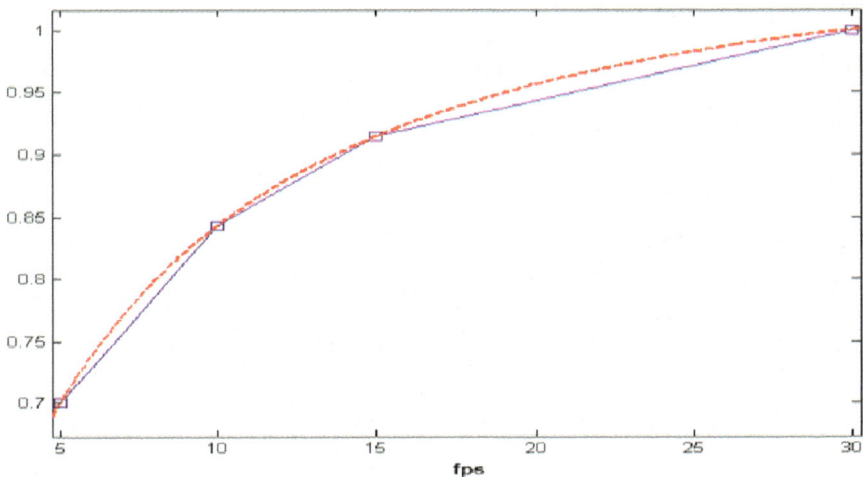

Fig. 4 Degradation of Quality based in Frame Rate

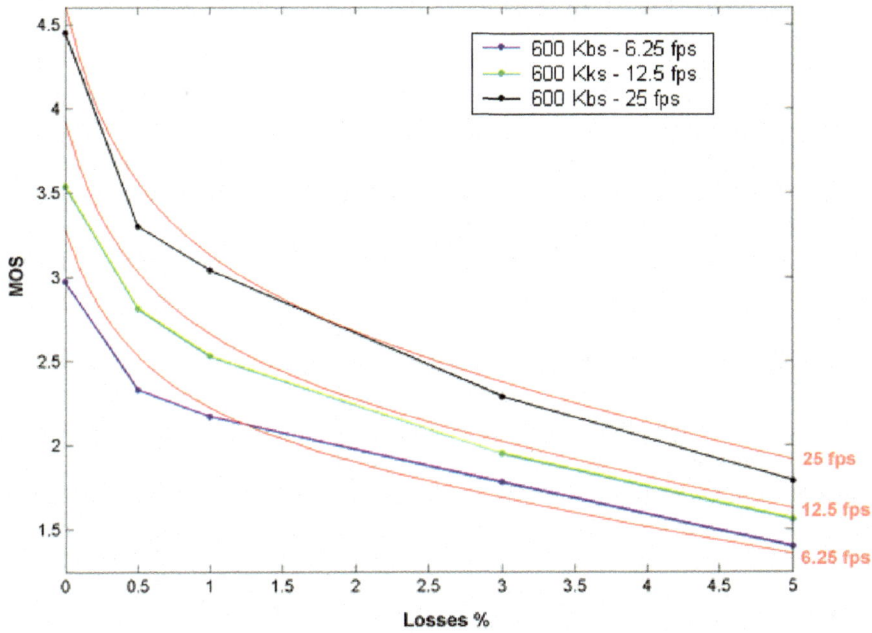

Fig. 5 Frame Rate Model Comparison

4.3.5 Impact of Delay

The delay is important when a high interactivity is required, e.g., video conference, whereas in scenarios like video streaming where a buffer space is created and as long as the data can be downloaded as fast as it is used up in playback, the influence of this parameter will practically disappear. Therefore, we have not further considered this factor.

4.3.6 Model Proposed

The quality perceived by the user will be obtained mainly from the values of the bit rate and of the packet loss within the network, this maximum value will be influenced by the frame rate of the video proportionally. Therefore, the final formula used to obtain the perceived quality of a video streaming is shown in Equation (6).

$$MOS\ (l,br\,,fps\,) = f_L(l,br\,) \cdot f_R(\,fps\,)$$

$$\forall\, l \in [\,0,10\,]\%$$

$$\forall\, br \in [\,150\,,1500\,]kbs$$

$$\forall\, fps \in [\,5,30\,]$$

(6)

4.3.7 Adding MPEG-21 Support Enabling Interoperable Cross-Layer Interactions

In order to provide interoperability when well established layers are broken up, we propose to utilize MPEG-21 DIA tools for describing functional dependencies between the layers. In particular, QoS mappings as described in Section 3.1 – possibly ranging across well-defined network layers – are instantiated with the MPEG-21 DIA tools introduced in Section 2.2. In this section we will show how the proposed QoS model as defined in Equation (6) can be instantiated using interoperable description formats according to the three-step approach described in [7].

The functional dependencies of the MOS function are described using MPEG-21 DIA AdaptationQoS' stack functions, i.e., XML-based reverse polish notation (RPN), given a range of possible content frame rate and bit-rate combinations (expressed via a so-called look-up table) with the objective to maximize the MOS. The context in which the multimedia content is consumed is characterized through the packet loss measured at the receiving terminal and communicated towards the multimedia content providing entity using an MPEG-21 DIA UED. In addition to the UED, the constraints of the probe – as indicated by Equation (6) – are expressed by an MPEG-21 DIA UCD utilizing limitation constraints which are attached to the UED.

Excerpts of the aforementioned interoperable descriptions are shown in Listing 1. Note that the complete descriptions can be found elsewhere [30].

Listing 1 MPEG-21 DIA Description Excerpts

```xml
AdaptationQoS Stack Function for MOS (aqos.xml):
<!-- Stack Function for MOS calculation -->
<Module xsi:type="StackFunctionType" iOPinRef="MOS">
  <StackFunction>
    <Argument xsi:type="InternalIOPinRefType"
        iOPinRef="F_FRAMERATE"/>
    <Argument xsi:type="InternalIOPinRefType"
        iOPinRef="F_PACKETLOSS"/>
    <!-- multiply -->
    <Operation operator=":SFO:18"/>
  </StackFunction>
</Module>
UCD maximizing the MOS (ucd_provider.xml):
<OptimizationConstraint optimize="maximize">
  <Argument xsi:type="ExternalIOPinRefType"
      iOPinRef="aqos.xml#MOS"/>
</OptimizationConstraint>
UED (ued.xml):
<Network xsi:type="NetworkType">
  <NetworkCharacteristic xsi:type="NetworkConditionType">
    <AvailableBandwidth average="1500000"/>
    <Error packetLossRate="0.03"/>
  </NetworkCharacteristic>
</Network>
UCD for probe constraints (ucd_probe.xml):
<!-- packet loss <= 0.1 (10%) -->
<LimitConstraint>
  <Argument xsi:type="SemanticalRefType"
      semantics=":AQoS:6.6.5.8"/>
  <Argument xsi:type="ConstantDataType">
    <Constant xsi:type="FloatType">
      <Value>0.1</Value>
    </Constant>
  </Argument>
  <Operation operator=":SFO:38"/>
</LimitConstraint>
```

The AdaptationQoS excerpt provides the multiplication of the functions representing the MOS vs. Packet Loss and bit rate equation (cf. Equation (4)) and the frame rate model (cf. Equation (5)) respectively. Both functions are also represented as stack functions demonstrating the modular usage of this tool. The maximization of the MOS is indicated by an optimization constraint referencing the MOS IOPin of the AdaptationQoS description. The UED excerpt describes a network with 1500 kbps available bandwidth and three percent packet loss. Finally, the probes' UCD excerpt defines that only those packet losses smaller than 10 percent are allowed.

Once these descriptions are available, they can be fed into an adaptation decision-taking engine (ADTE). In our work we rely on an existing implementation [11]. In particular, the implementation is generic in a sense that the core is independent of the actual description format and solves a mathematical optimization problem by restricting the solution space in order that the limitation and optimization constraints are fulfilled. In our example, the ADTE will assign values to the AdaptationQoS' IOPins, which can be used to adjust the bit-rate and frame rate according to the measured packet loss while maximizing the MOS.

5 Conclusions

In this chapter we have proposed an architecture allowing for the end-to-end management of heterogeneous environments enabling QoS/QoE. On top of this architecture the ENTHRONE Integrated Management Supervisor (EIMS) specifies subsystems (i.e., EIMS Managers) providing various functionalities required for modern end-to-end media delivery chains featuring a cross-layer QoS/QoE adaptation concept. Although cross-layer interactions violate the traditional protocol hierarchy and traditional isolation model, interoperability is preserved through the adoption of standardized description formats based on MPEG-7 and MPEG-21. Additionally, the interfaces to the EIMS Managers are defined using Web Services and interoperable payload formats, also based on MPEG-7 and MPEG-21.

The second part of this chapter focuses on the problem on how to measure QoS/QoE degradations at the terminal as perceived by the user without requiring cost-intensive subjective test. Therefore, a detailed QoS/QoE model for video transmission is proposed following the philosophy of the ITU-T's E-model for audio and taking into account the impact on frame rate, bandwidth, and packet loss. In order to use this model within the EIMS architecture we demonstrated its translation into interoperable description formats compliant to MPEG-21.

Acknowledgments. This work is supported by the European Commission in the context of the ENTHRONE project (IST-1-507637). Further information is available at http://www.ist-enthrone.org. In particular, the authors would like to thank Michael Ransburg, Ingo Kofler, Hermann Hellwagner (Klagenfurt University), Pedro Souto, Pedro Carvalho, Hélder Castro (INESC Porto), Mamadou Sidibé, Ahmed Mehaoua, Li Fang (CNRS-PRISM), Adam Lindsay, Michael Mackay (University of Lancaster), Artur Lugmayr (Tampere University of Technology), Bernhard Feiten (T-Systems), Víctor H. Ortega (Tecsidel), and José M. González (Telefónica I+D).

References

1. Guenkova-Luy, T., Kassler, A., Mandato, D.: End-to-End Quality of Service Coordination for Mobile Multimedia Applications. IEEE Journal on Selected Areas in Communications 22(5), 889–903 (2004)
2. Network Neutrality,
   http://en.wikipedia.org/wiki/Network_neutrality
   (accessed October 5, 2009)

3. Kilkki, K.: Quality of Experience in Communications Ecosystem. Journal of Universal Computer Science 14(5), 615–624 (2008)
4. Quality of Experience, http://en.wikipedia.org/wiki/Quality_of_experience (accessed October 5, 2009)
5. Pereira, F., Smith, J.R., Vetro, A. (eds.): Special Section on MPEG-21. IEEE Transactions on Multimedia 7(3), 397–479 (2005)
6. Burnett, I., Pereira, F., Van de Walle, R., Koenen, R. (eds.): The MPEG-21 Book. Wiley, Chichester (2006)
7. Kofler, I., Timmerer, C., Hellwagner, H., Ahmed, T.: Towards MPEG-21-based Cross-layer Multimedia Content Adaptation. In: Proc. 2nd International Workshop on Semantic Media Adaptation and Personalization (SMAP 2007), London, UK (2007)
8. Negru, O.: End-to-End QoS through Integrated Management of Content, Networks and Terminals: The ENTHRONE Project. In: Proc. 1st International Workshop on Wireless Internet Services (WISe 2008), Doha, Qatar (2008)
9. Vetro, A., Timmerer, C.: Digital Item Adaptation: Overview of Standardization and Research Activities. IEEE Transactions on Multimedia 3(7), 418–426 (2005)
10. Mukherjee, D., Delfosse, D., Kim, J.-G., Wang, Y.: Optimal Adaptation Decision-Taking for Terminal and Network Quality-of-Service. IEEE Transactions on Multimedia 3(7), 454–462 (2005)
11. Kofler, I., Timmerer, C., Hellwagner, H., Hutter, A., Sanahuja, F.: Efficient MPEG-21-based Adaptation Decision-Taking for Scalable Multimedia Content. In: Proc. 14th SPIE Annual Electronic Imaging Conference – Multimedia Computing and Networking (MMCN 2007), San Jose, CA, USA (2007)
12. ENTHRONE Web site, http://ist-enthrone.org/ (accessed October 5, 2009)
13. Burnett, I.S., Davis, S.J., Drury, G.M.: MPEG-21 Digital Item Declaration and Identification — Principles and Compression. IEEE Transactions on Multimedia 3(7), 400–407 (2005)
14. Chernilov, A., Arbel, I.: Content Handling and Content Protection under E2E QoS Distribution. In: Proc. International Conf. on Automation, Quality and Testing, Robotics (AQTR 2008), Cluj-Napoca, Romania (2008)
15. Jannach, D., Leopold, K., Timmerer, C., Hellwagner, H.: A Knowledge-based Framework for Multimedia Adaptation. Applied Intelligence 2(24), 109–125 (2006)
16. Sidibé, M., Mehaoua, A.: Service Monitoring System for Dynamic Service Adaptation in Multi-domain and Heterogeneous Networks. In: Proc. 9th International Workshop on Image Analysis for Multimedia Interactive Services (WIAMIS 2008), Klagenfurt, Austria (2008)
17. Koumaras, H., Kourtis, A., Lin, C.-H., Shieh, C.-K.: A Theoretical Framework for End-to-End Video Quality Prediction of MPEG-based Sequences. In: Proc. 3rd International Conference on Networking and Services (ICNS 2007), Athens, Greece (2007)
18. Shao, B., Renzi, D., Mattavelli, M., Battista, S., Keller, S.: A Multimedia Terminal Supporting Adaptation for QoS Control. In: Proc. 9th International Workshop on Image Analysis for Multimedia Interactive Services (WIAMIS 2008), Klagenfurt, Austria (2008)
19. Lugmayr, A.: The ENTHRONE 2 Metadata Management Tool (MATool). In: Proc. 1st International Workshop on Wireless Internet Services (WISe 2008), Doha, Qatar (2008)

20. Carvalho, P., Andrade, M.T., Alberti, C., Castro, H., Calistru, C., de Cuetos, P.: A unified data model and system support for the context-aware access to multimedia content. In: Proc. International Workshop on Multimedia Semantics - The role of metadata, Aachen, Germany (2007)
21. Gruhne, M., Tous, P., Döller, M., Delgado, J., Kosch, H.: MP7QF: An MPEG-7 Query Format. In: Proc. 3rd International Conference on Automated Production of Cross Media Content for Multi-channel Distribution (AXMEDIS 2007), Barcelona, Spain (2007)
22. Mehaoua, A., Fang, L.: Multicast Provisioning and Multicast Service Management. In: ENTHRONE Workshop at WIAMIS 2008, Klagenfurt, Austria (2008)
23. Pereira, F., Burnett, I.: Universal multimedia experiences for tomorrow. IEEE Signal Processing Magazine 20(2), 63–73 (2003)
24. van der Schaar, M., Sai Shankar, N.: Cross-Layer Wireless Multimedia Transmission: Challenges, Principles, and New Paradigms. IEEE Wireless Communications 3(4), 55–58 (2005)
25. Klaue, J., Rathke, B., Wolisz, A.: EvalVid - A Framework for Video Transmission and Quality Evaluation. In: Proc. 13th International Conference on Modeling Techniques and Tools for Computer Performance Evaluation, Urbana, Illinois, USA (2003)
26. Brill, M.H., Lubin, J., Costa, P., Pearson, J.: Accuracy and cross-calibration of video-quality metrics: new methods from ATIS/T1A1. In: Proc. of the International Conference on Image Processing (ICIP), Rochester, New York, USA (2002)
27. Ruiz, P.: Seamless Multimedia Communications in Heterogeneous Mobile Access Network. In: Proc. of the Terena Networking Conference, Rhodes, Greece (2004)
28. ENTHRONE Public Survey, http://www.enthrone2.tid.es/ (accessed October 5, 2009)
29. Apteker, R.T., Fisher, J.A., Kisimov, V.S., Neishlos, H.: Video acceptability and frame rate. IEEE Multimedia 3(2), 32–40 (1995)
30. Complete MPEG-21 DIA Descriptions Enabling Interoperable Cross-Layer Interactions,
http://www-itec.uni-klu.ac.at/~timse/WISe08/wise08.zip
(accessed October 5, 2009)

Chapter 21
Quality-Driven Coding and Prioritization of 3D Video over Wireless Networks

Sabih Nasir, Chaminda T.E.R. Hewage, Zaheer Ahmad, Marta Mrak,
Stewart Worrall, and Ahmet Kondoz

Abstract. The introduction of more affordable and better quality 3D displays has increased the importance of designing efficient transmission systems for 3D video services. For achieving the best possible 3D video quality, it is necessary to take into account human perception issues to avoid potentially disturbing artifacts. These perception issues are considered in the design of the cross layer video adaptation and transmission system proposed in this work. One of the most popular formats for representing 3D video is color plus depth, where a 2D color video is supplemented with a depth map. The depth map represents the per pixel distance from the camera and can be used to render 3D video at the users terminal. The proposed scheme uses this depth information to segment the video into foreground and background parts. The foreground and background are then coded as separate video streams, whose data is prioritized according to its influence on video quality at the decoder. Received video quality is estimated by modeling the effect of packet loss and subsequent concealment. A home gateway scenario with IEEE 802.11e is simulated. The proposed 3D video distribution system exploits the different priority access classes to achieve high quality 3D video even with significant levels of packet loss, and high network loads. The simulations show that gains in PSNR of 1-3dB can be achieved, depending on the amount of activity within a particular video sequence. Subjective results are also obtained to demonstrate that the performance improvements are perceptible.

1 Introduction

Transmission of video data over wireless links such as Wi-Fi has seen rapid growth in recent years. With the provision of 3D video content and advances in 3D video

Sabih Nasir · Chaminda T.E.R. Hewage · Zaheer Ahmad · Marta Mrak · Stewart Worrall ·
Ahmet Kondoz
Centre for Vision, Speech and Signal Processing
University of Surrey, Guildford, GU2 7XH, UK
e-mail: {s.nasir,e.thushara,zaheer.ahmad,m.mrak,s.worrall,
a.kondoz}@surrey.ac.uk

displays [1, 2], a new era of 3D video has started. On one hand, new methods are being proposed to generate 3D video content in different forms, and on the other hand compression techniques for 3D video data are being developed [3, 4]. In the entertainment sector, 3D video is being used in filming, 3D video animation, and games. In the security sector, surveillance companies and authorities are using 3D video cameras to capture and analyze 3D video of sensitive locations such as airports, railway stations, public gatherings, office buildings, and car parks. High definition 3D images are also used in medical research facilities to aid analysis.

Recently, 3D home cinema for entertainment has become a reality. Service providers and high-tech industry leaders are making alliances to provide consumers with high quality, yet affordable in-home 3D entertainment [5]. A large number of TV users are already enjoying stereoscopic 3D content. The growth in sales volume of 3D LCDs, and 3D enabled DLP HDTVs is already attracting the interest and attention of many of the key players in video communications. A typical home entertainment scenario will have 3D high definition viewing devices, broadband connection for delivering multimedia content form the server to the user's premises, and Wireless Local Area Network (WLAN) for transmission of video and other data to various terminals inside the home.

Video data transmission over wireless networks has always been a challenging problem due to the highly variable nature of wireless networks. Packets may be dropped during transmission or may only reach the destination after significant delay. The effects of errors propagate into succeeding frames in almost every form of video coding process. Various schemes exist in the literature that mitigate the effects of errors during transmission of 2D video data [6, 7]. However, the transmission of compressed 3D video data still needs to be analyzed for various wireless conditions. In this work, 3D video transmission over wireless networks is considered in a home entertainment scenario.

Video data passes through a number of network layers during its distribution. Recent research findings, [8], have demonstrated that sub optimal perceptual quality is achieved if encoding algorithms at higher level layers do not consider the techniques applied at the lower level layers, e.g. scheduling, routing, etc. However, adapting the coding and transmission strategies jointly across the layers helps in maintaining a consistently high quality at the receiving end. Various cross layer design approaches have been suggested for optimization of 2D video transmission. A weighted fair scheduling algorithm based on adaptive rate control is presented in [9]. An adaptive cross layer video multicast streaming algorithm for multi-rate WLANs has been proposed in [10], which optimizes the data rate at the physical layer for every multicast receiver according to its perceived channel conditions. However, these schemes have not considered any of the content attributes for optimization.

Our main objective is to design a cross layer optimisation approach in order to achieve efficient 3D video transmission over wireless networks. It is assumed that in most videos, objects closer to the camera are more important for perceptual quality compared to those that lie farther away in the video scene. Two reasons can be given for this assumption. First, objects closer to the camera may be rendered so that they appear in front of the display, and will thus draw most of the viewers attention. Secondly, objects closer to the camera undergo more warping during the 3D stereoscopic rendering process. Any errors, such as blocking and mismatch between color and depth map structures, will therefore be magnified for objects closer to the camera. The proposed solution uses the depth information for image segmentation at the application layer, and then at the transmission layer, objects are prioritized depending on their perceptual importance in the scene. Fig. 1 presents one such example. The background object in left image has been quantized by twice the factor as used for quantization in the right image. However, both images appear to have almost similar visual quality.

Fig. 1 Visual attention example. Left image has twice as coarsely quantized background object as in right image.

The chapter is structured as follows. Section 2 presents background information about the issues addressed in the design of the proposed solution. In section 3, a cross layer design for optimized 3D video communication is proposed. Section 4 gives experiments details and results. Conclusions are provided in Section 5.

2   3D Video Transmission

There are numerous formats to represent 3D video, and one of the most popular is video plus depth. This format is described in the MPEG-C, Part 3 standard, and is used in a number of commercially available 3D displays [5]. The basic principles for coding of 2D video can be applied to 3D videos. However, 3D video presents new challenges for transmission as it carries extra depth information along with color and texture. The depth information is required for content rendering at the 3D viewable display devices. This section discusses different aspects of the 3D video transmission system and highlights the perceptual based optimization possibilities for 3D video content transmission over WLAN.

2.1   3D Video Fundamentals

According to the classification of MPEG-3DAV (Motion Picture Expert Group-3D Audio Visual), three scene representations of 3D video have been identified, namely Omni-directional (panoramic) video, interactive multiple view video (free viewpoint video) and interactive stereo video [11]. Stereoscopic video is the simplest form of 3D video and can be easily adapted in communication applications with the support of existing multimedia technologies. Stereoscopic video renders two views for each eye, which facilitates depth perception of the scene. There are several techniques to generate stereoscopic video content including dual camera configuration, 3D/depth-range cameras and 2D-to-3D conversion algorithms [12]. Stereoscopic capture using a stereo camera pair is the simplest and most cost effective way to obtain stereo video, compared to other technologies available in the literature. In this case, depth information can be extracted from the video obtained form stereo camera setup. The latest depth-range camera generates a color image and a per-pixel depth image of a scene. An example of the color and the depth image is shown in Fig. 2. This depth image with its corresponding color image can be used to generate two virtual views for the left and right eye using the Depth-Image-Based Rendering (DIBR) method described in [13]. The depth information is essentially required for 3D content display; therefore, it is always transmitted along with the color information for 3D video communication.

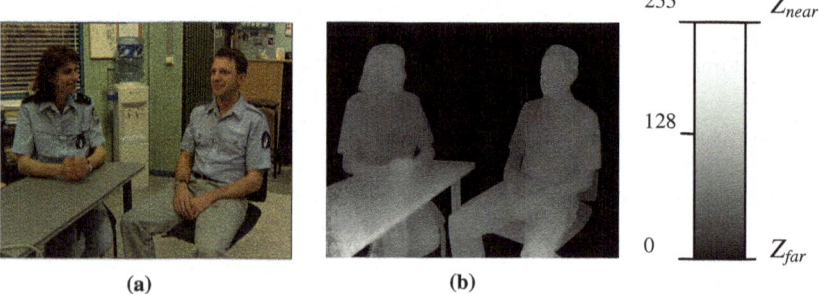

(a)   (b)

Fig. 2 Interview sequence a) Colour image, b) Per-pixel depth image. The depth images are normalized to a near clipping plane Z_{near} and a far clipping plane Z_{far}.

The two virtual views for an image of 8 bits per pixel color depth are generated using (1).

$$P_{pix} = -x_B \frac{N_{pix}}{D} \left[\frac{m}{255} (k_{near} + k_{far}) - k_{far} \right] \tag{1}$$

Here, N_{pix} and x_B are the number of horizontal pixels of the display and eye separation, respectively. The depth value of the image is represented by the N-bit value m. k_{near} and k_{far} specify the range of the depth information respectively behind and in front of the picture, relative to the screen width N_{pix}. D represents the viewing distance.

2.2 Perceptually Optimized Coding

The perceptual importance of scene elements can be utilized to code and transmit video content more efficiently. For example, the foreground objects of a video frame can be coded with finer details than the background information. Perceptually optimized coding methods (e.g. ROI) are a common topic in 2D video coding research, and can provide improved perceptual quality at reduced bitrates. For example, the dynamic bit-allocation methods proposed in [14] and [15] utilize the response of the human visual system to allocate bits for block-based and object-based coding strategies respectively. Mixed-resolution coding of stereoscopic video is also based on the response of the human visual system (e.g. the binocular suppression theorem [16]). Reduced temporal-spatial coding of the right image sequence has achieved overall bit-rate reduction, with no effect on the perceived quality of stereoscopic video [17]. Therefore, the response of the human visual system can be utilized to encode 2D/3D video more efficiently. Consequently, compression schemes have been designed that use coarsely compressed depth images [18] and reduced resolution depth images [19] in order to optimize bit-rate.

The transmission of 3D video in particular can be prioritized based on the perceptual importance of each component. For example, in the case of color plus depth 3D video, the color image sequence which is directly viewed by the human viewers can be sent over a more protected channel than the depth sequence. However the selection of priority levels should be made after careful observation of their effect on the reconstructed quality of 3D video. For example, human perception is used to decide the transmission strategy for 3D objects in [20]. In the presented framework, 3D video content is divided into segments, which are then coded and transmitted based on their individual perceptual importance to achieve enhanced visual quality at the receiving end. Perceptual importance is derived for these segments (background objects, foreground objects and depth information) by their expected transmission distortion and corresponding distance from the viewing point.

2.3 Quality of Service for Wireless Local Area Networks

In order to enable wireless transmission of 3D videos in home environments, a new standard IEEE 802.11e [21] is used. It has been selected because it provides the Quality-of-Service (QoS) support for demanding multimedia applications with stringent requirements. IEEE 802.11e supports two medium access mechanisms, namely, controlled channel access and contention-based channel access, referred to as Enhanced Distributed Channel Access (EDCA). EDCA provides the MAC layer with per-class service differentiation. QoS support is realized with the introduction of Traffic Categories (TCs) or Access Classes (ACs). With this approach, frames are delivered through multiple back off instances within one station. The implementation of legacy Distributed Coordination Function (DCF) and EDCA with different traffic classes and independent transmission queues are shown in Fig. 3. The priority levels for each TC can be differentiated based on the parameters selected for Contention Window (CW) size (e.g. CW_{min}, CW_{max}), Arbitrary

Inter Frame Space (AIFS) number and re-transmission limit [22]. Depending on the traffic class assigned, the particular traffic will undergo different packet-dropping probabilities and delay constraints. Therefore, the proposed transmission solution with IEEE 802.11e assigns higher priority for important parts of the video bit-stream whereas less important components are assigned a lower priority traffic class.

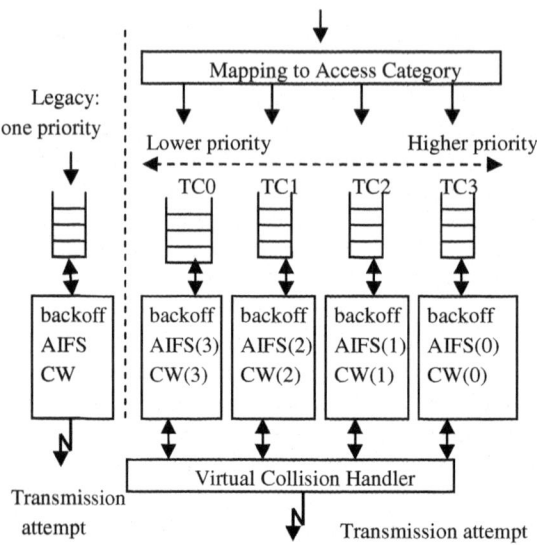

Fig. 3 Virtual back off of Traffic Categories (TCs) with DCF (left) and EDCA (right) methods

2.4 Robust Transmission

Generic error resilience techniques and channel protection schemes can make the video data more resilient to channel degradations. If they are separately optimized, the performance gain is limited. A significant improvement in performance has been demonstrated by the use of adaptive optimization of source and channel codes as in joint source-channel coding approaches [23].

These techniques can be divided into two areas: channel-optimized source coding and source-optimized channel coding. In the first approach, the channel-coding scheme is fixed and the source coding is designed to optimize the perform-ances. In the case of source-optimized channel coding, the optimal channel code is derived for a fixed source code. The application of source optimized channel cod-ing for video transmission in error prone propagation environments is considered in [24, 25]. However, the derivation of optimal channel coding at the physical link layer for fixed source coding at the application layer requires modification of all underlying network protocol layers in order to avoid redundancy for efficient resource utilization [25]. This problem can be resolved by partitioning of video bit stream and prioritizing its parts by sending them via multiple bearers with different characteristics, such as different channel coding, and modulation schemes. In order to implement such approach, it is necessary to separate the encoded bit stream optimally into a number of sub-streams, such as in [26].

Joint source-channel coding principles have been adopted here for achieving optimized transmission. Different segments of the 3D video bit stream are prioritized to provide maximum security to the most vulnerable parts. The encoded data is separated into a number of sub-streams based on the relative importance of different video packets, which is calculated using the estimated perceived distortion of packets at the encoder. The importance level has a further biasing factor of depth information related to that particular object. The next section explains the proposed prioritization algorithm.

3 Cross-Layer Design for Optimized 3D Video Transmission`

Cross layer design techniques for jointly adapting the coding/transmission techniques across all network layers have demonstrated considerable performance gains for multimedia applications over wireless networks [8]. This approach is followed in this work for optimized transmission of 3D video content over WLAN.

3.1 Design of Cross Layer Video Communications System

In a typical home entertainment scenario, a group of users may simultaneously access a variety of services, e.g. VoIP, data downloads, and video streaming. The WLAN router selects the appropriate channel for transmission of voice, video or data packets. Such a home WLAN enabled environment is shown in Fig. 4. 3D videos are available at the source server. The objective is to prioritize data across various layers of transmission chain for optimized 3D video content (color plus depth) transmission. A segmentation module extracts background and foreground information from the 3D video color stream. In order to separate the foreground from the background for streaming application, we need faster segmentation solution. Details of the segmentation process used in this work are provided in the next subsection.

Segmented data is then prioritized based upon its estimated distortion at the receiver. The distortion estimation module calculates optimal QP values for the three streams (the depth stream, the background object stream and the foreground object stream) for different network conditions to minimize distortion. The module also performs mapping of each video packet into one of the streams at the application layer level according to their expected distortion, and the error rate of the communication channel over which they are to be transmitted. The selection procedure is discussed in subsection 3.3.

One important aspect of this cross layer design is that the information of depth is also considered for the prioritization process. The distortion cost for an object that is farther from the camera or the viewing point in a video scene is adjusted according to its depth value. This allows prioritization of packets that are closer to the viewing point as they are assumed to contribute more to the perceived visual quality.

In the home entertainment scenario, the transcoder receives three segmented, encoded video streams. The transcoder also receives feedback from the WLAN transmitter about the channel conditions. The function of transcoder is to tune the data rate of the three streams based upon the channel condition. It changes the quality of the video by selecting the appropriate QP value from the available QP-distortion set table. This table information is received by the transcoder in the form of signaling form the source provider. The transcoder operates at the compressed domain level to avoid excessive complexity burden as suggested in [27]. Existing transcoders can modify the QP with very minor degradation in quality compared to directly compressing the video at the modified QP.

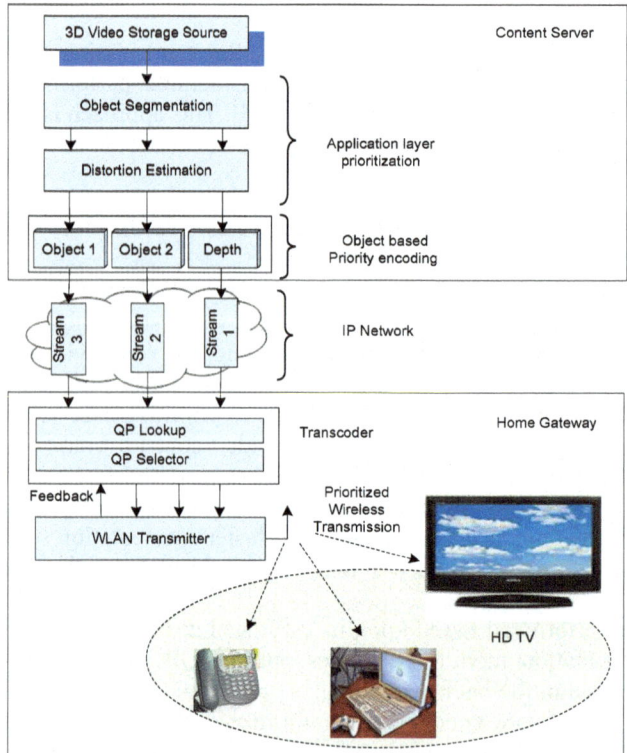

Fig. 4 Schematic diagram of the proposed cross layer multimedia transmission system for the home entertainment environment

3.2 3D Video Object Segmentation for Packet Prioritization

Shape information is necessary to separate foreground and background objects in the scene. In the case of color plus depth video, joint segmentation of the color and depth map can be performed to generate the shape information. The study carried out in [28] describes the generation of shape information using joint segmentation of color plus depth images in cluttered backgrounds or with low quality video input. In the presented work, a simple segmentation algorithm is applied to generate shape data based on the threshold levels of the depth map sequence. Depth pixel values greater than the mean pixel value of that depth map frame, D_{av} are set to the maximum pixel value of 255. Otherwise the pixel value is set to 0. The shape information generated by using this method is noisy. Therefore, objects which contain few pixels are removed and merged with surrounding area.

3.3 Quality-Driven Coding

A video object is made of rectangular or arbitrarily shaped sets of pixels of the video frame, capable of representing both natural and synthetic content types, e.g. a talking person without the background, or any graphics or text [29]. The source video frame may therefore, consist of a number of video objects, each of which can be separately encoded and transmitted into separate sub streams in an object based video coding scenario.

Paarticular optimization schemes, depending on the underlying network protocols, can be applied to separate the object's sub streams for prioritized communication. For instance, UEP could be used if different priority sub-carriers would be available for transmission of sub streams. The 802.11e protocol, for multimedia transmission with different QoS over WLAN, supports traffic classes with different priority levels, as described in section 2.3.

Video objects are prioritized on the basis of their expected distortion estimates and their relative depth in the scene. Each of the segmented objects is parsed at the packet level for estimating the distortion expected during its transmission. Let the input video sequence consists of M video frames. Each video frame is separated into L number of objects. Each object is coded separately using its own segmentation mask and is divided into N number of video packets. If expected distortion due to the corruption of nth packet of mth frame of lth video object is $\alpha(l, m, n)$, then the total expected distortion $E(D_{l,m,n})$ of the mth frame becomes

$$E(D_{l,m,n}) = \sum_{l=0}^{L-1} \sum_{n=0}^{N-1} \alpha(l, m, n) \tag{2}$$

The optimization problem is to minimise $E(D_{l,m,n})$. This is achieved by reformatting the bit-streams according to their importance. Based on the distortion estimates coupled with depth information, each packet is assigned an importance level I_n, which is used to reallocate packets with higher importance to channel with higher protection and vice-versa. The importance factor is calculated by multiplying the expected distortion estimate for a packet with its cumulative depth factor, CD.

$$I_n = \alpha(l,m,n) \cdot CD(l,m,n) \tag{3}$$

The cumulative depth factor represents the proportional effect of depth perception for every packet under the assumption that significant focal attention of the viewer is received by nearer objects in the video scene as compared to distant objects. It reduces the importance level of distant objects to give priority to nearer objects. It is calculated as follows.

$$CD(l,m,n) = \begin{cases} D_{av}\big/100, & \text{if } D_{av} \langle D_{th} \\ 1, & \text{otherwise} \end{cases} \tag{4}$$

where D_{av} is the average depth of the pixels in that packet. The depth threshold variable, D_{th} prioritizes the content according to its depth information. The optimal value for this variable is related to the average depth of the whole video frame, which would differ from sequence to sequence. However, it has been observed from experiments that pixels with depth values less than 70 do not contribute significantly to perceptual quality; hence a threshold value of 70 would be a good estimate for sub optimal separation of video packets. It must be noted that depth information represents the distance of a pixel from the viewing point calibrated in gray scale levels. The scale values range from 255 for nearest point to 0 for farthest point. The distortion estimation model is discussed in the next section.

3.4 Optimized Transmission and Decoding

This section describes the estimation of the distortion due to corruption of data in a video packet during its transmission.

The video quality is affected by the quantisation distortion $E(Q^D)$ and channel distortion. Channel distortion is composed of concealment distortion and distortion caused by error propagation over frames coded using temporal prediction. Concealment distortion depends on the concealment techniques applied at the decoder. If errors in the video data are concealed with concealment data from the previous frames, then it is called temporal concealment and distortion caused by such a scheme is called as temporal concealment distortion $E(D^T)$. Errors can propagate in two ways, either in the temporal direction or in the spatial direction. Frame to frame error propagation through motion prediction and temporal concealment is called temporal domain error propagation f^p. The distortion estimation model used in this work has similarities with the method proposed in [30]. However, the distortion induced due to error propagation is calculated differently and an adaptive intra refresh technique [31] is used to minimize the error propagation factor, where a fixed number of macroblocks in every P frame are randomly selected and 'intra' coded. Previous research has suggested that spatial error concealment is less important than the temporal error concealment [32] so in this work only temporal concealment is considered.

Taking the video packet as the base unit, the expected frame quality can be written as:

$$E(Q_{l,m}) = 10 \cdot \log_{10} \left(g / \sum_{n=0}^{N-1} E(D_{l,m,n}) \right) \tag{5}$$

where $E(Q_{l,m})$ is the expected quality of mth frame of the lth object, $E(D_{l,m,n})$ is the expected distortion of nth video packet and N is the total number of video packets. g is a constant defined by the color depth and by the size of given composite video frame, $g = c^2 \cdot w \cdot h$, where c is the color depth factor, w and h are the width and height of the video frame, respectively. $E(D_{l,m,n})$ can be written as:

$$E(D_{l,m,n}) = E(D_{l,m,n}^Q) + \rho^e(l,m,n) \cdot E(D_{l,m,n}^T) + f^{tp}(l,m,n) \tag{6}$$

Calculation of each term shown in (6) depends on the implemented concealment techniques, and the applied video coding scheme and its parameters. Quantization distortion is computed from the squared difference of original and reconstructed luminance values for every macroblock in the particular video packet.

The probability $\rho^e(l, m, n)$ of receiving a video packet with errors depends on the channel's Bit Error Rate (BER) and size of the video packet. If ρ^b is the BER of the transmission channel then $\rho^e(l, m, n)$ is given as:

$$\rho^e(l,m,n) = 1 - (1 - \rho^b)^{p_b(l,m,n)} \tag{7}$$

where p_b represents the size of the packet in bits.

The extent of temporal concealment distortion depends on the algorithm used for the temporal concealment. Here a simple concealment approach is followed, where macroblocks in the corrupted video packet are replaced by data of the corresponding macroblocks of the previous frame. For the estimation it is assumed that the neighbouring video packets and reference frames are received correctly.

Therefore, the concealment distortion is given as:

$$E(D_{l,m,n}^T) = \left\| Y_{l,m,n} - Y'_{l,m-1,n} \right\| \tag{8}$$

which is the squared difference of the luminance components of macroblocks of reconstructed part of current frame $Y_{m,n}$ and the same spatial area in the previous frame $Y'_{m-1,n}$.

The temporal error propagation due to MB mismatch between adjacent video frames is quantified by the term $f^{tp}(l, m, n)$ in (3), which is computed as:

$$f^{tp}(l,m,n) = (1 - \rho^e(l,m,n)) \cdot p_{l,m-1}^{tp} \cdot$$
$$\sum_{k=0}^{K-1} \left[\rho^e(l,m-1,n) \cdot E(D_{k,l,m-1,n}^T) \right] + (1 - \rho^e(l,m-1,n)) \cdot p_{l,m-2}^{tp} \tag{9}$$

where K is the total number of macroblocks in the frame. The summation in (9) represents the error propagation through the MBs. $p_m^{tp} = 1 - (1 - \rho^b)^{F_m}$ quantifies the fraction of distortion of the mth reference video frame, which should be considered in the propagation loss calculation and F_m is the size of the frame area contained in the observed packet.

Packets are then distributed on the basis of the estimated distortion to the respective bit-streams. Packets with higher expected distortion are put on the sub-channel with higher protection.

4 Experimental Results

This section explains the experimenting environment used to verify the model presented in the earlier sections and their results.

4.1 Experimental Setup

The performance of the proposed scheme has been tested using a simulated WLAN environment, implemented with the Network Simulator 2 (NS-2) platform. NS-2 is a widely used open source network simulation tool developed at the UC Berkley [33]. The specific version used for these experiments has been built upon version 2.28 with an 802.11e EDCA extension model implemented [34].

A wireless scenario with six nodes is considered. Four different Access Classes (AC), namely AC_VO, AC_VI, AC_BE, and AC_BK are employed for voice, video, best-effort and background traffic respectively. The AC_VO has the highest access priority whereas AC_BK class traffic has the least access priority. The 3D video streaming from Access Point (AP) to Node 2 is carried over three traffic flows, one each for the higher priority stream, the lower priority stream and the depth information. After considering the perceptual importance of each 3D video stream, they are allocated to the AC_VO, AC_BE, and AC_BK access classes. Due to the allocation of different access priorities for 3D video streams, they are subjected to different Average Packet Loss Rates (Av PLRs). The services used by each node and their access classes are listed in Table 1.

Table 1 Data streams over WLAN and their priority classes

Stream	Service	Access Class (AC)
1	Voice (flows from AP to Node 1)	AC_VO
2	Foreground object stream of 3D video (flows from AP to Node 2 over UDP)	AC_VI
3	Background object stream of 3D video (flows from AP to Node 2 over UDP)	AC_BE
4	Depth map stream of 3D video (flows from AP to Node 2 over UDP)	AC_BK
5	2D video streaming (flows from AP to Node 3 over UDP)	AC_VI
6	FTP stream with 1500 byte MTUs (flows from Node 4 to Node 5 over TCP)	AC_BE
7	FTP stream with 256 byte MTUs (flows from Node 5 to Node 6 over TCP)	AC_BK

Different prioritization levels for each class are obtained through changing the Contention Window (CW) parameters such as CW_{min}, CW_{max} and CW_{offset} of each station. A total simulation time of 30 seconds is considered. The resultant Av PLRs for each of the video streams is presented in Table 2.

Table 2 Packet loss rates for each testing point

Obj/Case	A	B	C	D	E
Depth	0.10	0.15	0.20	0.30	0.40
Obj 1	0.01	0.01	0.05	0.05	0.05
Obj 2	0.10	0.15	0.20	0.30	0.40

Video sequences are encoded using the MPEG-4 MoMuSys codec, which supports binary shape coding. Four standard 3D test sequences 'Orbi', 'Interview', 'Ballet' and 'Break dance' have been used for experiments. The 'Orbi' and 'Interview' sequences are of standard definition (720 x 576) resolution, while the 'Ballet' and 'Break dance' sequences are of high definition (1024 x 768) resolution. For these last two multi-view sequences, the fourth camera view and the corresponding depth map computed from stereo are utilized in this experiment [35].

Each sequence is segmented into foreground and background objects, using the segmentation technique discussed in section 3B, and each object is encoded separately. The prioritization module then estimates the expected distortion and the packet allocation to individual bit streams is changed accordingly. In the WLAN scenario, each stream is classified as a separate traffic flow from the access point to the receiver. Experiments have been conducted for a range of channel conditions and quality is compared with standard MPEG-4 object based video coding.

4.2 Performance Evaluation

The algorithm's performance is shown in terms of average frame PSNR vs AvPLR in Fig. 5. 90 frames of the 'Orbi' and 'Interview' sequence, and 70 frames of the 'Ballet' and 'Break dance' sequence have been used to make one test sequence of each type. Each point of the graph has been taken as the average of 30 runs of the decoder for each sequence. Each sequence consists of three sub bit streams, one for the depth information and the other two for prioritized objects. The streams are transmitted on different transmission flows. Av PLRs for each flow in different test cases is shown in Table 2.

The performance gain for the 'Break dance' and 'Ballet' sequences is significantly higher than other sequences, which is due to the nature of the video content. The 'Interview sequence', for example, has very little movement in it and the general perception of the sequence is static, which, therefore, leaves very little room for optimization. This is demonstrated by fairly high PSNR values even for very high packets error rates for the non prioritized streams as well. The 'Break dance' sequence on the other hand has very quick movement in it and video quality is highly vulnerable to transmission errors. A significant gain in performance is observed with this sequence, which makes the scheme suitable for use with sensitive data transmission.

The QP values used for QP optimization are 12, 16, and 20 for the background object, and 12, 14, and 16 for the foreground object. The optimal value of QP for each object is used to encode that particular frame. The QP values used for standard MPEG-4 that is used in comparison are 10 for I frames and 12 for P frames.

Fig. 6 shows selected frames from simulations with test case E. The selected frames provide a visual indication of the differences seen in the objective results.

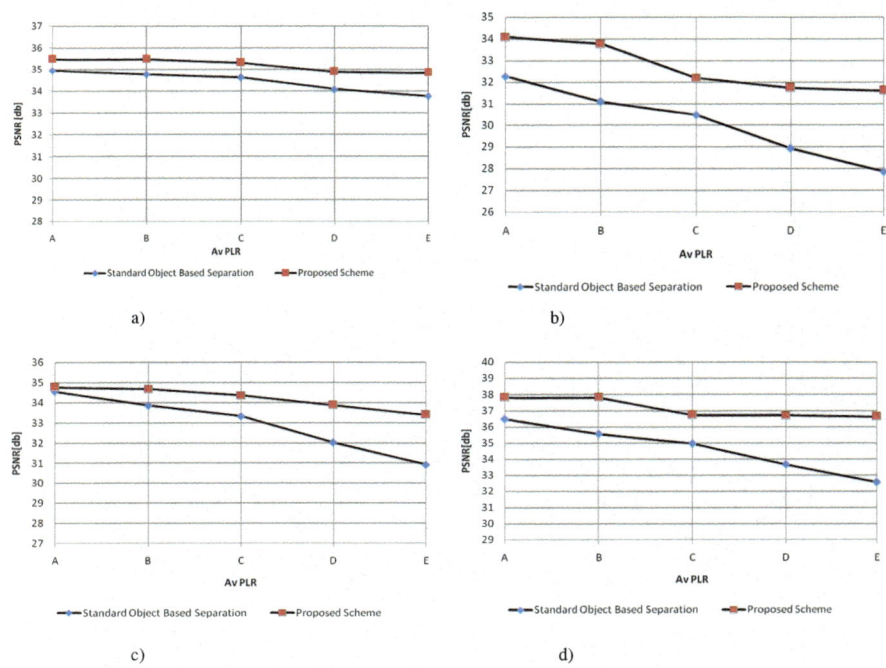

Fig. 5 Performance comparison for test sequences a) 'Interview'. b) 'Break dance'. c) 'Orbi', d) 'Ballet'.

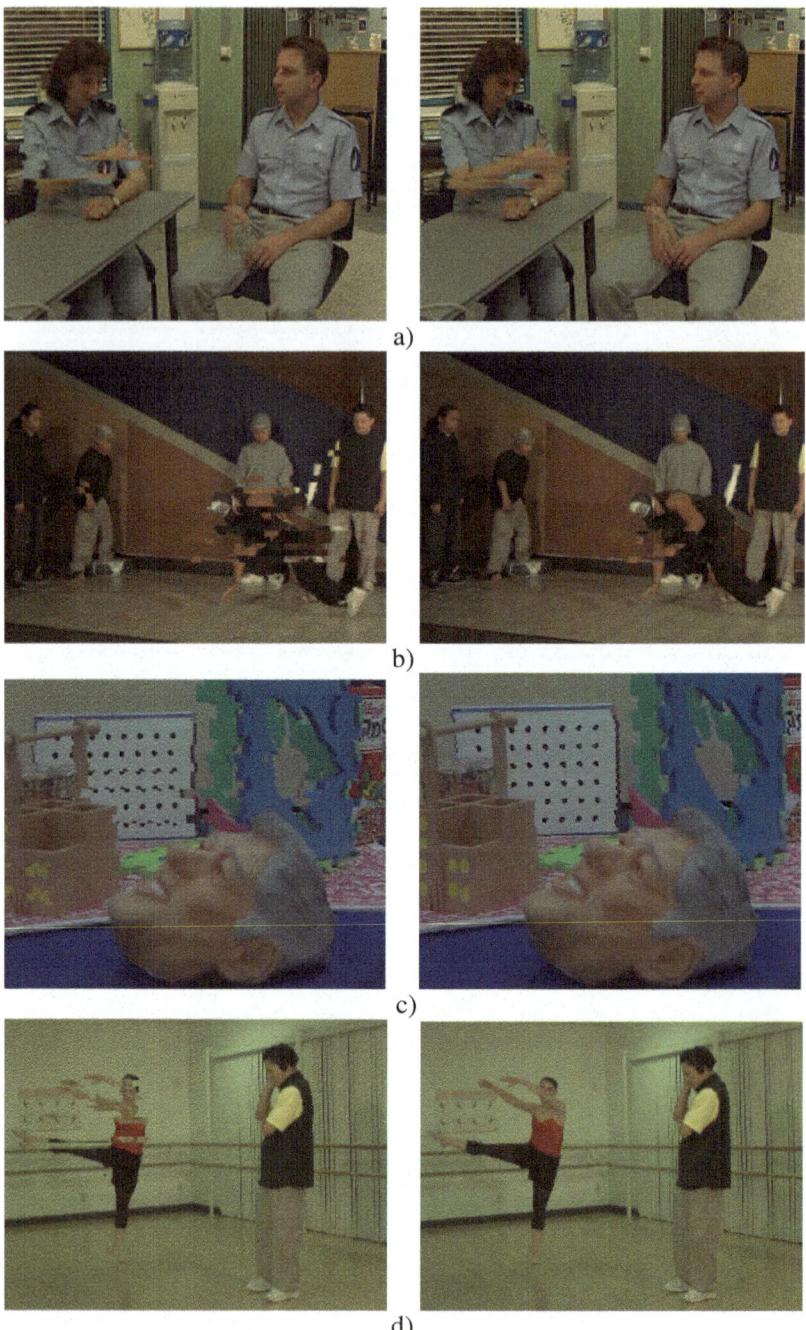

Fig. 6 Subjective performance comparison for test sequences a) 'Interview', b) 'Break dance', c) 'Orbi', d) 'Ballet' (frames taken for test case E) – left image shows normal object based coding, while right hand column shown the effect of the proposed scheme.

4.3 Subjective Quality Results

To determine whether the performance gains shown in the objective results are significant, the quality of the proposed method has been evaluated subjectively for the 'Interview' sequence. This sequence is chosen for subjective tests because the gains in objective quality are the lowest of the four tested sequences. The test used the single stimulus method described in ITU-R BT.1438 recommendation, which specify the quality evaluation procedure for stereoscopic video applications. A 42" Philips multi-view auto-stereoscopic display was used in the experiment to display the stereoscopic material to 16 subjects. Subjects rated the test sequences for the overall perceived image quality. The Mean Opinion Scores (MOS) have been calculated for each test sequence after averaging the opinion scores across all subjects.

Table 3 shows the subjective test result for the proposed prioritization scheme compared to the non-prioritized case. The MOS difference is greater than the 95% Confidence Interval (CI). This shows that a perceptible difference in quality can be observed with the proposed technique, even for sequences that exhibit a small gain in objective quality.

Table 3 Subjective Comparison for the Interview Sequence for Test Case A, D

	Un Prioritized ± CI	Prioritized ± CI
Err Free	3.59 ± 0.392	3.59 ± 0.392
Case A	2.67 ± 0.324	3.54 ± 0.352
Case D	2.06 ± 0.308	2.83 ± 0.376

5 Conclusions

A novel object based prioritization scheme for optimized transmission of 3D color plus depth video content over wireless networks is proposed in this work. A 3D home entertainment scenario is considered and an efficient channel resource allocation scheme is implemented for improved visual quality. Video content is segmented into video objects, based on the depth map, and the expected distortion is then calculated at the video packet level using the conventional 2D distortion parameters plus the effective contribution of depth information to distortion for each object. Data prioritization is achieved by rearranging the video packets into different bit streams, according to estimates of their distortion. These bit streams are then transmitted over prioritized communication links with different QoS facilities supported by the WLAN communication standard, IEEE 802.11e. The experiments carried out to evaluate the proposed scheme have demonstrated significant quality improvement by providing PSNR gain of 1-3dB at higher error rates for test sequences with rapid motion content. The proposed scheme is quite generic in implementation and can be adapted for use with the latest coding standards, as well for enhanced perceptual video quality.

References

1. Takada, H., Suyama, S., Date, M., Ohtani, Y.: Protruding apparent 3D images in depth-fused 3D display. IEEE Trans.: Consumer Electronics 54, 233–239 (2008)
2. Lee, C.M.G., Travis, A.R.L., Lin, R.: Flat-panel autostereoscopic 3D display. IET: Optoelectronics 2, 24–28 (2008)
3. Stoykova, E., Alatan, A.A., Benzie, P., Grammalidis, N., Malassiotis, S., Ostermann, J., Piekh, S., Sainov, V., Theobalt, C., Thevar, T., Zabulis, X.: 3-D Time-Varying Scene Capture Technologies—A Survey. IEEE Trans.: Circuits and Sys. for Video Tech. 17, 1568–1586 (2007)
4. Seung-Ryong, H., Toshihiko, Y., Kiyoharu, A.: 3D Video Compression Based on Extended Block Matching Algorithm. In: Proc. IEEE Int. Conf. on Image Processing (2006)
5. 3D@Home Consortium, http://www.3dathome.org/ (accessed June 02, 2009)
6. Chang, Y.C., Lee, S.W., Komyia, R.: A fast forward error correction allocation algorithm for unequal error protection of video transmission over wireless channels. IEEE Trans.: Consumer Electronics 54, 1066–1073 (2008)
7. Sprljan, N., Mrak, M., Izquierdo, E.: A fast error protection scheme for transmission of embedded coded images over unreliable channels and fixed packet size. In: Proc. IEEE International Conference on Acoustics, Speech, and Signal Processing (2005)
8. Van Der Schaar, M., Sai Shankar, N.: Cross-layer wireless multimedia transmission: challenges, principles, and new paradigms. IEEE Trans.: Wireless Comm. 12, 50–58 (2005)
9. Qiuyan, X., Xing, J., Hamdi, M.: Cross Layer Design for the IEEE 802.11 WLANs: Joint Rate Control and Packet Scheduling. IEEE Trans.: Wireless Comm. 6, 2732–2740 (2007)
10. Villalon, J., Cuenca, P., Orozco-Barbosa, L., Seok, Y., Turletti, T.: Cross-Layer Architecture for Adaptive Video Multicast Streaming over Multi-Rate Wireless LANs. IEEE Trans.: IEEE Journal on Selected Areas in Communications 25, 699 (2007)
11. Smolic, A., Kimata, H.: Applications and requirements for 3DAV. ISO. IEC JTC1/SC29/WG11 (2003)
12. Meesters, L.M.J., Ijsselsteijn, W.A., Seuntiens, P.: Survey of perceptual quality issues in threedimensional television systems. In: Proc. SPIE (2003)
13. Fehn, C.: A 3D-TV approach using depth-image-based rendering (DIBR). In: Proc. VIIP, Spain (2003)
14. Chen, Z., Han, J., Ngan, K.N.: Dynamic Bit Allocation for Multiple Video Object Coding. IEEE Trans.: Multimedia 8, 1117–1124 (2006)
15. Chih-Wei, T., Ching-Ho, C., Ya-Hui, Y., Chun-Jen, T.: Visual sensitivity guided bit allocation for video coding. IEEE Trans.: Multimedia 8, 11–18 (2006)
16. Perkins, M.G.: Data compression of stereopairs. IEEE Trans.: Comms. 40, 684–696 (1992)
17. Aksay, A., Bilen, C., Kurutepe, E., Ozcelebi, T., Akar, G.B., Civanlar, M.R., Tekalp, A.M., Ankara, T.: Temporal and spatial scaling for stereoscopic video compression. In: Proc. IEEE EUSIPCO (2006)

18. Hewage, C.T.E.R., Worrall, S.T., Dogan, S., Kondoz, A.M.: Prediction of stereoscopic video quality using objective quality models of 2-D video. IET Electronics Letters 44, 963–965 (2008)

19. Ekmekcioglu, E., Mrak, M., Worrall, S.T., Kondoz, A.M.: Edge adaptive upsampling of depth map videos for enhanced free-viewpoint video quality. IET Electronics Letters 45(7), 353–354 (2009)

20. Cheng, I., Basu, A.: Perceptually optimized 3D transmission over wireless networks. In: Proc. International Conference on Computer Graphics and Interactive Techniques ACM SIGGRAPH, California (2005)

21. Draft Supplement to STANDARD FOR Telecommunications and Information Exchange Between Systems-LAN/MAN Specific Requirements - Part 11: Wireless Medium Access Control (MAC) and Physical Layer (PHY) specifications: Medium Access Control (MAC) Enhancements for Quality of Service. IEEE 802.11e/Draft 4.2, IEEE 802.11 WG (2003)

22. Deng, D.J., Chang, R.S.: A Priority Scheme for IEEE 802.11 DCF Access Method. IEICE Trans.: Comms. 82, 96–102 (1999)

23. Bystrom, M., Stockhammer, T.: Dependent source and channel rate allocation for video transmission. IEEE Trans.: Wireless Comm. 3, 258–268 (2004)

24. van Dyck, R.E., Miller, D.J.: Transport of wireless video using separate, concatenated, and jointsource-channel coding. Proceedings of the IEEE 87, 1734–1750 (1999)

25. Gharavi, H., Alamouti, S.M.: Video Transmission for Third Generation Wireless Communication Systems. Journal of research: National institute of standards and technology 106, 455–470 (2001)

26. Nasir, S., Worrall, S.T., Mrak, M., Kondoz, A.M.: An unequal error protection scheme for object based video communications. In: Proc. IEEE International Symposium on Consumer Electronics, Portugal (2008)

27. Dogan, S., Cellatoglu, A., yguroglu, M., Sadka, A.H., Kondoz, A.M.: Error-resilient video transcoding for robust internetwork communications using GPRS. IEEE Trans.: Circuits and Sys. for Video Tech. 12, 453–464 (2002)

28. Ma, Y., Worrall, S.T., Kondoz, A.M.: Automatic Video Object Segmentation Using Depth Information and an Active Contour Model. In: Proc. Int. Conf. on Image Processing, San Diego, USA (2008)

29. ISO: Overview of the MPEG-4 Standard, ISO/IEC JTC/SC29/WG11 N4668 (2002)

30. Il-Min, K., Hyung-Myung, K.: An optimum power management scheme for wireless video service in CDMA systems. IEEE Trans.: Wireless Comm. 2, 81–91 (2003)

31. Worrall, S.T., Fabri, S.N., Sadka, A.H., Kondoz, A.M.: Prioritisation of Data Partitioned MPEG-4 Video over Mobile Networks. ETT: Europ. Trans. on Telecom. 12, 169–174 (2001)

32. Wang, Y., Wenger, S., Wen, J., Katsaggelos, A.K.: Error resilient video coding techniques. IEEE Signal Pro. Magazine 17, 61–82 (2000)

33. Fall, K., Varadhan, K.: The ns manual,
    http://www.isi.edu/nsnam/ns/doc/index.html
    (accessed June 02, 2009)

34. Ni, Q.: IEEE 802.11e NS2 Implementation,
    http://www.sop.inria.fr/planete/qni/Research.html
    (accessed June 02, 2009)

35. Zitnick, C.L., Kang, S.B., Uyttendaele, M., Winder, S., Szeliski, R.: High-quality video view interpolation using a layered representation. ACM Transactions: Graphics (TOG) 23, 600–608 (2004)

Chapter 22
Scalable Indexing of HD Video

Jenny Benois-Pineau, Sandrine Anthoine, Claire Morand,
Jean-Philippe Domenger, Eric Debreuve, Wafa Bel Haj Ali, and Paulo Piro

Abstract. HD video content represents a tremendous quantity of information that all types of devices can not easily handle. Hence the scalability issues in its processing have become a focus of interest in HD video coding technologies. In this chapter, we focus on the natural scalability of hierarchical transforms to tackle video indexing and retrieval. In the first part of the chapter, we give an overview of the transforms used and then present the methods which aim at exploring the transform coefficients to extract meaningful features from video and embed metadata in the scalable code-stream. Statistical global object-based descriptor incorporating low frequency and high-frequency features is proposed. In the second part of the chapter, we introduce a video retrieval technique based on a multiscale description of the video content. Both spatial and temporal scalable descriptors are proposed on the basis of multi-scale patches. A statistical dissimilarity between videos is derived using Kullback-Leibler divergences to compare patch descriptors.

1 Introduction

HD video content represents a tremendous quantity of information that cannot be handled by current devices without adapting the processing chain. There is thus a need to develop new content-based indexing methods adapted to 1) the high quality and complexity of the HD video content and 2) the fact that such a content will be accessed through heterogeneous networks. In particular, scalability is a most

Jenny Benois-Pineau · Claire Morand · Jean-Philippe Domenger
LaBRI, UMR CNRS/ University of Bordeaux 1, 351 Cours de la Liberation, F-33405 Talence, France
e-mail: {jenny.benois,morand,domenger}@labri.fr

Sandrine Anthoine · Eric Debreuve · Wafa Bel Haj Ali · Paulo Piro
I3S, University of Nice Sophia-Antipolis/ CNRS, 2000 route des lucioles, 06903 Sophia-Antipolis, France
e-mail: {anthoine,debreuve,benelhaj,piro}@i3s.unice.fr

desirable property in the developed technology. It stands for the adaptability of the data encoding and delivery process to different temporal and spatial resolutions that may be imposed by specific network properties. It has become a focus of interest in HD video coding technologies which led to coding standards such as SVC, H.264 or (M)JPEG2000 [1] and is also of interest in post-processing technologies. Hierarchical transforms (e.g. wavelet transforms) are not only efficient tools to describe and compress video content but they also naturally yield a scalable description of this content. They are thus natural candidates to help defining scalable indexing methods.

In this chapter, we present the first research works on scalable HD video indexing methods in the transformed domain. The first method extracts information directly from the compressed video-stream while the second deals with raw data. The standards in video coding (e.g. SVC etc.) indeed use a hierarchical transform to compress the data. The design of descriptors directly extracted from this domain thus ensures the scalability of the proposed method while allowing for a coherent and fast processing of the data. In this framework, we propose two video indexing and retrieval methods.

The first part of the chapter focuses on indexing in the compressed domain. We give an overview of the transforms used and then present the methods which aim at exploring the transform coefficients to extract from the video stream, at different levels of decomposition, meaningful features such as objects [2] or visual dictionaries [3]. The resulting descriptors will be used for video partitioning and retrieval. First we will introduce a system based on Daubechies wavelets and designed for joint indexing and encoding of HD content by JPEG2000-like encoders. Here we will present an overview of emerging methods such as [3] and develop our previous contributions [2, 4].

In the second part of this chapter, we describe a video indexing method that builds up on a hierarchical description of the decoded data. Spatial and respectively temporal descriptors of the video content are defined, that rely on the coherence of a wavelet description of the key-frames and respectively the motion of blocks in the video. The method builds on the redundancy of the descriptors (induced, in part, by the transform used which is the Laplacian pyramid) to statistically compare two videos. The invariance properties of the descriptors as well as the statistical point of view allow for some robustness to geometric and radiometric alterations.

2 HD Content Indexing in the Compressed Domain

2.1 Scalable Compression Standards

The scalability of a representation of the video content is the property which has been introduced in multimedia standards since MPEG2. It means that various temporal and spatial resolutions of a video stream and also different qualities of videos can be decoded from the same code-stream. In the first case, this is a temporal scalability, in the second, spatial scalability and finally, the SNR scalability is concerned. Two latest standards designed for HD video and film content have this property: H.264 and motion JPEG2000 known as (MJPEG2000 [1]. While H.264 has been designed for HD TV and continues the principles of previous standards in the sense that the transform used (Integer Block Transform) is a variation of a DCT, which does not have the property of scalability, (M)JPEG2000 standard has this property naturally, due to the scalable nature of the transform used: the Discrete Wavelet Transform (DWT).

(M)JPEG2000 is a part of JPEG2000 standard for motion sequences of images. Nevertheless, contrary to H. 264 it does not contain motion information, each frame being encoded in an intra-frame mode by JPEG2000. In the following, we give the insights to JPEG2000 standard [5].

2.1.1   (M)JPEG2000 Standard

Initiated in march 1997 and becoming international ISO standard in December 2000, the standard JPEG2000 exhibited a new efficiency with regard to specifically high-resolution (HD) images. The specifications of DCI (Digital Cinema Initiative, LLC [6]) made (M)JPEG2000 the digital cinema compression standard. (M)JPEG2000 is the extension of the standard JPEG2000 for videos: each frame in the video sequence is considered separately and encoded with JPEG2000. Furthermore (M)JPEG2000 is becoming the common standard for archiving [7] cultural cinematographic and video heritage with the greater quality/compression compromise than previously used solutions. The JPEG2000 standard follows the ideas initially proposed in MPEG4 [8] for object-based coding, namely the possibility to encode more precisely Regions of Interest (ROI) in each frame or in a single image. The industrial reality in the usage of this advanced feature in JPEG2000 turned to be pretty much the same as with MPEG4. Despite the very rich research work proposing various methods for extraction of ROI (e.g. [9, 10]), the most commonly used JPEG2000 limits to encoding the whole frame. More precisely, an image frame is modeled as a set of tiles on which the coding process performs independently as depicted in Figure 1, a frame being considered as a single tile.

The core of the standard is the DWT which in case of lossy compression is realized by High-Pass and Low-Pass filters designed for zero-mean signals. This is why the Level offset is necessary at the pre-processing step. Furthermore, the standard operates on YCrCb color system, hence if the source is in RGB, a linear transform has to be applied. Then the resulting frame undergoes the DWT which we describe below. The transform coefficients are quantized to reduce the quantity of information and entropic coding known as EBCOT (Embedded Block Coding with Optimized Truncation) is performed on these quantized values. At the first step (Tier 1) context modeling is realized, at the second step (Tier 2) the bit allocation for output bit stream is performed.

The decoder proceeds in an inverse order to decode the original frame. In the lossy scheme the original pixel values cannot be recovered, but the quantization matrix is designed in a way to take into account psycho-visual properties of Human

Fig. 1 Simplified Block-
Diagram of JPEG2000
frame encoding without
ROI

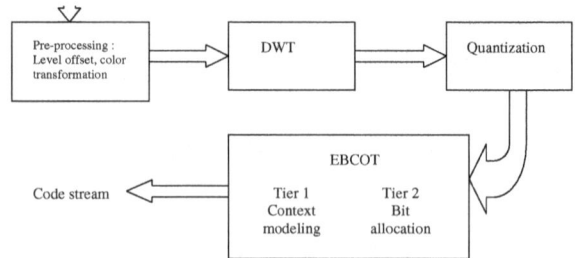

Fig. 1 Simplified Block-Diagram of JPEG2000 frame encoding without ROI

Visual System (Part 2 of the standard). Hence the degradations on decoded HD video frames could not be perceived.

As most of digital HD content is now available in compressed form, the compressed data are very much attractive to use directly for analysis and indexing purposes. This was the case for instance in [11], where the Rough Indexing paradigm was proposed to fulfill all mid-level indexing tasks such as camera motion identification, scene boundary detection and meaningful object extraction from MPEG2 compressed streams. The motivation of the earlier work in compressed domain was mainly in saving computational power and re-using already available low-resolution information. In the case of JPEG2000, this is the hierarchical nature of DWT which is in the focus, as it can allow analysis and indexing of image content at various spatial resolutions. Hence in order to give understanding of the data in the compressed domain to be used for this purpose we will briefly introduce the DWT used in JPEG2000 standard.

2.1.2   DWT in JPEG2000

We will first limit ourselves to the presentation of Wavelet Transform in the 1D continuous case. A wavelet is a waveform function localized and sufficiently regular. These properties are expressed by the following

$$\psi \in L^1 \cap L^2 \text{ and } \int_0^{+\infty} \frac{|FT[\psi](\omega)|^2}{|\omega|} d\omega = \int_{-\infty}^0 \frac{|FT[\psi](\omega)|^2}{|\omega|} d\omega < +\infty \qquad (1)$$

Where L_1 is the space of integrable functions on $\mathbb{R}$ and L_2 is the space of square-integrable functions on $\mathbb{R}$, $FT[\psi]$ is the Fourier transform of ψ and ω is the frequency. From this unique function called "mother wavelet" it is possible to build a basis for analysis of a function f from inner product space L_2 by translation and scaling of the mother wavelet:

$$\psi_{h,\tau}(t) = \frac{1}{\sqrt{h}} \psi\left(\frac{t-\tau}{h}\right), \, h \in \mathbb{R}^+, \, \tau \in \mathbb{R} \qquad (2)$$

The analysis of f consists in computing the coefficients $w_f(h, \tau)$ of the projection of f :

$$w_f(h, \tau) = \int_{\mathbb{R}} f(t) \overline{\psi}_{h,\tau}(t) dt, \tag{3}$$

where $\overline{\psi}$ is the complex conjugate of ψ. The synthesis can be performed under conditions of admissibility in Eq. (1) as

$$f(t) \overset{L_2}{=} \frac{1}{K_\psi} \int_{\mathbb{R}^+ \times \mathbb{R}} f(t) \psi_f(h, \tau) \frac{dh d\tau}{h^2} \tag{4}$$

where K_ψ is the common value of the integrals in Eq. (1). Under certain conditions it is possible to built an orthonormal wavelet basis. Nevertheless, the basis functions are often difficult to construct. Hence bi-orthogonal wavelets are considered such that two bases, the direct $B = \{e_i\}$ and the dual $\tilde{B} = \{\tilde{e}_i\}$, satisfying condition of duality $(e_i, \tilde{e}_j) = \delta_{ij}$, serve for analysis and synthesis respectively.

In JPEG2000, bi-orthogonal wavelets are used. Image and video compression are applied to the discrete signals, hence instead of continuous case, a discrete wavelet transform (DWT) has to be performed. In this transform the wavelets are defined on discretely sampled space, for instance a dyadic case can be considered with

$$h = 2^k, \tau = l2^k, (k, l) \in \mathbb{Z}^2 \tag{5}$$

This transform allows re-covering good approximations converging to f

$$f(t) = \sum_{k \in \mathbb{Z}} \sum_{l \in \mathbb{Z}} w_f(2^k, l2^k) \, \psi_{2^k, l2^k}(t) \tag{6}$$

Mallat [12] showed that the DWT could be computed with a bank of filters. This is the way the DWT is realised in JPEG2000 with Daubechies filters. Figure 2 depicts the analysis process applied to an image where the arrows correspond to the sub-sampling by a factor of 2 and squares depict discrete convolution operation. The resulting subbands are denoted as LL for low-pass filtering results on lines and columns, LH for consecutive low and high-pass filtering, HL for filtering in the inverse order and HH for two consecutive high-pass filtering steps. The subbands HL, LH and HH are called "High Frequency" (HF) subbands while LL is called "Low Frequecy" (LF)subband.

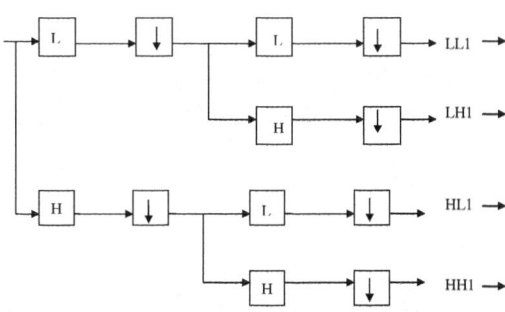

Fig. 2 One level wavelet analysis on an image in JPEG2000 standard

Table 1 Coefficients of 9/7 Daubechies filters

Analysis Filters			Synthesis Filters		
i	Low-Pass Filter	High-Pass Filter	i	Low-Pass Filter	High-Pass Filter
0	0.6029490182363	1.115087052	0	1.115087052	0.6029490182363
± 1	0.2668641184428	-0.591271763114	± 1	-0.591271763114	0.2668641184428
± 2	-0.07822326652898	-0.05754352622849	± 2	-0.05754352622849	-0.07822326652898
± 3	-0.01686411844287	0.09127176311424	± 3	0.09127176311424	-0.01686411844287
± 4	0.02674875741080		± 4		0.02674875741080

Fig. 3 An example of 2-level wavelet decomposition of and HD Video Frame. LABRI corpus

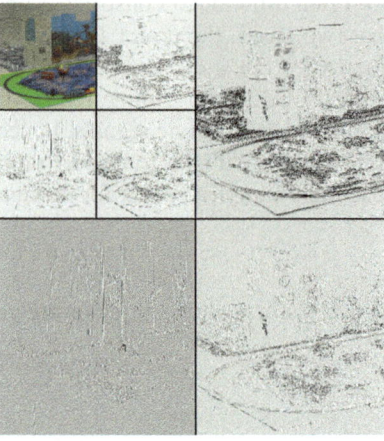

The coefficients of Daubechies analysis /synthesis filters 9/7 used for lossy compression are given in table 1. The decomposition process can be re-iterated on the LL subband thus resulting in a wavelet pyramid with K decomposition levels. Figure 3 contains an example of 2-level decomposition of an HD 1080p video frame.

Hence, if the HD video is encoded in the JPEG2000, the wavelet pyramid obtained after a partial decoding and de-quantizing of wavelet coefficients without inverse transform contains low-frequency and high-frequency information at several levels. In the following we will present its exploitation in the scalable indexing and retrieval of video content.

2.2 Scalable Extraction of Low and Mid-level Features from Compressed Streams

In the overall set of problems to be resolved in the task of video indexing one can distinguish:

- temporal partitioning, linear or non-linear, into semantically homogeneous sequences such as video shots and video scenes. This partition can be used for various tasks such as generation of video summaries [13], video retrieval with query-by-clip scenarios or query by key-frame, navigation in video content.

- spatio-temporal partitioning, that is extraction of meaningful objects from video in order to detect events of interest related to objects or to realize object-based queries on video databases [14].

In the first case, the key feature to be extracted from video is represented by shot boundaries which are considered as a "mid-level" semantic feature compared to high level concepts appealing to human-like interpretation of video content. A shot is a sequence of frames corresponding to one take of the camera. Very vast literature has been devoted to the problem of shot boundary detection in the past. This was the subject of TRECVid competition [15] in 2001–2007 and various industrial solutions have come from this intensive research. Shot boundaries are the most natural metadata which can be automatically generated and allow for a sequential navigation in a coded stream. Scene boundaries correspond to the changes of the content with more semantic interpretation, they delimit groups of subsequent shots which convey the same editing ideas. In the framework of scalable indexing of HD video we wish to analyze a new trend for efficient video services: embedded indexing in Scalable Video coding. This is namely one of the focuses of JPSearch initiative [16]: embedding of metadata into the data encoded in the JPEG2000 standard. Hence the latest research works link content encoding and indexing in the same framework be it images or videos [3].

2.2.1 Embedded Mid Level Indexing in a Scalable Video Coding

The coding scheme [3] inspired by JPEG2000 scalable architecture allows a joint encoding of a video stream and its temporal indexing such as shot boundaries information and key pictures at various quality levels. The proposed architecture of scalable code-stream consists of a sequences of Groups of Pictures (GOPs) corresponding to a video shot and containing its proper source model which at the same time represents a content descriptor.

The video stream architecture is depicted in Figure 4. The authors propose using Vector Quantization (VQ) to build a visual code-book VC for the first key-picture of each shot. Here the encoding follows a traditional scheme in VQ coding: the key-picture is split into blocks forming vectors in the description space. Then an accelerated K-means algorithm is applied resulting in a VC of pre-defined dimension. The key-picture is then encoded with this code-book and the error of vector quantization is encoded using a JPEG2000 encoder. For all following key-pictures, the code-book obtained is applied for their encoding. If the coding distortion for a given key-picture I_{ij} in the shot S_i encoded with VC_i is higher than a pre-defined

Fig. 4 Temporal decomposition of a video stream

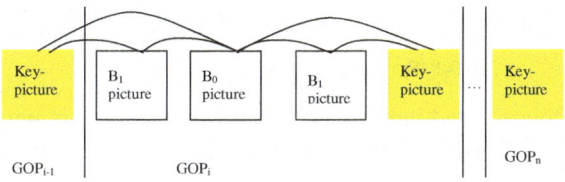

threshold, then it means that I_{ij} cannot be efficiently encoded with VC_i and hence most probably belongs to the next shot. The visual code-book VC_i is entropy-encoded and embedded in the code stream. All the B-frames of a given GOP are encoded applying a conventional motion estimation and compensation method.

In this approach, the visual code-book plays two roles: on the one hand, it is used for decoding the key frame; on the other hand, it is a low-level content descriptor. Indeed, in their previous work [17] the authors showed that visual code-books are good low-level descriptors of the content. Since this pioneer work visual code books have been applied in specific video description spaces [18] and become very popular in visual content indexing [19].

Assuming the visual content to be quantized using a VC, the similarity between two images and/or two shots can be estimated by evaluating the distortion introduced if the role of two VC is exchanged. The authors propose a symmetric form:

$$\phi(S_i, S_j) = \left| D_{VC_j}(S_i) - D_{VC_i}(S_i) \right| + \left| D_{VC_i}(S_j) - D_{VC_j}(S_j) \right| , \tag{7}$$

where $D_{VC_j}(S_i)$ is a distortion of encoding a shot (or key-frame) S_i with the visual code-book D_{VC_j}.

Visual distortion is understood as for usual case of VQ encoding:

$$D_{VC_j}(S_i) = \frac{1}{N_i} \sum_{p=1}^{N_i} ||v_p - c_j(v_p)||^2 \tag{8}$$

Here v_p is the visual vector representing a key-frame pixel block, N_i is the number of blocks in the key-frame (or whole shot S_i), $c_j(v_p)$ is the vector in a code-book closest to v_p in the sense of Euclidean distance.

The symmetric form in Eq. (7) is a good measure of dissimilarity between shots. Hence the temporal scalability of video index can be obtained by grouping shots on the basis of Eq. (7).

Thus two scalable mid-level indexes are implicitly embedded in a scalable code-stream: the two different code-books for the subsequent groups of GOPs indicate a shot boundary, a scene boundary can be obtained when parsing and grouping shots with Eq. (8).

On the other hand decoding of visual code-books and representation of key-frames only with code-words supplies the base level of spatial scalability. The enhancement levels can be obtained by scalable decoding of VQ error on key-frames.

This scheme is very interesting for the HD content: a quick browsing does not require decoding full HD, but only the base layer can be used to visualize the frames. Reduced temporal resolution can also be achieved when parsing.

2.2.2 Object-Based Mid-level Features from (M)JPEG2000 Compressed Stream

Object-based indexing of compressed content remains one the most difficult problems in the the vast set of indexing tasks be it Low Definition, Standard Definition or High Definition video and film content. The main stream research in indexing and retrieval of video content nowadays avoids the complex, ill-posed "chicken and egg" problem of extracting meaningful objects from video. It focuses on local features such as SIFT descriptors proposed by Lowe [20]. Hence, in the paper entitled "Unsupervised Object Discovery: A comparison", [21], where the authors search for images containing objects, one can read "Images are represented using local features". Pushing this reasoning to its extreme end, we come to the famous cat illusion and make a "bottom-up" effort in visual content understanding. At the same time, the strong effort of the multimedia research community related to the elaboration of MPEG4, MPEG7 [22] and JPEG2000 (part 1) standards was devoted to the development of automatic segmentation methods of video content to extract objects. Here the approach is just the contrary: first an Entity has to be extracted and then a description (sparse, dense, local or global) of it can be obtained. The results of these methods, e.g. [23, 24, 25], while not always ensuring an ideal correspondence of extracted object borders to visually observed contours, were sufficiently good for fine-tuning of encoding parameters and for content description.

Hence, we are strongly convinced that the paradigm consisting of segmenting objects first and then representing them in adequate feature spaces for object based indexing and retrieval of video remains the promising road to the success and a good alternative for local modeling of content by feature points. In the context of scalable HD content, the object extraction process has to be adapted to the multiple resolutions present in code-stream. It has to supply mid-level, object-based features corresponding to each resolution level.

In [26] we proposed a full solution for mid-level global feature extraction for generic objects in (M)JPEG2000 compressed content by an approach operating directly on the Daubechies 9/7 pyramid of a HD compressed stream. The underlying assumptions of the method are as follows : i) we suppose that generic objects can be "discovered" in video when the magnitude of object local ego-motions sufficiently differs from the global motion, that of the camera ii) the high-frequency information contained in HF subbands at each level of the wavelet pyramid can be efficiently reused for delimiting objects boundaries, iii) both LF and HF subbands are necessary to convey global object features.

According to our indexing paradigm, the first step consists of extraction of objects from a compressed stream. The overall strategy follows fruitful ideas of cooperative motion-based and color-based spatio-temporal video object segmentation [11]. Here the areas of local motion have to be identified in video frames first. They form the so-called motion masks M_t^k at the lowest resolution level ($k = K - 1$) of K-level Daubechies pyramid. Then a color-based segmentation of the low frequency LL^k subband has to be fulfilled on the whole subband. Finally motion masks and segmentation map are merged by majority vote resulting in object masks $O_t^k = \{O_{t,i}^k, \, i = 1..n(k)\}, k = K - 1$. Objects at the top of the pyramid corresponding to the lowest scalability level are thus extracted.

The object masks obtained are then projected on the higher resolution levels using the wavelet location principle (see Figure 5) allowing for establishing direct

Fig. 5 Location principle in
Daubechies Pyramid

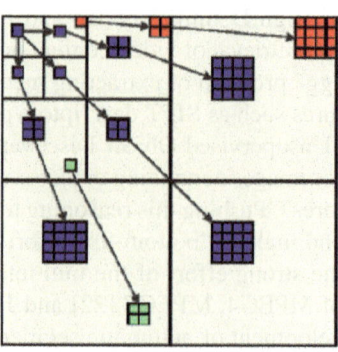

correspondences between a coefficient of the current level and four coefficients at
the lower level.

Then the projected segmentation is refined using motion masks obtained at lower
level and low and high-frequency coefficients contained in the subbands of the current level of the pyramid.

Colour segmentation of LL subband at the top of the pyramid is performed by
a known morphological approach. Its application does not really differ from previously studied morphological segmentation by a simplified watershed adapted to low
resolution version of compressed video frames [11].

It is in the detection of motion masks and refinement of the segmentation across
wavelet pyramid, that the nature of wavelet subbands is truly used. Hence we will
focus on these aspects in the following.

Motion estimation in the Wavelet domain

The problem of motion estimation in the wavelet domain has been extensively studied for the development of motion-compensated wavelet encoders [26]. Due to the
shift variant nature of DWT, direct band-to-band motion estimation by classical
Block Matching (BM) fails to give sufficiently reliable information when used on
the lower resolution levels, especially for the HF subbands. Several methods have
been developed to limit the effects of the shift-variance of the wavelet transform.
One of the possible approaches was to estimate motion on the LL subband and
motion-compensate HF subbands with the estimated vectors. In order to limit the
shift-variance, the matching is done between the current frame and the translated
versions of the reference frame in wavelet domain [27], other approaches consist
of estimating motion in Overcomplete DWT (ODWT) without sub-sampling [28].
Nevertheless in JPEG2000, the Daubechies pyramid is already imposed by the standard. Hence for the sake of separation of motion masks from the background motion,
the estimation between the LL subbands of each level of the pyramid and regularization with robust global motion estimation is proposed.

The first step consists in estimating motion vectors on the block-basis on the
lowest $k - th$ resolution level minimizing "Mean Absolute Difference" criterion in
Eq. (9) when initializing them by zero motion.

$$MAD_B(dx, dy) = \frac{1}{|B|} \sum_{(x,y) \in B} \left| LLY^K(x, y, t) - LLY^K(x + dx, y + dy, t - dt) \right| \quad (9)$$

Here LLY^K is the Y-component of the low frequency subband and B is the considered block. Estimation with pixel accuracy turns to be better, than half pixel because of shift-variance of wavelets. Then the global affine six parameters motion model is estimated by robust weighted least squares:

$$\begin{cases} dx(x, y) = a_1 + a_2 x + a_3 y \\ dy(x, y) = a_4 + a_5 x + a_6 y \end{cases} \quad (10)$$

The outliers with regard to this model with weak weights $w(B)$ form the motion mask M_t^K at the top of the pyramid and serve for extraction of objects O_t^K. When estimating the model of Eq. (10) the coefficients of the HF subbands are used in order to a priori exclude "flat areas" in a subband LL, which are not reliable for motion estimation. Here the standard deviation vector $\sigma(B) = (\sigma_{LH}, \sigma_{HL}, \sigma_{HH})^T$ is computed for each block. If its norm $||\sigma(B)||_\infty$ is less than a level-dependent threshold Th_σ^k, then the block is considered as "flat".

The projection of motion vectors to initialize the estimator at the lower levels of the pyramid is realized with location principle on the subband LL diadycally increasing block size and vector magnitudes. The outlier blocks, projected with this scheme are then split into smaller blocks in order to keep precise motion estimation in areas with proper motion. The motion model of Eq. (10) re-estimated at each level of the pyramid allows for improvement of PSNR measured on non-outliers up to 8% on average.

In filtering of outliers from blocks which follow the model of Eq. (10), the absolute difference between optimal values of MAD obtained when a block is compensated with its original vector and with Eq. (10) is computed. If it is greater than a threshold Th_{MAD}^k, than the "proper" motion of a block is confirmed. Otherwise, it is incorporated in the set of the blocks following the global motion, the same test is made for flat blocks. Figure 6 depicts the results of this filtering at the second resolution level of a Daubechies pyramid. The upper row represents the LL subband at level 2, the mid-raw is the result of outlier rejection by weighted least squares, the lower row is the result of filtering.

The merged motion masks and segmentation map at the top of the pyramid form extracted objects (see an example in Figure 7).

To form a scalable object-based descriptor, it is necessary to get extracted objects at all levels of the pyramid. The object masks extracted from the top of the pyramid have to be projected and refined at each level. If the projection across pyramid levels is naturally guided by wavelet location principle (Figure 5), fitting of object boundaries to the LL subband content at the lower pyramid levels is a problem per se. It is natural try to use already available contour information in HF subbands. This can be done in the framework of Markov Random Field (MRF) modeling.

Fig. 6 Results of filtering of flat blocks and outliers. Sequence "Claire" ©LABRI.

Fig. 7 Object extraction at the top of the pyramid by merging frame segmentation and motion masks. Sequence "Lancé trousse" ©LABRI.

MRF based adjustment of object borders with LL and HF subbands

When projected between two successive levels of the pyramid, one wavelet coefficient of a higher $W^k(i,j)$ level in each subband is transformed into four coefficients $W^{k-1}(2i,2j)$, $W^{k-1}(2i,2j+1)$, $W^{k-1}(2i+1,2j)$, $W^{k-1}(2i+1,2j+1)$ of lower level. To avoid this aliasing in resulting object border a morphological erosion of the projected mask is performed thus defining an uncertainty area. Then pixel-coefficients in it are assigned according to the minimal energy adapting the classical MRF formulation with Maximal A posteriori Probability (MAP) criterion. Gaussian multivariate distribution of LL coefficients in each segmented region is supposed. For a coefficient in LL subband at position p the label l is assigned according to the minimum of the energy:

$$l(p) = \arg \min_{l \in [1,L]} \left(U_1(l,p) + U_2(l,p) \right) \tag{11}$$

The first potential U_1 expresses the Gaussian assumption of distribution of the coefficients in LL subband:

$$U_1(l,p) = (p_{LL} - \mu_l)^T \Sigma_l^{-1}(p_{LL} - \mu_l) \tag{12}$$

with μ_l the mean color vector of a region and Σ_l its covariance matrix.

The second potential usually stands for regularity of segmentation maps. In the context of scalable segmentation in the wavelet domain it is completed with the contrast values expressed by HF coefficients:

$$U_2(l,p) = \sum_{c \in C_p} A\left(1 - \delta(l,l(c)) + (2\delta(l,l(c)) - 1)\,|HF|_n^c\right) \tag{13}$$

Here c are the cliques in the 8-connected neighborhood and $|HF|_n^c$ is the normalized HF coefficient from HL, LH or HH subband accordingly to the direction of the clique (horizontal, vertical or slant), A is a constant and δ is the Kronecker symbol.

Hence after such an adjustment at each level of the pyramid, meaningful objects are extracted. An illustration is given in Figure 8. Contrary to the main-stream approaches describing objects by local (e.g. SIFT) features, a global statistical descriptor of objects at each level of the pyramid is proposed. This descriptor is a pair of LL and HF histograms of wavelet coefficients extracted on object masks. For each object $O_{t,i}$ the descriptor is denoted by

$$H_{t,i} = \left\{ h_{LL}^k(O_{t,i}), h_{HF}^k(O_{t,i}),\ k = 0,\ldots,K-1 \right\} \tag{14}$$

Here h_{LL}^k is a normalized joint histogram of color wavelet coefficients of LL^k subband and h_{HF}^k is a normalized joint histogram of mean values of three coefficients in HF subbands. The descriptor therefore contains both LL and HF parts. A complex question of histogram quantization is resolved from general statistical considerations. The Sturges rule [29] is used relating the number of wavelet coefficients in the object mask and the number of bins in the resulting histogram b:

$$b^k = 1 + \log_2(|O_i^k|) \tag{15}$$

Fig. 8 Object extracted at various resolution levels in LL subbands of a wavelet pyramid. Sequence "Lancé trousse" ©LABRI according to [26].

The similarity measure for this descriptor is a linear combination of metrics for histogram comparison (e.g. Bhattacharya coefficient):

$$\rho_{sb}^k(O_{t,i}, O_{t,j}) = \sum_x \sqrt{h_{sb}^k(O_{t,i})(x)\, h_{sb}^k(O_{t,j})(x)} \qquad (16)$$

Here sb stands for LL or HF and x is the bin index. Finally, the similarity measure is expressed as

$$\rho(O_{t,i}, O_{t,j}) = \alpha \rho_{LL} + (1 - \alpha)\rho_{HF} \qquad (17)$$

2.3 On Scalable Content-Based Queries

A vast literature is devoted to image retrieval in large databases and much work on video retrieval is being done. In our previous work we specifically focused on retrieval of objects in video content [14]. In this chapter two questions in the object-based framework: query by clip and scalable queries are addressed.

The retrieval scenario considered consists of searching for a clip in a HD video database containing a query object. This scenario can for instance be used for detection of a fraudulent post-production, where an object is extracted from a video clip frame by frame and inserted into the background extracted from another sequence.

Let us consider a clip C_{DB} in a video database (DB). A set of objects masks $O_{DB} = \{O_{t,i}, t = t0, t0 + \Delta t, \ldots\}$ is extracted for each object at each level of the wavelet pyramid. The histogram features H_{DB} are computed and stored as metadata.

Let us then consider a query clip C_Q and histogram features H_Q of objects extracted from this clip. The user is invited to select an image $I_{t^*} \in C_Q$ in which the object extraction result is visually the most satisfactory.

We consider both mono-level and cross-level search. In the case of mono-level search, the descriptor H_Q^k at a given pyramid level k is compared to all the descriptors available in the DB at the same level. We call this query a "mono-level" query. Hence, a clip from the DB is the response to the query clip for a given resolution if at least one of its frames is a "good" response to the query. The "goodness" of a response is measured in comparison with a given threshold. This scenario is well adapted to the scalable script in the case when the query is not transmitted with full-resolution.

The "cross-level" search consists in comparison of a query descriptor extracted at a chosen resolution level k with descriptors in DB extracted at a specified resolution level. First of all, this type of query is interesting for a "light" processing at a client side. The query object can be extracted on low resolution levels of wavelet pyramid while the high resolution descriptors in the DB will be used for retrieval at server side. Inversely, if the high-resolution descriptors are available in the original clip (e.g. stored in the video archive), it can be compared with a low-resolution collection of videos when searching for a fraudulent low-quality video.

In [26] main stream retrieval consisting of matching of SIFT descriptors extracted on object masks and the global descriptor, i.e. a pair of wavelet histograms are compared. It turns out, that firstly the HF histogram is necessary ($0 < \alpha < 1$ in

Eq. (17)). Secondly, histogram-based descriptor turns out to be more efficient both in mono-level and cross-level retrieval. Figures 9 and 10 depict the curves of inter-polated average precision on HD video database produced in the framework of the

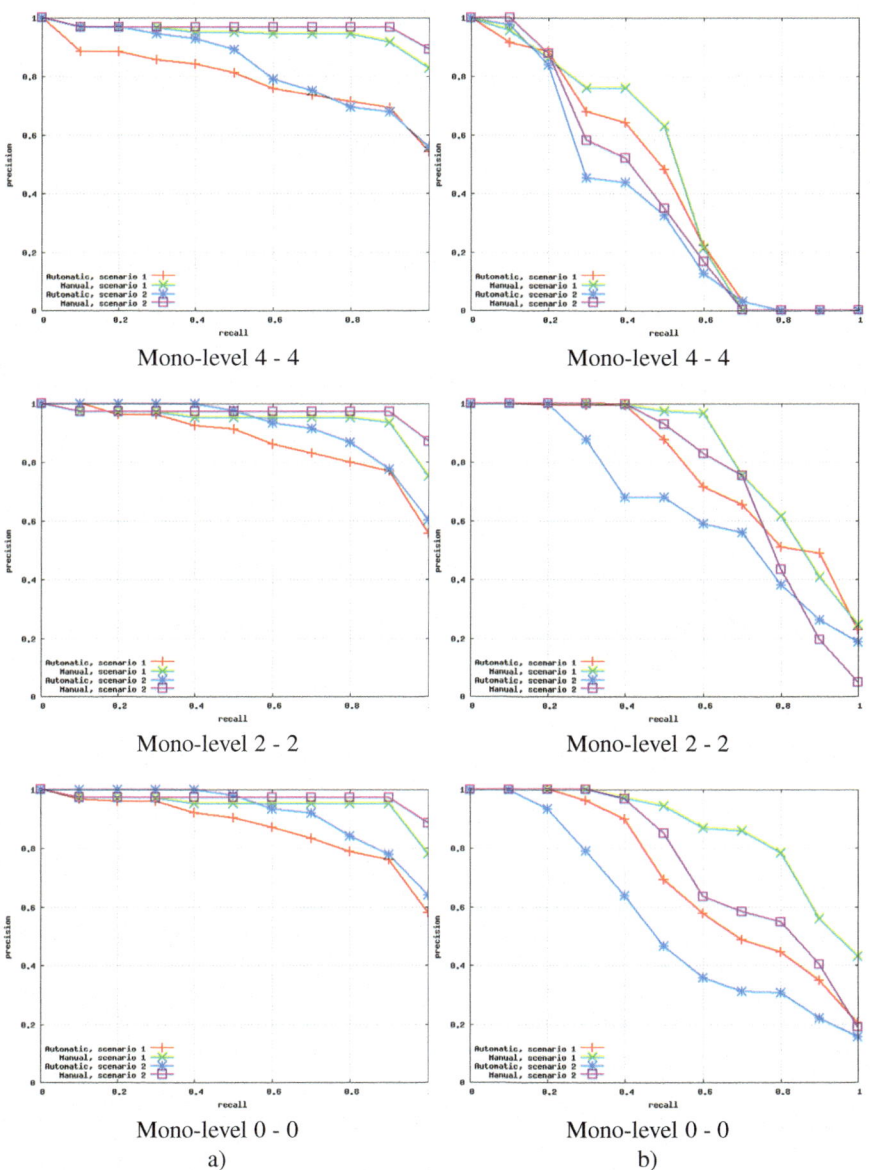

Fig. 9 Examples of mono-level queries. According to [26]. a) global object descriptors, b) SIFT descriptors)(Figure continued on Fig. 10).

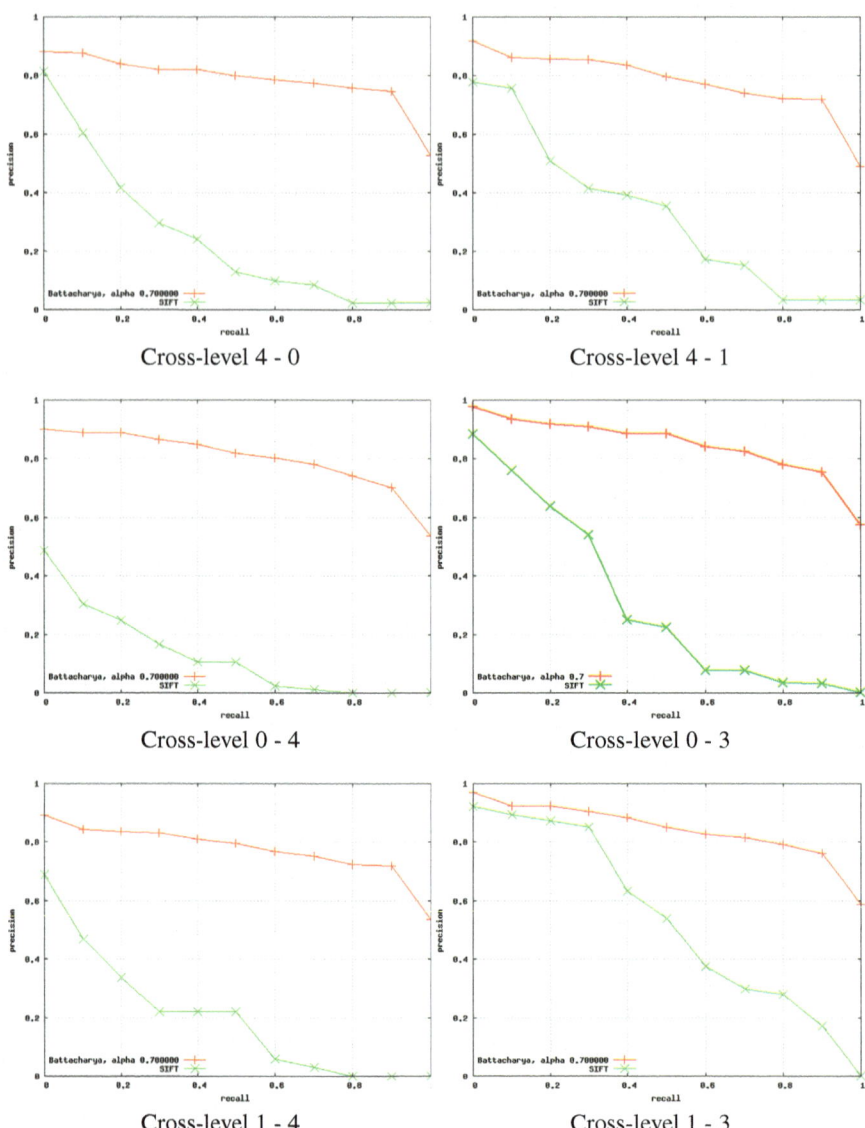

Fig. 10 Examples of cross-level queries. According to [26]. (Continuation of Fig. 9)

ICOS-HD French national research project[1] which contains affine-transformed versions of 17 HD video clips of 3 seconds. Two scenarii were designed, one with a fraudulent object and one with various sequences containing the same object. These two scenarios differ in the expected answers to the queries as summarized in

---

[1] ICOS-HD (Scalable Joint Indexing and Compression for High-Definition Video Content) is a research project funded by ANR-06-MDCA-010-03 (French Research Agency).

Table 2 Summarization of scenarios 1 and 2 settings

	Scenario 1	Scenario 2
Aim	Finding all the copies of an object in a DataBase	Finding objects similar to an object in a DataBase
Query		
Examples of answers to the query		
Examples of non-answer to the query		

table 2. (The detailed description of this corpus is presented in the following Section of this chapter.)

Both in case of manually and automatically extracted object masks the global histogram-based descriptors are more efficient in object-based queries. The methods of joint scalable coding/indexing or indexing on already encoded scalable stream presented in this chapter open an exciting perspective for efficient indexing of HD content. Indeed, the full HD resolution is not necesserily needed for fast browsing, summarizing and retrieval of HD video content. According to the architecture of emerging information systems containing HD video, various levels of granularity can be accessed, indexed and retrieved. Wavelets, the hierarchical transform widely used for actual and future scalable HD standards is an excellent basis for this.

3   HD Content Indexing Using Patch Descriptors in the Wavelet Domain

In this part of the chapter, we present a method for comparing HD video segments statistically using a sparse multiscale description of the content. This description is both spatial and temporal and relies on the following concepts: 1) a sparse and multiscale transform of the video content; 2) a local patch description obtained by grouping spatially or temporally coherent pieces of information; 3) the multiple occurrences of similar patches throughout the video lead to a global statistical description of the video content that is robust to the usual geometric or radiometric video transformations. The comparison of these descriptors is naturally done in a statistical fashion. The global dissimilarity proposed is a weighted combination of Kullback-Leibler divergences between the probability densities of the different kinds of patches. The estimation of the dissimilarity is done non-parametrically in a k-th nearest neighbor context, which enables us to cope with the high-dimensionality of the probability density functions at stake. This method is designed to compare short video segments (e.g. Groups of Pictures (GOPs) of eight frames), with the understanding that to compare larger videos we sum up the dissimilarities between their consecutive GOPs. In the sequel, we describe how to extract the video description from GOPs and how to estimate the proposed dissimilarity. Finally, we report results obtained on content-based queries experiments.

3.1   Sparse Multiscale Patches and Motion Patches Descriptors

Our description of a GOP extracts separately a spatial information relative to the scene and a temporal information relative to the motion within the GOP. The spatial information is extracted from the first frame of the GOP while the temporal information is extracted from the motion of blocks throughout the GOP.

3.1.1   Spatial Descriptors: Sparse Multiscale Patches (SMP)

A structure in an image I can be identified by the coherence (or correlation) of the multiscale coefficients of I around a particular location p and a particular spatial scale k. A patch of the *sparse multiscale patches* (*SMP*) description of an image I [30, 31] is a group of multiscale coefficients that vary simultaneously in presence of a spatial structure: these are coefficients of all color channels that are neighbors across scale and location.

More precisely, we write $w_{k,p}^{Ic}$ for the multiscale coefficient of channel c of image I at scale k and location p (this would be the dot product of channel c of image I with a waveform of scale k centered at location p). With the detail coefficients

Fig. 11 Building a patch of multiscale coefficients, for a single color channel image

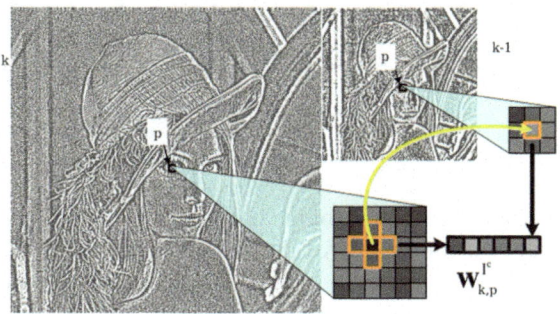

(i.e band-pass or high-pass subbands of the multiscale transform), we build inter-channel and interscale patches. To do so, in each color channel, we first group the coefficients of closest scale and location (see Fig. 11):

$$\mathbf{w}_{k,p}^{I^c} = \left(w_{k,p}^{I^c}, w_{k,p+(1,0)}^{I^c}, w_{k,p-(1,0)}^{I^c}, w_{k,p+(0,1)}^{I^c}, w_{k,p-(0,1)}^{I^c}, w_{k-1,p}^{I^c} \right); \qquad (18)$$

and then build interchannel patches $\mathbf{W}_{k,p}^{I}$ by concatenating the patches of the three color channels (YUV):

$$\mathbf{W}_{k,p}^{I} = \left(\mathbf{w}_{k,p}^{I^Y}, \mathbf{w}_{k,p}^{I^U}, \mathbf{w}_{k,p}^{I^V} \right). \qquad (19)$$

With the approximation coefficients (i.e the low-pass subband of the multiscale transform), we build interchannel and intrascale patches by concatenating across channels the 3 by 3 neighborhoods of the low-frequency coefficients (making patches of length 27). We denote by $\mathbf{W}_{k,p}^{I}$ either a low-pass or a high-pass or band-pass patch. We use the Laplacian pyramid as the multiscale transform of the images for its low redundancy and near invariance properties.

The multiscale patches description obtained is the set of all patches $\mathbf{W}_{k,p}^{I}$ for all scales k and locations p. It is said to be sparse because 1) the set of patches of large energy (sum of squared coefficients) is a small subset of the set of all multiscale patches $\{\mathbf{W}_{k,p}^{I}\}_{0 \leq k \leq K-1, p \in \mathbb{Z}}$ and 2) this small subset describes well the content of the image (this is a sparsity property inherited from the sparsity of the multiscale decomposition: a small group yields a good representation). We select the so-called *sparse multiscale patches* by thresholding the energy level at each scale k and thus obtain spatial descriptors of a frame of the video (see Section 3.3 for specific values of thresholds).

3.1.2 Temporal Descriptors: GOP Motion Patches (GOP-MP)

To capture the motion information in a GOP, we also use the concept of patches built on coherent information. Here, the coherence is sought through time: the patches are made of motion vectors that follow motion through the GOP. One patch of the GOP Motion Patches (*GOP-MP*) description captures the temporal coherence within the GOP at a particular location $p = (x, y)$ by encoding the motion of the block centered

Fig. 12 Building a motion
patch

$$m(\text{x,y}) = \big(x, y, \mathbf{u}_{1,2}(\text{x,y}), \mathbf{u}_{2,3}(\text{x,y}), \dots, \mathbf{u}_{n-1,n}(\text{x,y})\big) \tag{20}$$

at $p = (x, y)$ (the size of the blocks corresponds to the lowest scale of the multiscale decomposition used for the spatial description). More precisely, for a GOP of n consecutive frames $f_1, \dots, f_n$, we compute the following motion patches for each block of center (x,y):

where $\mathbf{u}_{n-1,n}(\text{x,y})$ is the apparent motion of the block centered at (x,y) from frame f_{n-1} to frame f_n (see Fig. 12). The motion vectors $\mathbf{u}$ are computed via a diamond-search block matching algorithm. For each GOP studied, we compute the motion patches $m(\text{x,y})$ for each block (x,y). Note that we include in the motion patch its location (x,y) so that each patch has length $2n$ (which is 16 for GOPs of 8 frames).

As is the case for spatial patches, in fact only a few motion patches effectively describe motion (sparsity). Thus, we select the significant motion patches by a thresholding that keeps only the patches having the largest motion amplitude (sum of squares of the $\mathbf{u}$ components in Eq. (20)). (The threshold value used in Section 3.3 is zero: the motion patches kept are those for which the motion amplitude is non-zero).

3.2 Using the Kullback-Leibler Divergence as a Similarity Measure

3.2.1 Motivation and Expression

As mentioned in Section 3, the comparison between two HD video segments is performed by statistically measuring the dissimilarity between their respective sets of (spatial and temporal) descriptors within the successive GOPs. Indeed, the scale and location of the descriptors extracted in each segment will not match in general even if the segments are visually similar. Therefore, a dissimilarity based on one-to-one distance measures is not adequate. Instead, it is more appropriate to consider each set of descriptors as a set of realizations of a multidimensional random variable characterized by a particular probability density function (PDF), and to measure the dissimilarity between these PDFs. Because the descriptors were defined in high-dimensional spaces, PDF estimation is problematic. The k-th nearest neighbor (kNN) framework provides interesting estimators in this context [32, 33, 34]. First, they are less sensitive to the curse of dimensionality. Second, they are expressed directly in terms of the realizations. Besides a PDF estimator, a consistent, asymptotically unbiased entropy estimator has been proposed [35, 36, 37]. To compare two PDFs in this framework, entropy-based measures then appear as a good option. We chose the Kullback-Leibler divergence because it proved to be successful in similar applications [38, 39]. Since the descriptors are heterogenous (SMPs, low-frequency patches, and motion patches), several such divergences will be combined.

Let us assume that the two video segments to be compared are both composed of a single GOP. One of them will be referred to as the query and denoted by G_Q, the other one as the reference, G_R. The dissimilarity D between G_Q and G_R is defined as

$$D(G_Q, G_R) = \alpha_s \underbrace{D_s(G_Q, G_R)}_{\text{spatial term}} + \alpha_t \underbrace{D_t(G_Q, G_R)}_{\text{temporal term}} \qquad (21)$$

where

$$\begin{cases} D_s(G_Q, G_R) = \displaystyle\sum_{0 \le k \le K-1} D_{\text{KL}}\big(p_k(G_Q) \| p_k(G_R)\big) \\ D_t(G_Q, G_R) = D_{\text{KL}}\big(p_m(G_Q) \| p_m(G_R)\big) \end{cases} \qquad (22)$$

The positive parameters α_s and α_t allow us to tune the relative influence of the spatial and temporal terms. The scale $k = 0$ is the coarsest scale of the decomposition corresponding to the low-pass subband. $p_k(G)$, respectively $p_m(G)$, denotes the PDF underlying the SMPs or low-frequency patches $\{\mathbf{W}_{k,p}^l, p\}$, respectively the motion patches $\{m_p, p\}$, extracted from the GOP G. Finally, D_{KL} denotes the Kullback-Leibler divergence.

The term of scale k in the sum D_s can be interpreted as a measure of how dissimilar local spatial structures are at this scale in the respective key frames of G_Q and G_R. Overall, D_s indicates whether some objects are present in both frames. Since the motion patches group together motion vectors and their respective location, the temporal term D_t not only tells about how the motions throughout the GOPs compare; it also tells (roughly) whether similar shapes move the same way in both GOPs.

Let us now see how the Kullback-Leibler divergences involved in the definition of D can be conveniently estimated from a set of realizations.

3.2.2 Estimation in the kNN Framework

Estimation of the Kullback-Leibler divergence between two PDFs p and q when these PDFs are *known* only through two respective sets of realizations U and V apparently requires prior estimation of the PDFs. Because the realizations are vectors of high-dimension (18 for high-pass and band-pass SMPs, 27 for low-pass patches, and 16 for motion patches), PDF estimation is afflicted with the curse of dimensionality [34]. Assuming that an accurate parametric model of the PDFs can be built anyway (for example, a mixture of Gaussians), an analytic expression of the divergence in terms of the model parameters exists only for some restricted cases such as mixtures composed of a unique Gaussian. Alternatively, an entropy estimator written directly in terms of the realizations has been proposed in the kNN framework [35, 36, 37]. Then, the Kullback-Leibler divergence being the difference between a cross-entropy and an entropy, the divergences D_{KL} involved in (22) can be expressed as functions of the sets of patches $\{\mathbf{W}_{k,p}^l, p\}$, for each scale k, and $\{m_p, p\}$.

To give an intuition of the kNN entropy estimator, let us mention that it can be considered as the combination of the kNN PDF estimator $\hat{p}$ with the Ahmad-Lin entropy approximation H_{AL} [40]

$$\begin{cases} \hat{p}(x) = \displaystyle\sum_{w \in W} \frac{1}{|W| v_d \rho_k^d(W,x)} \, \delta\left[|x-w| < \rho_k(W,x)\right] \\[4mm] H_{AL}(W) = -\dfrac{1}{|W|} \displaystyle\sum_{w \in W} \log p(w) \end{cases} \tag{23}$$

where x is an element of $\mathbb{R}^d$, W is a set of d-dimensional realizations whose underlying PDF is p, $|W|$ is the cardinality of W, v_d is the volume of the unit ball in $\mathbb{R}^d$, $\rho_k(W,x)$ is the distance between x and its k-th nearest neighbor among the elements of W, and $\delta(B)$ is equal to 1 if B is true and zero otherwise. Replacing p in H_{AL} with $\hat{p}$ leads to a (biased) kNN-based entropy estimator close to the unbiased version proposed in [35, 36, 37].

Subtracting the kNN entropy estimation from the kNN cross-entropy estimation leads to the following kNN Kullback-Leibler estimation:

$$D_{KL}(U\|V) = \log \frac{|V|}{|U|-1} + \frac{d}{|U|} \sum_{u \in U} \log \rho_k(V,u) - \frac{d}{|U|} \sum_{u \in U} \log \rho_k(U,u) \, . \tag{24}$$

3.3 Scalable Content-Based Queries with Patches Descriptors

In this section we assess the quality of the proposed GOP dissimilarity measure for the retrieval problem. The experiments were performed on video sequences from the ICOS-HD project database. After a brief description of the database, we analyze retrieval results based on spatial frame descriptors alone, temporal/motion descriptors alone, and both sets of descriptors combined together.

3.3.1 ICOS-HD Video Database

The ICOS-HD project provides a large database of both original full HD videos and edited versions. Each original sequence contains 72 Full HD frames (1920×1080 pixels) and has been manually split up into clips, such that the boundary between the clips roughly corresponds to a relevant motion transition. In addition, common geometric and radiometric deformations were applied to the original HD video sequences, thus obtaining different versions of each video clip.

For these experiments, we used ten video sequences (see some thumbnails in Figure 13). The deformations we considered are scaling and quality degradation by high JPEG2000 compression, for a totla of four different versions of each video clip:

- original Full HD (1920×1080 pixels), referenced as 1920 in the figures;
- two rescaled versions (960×540 pixels), referenced as 960;
- two JPEG2000 coded versions (low and very low quality) referenced as jpeg-ql and jpeg-q10.

Fig. 13 Thumbnails of
two video sequences. Left:
"Man in Restaurant", and
right: "Street with trees
and bicycle". Original HD
sequences ©Warner Bros
issued from the Dolby 4-4-4
Film Content Kit One.

As explained in Section 3.1, we used GOPs of 8 consecutive frames as basic units of video information to extract spatial and temporal descriptors for each clip. The spatial *SMP* descriptors were extracted from the first frame of each GOP using five resolution levels of the Laplacian pyramid as well as the low-frequency residual. The thresholds were set to keep $1/6$ of the patches at each scale, except for the lowest one where all patches were used. The temporal descriptors were extracted using a diamond-search block matching algorithm to estimate inter-frame motion vectors on 16×16.

3.3.2   Spatial Dissimilarity

We consider the task of retrieving the GOPs most similar to a query GOP. Hence all transformed versions of the query GOP itself are expected to be ranked first by the dissimilarity measure defined above. The dissimilarity measure D between a query GOP G_Q and a reference GOP G_R as defined in Eq. (21) is a combination of a spatial term D_s taking into account only spatial features and a temporal term D_t defined over temporal features. While the spatial descriptors are essentially useful for comparing statistical scene information of two video pieces, motion descriptors are expected to highlight similarities based on dynamical patterns like the movement of objects or persons in a scene. In order to appropriately choose the weighting factors α_1 and α_2 in Eq. (21), we studied the spatial and temporal parts of the measure separately first.

Firstly we considered only the spatial descriptors ($\alpha_1 = 1$, $\alpha_2 = 0$) to retrieve similar GOPs. The *SMP* descriptors prove to be crucial for distinguishing GOPs of the same video sequence as the query from those belonging to different video sequences. The results obtained are shown in Figure 14. In this figure each curve shows the dissimilarity between a fixed query GOP and all GOP from 2 clips of the same sequence and one clip of a different sequence in all possible versions. The query GOP is the first GOP of the first clip of either "Man in Restaurant" or "Street with Bicycle an Trees". A particular reference GOP is identified by the sequence, clip and version indicated in the middle rectangles of the figure, and by the GOP label on the x-axis, the 9 GOPs of a particular clip being ordered chronologically.

Even when frame transformations are applied - either rescaling and very lossy compression - all GOPs originating from the same video sequence sequence are far more dissimilar to the query. These results confirm that *SMP* descriptors are relevant for retrieving video scenes that share overall visual similarity with a query scene, and show in particular that the spatial part of the measure is robust to scaling and very lossy compression (spatial scalability).

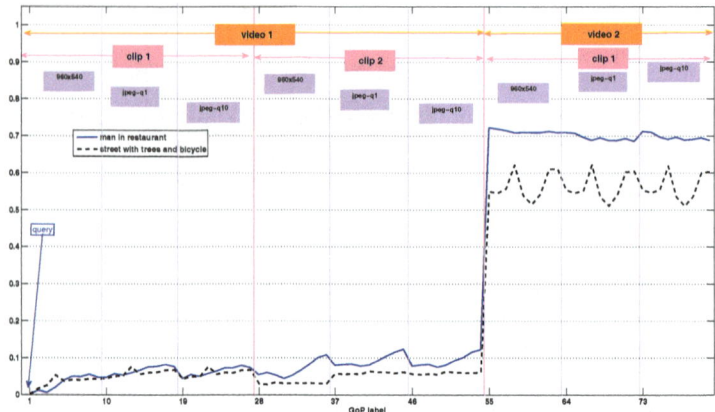

Fig. 14 GOP retrieval based on spatial descriptors. The query is GOP 1 from clip 1 of version 960 of "Man in Restaurant" (plain curve) or "Street with Trees and Bicycle" (dotted curve).

3.3.3 Temporal Dissimilarity

We now analyze the dissimilarity measure of Eq. (21) using only motion descriptors ($\alpha_1 = 0$, $\alpha_2 = 1$). Since the different clips of each sequence in our database differ from each other mainly with respect to motion information, this measure is expected to discriminate GOPs of different clips of the same video sequence. This is confirmed by the experimental results shown in Figure 15, which show the motion dissimilarity between a fixed query GOP and all GOPs of the two clips of the same sequence as

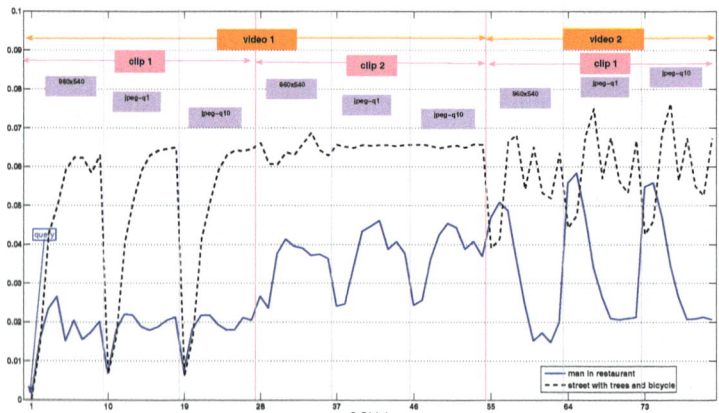

Fig. 15 GOP retrieval based on motion descriptors. The query is GOP 1 from clip 1 of version 960 of "Man in Restaurant" (plain curve) or "Street with Trees and Bicycle" (dotted curve).

Table 3 Mean and variance of the spatial and temporal dissimilarities

	Spatial term (across scenes)	Spatial term (within a scene)	Temporal term
Mean	122.8	12.1	3.7
Standard deviation	1.7	4.7	2.5

well as a clip of a different sequence in all versions (same labeling of the reference GOPs as for Fig. 14). As expected, the GOPs from the same sequence as the query and that are close in time to the query have far smaller dissimilarity values than those originating from the second clip. As previously, we note that the temporal part of the measure is robust to scaling and lossy compression (spatial scalability).

3.3.4 Spatio-Temporal Dissimilarity

Considering that the spatial term of the dissimilarity is able to differentiate video scenes and the temporal term allows us to characterize different motions within a single sequence, we expect that the combination of the two will enable us to globally compare two clips whether there are from the same sequence or not. The typical ranges and variances of the spatial and temporal similarities are quite different (see Table 3). As seen from the previous experiments, the spatial term is not discriminative within a scene but shows a clear discontinuity marking the difference between scenes, while the temporal term differentiates GOPs within a video. We thus rescale the temporal term to ensure that on average it modulates the spatial term within a scene without breaking the discontinuity across scenes. To do so, we

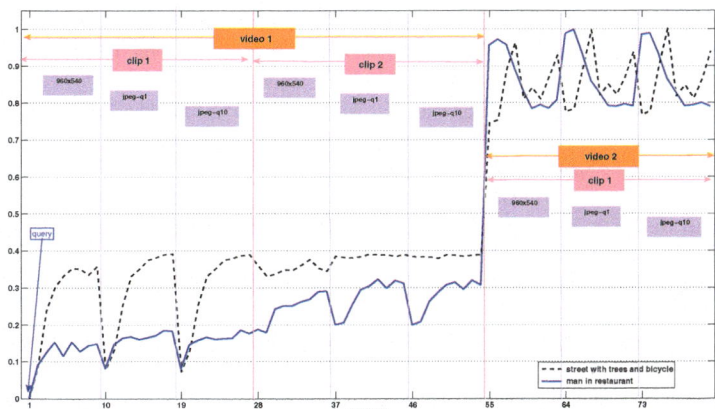

Fig. 16 GOP retrieval combining spatial (weight $\alpha_1 = 1$) and temporal (weight $\alpha_2 = 10$) dissimilarities. The query is GOP 1 from clip 1 of version 960 of "Man in Restaurant" (plain curve) or "Street with Trees and Bicycle" (dotted curve).

set $\alpha_1 = 1$, $\alpha_2 = 10$. The results displayed in Fig. 16 indeed show that the two clips within a sequence are discriminated independently of which degradation is applied to the reference GOP.

4 Conclusion and Perspectives

This chapter has presented an overview of two frameworks for scalable video indexing based on hierarchical decomposition. The first deals directly with the video content in the compressed domain, i.e. the domain defined by the given compression standard, while the second relies on a hierarchical decomposition of the decoded data. Both approaches have their own advantages and drawbacks. Techniques working in the compressed domain provide potentially faster processing since basic features such as the motion vectors are readily available; however, the features are computed so as to achieve the best quality-versus-compression ratio trade-off and might not ideally serve the purpose of indexing. Techniques processing decoded data, on the other hand, give full freedom to design ad-hoc multiscale descriptors; however they are significantly computationally slower due to the full decoding followed by the chosen hierarchical decomposition. Both approaches open an exciting perspective for efficient processing of HD content such as indexing but also fast browsing, summarizing and retrieval of HD video content by exploiting the various levels of granularity that can be accessed in new information systems.

References

1. JPEG 2000 image coding system: Motion JPEG 2000, ISO/IEC. 15444-3:2007. Information technology edn.
2. Morand, C., Benois-Pineau, J., Domenger, J.P., Mansencal, B.: Object-based indexing of compressed video content: From sd to hd video. In: International Conference on Image Analysis and Processing Workshops, Modena, Italy (2007)
3. Adami, N., Boschetti, A., Leonardi, R., Migliorati, P.: Embedded indexing in scalable video coding. In: 7th International Workshop on Content-Based Multimedia Indexing, Chania, Crete (2009)
4. Morand, C., Benois-Pineau, J., Domenger, J.P.: HD motion estimation in a wavelet pyramid in JPEG 2000 context. In: 5th International Conference on Image Processing, Chicago, IL, USA (2008)
5. JPEG 2000 image coding system: Core coding system, ISO/IEC 15444-1:2004. Information technology edn.
6. Digital Cinema Initiative, http://www.dcimovies.com/ (accessed November 9, 2009)
7. Pearson, G., Gill, M.: An evaluation of Motion JPEG 2000 for video archiving. In: Archiving, Washington, D.C., USA (2005)
8. Coding of audio-visual objects – Part 2: Visual (MPEG4), ISO/IEC 14496-2:2004. Information technology edn.
9. Wang, Y., Hannuksela, M., Gabbouj, M.: Error-robust inter/intra mode selection using isolated regions. In: Packet Video (PV), Nantes, France (2003)

10. Totozafiny, T., Patrouix, O., Luthon, F., Coutellier, J.M.: Dynamic background segmentation for remote reference image updating within motion detection JPEG 2000. In: IEEE International Symposium on Industrial Electronics (ISIE), Montreal, Canada (2006)

11. Manerba, F., Benois-Pineau, J., Leonardi, R.: Extraction of foreground objects from MPEG2 video stream in rough indexing framework. In: SPIE Storage and Retrieval Methods and Applications for Multimedia, San Jose, CA, USA (2004)

12. Mallat, S.: A theory for multiresolution signal decomposition: The wavelet representation. IEEE Transactions on Pattern Analysis and Machine Intelligence 11(7), 674–693 (1989)

13. Ćalić, J., Mrak, M., Kondoz, A.: Dynamic layout of visual summaries for scalable video. In: 6th International Workshop on Content-Based Multimedia Indexing, London, UK (2008)

14. Chevalier, F., Domenger, J.P., Benois-Pineau, J., Delest, M.: Retrieval of objects in video by similarity based on graph matching. Pattern Recognition Letters 28(8), 939–949 (2007)

15. TREC Video Retrieval Evaluation, http://www-nlpir.nist.gov/projects/trecvid/ (accessed November 9, 2009)

16. Dufaux, F., Ansorge, M., Ebrahimi, T.: Overview of JPSearch. A standard for image search and retrieval. In: 5th International Workshop on Content-Based Multimedia Indexing, Bordeaux, France (2007)

17. Saraceno, C., Leonardi, R.: Indexing audio-visual databases through a joint audio and video processing. International Journal of Imaging Systems and Technology 9(5), 320–331 (1998)

18. Benois-Pineau, J., Dupuy, W., Barba, D.: Recovering visual scenarios in movies by motion analysis and grouping of spatio-temporal signatures of shots. In: 2nd International Conference in Fuzzy Logic and Technology (EUSFLAT), Leicester, UK (2001)

19. Nilsback, M.E., Zisserman, A.: A visual vocabulary for flower classification. In: IEEE Comp. Soc. Conf. on Computer Vision and Pattern Recognition (CVPR), New York, NY, USA (2006)

20. Lowe, D.G.: Distinctive image features from scale-invariant keypoints. International Journal of Computer Vision 60(2), 91–110 (2004)

21. Tuytelaars, T., Lampert, C.H., Blaschko, M.B., Buntine, W.: Unsupervised object discovery: a comparison. International Journal on Computer Vision (2009) (Online First version)

22. Multimedia content description interface – Part 3: Visual (MPEG7), ISO/IEC 15938-3:2002. Information technology edn.

23. Benois-Pineau, J., Morier, F., Barba, D., Sanson, H.: Hierarchical segmentation of video sequences for content manipulation and adaptive coding. Signal Processing 66(2), 181–201 (1998)

24. Salembier, P., Marqués, F., Pardàs, M., Morros, J., Corset, I., Jeannin, S., Bouchard, L., Meyer, F., Marcotegui, B.: Segmentation-based video coding system allowing the manipulation of objects. IEEE transactions on circuits and systems for video technology 7(1), 60–74 (1997)

25. Jehan-Besson, S., Barlaud, M., Aubert, G.: A 3-step algorithm using region-based active contours for video objects detection. EURASIP Journal on Applied Signal Processing 2002(6), 572–581 (2002)

26. Morand, C., Benois-Pineau, J., Domenger, J.P., Zepeda, J., Kijak, E., Guillemot, C.: Scalable object-based video retrieval in HD video databases. Submitted to Signal Processing: Image Communication (2009)

27. Liu, Y., Ngi Ngan, K.: Fast multiresolution motion estimation algorithms for wavelet-based scalable video coding. Signal Processing: Image Communication 22, 448–465 (2007)
28. Liu, Y., Ngan, K.N.: Fast multiresolution motion estimation algorithms for wavelet-based scalable video coding. Signal Processing: Image Communication 22(5), 448–465 (2007)
29. Sturges, H.A.: The choice of a class interval. Journal of the American Statistical Association 21(153), 65–66 (1926)
30. Piro, P., Anthoine, S., Debreuve, E., Barlaud, M.: Image retrieval via kullback-leibler divergence of patches of multiscale coefficients in the knn framework. In: 6th International Workshop on Content-Based Multimedia Indexing, London, UK (2008)
31. Piro, P., Anthoine, S., Debreuve, E., Barlaud, M.: Sparse Multiscale Patches for Image Processing. In: Nielsen, F. (ed.) Emerging Trends in Visual Computing. LNCS, vol. 5416, pp. 284–304. Springer, Heidelberg (2009)
32. Fukunaga, K., Hostetler, L.: The estimation of the gradient of a density function, with applications in pattern recognition. IEEE Transactions on Information Theory 21(1), 32–40 (1975)
33. Fukunaga, K.: Introduction to statistical pattern recognition, 2nd edn. Academic Press Professional, Inc., San Diego (1990)
34. Terrell, G.R., Scott, D.W.: Variable kernel density estimation. The Annals of Statistics 20(3), 1236–1265 (1992)
35. Kozachenko, L.F., Leonenko, N.: On statistical estimation of entropy of random vector. Problems of Information Transmission 23(2), 95–101 (1987)
36. Goria, M., Leonenko, N., Mergel, V., Novi Inverardi, P.: A new class of random vector entropy estimators and its applications in testing statistical hypotheses. Journal of Nonparametric Statistics 17(3), 277–298 (2005)
37. Leonenko, N., Pronzato, L., Savani, V.: A class of Rényi information estimators for multidimensional densities. Annals of Statistics 36(5), 2153–2182 (2008)
38. Hero, A.O., Ma, B., Michel, O., Gorman, J.: Alpha-divergence for classification, indexing and retrieval. Tech. Rep. CSPL-328, University of Michigan (2001)
39. Do, M., Vetterli, M.: Wavelet based texture retrieval using generalized Gaussian density and Kullback-Leibler distance. IEEE Transactions on Image Processing 11(2), 146–158 (2002)
40. Ahmad, I., Lin, P.E.: A nonparametric estimation of the entropy for absolutely continuous distributions. IEEE Transactions on Information Theory 22(3), 372–375 (1976)

Chapter 23
Stereo Correspondence in Information Retrieval

Huiyu Zhou and Abdul H. Sadka

Abstract. Stereo correspondence is a very important problem in information retrieval. Optimal stereo correspondence algorithms are used to generate optimal disparity maps as well as accurate 3-D shapes from 2-D image inputs. Most established algorithms utilise local measurements such as image intensity (or colour) and phase, and then integrate the data from multiple pixels using a smoothness constraint. This strategy applies fixed or adaptive windows to achieve certain performance. To build up appropriate stereo correspondences, a global approach must be implemented in the way that a global energy or cost function is designed by considering template matching, smoothness constraints and/or penalties for data loss (e.g. occlusion). This energy function usually works with optimisation methods like dynamic programming, simulated annealing and graph cuts to reach the correspondence. In this book chapter, some recently developed stereo correspondence algorithms will be summarised. In particular, maximum likelihood estimation-based, segment-based, connectivity-based and wide-baseline stereo algorithms using descriptors will be introduced. Their performance in different image pairs will be demonstrated and compared. Finally, future research developments of these algorithms will be pointed out.

1 Introduction

Digital video cameras are widely used in our community, and the quantity of digital videos has significantly increased up to date. For the reuse and storage purpose, consumers have to retrieve a video from a large number of multimedia resources. To

Huiyu Zhou
Queen's University Belfast, Belfast, BT3 9DT, United Kingdom
e-mail: H.Zhou@ecit.qub.ac.uk

Abdul H. Sadka
Brunel University, Uxbridge, UB8 3PH, United Kingdom
e-mail: Abdul.Sadka@brunel.ac.uk

seek similar videos from a definite database, information retrieval systems have been established with promising performance in searching accuracy and efficiency, e.g. [1]. Many of these established systems attempt to search for videos that have been annotated with metadata *a priori* (e.g. [2]). Nevertheless, there are still a significant number of footages that have been recorded but not ever used [3]. These footages normally have not been properly annotated, and hence the retrieval can only be carried out according to the video contents rather than the annotated information.

Of the generic video contents, the need for the ability to retrieve 3-D models from databases or the Internet has gained dramatic prominence. Content-based 3-D model retrieval currently remains a hot research area, and has found its tremendous applications in computer animation, medical imaging, and security. To effectively extract a 3-D object, shape-based 3-D modelling (e.g. [4]) and similarity or dissimilarity (or distance) computation (e.g. [5]) are two of the main research areas. In this chapter, we review the algorithms that have been recently developed for the reconstruction of 3-D shapes from 2-D video sequences. This work is inspired by the fact that the estimation of 3-D shapes critically affects the retrieval quality of 3-D models. We believe that the introduction to these summarised approaches here will be used to effectively facilitate the application of 3-D model retrieval in the databases or Internet. However, this potential application is beyond the scope of the current report and omitted in the current report.

One of the commonly used strategies to recover 3-D shapes is the use of multiple view reconstruction. For example, Bartoli and Sturm [6] used Plucker coordinates to represent the 3-D lines in the scope of maximum likelihood estimation, and then they proposed an orthonormal representation to challenge the bundle adjustment problem. Zhou *et al.* [7] conducted co-planarity checks using cross-ratio invariants and periodic analysis of the triangular regions. Klaus *et al.* [8] presents a segment-based method to extract the regions of homogeneous colours, followed by local window based matching, plane fitting and disparity assignment. Similar approaches have been introduced in [9], [10]. Sun *et al.* [11] reported a stereo mathcing algorithm using Bayesian belief propagation. The stereo problem was solved by taking into account the three Markov random fields: a smooth field for depth/disparity, a line process for depth discontinuity and binary process for occlusion. An iterative RANSAC plane fitting strategy reported in [12] shows a maximum likelihood estimation approach. This technique enables one to obtain the best plane fitting to the generated 3-D points automatically rather than using empirical criteria, which is determined according to a limited number of image samples.

Regarding the non-linear surface reconstruction from motion, Laurentini reported the visual hull as the largest volume consistent with the contours that have been observed from several viewpoints [13]. This approach ignores the small details but capture the approximate shape of the scene. Roy and Cox [14] introduced a method using the graph flow theory to generalise the purely 1-D dynamic programming technique to the 2-D problem raised by disparity maps. Kolmogorov and Zabih [15] a graph cuts based general theory tp disparity maps in the multi-view context. Narayanan *et al.* [16] reconstructed several depth maps that are aggregated into a single structure. Hoff and Ahuja [17] constructed a disparity map by gathering the information of a few quadratic patches.

From the next section, we are going to briefly summarise several important established stereo matching algorithms. The performance of these schemes will be demonstrated in the evaluation section, where the characteristics of each algorithm can be clearly observed. First of all, we start from a flat surface detection algorithm [12] where an improved random sampling census algorithm is integrated for better estimates of planar surfaces. A segment based stereo matching algorithm using belief propagation and self adapting dissimilarity measure will be introduced [8]. Afterwards, a connectivity based stereo matching algorithm is introduced. This approach integrates the stereo correspondence with shape segmentation in order to reach higher accuracy than the classical approaches. Finally, a wide baseline stereo corresponding algorithm using local descriptors is presented. Evaluation of these established algorithms will be provided before conclusions and future work are given.

2   Maximum Likelihood Estimation for Flat Surface Detection

This planar determination algorithm starts with corner feature detection using two neighboring frames in a monocular video sequence. Given the epipolar geometry constraint, we then build up dense matching between these two groups of points of interest using the sum squared of differences (SSD) correlation method. Assuming a calibrated camera (used to collect this sequence), we then compute a depth map, based on the estimated disparity map. If there is only one single flat surface in the scene (this constraint can only be satisfied in a small image region in many applications), we can launch a RANSAC algorithm [18] to fit a plane to the available three-dimensional points. This RANSAC operation is iterated in an expectation-maximisation context for seeking global minimal errors, which is the main contribution of our work. The algorithmic flowchart is illustrated in Fig. 1. Note that the proposed strategy works in the presence of motion parallax. To retrieve planes from uncalibrated scenes, a fast multiple-view reconstruction strategy, based on the algorithm presented here, will be explored in a future work.

2.1   Estimation of a Depth Map

Before a plane fitting scheme starts, 3-D point sets need to be generated based on the 2-D image inputs. Of two neighboring images, we consider the later image is the shifted one from the previous image. Given a shift $(\triangle x, \triangle y)$ and an image point (x,y) in a previous frame, the auto-correlation function for similarity check across frames is defined as $c(x,y) = \sum_W [I(x_i,y_i) - I(x_i + \triangle x, y_i + \triangle y)]^2$, where $I(\cdot)$ denotes the image function and (x_i,y_i) are the image points in the window W (Gaussian) centred at (x,y). The shifted image can be approximated by a Taylor expansion as follows, $I(x_i + \triangle x, y_i + \triangle y) \approx I(x_i,y_i) + [I_x(x_i,y_i), I_y(x_i,y_i)] \begin{bmatrix} \triangle x \\ \triangle y \end{bmatrix}$, where $I_x(\cdot)$ and $I_y(\cdot)$ denote the partial derivations along x and y, respectively. Eventually, we have

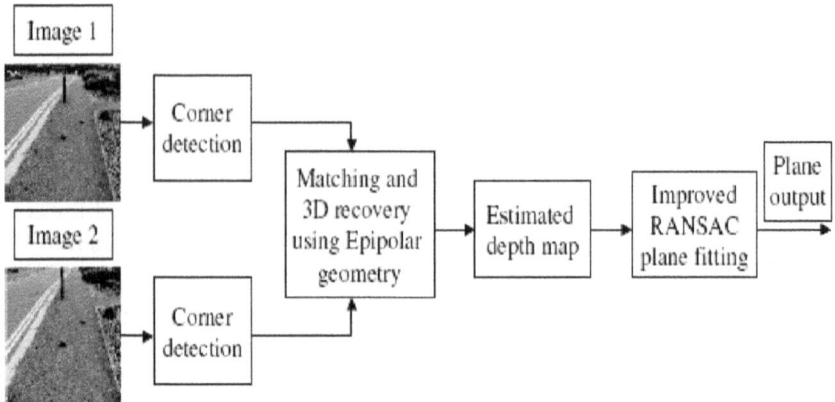

Fig. 1 Flowchart of the maximum likelihood estimation based flat surface detection algorithm

$c(x,y) = [\triangle x, \triangle y] C(x,y) \begin{bmatrix} \triangle x \\ \triangle y \end{bmatrix}$, where $C(x,y)$ represents the intensity structure of the local neighborhood. Let λ_1 and λ_2 be two eigenvalues of matrix $C(x,y)$. A corner point can be detected if $\min(\lambda_1, \lambda_2)$ is larger than a pre-defined threshold.

Once holding the points of interest, we then apply the sum squared of differences correlation method to match these corner features. Using the matched features, we exploit the well-established epipolar constraints to further refine the correspondence of features. The camera parameters are then used for recovering the scene geometry [7]. As an example, Fig. 2(a) and (b) show the original images superimposed by the extracted corner features using the Harris corner detector [19], (c) is the disparity map and (d) refers to the estimated depth map according to the relationship: $D = fd/z$, where D is depth to be computed, f focal length, d introcular distance and z estimated disparity.

2.2 Iterative RANSAC Planar Surface Detection

RANSAC planar estimation is supposed to effectively work in the presence of data outliers. This method starts from fitting a plane to a set of 3 points (considered as inliers) randomly selected from the matched corner features. Other image points are then evaluated using the Euclidean distances between these 3-D points and the fitted plane. If the points fall in a pre-defined region, then they will be classified as inliers. Otherwise, the points will be removed from the consideration of coplanarity. These steps are repeated until a count limit is reached. In a classical RANSAC plane fitting approach, the iteration is terminated by either a user-specified number or the number of outliers falling below a pre-defined threshold. This heuristic trick cannot handle general situations, where either under- or over-estimation usually appears.

(a)   (b)

(c)   (d)

Fig. 2 Estimation of disparity and depth maps: (a) and (b) feature extraction, (c) disparity map and (d) depth map

We here intend to find a strategy to achieve maximum likelihood estimation to the flat surfaces. Let N independent samples be represented as $\mathscr{X} = \mathbf{x}_1, ..., \mathbf{x}_N$ ($N \geq 30$ denoting a part of the overall image points), the probability density function $p(\mathbf{x})$ (Euclidean distance between the selected 3-D points and the fitted plane) and a Gaussian exits as $\mathscr{N}(\mathbf{x}, \theta, \mathbf{r})$, where θ and $\mathbf{r}$ stand for a fraction of the inliers of the estimated plane and the relationship between the samples and the inliers, respectively. To obtain a maximum likelihood estimation of θ and $\mathbf{r}$, we can maximise the likelihood function $\Pi_{i=1}^{N} p(\mathbf{x}_i)$. The object function can be generalised as $f(\theta, \mathbf{r}) = \sum_{i=1}^{N} \omega_i \mathscr{N}(\mathbf{x}_i, \theta)$, where ω_i are weight factors and will be determined when we carry out similarity measurements. Based on the Jensen's inequality, we have an alternative object function as $\log f(\theta, \mathbf{r}) \geq \sum_{i=1}^{N} \log \left(\frac{\omega_i \mathscr{N}(\mathbf{x}_i, \theta, \mathbf{r})}{q_i} \right)^{q_i}$, where q_i is a non-negative constant that satisfies $\sum_{i=1}^{N} q_i = 1$.

Considering the current estimation θ_k and $\mathbf{r}_k$ (k indicates current state), we iterate the following E and M stages via the expectation-maximisation (EM) algorithm [20]:

(1) E-stage: Assuming that θ_k and $\mathbf{r}_k$ are fixed, we expect to obtain q_i that maximises the right hand side of the object function. The solution is expressed as: $q_i = \frac{\omega_i \mathcal{N}(\mathbf{x}_i, \theta_k, \mathbf{r}_k)}{\sum_{i=1}^{N} \omega_i \mathcal{N}(\mathbf{x}_i, \theta_k, \mathbf{r}_k)}$.

(2) M-stage: Considering q_i as constants, we maximise the right side of the object function with respect to θ and $\mathbf{r}$. The inlier fraction θ is solved by $\theta_{k+1} = \frac{\sum_{i=1}^{N} \mathbf{x}_i \omega_i \mathcal{N}(\mathbf{x}_i, \theta_k, \mathbf{r}_k)}{\sum_{i=1}^{N} \omega_i \mathcal{N}(\mathbf{x}_i, \theta_k, \mathbf{r}_k)}$, where $\mathbf{r}$ is updated according to the following equation $\mathbf{r}_{k+1} \propto \sum_{i=1}^{N} q_i(\mathbf{x}_i - \theta_k)(\mathbf{x}_i - \theta_k)^T$. This E-M iteration will terminate if and only if $|\bar{\theta}_{m+1} - \bar{\theta}_m|$ is less than a pre-defined threshold ($\bar{\theta}_m$ denotes an averaged θ in group m). In other words, the difference between two distributions instead of two consecutive samples is used as a stopping criterion.

(a)   (b)

Fig. 3 Estimated ground planes (in red color and hereafter) by (a) the improved RANSAC method, and (b) a classical RANSAC technique with the constraint where the number of outliers falls below a pre-defined threshold.

Fig. 3 illustrates the estimated ground planes, highlighted by red color, using two different techniques. It is observed that the proposed scheme leads to more accurate coplanar determination. For example, Fig. 3(a) shows that the points on the stones (in the image centre) have been correctly identified to be over the ground plane by the proposed approach. At the same time, the classical RANSAC plane fitting approach fails to do so (Fig. 3(b)). This indicates that the proposed algorithm can be used to accurately recover 3-D shapes from 2-D image pairs.

2.3 Case Study

We conduct a few more experiments to demonstrate how the iterative RANSAC plane fitting scheme performs in the extraction of flat surfaces, particularly ground planes. The performance of the proposed method is compared to that of the classical RANSAC plane fitting scheme with the constraint where the number of outliers falls

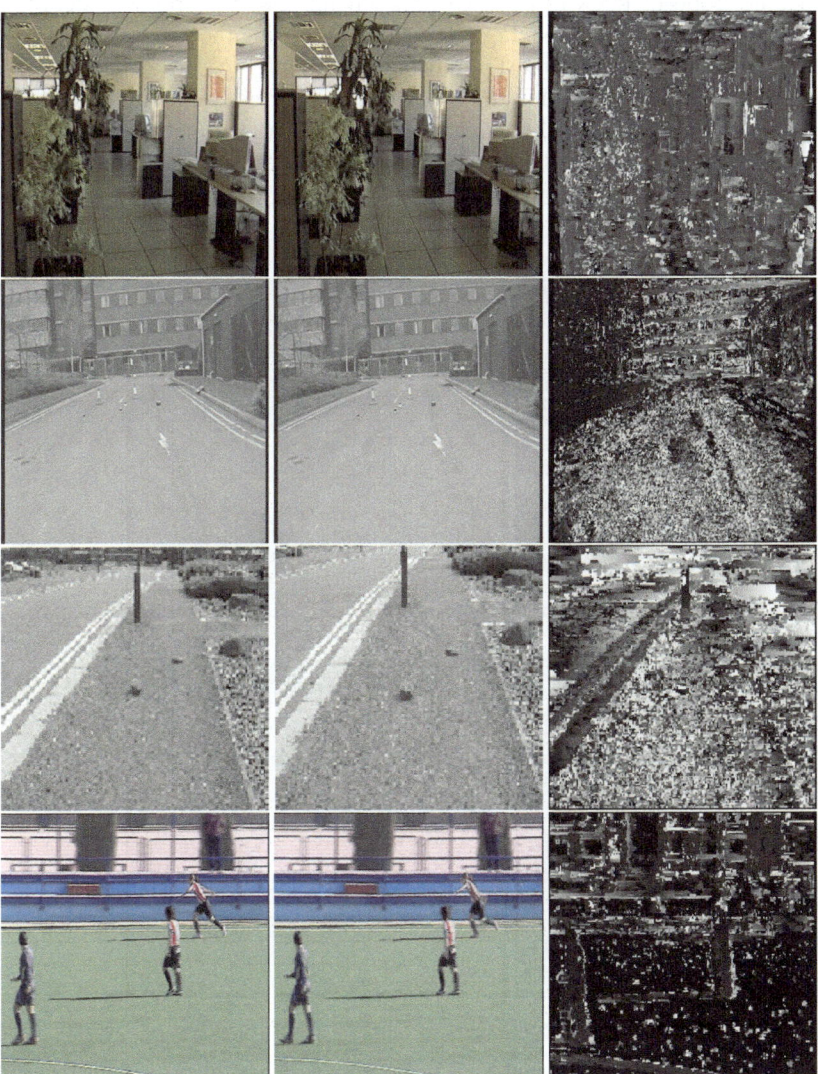

Fig. 4 Four test pairs and their corresponding disparity maps, where 1st- and 2nd-column are original images and 3rd-column is the disparity map

below a pre-defined threshold. Four image pairs and their corresponding disparity maps have been obtained and illustrated in Fig. 4. The proposed algorithm has been able to outline the actual planar areas.

We further demonstrate the performance of the proposed algorithm in planar surface detection that usually is the measure of corresponding accuracy. Fig. 5 illustrates two neighboring image frames of a test sequence namely "campus", superimposed by the detected corner features (see Fig. 5(a) and (b)). It exhibits in Fig. 5(c) and (d) that the proposed RANSAC plane fitting scheme results in optimal outcomes of flat surface fitting. For example, Fig. 5(c) shows that using the proposed method we are able to correctly identify most points on the ground. Fig. 5(d) denotes a significant number of points on the buildings have been incorrectly classified to be on the ground plane by the classical technique. Meanwhile, the points on the ground plane shown on Fig. 5(d) are less dense than those of Fig. 5(c), which is an issue in the classical RANSAC method.

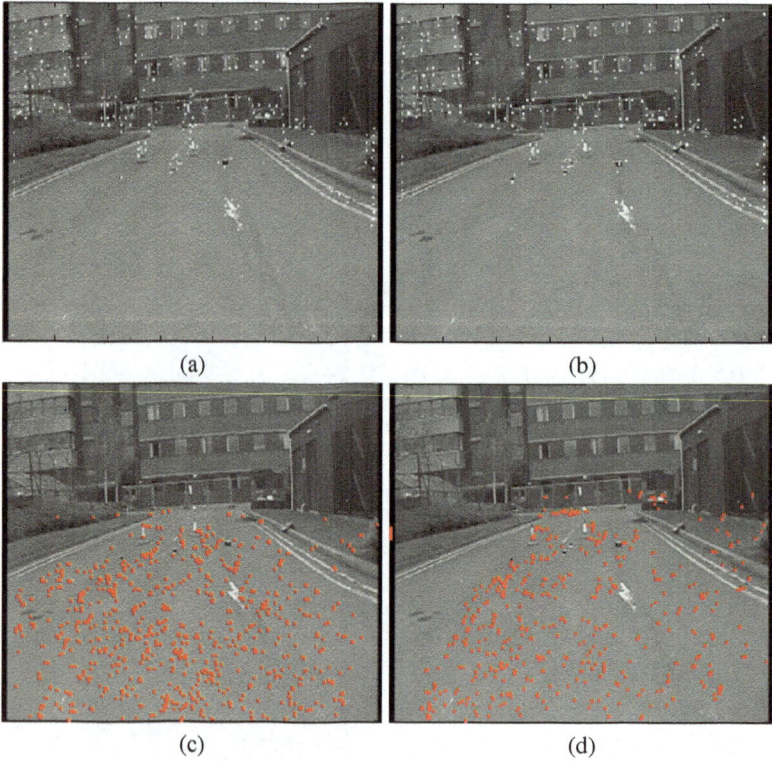

(a)   (b)

(c)   (d)

Fig. 5 Examples of the estimated ground plane in sequence "campus" by two different methods: (a) and (b) feature extraction, (c) outcome of the proposed method, and (d) outcome of the classical method

3 Segment Based Stereo Matching

Segment based methods normally consist of four steps to obtain the estimated surfaces, whose algorithm is illustrated in Fig. 6. First of all, the regions of homogeneous colour are detected by using a colour segmentation algorithm. This is followed by a local window based matching method that is utilised to estimate the disparities across two groups of image points. A plane fitting is then applied to generate disparity planes that indicate the feature points. Finally, an optimal disparity plane is obtained using a greedy optimisation algorithm.

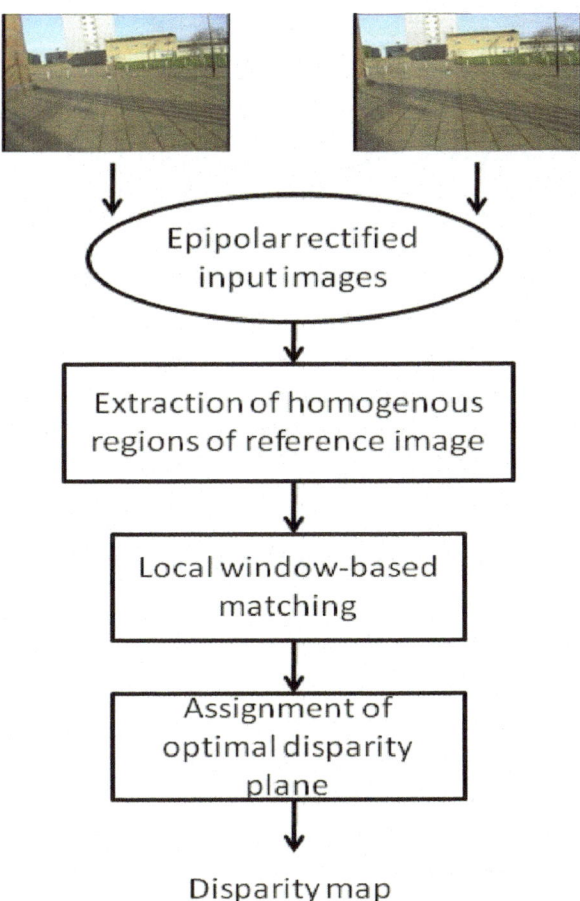

Fig. 6 Flowchart of the segment based stereo matching algorithm

3.1   Colour Segmentation

Considering a pair of colour images that can be used to extract a 3-D structure, we
mainly focus on the region edges where most likely embed depth discontinuities. To
extract homogeneous regions mean shift based colour segmentation [21] is applied
to search for a maxima in a density function. This process is demonstrated in Fig. 7,
where (a) and (b) are original colour images, and (c) is the colour segmentation by
mean shift.

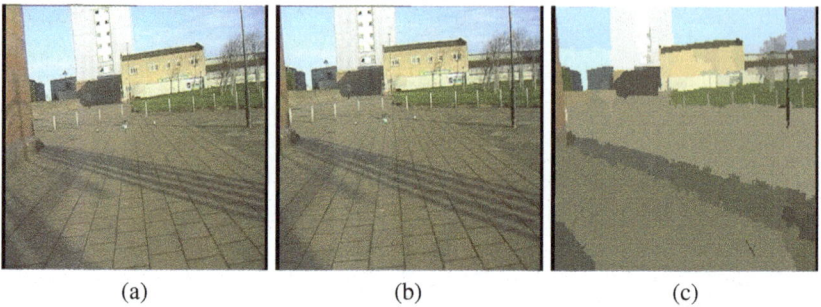

| (a) | (b) | (c) |

Fig. 7 Colour images and the segmentation by mean shift: (a) Left image, (b) right colour
and (c) segmentation result

A surface comprises a number of patches that can be represented by a disparity plane: $d = c_1 x + c_2 y + c_3$, where (x, y) refers to image pixel coordinates, and
(c_1, c_2, c_3) are used to determine a disparity d. Without further process, the available
disparity planes will be redundant and sometimes appear to be "noisy". A number of
approaches can be used to reduce the noise. Klaus *et al.* [8] utilised a self-adapting
dissimilarity measure that integrates the sum of absolute intensity differences (SAD)
and a gradient based measure which is defined as

$$F_{SAD}(x, y, d) = \sum_{(i,j) \in N(x,y)} I_1(i, j) - I_2(i + d, j) \tag{1}$$

and

$$F_{GRAD}(x, y, d) = \sum_{(i,j) \in N_x(x,y)} |\nabla_x I_1(i, j) - \nabla_x I_2(i + d, j)|$$
$$+ \sum_{(i,j) \in N_y(x,y)} |\nabla_y I_1(i, j) - \nabla_y I_2(i + d, j)|, \tag{2}$$

where $N(x, y)$ is a 3×3 window surrounding position (x, y). $N_x(x, y)$ is a window
without the rightmost column, $N_y(x, y)$ is a window without the lowest row, ∇_x is
the forward gradient to the right and ∇_x is the forward gradient to the left.

An optimal weight ω between F_{SAD} and F_{GRAD} can be used to maximise the
number of reliable correspondences that are handled by a cross-checking scheme in

line with a winner-take-all strategy (the disparity is determined in the presence of the lowest matching cost). The dissimilarity measure finally can be produced using the following formula:

$$F(x,y,d) = (1 - \omega) \times F_{SAD}(x,y,d) + \omega \times F_{GRAD}(x,y,d). \tag{3}$$

3.2 Estimation of Disparity Planes

Once the disparity planes have been processed, then we may find appropriate disparity planes to represent the scene structure. A robust solution is applied to estimate the parameters. First of all, the horizontal slant is computed using the reliably estimated disparities that fall in the identical line. The derivation $\frac{\partial d}{\partial x}$ is conducted and used to determine the horizontal slant by applying convolution with a Gaussian kernel.

Secondly, the vertical slant is calculated using a similar way to the above approach. Thirdly, the determined slant is used to obtain a robust estimation of the centre of the disparity pitch. The disparity map obtained according to the previous descriptions is not good enough in terms of accuracy. A matching procedure for each "segment to plane" assignment is used as follows:

$$F_{SEG}(S,P_e) = \sum_{(x,y) \in S} F(x,y,d), \tag{4}$$

where P_e is a disparity plane that defines the disparity d. This equation is iteratively used to find the segments with the minimum matching cost, and all the segments go over this process.

The final stage of this segment based stereo matching is to search the solution to the segment-to-disparity plane assignment. This in fact is a minimisation problem that satisfies

$$E(f) = E_{data}(f) + E_{smooth}(f), \tag{5}$$

where

$$\begin{cases} E_{data}(f) = \sum_{s \in R} F_{SEG}(s, f(s)) \\ E_{smooth}(f) = \sum_{\forall (s_i, s_j) \in S_N | f(s_i) \neq f(s_j))} \Omega(s_i, s_j) \end{cases} \tag{6}$$

where f is a labeling function, S_N is a set of adjacent segments and Ω is a discontinuity penalty. An optimal labeling with minimum energy is approached using the Loopy belief propagation algorithm [22]. This optimisation is illustrated in Fig. 8, where (a) indicates the pixel disparity map and (b) is the optimisation of (a). To further demonstrate the performance of the colour segment based stereo approach, we use three pairs of images for the estimation of disparity maps, which is revealed in Fig. 9. It is observed that this proposed algorithm can effectively handle the scenarios that possess less clutters but fails in complex scenes.

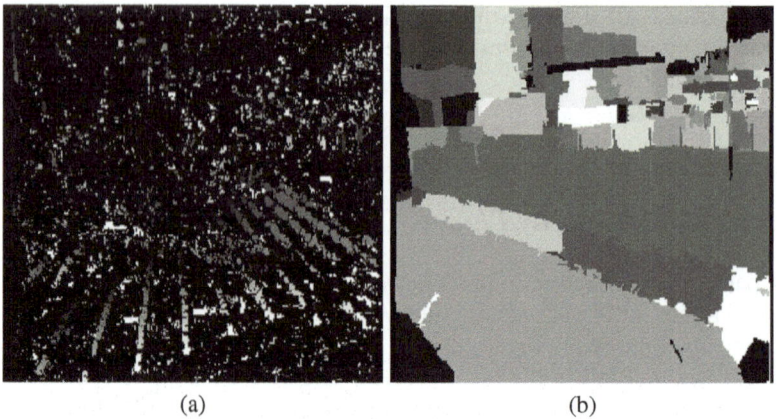

(a)   (b)

Fig. 8 Disparity maps: (a) pixel-wise, and (b) final disparity map

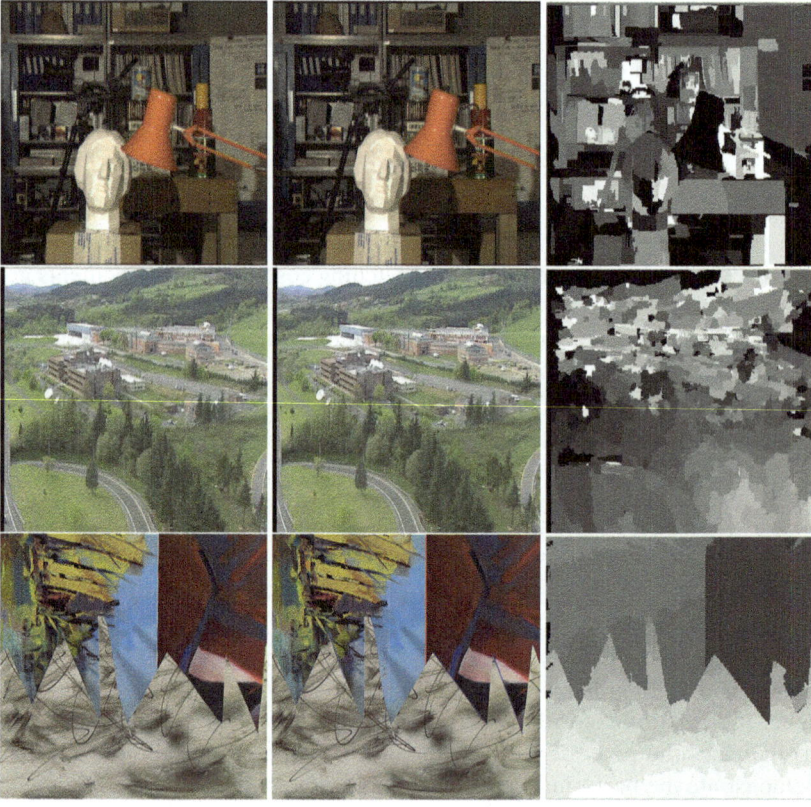

Fig. 9 Three test pairs and their corresponding disparity maps, where 1st- and 2nd-column are original images and 3rd-column is the disparity map

4 Connectivity Based Stereo Correspondence

Corresponding two images of a scene involves the selection of local metric, e.g. intensity or colour. However, image matching only based on the available local information will not be enough. This is due to the fact that colour repetition and redundancy exit everywhere. To reduce the effects of this uncertainty, the pixel characteristics must be used along with additional assumptions or constraints, e.g. continuity or smoothness. Prior knowledge of these constraints can be dominant in the estimation of patches. For example, depth discontinuities (edges or connective parts of inhomogeneous regions) will be determined if and only if smoothness is enforced. This also brings an interesting question, "has a shape anything to with the stereo correspondence"? The answer is yes. In fact, if we know where to find a shape, then the segmentation and correspondence of the associated image areas will be achieved without any problem, and vice versa. Unfortunately, this kind of prior knowledge is unavailable in all the time.

To effectively solve this problem, Ogale and Aloimonos [23] treat the disparity map of a real scene as a piecewise continuous function, where the images are described with the minimum possible number of pieces (segmentations). This piecewise continuous function is approximated by piecewise consistency. The role of shape in establishing correspondence is also discussed in their report. Particularly, the relation of the image correspondence and the segmentation is un-separated. The authors also emphasize on the geometric effects that were raised regarding the correspondence of a horizontally slanted surface. This is because the uniqueness constraint used to find the one-to-one correspondence does not hold in the presence of horizontally slanted surfaces and hence one against many matches will be observed.

The proposed algorithm presented in [23] is summarised as follows, given that the two images has shifts $\sigma_x \in \{\sigma_1, \sigma_1, ..., \sigma_k\}$:

Step 1: Shift the left image I_L horizontally by σ_x and then generate a new image I'_L. Then match I'_L with I_R.

Step 2: Investigate the closeness of the pixel (x, y) and its vertical neighbor $(x, y-1)$.

Step 3: Build up connected components using the vertical connections from Step 2.

Step 4: Determine the weights of the connected components.

Step 5: If the connected components surrounding the image pixel cause larger shifts, then the estimated left/right disparity maps must be updated by taking into account the uniqueness constraint.

One simple scanline algorithm was used to deal with the horizontal slant that leads to the violation of the uniqueness constraint in the correspondence. Assume that we have a pair of scanlines $I_L(x)$ and $I_R(x)$. Horizontal disparities $\triangle_L(x)$ are assigned to the left scanline within RANGLE OF $[\triangle_1, \triangle_2]$, and $\triangle_R(x)$ to the right scanline with the range $[-\triangle_1, -\triangle_2]$. The left scanline consists of the functions $m_L(x)$ and $d_L(x)$ and the right scanline has the functions $m_R(x)$ and $d_R(x)$. Two image points x_L and x_R must satisfy the following formula:

Fig. 10 Four test pairs and their corresponding disparity maps, where 1st- and 2nd-column are original images and 3rd-column is the disparity map.

$$x_R = m_L(x_L) \cdot x_L + d_L(x_L), \tag{7}$$

and

$$x_L = m_R(x_R) \cdot x_R + d_R(x_R). \tag{8}$$

Since

$$\begin{cases} m_R(x_R) = \frac{1}{m_L(x_L)} \\ d_R(x_R) = -\frac{d_L(x_L)}{m_L(x_L)} \end{cases} \tag{9}$$

Therefore, the disparity map can be estimated as follows:

$$\begin{cases} \triangle_L(x_L) = (m_L(x_L) - 1) \cdot x_L + d_L(x_L) \\ \triangle_R(x_R) = (m_R(x_R) - 1) \cdot x_R + d_R(x_R) \end{cases} \tag{10}$$

where the functions m_l and m_R are the horizontal slants that enable line segments on two scanlines to match.

In the occurance of the horizontal slants, we shall have a developed algorithm for the disparity estimation:

Step 1. For all $m_L \in M$, $\triangle_L \in [\triangle_1, \triangle_2]$,
(a) Stretch I_L by m_L to get I_L'.
(b) Define a range for d_L using the given range for $\triangle_L$.
(c) For every d_L, match I_L' and I_R. Then find connected matching segments and their sizes; update correspondence map while enforcing the uniqueness constraint.
Step 2. For all $m_R \in M$, $\triangle_R \in [-\triangle_2, -\triangle_1]$,
Similar approaches to the above.
3. $m_L = m_R = 1$
(a) For every $d_L \in [\triangle_1, \triangle_2]$, match I_R and I_L and find connected matching segments and their sizes; update correspondence map using the uniqueness constraint.

If there is any vertical slant in the view, similar approaches to the case of horizontal slants can be considered. It is worthy to point out that when a higher order model of shapes is met, there will not be any established algorithm for this sort of problems yet. To demonstrate the performance of this connectivity based stereo correspondence algorithm, Fig. 10 denotes 4 pairs of images and their disparity maps. Is is observed that the performance of this connectivity based approach cannot be maintained due to the image clutters.

5 Wide Baseline Stereo Correspondence Using Local Descriptors

A number of short baseline stereo matching algorithms have been established with reasonable performance [24],[25],[26]. Due to the large distance and orientation change wide baseline stereo matching is more challenging and many application problems are related to this wide baseline issue.

The developed wide baseline methods intend to use small correlation windows or point-wise similarity measures [15],[27]. But these algorithms abruptly loose their capability in the presence of light changing [28]. Local image descriptors, e.g. SIFT [29] and GLOH [30], have been commonly used in dense matching, where the matching process can be efficiently and effectively achieved. For example, [31] reported the propagation of the disparity maps of the matched features to their neighbours.

Tola *et al.* reported a new approach based on the local descriptors [28]. This strategy comes up with a new descriptor that retains the robust features of SIFT and GLOH. The descriptor is then used for dense matching and view-based synthesis using stereo-pairs. The kernel of the technique is that computational complexity can be significantly reduced without sacrificing the performance by convolving orientation maps to compute the histogram bin values.

A vector here is made of values from the convolved orientation maps located on concentric circles centered at the pixel location. Let $\mathbf{h}_\Sigma(u,v)$ be the vector at (u,v) after the convolution by a Gaussian kernel of standard deviation Σ:

$$\mathbf{h}_\Sigma(u,v) = [\mathbf{G}_1^\Sigma(u,v), ..., \mathbf{G}_8^\Sigma(u,v)]^T, \tag{11}$$

where $\mathbf{G}_1^\Sigma(u,v), ..., \mathbf{G}_8^\Sigma(u,v)$ are the convolved orientation maps. These vectors are normalised to unit norm $\tilde{\mathbf{h}}$ so that they represent the pixels near occlusions as correct as possible. The propsoed descriptor $\mathbf{D}(u_0, v_0)$ for location (u_0, v_0) is then defined as a concatenated $\mathbf{h}$ vectors:

$$\mathbf{D}(u_0, v_0) = [\tilde{\mathbf{h}}_{\Sigma_1}^T(u_0, v_0), \tilde{\mathbf{h}}_{\Sigma_1}^T(\mathbf{l}_1(u_0, v_0, R_1)), ..., \tilde{\mathbf{h}}_{\Sigma_1}^T(\mathbf{l}_N(u_0, v_0, R_1)), ..., \tilde{\mathbf{h}}_{\Sigma_3}^T(\mathbf{l}_N(u_0, v_0, R_3))]^T, \tag{12}$$

where $\mathbf{l}_j(u, v, R)$ is the location with distance R from (u,v) in the j direction that has N values. Once these feature have been obtained using the descriptors, similarities across images will be measured using the graph-cut-based reconstruction algorithm presented in [32]. An occlusion map is used to handle occlusions using EM and binary masks that have been redefined to enforce the spatial coherence of the occlusion map. Fig. 11 shows an example where shadows may affect the estimation of the disparity maps. But in this example, the proposed algorithm handles this situation well and the shadow area has not been false-positive. In Fig. 12, the proposed algorithm leads details (human statues) to being explicitly separated. Finally, Fig. 13 demonstrates that three images can be combined to improve the outcome generated using two images.

Fig. 11 Test triples and the corresponding depth map, where 1st- and 2nd-column are original images and 3rd-column is the depth map

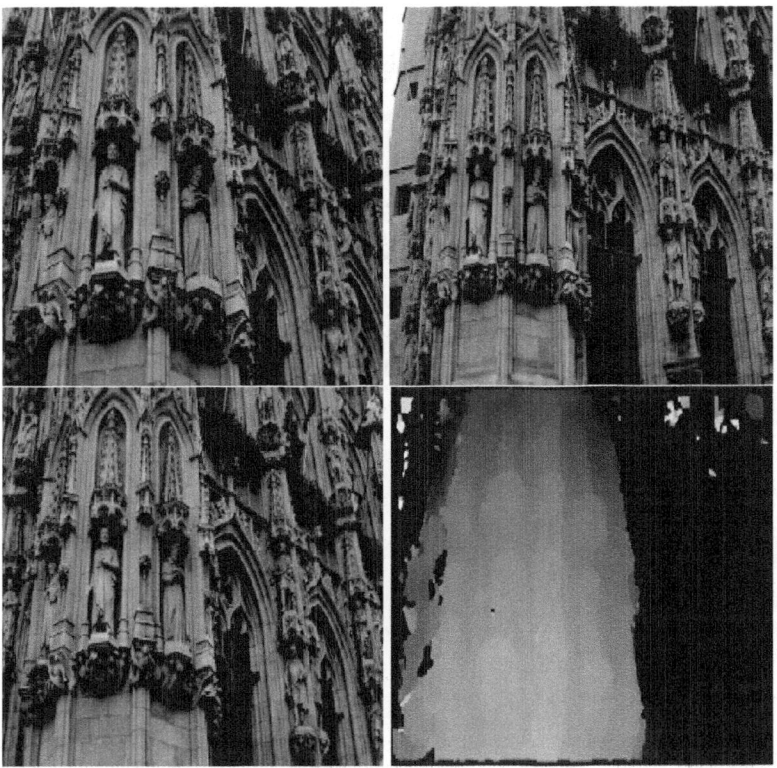

Fig. 12 Three test images and the corresponding depth map, where the bottom right is the depth map and the remainder refers to the original images

6 Evaluation of Different Methods

In the previous sections, we briefly summarised the principles and characteristics of a few classical and typical stereo matching algorithms. To evaluate their performance in information retrieval, we in this section compare these algorithms using the RUSHES database that was designed for video retrieval [3]. This database consists of 134 raw videos and has about 14 hours' length in total, provided by Spanish communication group Euskal Irrati Telebista (EITB). The videos used in this database include various contents, e.g. interviews, football matches, aerial views, shopping and rowing, etc. 3064 key frames have been extracted from the overall videos.

To illustrate the comparisons we here use exemplar image pairs taken from the database. Fig. 14 shows five image pairs that denote different image backgrounds. For example, 1st row shows an aerial view of a building from a helicopter, while row 4 reveals shots from a moving hand-held video camera. These scenes are challenging to the existing stereo correspondence algorithms in the sense that a number

Fig. 13 Three test images and the corresponding depth map, where the bottom right is the depth map and the remainder refers to the original images

of objects appear in the background as well as the foreground. The algorithms involved in the comparison comprise (1) Zhou's algorithm [12], (2) Klaus's algorithm [8], (3) Ogale' algorithm [23] and (4) Tola' algorithm [28].

Fig. 15 illustrates the outcomes of different stereo matching algorithms. For example, it has been observed from row 1 that Klaus's algorithm allows the building's details to be significantly presented. However, we observe that Zhou's and Tola' algorithms have the best performance in row 3's outcomes due to the explicit details. In the row, the human shapes can be noticed in the extracted disparity map by Zhou's algorithm, while the other algorithms seems to loose details. These comparison results demonstrate that the behaviors of individual algorithms may vary in different images. Therefore, more broad and further studies on better performance of stereo reconstruction are necessarily required in the community.

Fig. 14 Illustration of exemplar test image pairs from the RUSHES database

Fig. 15 Performance comparison of various stereo correspondence algorithms in the RUSHES image pairs, where rows are the outputs of rows shown in Fig. 14 and columns indicate the results of Zhou's, Klaus's, Ogale's and Tola's algorithms.

7 Conclusions and Future Work

We have summarised several techniques for effective stereo correspondence from 2-D images. These systems can reasonably deal with the common problems such as wide-baseline, clutters, and light changing. In the maximum likelihood estimation approach, flat surfaces are extracted from the scenes by analysing the video content, e.g. correspondence and 3-D recovery. An iterative RANSAC plane fitting scheme was also presented. In the segment based stereo algorithm, the image patches look neat in most cases. This is due to the colour segmentation before the disparity maps are estimated. In the connectivity based scheme, the stereo correspondence is integrated with shape segmentation. The shape segmentation is used to enhance the performance of estimating the disparity maps. The last one is the wide-baseline stereo strategy incorporating local descriptors. This algorithm can effectively handle some significant wide-baseline cases. In spite of their success, these systems also reveal their weakness in certain circumstances. One of the disadvantages is that, in many cases, the local noise still evidently appears and somehow affects the structure representation of the scene or objects. This weakness may be tackled if prior knowledge of these details can be used after necessary training of a local classifier.

Acknowledgements. This work was supported by European Commission under Grant FP6-045189-STREP (RUSHES).

References

1. Sivic, J., Zisserman, A.: Video google: a text retrieval approach to obejct matching in videos. In: Proc. of Ninth IEEE International Conference on Computer Vision, pp. 1470–1477 (2003)
2. Davis, M.: An iconic visual language for video annotation. In: Proc. of IEEE Symposium on Visual Language, pp. 196–202 (1993)
3. Rushes project deliverable D5, requirement analysis and use-cases definition for professional content creators or providers and home-users (2007),
   http://www.rushes-project.eu/upload/Deliverables/
   D5_WP1_ETB_v04.pdf
4. Tangelder, J., Veltkamp, R.: A survey of content based 3d shape retrieval methods. In: Proc. of International Conference on Shape Modeling, pp. 145–156 (2004)
5. Ohbuchi, R., Kobayashi, J.: Unsupervised learning from a corpus for shape-based 3d model retrieval. In: Proc. of the 8th ACM international workshop on Multimedia information retrieval, New York, NY, USA, pp. 163–172 (2006)
6. Bartoli, A., Sturm, P.: Structure-from-motion using lines: Representation, triangulation, and bundle adjustment. Computer Vision and Image Understanding 100(3), 416–441 (2005)
7. Zhou, H., Wallace, A., Green, P.: A multistage filtering technique to detect hazards on the ground plane. Pattern Recognition Letters 24(9-10), 1453–1461 (2003)
8. Klaus, A., Sormann, M., Karner, K.: Segment-based stereo matching using belief propagation and a self-adapting dissimilarity measure. In: Proc. of 18th International Conference on Pattern Recognition, pp. 15–18 (2006)

9. Deng, Y., Yang, Q., Lin, X., Tang, X.: A symmetric patch-based correspondence model for occlusion handling. In: Proc. of the Tenth IEEE International Conference on Computer Vision, pp. 1316–1322 (2005)

10. Tao, H., Sawhney, H., Kumar, R.: A global matching framework for stereo computation. In: Proc. of Ninth IEEE International Conference on Computer Vision, pp. 532–539 (2001)

11. Sun, J., Zheng, N.N., Shum, H.Y.: Stereo matching using belief propagation. IEEE Transactions on Pattern Analysis and Machine Intelligence 25(7), 787–800 (2003)

12. Zhou, H., Sadka, A., Jiang, M.: 3d inference and modelling for video retrieval. In: Proc. of Ninth International Workshop on Image Analysis for Multimedia Interactive Services, pp. 84–87 (2008)

13. Laurentini, A.: The visual hull concept for silhouette-based image understanding. IEEE Transactions on Pattern Analysis and Machine Intelligence 16(2), 150–162 (1994)

14. Roy, S., Cox, I.: A maximum-flow formulation of the n-camera stereo correspondence problem. In: Proc. of the Sixth International Conference on Computer Vision, p. 492 (1998)

15. Kolmogorov, V., Zabih, R.: Multi-camera scene reconstruction via graph cuts. In: Heyden, A., Sparr, G., Nielsen, M., Johansen, P. (eds.) ECCV 2002. LNCS, vol. 2352, pp. 82–96. Springer, Heidelberg (2002)

16. Narayanan, P., Rander, P., Kanade, T.: Constructing virtual worlds using dense stereo. In: Proc. of the Sixth International Conference on Computer Vision, p. 3 (1998)

17. Hoff, W., Ahuja, N.: Surfaces from stereo: Integrating feature matching, disparity estimation, and contour detection. IEEE Trans. Pattern Anal. Mach. Intell. 11(2), 121–136 (1989)

18. Fischler, M., Bolles, R.: Random sample consensus: a paradigm for model fitting with applications to image analysis and automated cartography. Comm. ACM 24, 381–395 (1988)

19. Harris, C., Stephens, M.: A combined corner and edge detector. In: Proc. of Alvey Vision Conference, pp. 47–152 (1988)

20. Dempster, A., Laird, N., Rubin, D.: Maximum likelihood from incomplete data via the em algorithm. Journal of the Royal Statistical Society 39(1), 1–38 (1977)

21. Comaniciu, D., Meer, P.: Mean shift: A robust approach toward feature space analysis. IEEE Transactions on Pattern Analysis and Machine Intelligence 24(5), 603–619 (2002)

22. Felzenszwalb, P., Huttenlocher, D.: Efficient belief propagation for early vision. Int. J. Comput. Vision 70(1), 41–54 (2006)

23. Ogale, A., Aloimonos, Y.: Shape and the stereo correspondence problem. Int. J. Comput. Vision 65(3), 147–162 (2005)

24. Brown, M., Burschka, D., Hager, G.: Advances in computational stereo. IEEE Transactions on Pattern Analysis and Machine Intelligence 25(8), 993–1008 (2003)

25. Scharstein, D., Szeliski, R.: A taxonomy and evaluation of dense two-frame stereo correspondence algorithms. Int. J. Comput. Vision 47(1-3), 7–42 (2002)

26. Baker, H.: Depth from edge and intensity based stereo. Ph.D. thesis, Champaign, IL, USA (1981)

27. Strecha, C., Tuytelaars, T., Gool, L.V.: Dense matching of multiple wide-baseline views. In: Proc. of the Ninth IEEE International Conference on Computer Vision, p. 1194 (2003)

28. Tola, E., Lepetit, V., Fua, P.: A fast local descriptor for dense matching. In: CVPR (2008)

29. Lowe, D.: Distinctive image features from scale-invariant keypoints. Int. J. Comput. Vision 60(2), 91–110 (2004)

30. Mikolajczyk, K., Schmid, C.: A performance evaluation of local descriptors. IEEE Trans. Pattern Anal. Mach. Intell. 27(10), 1615–1630 (2005)
31. Yao, J., Cham, W.K.: 3d modeling and rendering from multiple wide-baseline images by match propagation. Signal Processing: Image Communication 21(6), 506–518 (2006)
32. Boykov, Y., Veksler, O., Zabih, R.: Fast approximate energy minimization via graph cuts. IEEE Transactions on Pattern Analysis and Machine Intelligence 23(11), 1222–1239 (2001)

Author Index

Printed by Printforce, the Netherlands